Universal Wireless Personal Communications

Universal Wireless Personal Communications

Ramjee Prasad

Artech House

Boston · London

Library of Congress Cataloging-in-Publication Data
Prasad, Ramjee
Universal wireless personal communications / Ramjee Prasad.
p. cm. — (Artech House mobile communications library)
Includes bibliographical references and index.
ISBN 0-89006-958-1 (alk. paper)
1. Wireless communication systems. 2. Mobile communication systems.
3. Personal communication service systems. 4. Cellular radio. I. Title.
II. Series.
TK5103.2.P73 1998
621.382—dc21 98-20073
CIP

British Library Cataloguing in Publication Data
Prasad, Ramjee
Universal wireless personal communications. — (Artech House mobile communications library)
1. Personal communication service systems
I. Title
621.3'845

ISBN 0-89006-958-1

Cover design by Lynda Fishbourne

International Standard Book Number: 0-89006-958-1
Library of Congress Catalog Card Number: 98-20073

10 9 8 7 6 5 4 3 2 1

To
my fatherly brother
Surendra Nath Sinha,

and my motherly sisters
Kunti Devi and Gaytri Devi,

who inspired me to reach the goal by conquering the mind,
I respectfully dedicate this book.

Contents

Preface

बन्धुरात्मात्मनस्तस्य येनात्मैवात्मना जितः ।
अनात्मनस्तु शत्रुत्वे वर्तेतात्मैव शत्रुवत् ॥

bandhur ātmātmanas tasya
yenātmaivātmanā jitah
anātmanas tu śatrutve
vartetātmaiva śatru-vat

For him who has conquered the mind, the mind is the best of friends; but for one who has failed to do so, his mind will remain the greatest enemy.

The Bhagavad Gita (6.6)

This year, 1998, is the "harvest year" of the full-fledged wireless (mobile) garden that I started in 1988 with the help of colleagues and masters, doctoral, and postdoctoral candidates of the Telecommunications and Traffic-Control Systems Group of Delft University of Technology (DUT), The Netherlands. A gardener feels happy when he finds his garden full of a variety of beautiful flowers; similarly I feel extremely happy in seeing the output of the wireless (mobile) garden in the form of this book *Universal Wireless Personal Communications*, which covers all the aspects of wireless personal communications from its first generation to its future fourth generation.

The first three flowers appeared in *Electronics Letters*, Vol. 24, August 1988; in *Proceedings of Canadian Conference on Electrical and Computer Engineering*, Vancouver, November 1988; and *IEEE Transactions of Vehicular Technology*, Vol. 37, November 1988, jointly with Jens Arnbak. These articles were in the area of random multiple access protocols in the wireless environment, cellular mobile communications, and wireless channel modeling. The number of flowers started growing rapidly in various areas, covering all aspects of wireless communications.

The idea of converting our research contributions into a book came to me in 1993 when I gave a course on mobile radio systems for the Advanced Studies in Electrical Engineering (ASEE) program. I realized that there was a shortage of technical material for this course. The teaching material, which I prepared in 1993, was revised during lecture preparations in 1994, 1995, 1996, and 1997. Until now, no such book was available in the market. I have attempted to fill this gap in the literature. I plan to use some chapters in this book for two courses, "Cellular Mobile Communications" and "Personal Communications Network," which I offer to prefinal and final year students studying for M.Sc. degrees in electrical engineering.

I wrote this book for the following types of readers and with the objective to have a single-source reference that offers in-depth coverage of the fundamental and advanced topics associated with wireless personal communications, from basic cellular systems, to third-generation mobile technologies to broadband multimedia and satellite communication:

- Undergraduate, postgraduate, and Ph.D. candidates;
- Lecturers and professors;
- Scientists;
- Engineers;
- Top technocrats.

Universal Wireless Personal Communications is the first book that covers a broad range of topics such as, radio propagation aspects, basic cellular communications including macrocellular, microcellular, and picocellular systems, adaptive equalization, multiple access protocols with CDMA concepts and random multiple access protocols in a hostile wireless environment, dynamic channel assignment, several aspects of future public land mobile telecommunication systems/international mobile telecommunications-2000/universal mobile telecommunication systems (FPLMTS/IMT-2000/UMTS), and broadband multimedia communications including orthogonal frequency division multiplexing (OFDM), OFDM-based asynchronous transfer mode (ATM) network and multicarrier-CDMA (MC-CDMA).

Chapter 1 presents a general introduction to wireless personal communications. A part of this chapter on UMTS is based on the contribution of Tero Ojanpera from Nokia Research Center, Irving, Texas. Tero is also busy working with me on his Ph.D. studies. Chandan K. Chatterjee from Orange Telecom Software, The Hague, The Netherlands has also contributed to this chapter in introducing wireless in local loop (WILL).

Chapters 2 to 9 deal with the basic knowledge of wireless (mobile) communications every undergraduate and postgraduate student, as well as practicing engineers, MUST learn. Chapters 2 to 5 are based on the research contributions of Adriaan Kagel, who was my colleague and is now retired; Aysal Safak, who finished her Ph.D. under my supervision; Gerard Janssen and Han Reijmers who are colleagues and are also finishing their Ph.D studies with me; and several masters students, particularly Willem Hollemans, Aart de Vos, Frank van der Wijk, Patrick A. Stigter, Aart J. 'T Jong, and Eric Wesselman. Gerard Janssen and Frank van der Wijk have contributed significantly toward the completion of Chapter 6. Chapter 7 is based on the research contributions of Aleksandar Zigic who finished his Ph.D. under my supervision, and Mark Krapels, who finished his M.Sc. in electrical engineering under the supervision of Gerard Janssen and myself. Chapter 8 is based on the research contributions of my predoctoral student Michel Jansen and doctoral student Casper van den Broek. Chapter 9 is based on the research contributions of Jean-Paul Linnartz, who is a colleague and also worked with me during his Ph.D. studies; Jos Nijhof, who is a colleague, Neeli Rashmi Prasad from Libertel, The Netherlands, and several master and Ph.D. students, particularly Marcel Kerkhoff, Aart J. 'T Jong, C.-Y. Liu, Iguh Widipangestu, Kasibhatia Sastry, Casper van den Broek, Michel Janssen, Caspar Wijffels, Howard Sewberath Misser, and Huub van Roosmalen. Random multiple access protocols and CDMA are some of the strongest research areas in DUT. My book *CDMA for Wireless Personal Communications*, is authored by me and published in 1996 by Artech House. Another book on wideband CDMA is under preparation and expected to be published in 1998 jointly with Tero Ojanpera.

Chapters 10 to 14, 17, and 18 present open problems for scientists and engineers. These chapters are also very useful for Ph.D. candidates. The research contributions of Dusan Matic, who is busy with his Ph.D. studies under my supervision, and Hiroshi Harada, who was a postdoctoral student with me during 1996–97, from Communications Research Laboratory, Tokyo, Japan are used in Chapter 10 in the field of OFDM. Currently, research in the field of OFDM is one of the active research topics in DUT. Homayoun Nikooker, a postdoctoral student with me from Iran, and Klaus Witrisal, a young researcher with me from Austria, are actively involved in the development of an OFDM system. A book on OFDM is under preparation jointly with Richard D.J. van Nee from Lucent Technology, The Netherlands. Multicarrier-CDMA is discussed in Chapter 11, using the research contributions of Shinsuke Hara, a postdoctoral student with me during 1995-96 from the University of Osaka, Japan, and Frans Kleer, who finished his M.Sc. in electrical engineering under the supervision of Shinsuki Hara. Chapter 12 summarizes much of the standard results in diversity theory using the research contributions of Jean-Paul van Deursen, who finished his M.Sc. in electrical engineering studies under the supervision of Kiwi Smit of Ericsson, Business Mobile Networks B.V., The Netherlands, Adriaan Kegel, Michel Jansen, and myself. A basic concept of millimeter-wave communications is introduced in Chapter 13 using the research contributions of Barend van Lieshout, who completed his M.Sc. in electrical engineering studies under the supervision of Gererd Janssen and myself. Currently, DUT is busy in developing a propagation model using measured data at 60 GHz. Dynamic channel assignment is introduced in Chapter 14 using the research contributions of Dirk Sparreboom, who is a colleague, Frank Brouwer of Ericsson Business Mobile Networks B.V., The Netherlands, Jan B. Punt who is a colleague, and S.Z. Beheshti Babanari, who finished his M.Sc in electrical engineering studies under the supervision of Dirk Sparreboom, Homayoun Nikooker, and myself. Chapter 17 proposes an OFDM-based wireless ATM transmission system using the research contributions of Hiroshi Harada. Chapter 18 is based on the research contributions of John Farserotu from NATO-C3 Agency, The Hague, The Netherlands, who is busy in finishing his Ph.D. studies as a part-time researcher under my supervision.

Chapters 15, 16, and 18 were written from the viewpoint of top technocrats from industries, government departments, and policy-making bodies. Of course, this information may be interesting to others as well. Air interface multiple access schemes for third-generation wireless personal communications are discussed using the research contributions of Tero Ojanpera. A general overview of wireless broadband multimedia communications is presented in Chapter 16 using the research contributions of Luis M. Correia from Technical University of Lisbon, Portugal, and Anand Raghwa Prasad from Tokyo Research Center, Uniden Co. Japan.

Thus, this book can be used as a textbook for undergraduate and postgraduate students. It can also be used by scientists, engineers, lecturers, and Ph.D. candidates. Finally, it can be very useful to top technocrats. Keeping the above points into mind, some chapters are written in great depth and some chapters present general information. Wherever the book presents technical opinions, they represent the technical views of the author.

This book is based on the contributions of several researchers who were or are actively involved in the development of the wireless (mobile) garden under my gardenership. As a gardener, I have tried my best to provide enough water and energy to make each flower in the form of a chapter quite complete in itself. I would greatly appreciate it if readers provide extra water and energy in improving the quality of the flowers— in improving the text and pointing out errors.

Acknowledgments

I would like to express my heartfelt gratitude to colleagues and students without whom this book would have never been completed. Jens Arnbak, with whom I started the book, gave me initial support. Adriaan Kegel, Jos Nijof, Dirk Sparreboom, Han Reijmens, Jean-Paul Linnartz (currently with Philips National Laboratory, Eindhoven, The Netherlands), and Gerard Janssen supported me in supervising many graduate students whose results are used in this book. Jayati Chatterjee gave her support in preparing the type script of the book.

Tero Ojanpera from Nokia Research Center, Chandan K. Chatterjee from Orange Telecom Software, Kiwi Smit and Frank Brouwer from Ericsson Business Mobile Networks B.V., John Farserotu from NATO-C3 Agency , Neeli Rashmi Prasad from Libertel and Anand Roughhew Prasad from Tokyo Research Center, Uniden Co. are deeply acknowledged for their valuable contributions. The postdoctoral research scholars Shinsuke Hara from the University of Osaka, Japan, Hiroshi Harada from the Communications Research Center, Tokyo, Japan, and Homayoun Nikooker from Iran are acknowledged for their contributions. Alexander Zigic, Aysal Safak, Gerard Janssen, Casper van den Broek, Tero Ojanpera, John Farserotu, and Dusan Matic are thanked for use of their doctoral research results in the book. Michel G. Jansen and Frank van der Wijk, predoctoral students, are especially thanked for their valuable contributions.

Last but not the least, the following master students are greatly acknowledged in generating several new results which are used throughout in this book: Willem Hollemans, Aart de Vos, Patrick A. Stigter, Aart J. 'T Jong, Eric Wesselman, Mark Krapels, Marcel Kerkchoff, C.-Y. Liu, Iguh Widipangostu, Caspar Wijffels, Howard S. Misser, Huub van Roosmelaan, Jean-Paul van Deursen, Barend van Lishout, Jan B. Punt, S.Z. Beheshti Babanari, Suzane 'Grujev, M.B. Loog, H.I. Cakil, M.A. Visser, and to all my master students whose research contributions are used in this book in some form or other.

Chapter 1
Introduction

One hundred years ago, the notion of transmitting information without the use of wires must have seemed like magic. Marchese Guglielmo Marconi made it possible. In 1896, the first patent for wireless communication was granted to him in the United Kingdom. In fact, he demonstrated the first wireless communication system in 1897 between a land-based station and a tugboat. Since then, unbelievable, extraordinary, and rapid developments in the field of wireless communication have been taking place that will shrink the world into a global communication village (GCV) by 2010. A GCV system will provide communication services from one person to another in any place at any time without any delay (except propagation and system delay), in any form through any medium by using one pocket-sized unit at minimum cost, with acceptable quality and security through using a personal telecommunication reference number/or personal identification (name, city, and country) [1–3].

1.1 AN OVERVIEW

This book deals with wireless communications in terms of RADIO technology only. Other wireless technology, such as infrared (IR), is not discussed.

The wireless era is divided in three periods: (a) pioneer era, (b) precellular era, and (c) cellular era, as shown in Table 1.1.

Table 1.1

Wireless Era

Pioneer Era

1860s	James Clark Maxwell's EM waves postulates
1880s	Proof of the existence of EM waves by Heinrich Rudolf Hertz
1890s	First use of wireless and first patent of wireless communications by Gugliemo Marconi
1905	First transmission of speech and music via a wireless link by Reginald Fessenden
1912	Sinking of the Titanic highlights the importance of wireless communication on the seaways; in the following years marine radio telegraphy is established

Precellular Era

1921	Detroit Police Department conducts field tests with mobile radio
1933	In the United States, four channels in the 30–40 MHz range
1938	In the United States, ruled for regular services
1940	Wireless communication is stimulated by World War II
1946	First commercial mobile telephone system operated by the Bell system and deployed in St. Louis
1948	First commercial fully automatic mobile telephone system is deployed in Richmond, United States
1950s	Microwave telephone and communication links are developed
1960s	Introduction of trunked radio systems with automatic channel allocation capabilities in the United States
1970s	Commercial mobile telephone system operated in many countries (e.g., 100 million moving vehicles on U.S. highways, "B-Netz" in West Germany)

Cellular Era

1980s	Deployment of analog cellular systems
1990s	Digital cellular deployment and dual mode operation of digital systems
2000s	Future public land mobile telecommunication systems (FPLMTS)/ international mobile telecommunications-2000 (IMT-2000)/ universal mobile telecommunication systems (UMTS) will be deployed with multimedia services
2010s	Wireless broadband communications will be available with B-ISDN and ATM networks
2010s +	Radio over fiber (such as fiber-optic microcells)

During the pioneer era, much fundamental research and development in the field of wireless communications took place. If we look back in history, we find that wireless communications began with the postulates of electromagnetic waves by James Clark Maxwell during the 1860s in England, the demonstration of the existence of these waves by Heinrich Rudolf Hertz in the 1880s in Germany, and the invention of wireless telegraphy by Marconi during the 1890s first in Italy and then in the United Kingdom [4,5]. During the 1890s, eminent scientists like Jagdish Chandra Bose in India, Oliver Lodge in England, and Augusto at Bolgna University were busy studying the fundamental nature of electromagnetic waves. In September 1895, Marconi could say that he had a wireless telegraph system with a potentially useful range, unaffected by natural obstacles [5]. After that event, a number of developments in the field of wireless communications took place, as seen in Table 1.1.

The precellular era began with the first land mobile wireless telephone system was installed in 1921 by the Detroit Police Department for police car wireless dispatch in the 2-MHz frequency band [6,7].

The cellular era started with the operation of analog cellular systems in 1980s. In fact, the successful operation of analog cellular systems gave the birth to the concept of the GCV. A family tree for the GCV system is shown in Figure 1.1.

The GCV has been evolving since the birth of the first-generation analog cellular system. Table 1.2 shows the summary of analog cellular radio systems [8–10]. Various standard systems were developed worldwide: advanced mobile phone service (AMPS) in the United States, Nordic mobile telephones (NMT) in Europe, total access communication systems (TACS) in the United Kingdom, Nippon Telephone and Telegraph (NTT) in Japan, and so on. The first AMPS cellular telephone service commenced operation in Chicago on October 13, 1983 [11,12]. In Norway, NMT-450 was launched in 1981 and in 1984 the network was already overloaded in Oslo, as 35% of all traffic occurred within a radius of 25 km from the center of Oslo [13]. The situation was similar in other Nordic capitals. In 1983, the NMT Committee had started the development of a new system NMT-900, C-450 was launched in Germany and Portugal, a modified version of TACS known as JTACS/NTACS was installed in Japan, radio telephone mobile system (RTMS) began in Italy, and Radiocomm started in France. All the first-generation systems used frequency modulation (FM) for speech and frequency shift keying (FSK) for signaling and the access technique used was frequency division multiple access (FDMA). Details of these systems are given in Table 1.2. The first-generation analog cellular system penetrated up to 10% of the calls in North America, Western Europe, and Japan. Capacity and quality were the major problems. In addition, systems were not compatible.

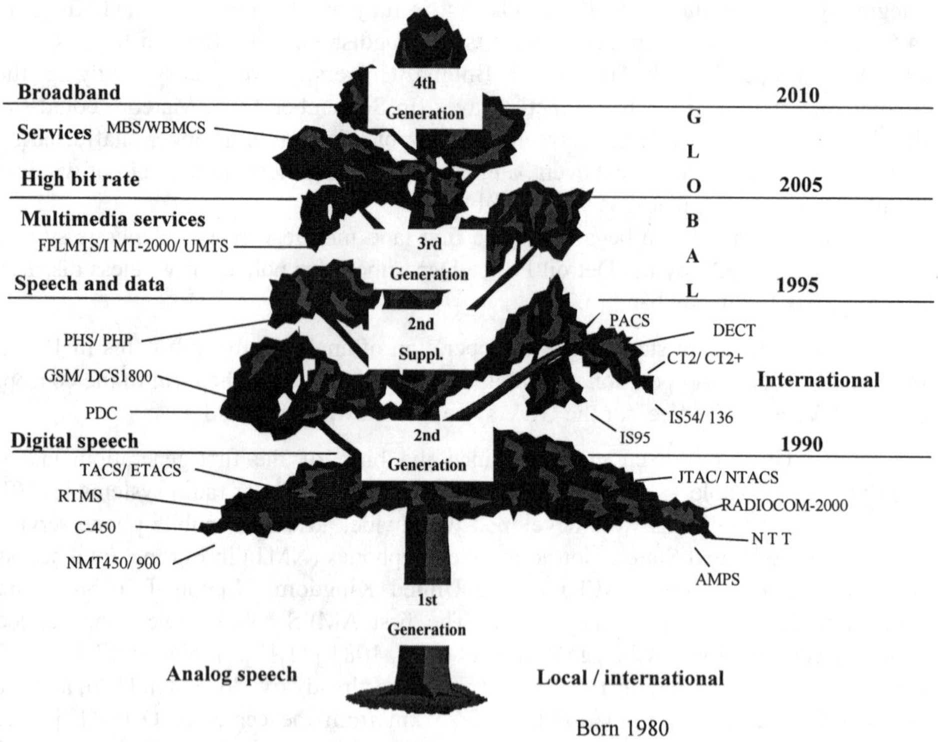

Figure 1.1 Family tree of the GCV system. Branches and leaves of the GCV family tree are not shown in chronological order.

Advancement in digital technology gave birth to Pan-European digital cellular mobile (DCM) Groupe Spéciale Mobile (GSM)/digital cordless system (DCS)-1800 systems in Europe, personal digital cellular (PDC) systems in Japan, and interim-standard (IS)-54/136 and IS-95 in North America, which are the second-generation systems. A summary of digital cellular radio systems is shown in Table 1.3 [8–10]. Additional information can be found in [13–20].

Table 1.2

Summary of Analog Cellular Radio Systems

AMPS, NMT-450, NMT-900, TACS, and ETACS

System	AMPS	NMT-450	NMT-900	TACS	ETACS
Frequency range (mobile TX/base TX) (MHz)	824-849/ 869-894	453-457.5/ 463-467.5	890-915/ 463-467.5	890-915/ 935-960	872-905/ 917-950
Channel spacing (kHz)	30	25	12.5*	25	25
Number of channels	832	180	1999	1000	1240
Region	The Americas, Australia, China, Southeast Asia	Europe	Europe, China, India, Africa	United Kingdom	Europe, Africa

C-450, RTMS, Radiocom-2000, JTACS/NTACS, and NTT

System	C-450	RTMS	Radiocom-2000	JTACS/NTACS	NTT
Frequency range (mobile TX/base) (MHz)	450-455.74/ 460-465.74	450-455/ 460-465	165.2-168.4/ 169.8-173 192.5-199.5/ 200.5-207.5 215.5-233.5/ 207.5-215.5 414.8-418/ 424.8-428	915-925/ 860-870 898-901/ 843-846 918.5-922/ 863.5-867	925-940/ 870-855 915-918.5/ 860-863.5 922-925/ 867-870
Channel spacing (kHz)	10*	25	12.5	25/12.5* 25/12.5* 12.5*	25/6.25* 6.25* 6.25*
Number of channels	573	200	256 560 640 256	400/800 120/240 280	600/2400 560 480
Region	Germany, Portugal	Italy	France	Japan	Japan

* Frequency interleaving using overlapping or "interstitial" channels, the channel spacing is half the nominal channel bandwidth.

Table 1.3

Summary of Digital Cellular Radio Systems

Systems	GSM DCS-1800	IS-54	IS-95	PDC
Frequency range (base Rx/Tx, MHz)	GSM: Tx: 935-960 Rx: 890-915 DCS - 1800: Tx: 1805-1880 Rx: 1710-1785	Tx: 869-894 Rx: 824-849	Tx:869-894 Rx: 824-849	Tx: 810-826 Rx: 940-956 Tx:1429-1453 Rx: 1477-1501
Channel spacing (kHz)	200	30	1250	25
Number of channels	GSM: 124 DCS - 1800: 375	832	20	1600
Number of users per channel	GSM: 8 DCS - 1800: 16	3	63	3
Multiple access	TDMA/FDMA	TDMA/FDMA	CDMA/FDMA	TDMA/FDMA
Duplex	FDD	FDD	FDD	FDD
Modulation	GMSK	$\pi/4$ DQPSK	BPSK/QPSK	$\pi/4$ DQPSK
Speech coding and its rate (Kbps)	RPE-LTP 13	VSELP 7.95	QCELP 8	VSELP 6.7
Channel coding	1/2 Convolutional	1/2 Convolutional	Uplink 1/3 Downlink 1/2 Convolutional	9/17 Convolutional
Region	Europe, China, Australia, Southeast Asia	North America, Indonesia	North America, Australia, Southeast Asia	Japan

Time division multiple access (TDMA) is used as the access technique, except for IS-95, which is based on code division multiple access (CDMA). The second-generation systems provide digital speech and short message services. These services have penetrated to more than 20% of the call population. GSM has become deeply rooted in Europe and in more than 70 countries worldwide. In fact, GSM has become the de facto world standard. GSM now stands for "global systems for mobile communications." DCS-1800 is also spreading outside Europe to East Asia and some South American countries.

The development of new digital cordless technologies gave birth to the second-supplement-generation systems, namely, personal handyphone systems (PHS, formerly PHP) in Japan, digital enhanced (formerly European) cordless telephone (DECT) in Europe, and personal access communication services (PACS) in North America. Table

1.4 shows the features of the second-supplement-generation systems [8–10], which have increased the call penetration depth up to 30% and introduced many new services.

Table 1.4 shows also the description of CT2 and CT2+. A detailed description of CT2 can be found in [21,22]. CT2+ is a Canadian enhancement of the CT2 common air interface (CAI).

Table 1.4

Summary of Digital Cordless Systems

System	CT2/CT2+	DECT	PHS	PACS
Frequency range (Base Rx/Tx, MHz)	CT2: 864-868 CT2+: 944-948	1880-1990	1895-1918	TX: 1850-1910 Rx: 1930-1990
Channel spacing (kHz)	100	1728	300	300
Number of channels	40	10	77	96
Number of users per channel	1	12	4	8
Multiple access	FDMA	TDMA/FDMA	TDMA/FDMA	TDMA/FDMA
Duplex	TDD	TDD	TDD	FDD
Modulation	GFSK	GFSK	$\pi/4$ DQPSK	$\pi/4$ DQPSK
Speech coding	ADPCM 32	ADPCM 32	ADPCM 32	ADPCM 32
Channel coding	None	CRC	CRC	CRC
Region	Europe, Canada, China, Southeast Asia	Europe	Japan, Hong Kong	United States

In recent years DECT, PHS, and PACS/wireless access communications systems (WACS) have been introduced to provide cost-effective wireless in local loop (WILL) [23–29].

Little did the author of the phrase, "Where there is a WILL there is a way" realize that telecommunications developers and researchers would apply the phrase literally to the telecommunications field. As we know, for extending telephone service to a user conventionally, underground physical cables are laid to the user's premises. Such cable laying has its own problems in terms of time required, costs, logistics and so forth. In many places, it is impossible to lay the cable because of narrow and congested by-lanes or solid rocky terrain. Wireless overcomes all such barriers and WILL is seen as a solution to all these problems and much more. Thus, where there is a WILL, a way will be found to reach the user. WILL is also popularly known as wireless local loop (WLL) or radio local loop (RLL).

The term local loop stands for the medium that connects the equipment on the user's premises with telephone switching equipment.

There are psychological and technical challenges that WILL faces in becoming acceptable to users. Since it replaces copper cable for connecting the user with the local exchange, the user may be apprehensive about reliability, privacy, and interference with wireless appliances like radio and television (man-made noise) and other WILL users. On the other hand, WILL must prove itself at least as good, if not better, than the services provided by physical cable. It should be able to carry and deliver voice, data, state-of-the-art multimedia (with some add-ons), leased-line connectivity, dial-up, Internet connectivity, high-speed data, and so forth, as efficiently as the age-old fixed line does. The onus rests on researchers and developers to prove that it is so. Only time will tell.

Fixed radio access systems are already being deployed in over 50 countries. Recently, the Indian Government has enjoined upon the prospective basic service providers to provide the local loop through WILL. In Germany since 1990, over 50,000 lines have been provided using Ericsson and Nokia systems. In Hungary, the Czech Republic, and Russia, over 300,000 lines are in the process of deployment. In Spain, the WILL-related users have grown to over 800,000. The figures will keep rising as confidence in WILL grows.

In Japan, there were already 3.56 million users in August 1996 who were using the PHS system, which gives radio access. PHS is also spreading to other countries; Australia, Hong Kong, Indonesia, Singapore, and Thailand have already decided to adopt it, and field trials have started in Malaysia. In Indonesia, about 100,000 users lines and an equal amount of fixed telephone terminals are being installed for WILL application. Mexico has proposed to connect 1.5 million new users through wireless. WILL has already made entry into Columbia.

China, India, Indonesia, and the Philippines are having a hard look at the possibility of using WILL in a big way to clear their backlogs of telephone lines.

WILL holds great promise toward fulfilling the requirement of a huge visible and pent-up demand for fast telephone connections. With WILL becoming a commonplace item in the communication arena, it will come under pressure to quench the thirst of the user for more and better services. On the other hand, with the advance of technology, wireless is offering better services in terms of quality and capacity, improved and miniaturized chips resulting in smaller, lighter and lesser power consuming equipment, more advanced digital signal processing capabilities and more interference-free working. The end user, however, does not care whether it is TDMA,

FDMA, CDMA, or American or European standard. What concerns the user is how much the service and calls will cost and what is the quality? To that extent, whatever the proponents have to say in support of their products, the user will remain king.

Although the second-generation and its supplement have covered local, national, and international services, they still have one major drawback in terms of a universal service facility. In addition to the system discussed in this section, wireless data systems and wireless local area networks (WLANs) are also very important in the field of wireless communications [30,31].Some features of wide area wireless packet data systems are shown in Table 1.5 [14].

Table 1.5

Summary of Wide-Area Wireless Packet Data Systems

System	CDPD	RAM Mobile (Mobitex)	ARDIS (KDT)	Metricom (MDN)
Data rate	19.2 Kbps	8 Kbps [19.2 Kbps]	4.8 Kbps [19.2 Kbps]	~ 76 Kbps
Modulation	GMSK BT = 0.5	GMSK	GMSK	GMSK
Frequency	~ 800 MHz	~900 MHz	~ 800 MHz	~915 MHz
Channel spacing	30 KHz	12.5 KHz	25 KHz	160 KHz
Status	1994 service	Full service	Full service	In service
Access means	Unused AMPS channels	Slotted Aloha CSMA		FH SS (ISM)
Transmit power			40W	1W

Note: Data in square brackets [ ] indicates proposed.

CDPD : Cellular digital packet data.
MDN : Micro-cellular data network.
ARDIS : Advanced radio data information service.

Advanced radio data information service (ARDIS) and RAM mobile data (RMD) are the earliest and best known systems in North America. Cellular digital packet data (CDPD) is a new wide area packet data network. The general packet radio service (GPRS) standard was developed to provide packet data service over the GSM infrastructure.

The third generation should be deployed by the year 2003 via universal wireless personal communications (UWPC) systems, which will provide universal speech services and local multimedia services. It is expected that the third-generation system will penetrate up to 50% of the telecommunication services population. The third-generation personal communication systems are in the process of development worldwide by the International Telecommunications Union (ITU) within the framework of the future public land mobile telecommunications systems (FPLMTS)/international mobile telecommunications-2000 (IMT-2000) activities. In Europe, this is supported by the universal mobile telecommunications system (UMTS) program within the European community. Both the FPLMTS and UMTS programs are tightly related and expected to lead to consistent and compatible systems. Figure 1.2 shows the possible configuration for UMTS subnetworks and fixed networks.

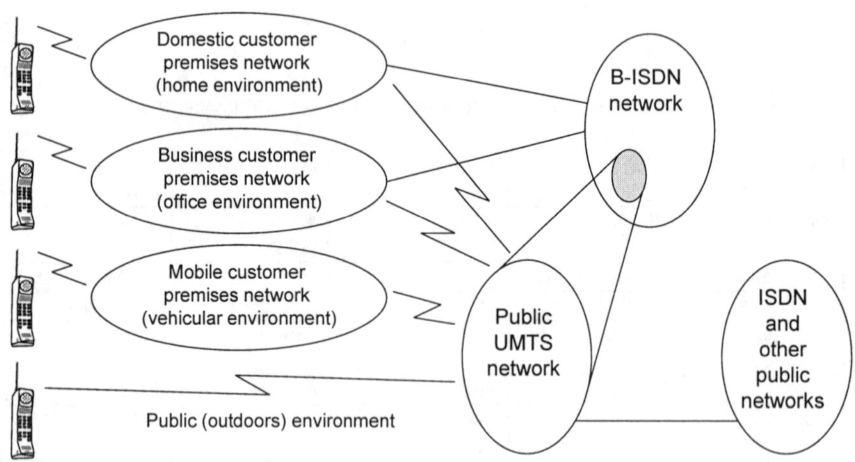

Figure 1.2 UMTS possible service configuration.

UMTS is intended to provide a wide range of mobile services to users via a range of mobile terminals that enable the use of the pocket telephone in almost any location, indoor or outdoor, in the home, office, or street. North America and Japan are also equally engaged in developing third-generation systems in the UMTS direction. Looking at research trends going on in Europe, Japan, and North America, it appears that the future access scheme will be either CDMA or one of the hybrid multiple access schemes based mainly on the combination of CDMA and TDMA/FDMA [1,31]. This subject is discussed in detail in Chapter 15.

The only drawback is the lack of rural (modern) communication coverage, particularly in developing countries. Hopefully, this will be solved at the beginning of the 21st century. Land-mobile satellite communications could solve this problem.

Figure 1.3 shows a GCV integrating terrestrial pico-, micro-, and macrocellular systems and satellite networks. An integrated satellite-terrestrial cellular system is the concept of GCV; that is, UWPC.

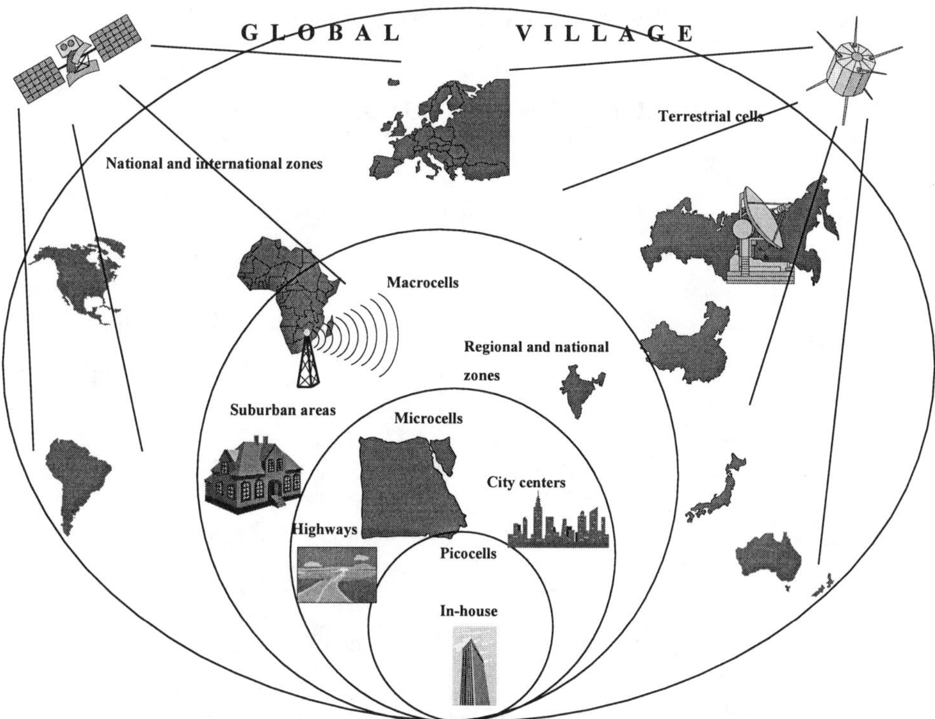

Figure 1.3 View of a future GCV.

The cellular systems are classified into three categories [32], depending on the size of the cells, as follows: (1) macrocells of 2- to 20-km diameter with antennas radiating power in the range of 0.6 to 10W from the top of tall buildings; (2) microcells of 0.4- to 2-km diameter with antennas at street-lamp elevation and radiating power less than 20 mW; and (3) picocells of much smaller size (20m to 400m diameter) especially suited for indoor radio communications (e.g., offices, research laboratories, hospitals, modern factories, and university campuses) with antennas placed on top of a bookshelf and radiating power on the order of a few milliwatts.

Research and development (R & D) is also going on in Europe, Japan, and North America for the fourth-generation mobile broadband systems (MBS) and wireless broadband multimedia communications systems (WBMCS). WBMCS is expected to provide its users customer premises services from information rates exceeding 2Mbps. A detailed description of WBMCS is presented in Chapter 16.

1.2 UNIVERSAL WIRELESS PERSONAL COMMUNICATIONS

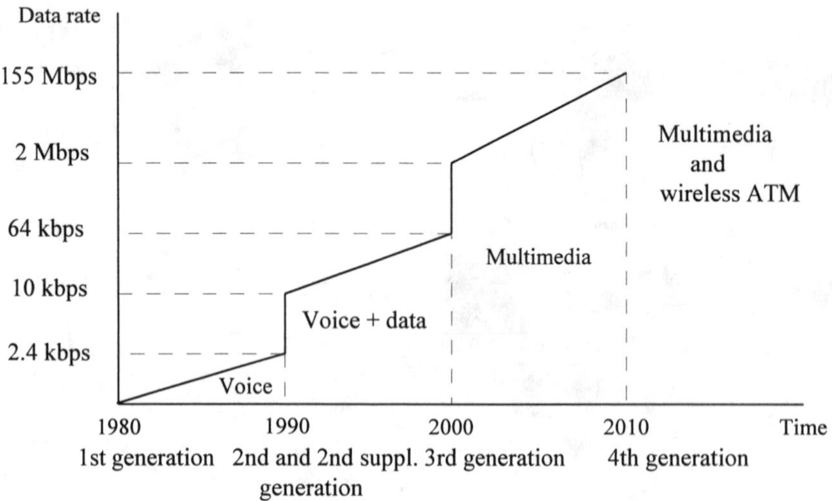

Figure 1.4 Worldwide service and time evolution.

Figure 1.4 shows the worldwide service evaluation [33]. Universities, research laboratories, and industries are jointly busy in exploring the following challenges to achieve the universal wireless personal communications (UWPC) systems:

- Multiple access protocols (CDMA?, TDMA?, or Hybrid?);
- Interfacing, internetworking, and integration (ATM, fiber, air, fixed, macrocells, microcells, picocells, and hypercells);

- System development (baseband, terminals, and antennas);
- Technology (low power, size, and cost);
- Frequency bands (62–63 GHz millimeter waves);
- Multimedia communications (MMC);
- Satellite (frequency allocation, channel characterization, radio access, etc.).

Significant advances have been achieved in meeting these challenges, which can be found in the published literature. It is worth mentioning that the European Community, Japan, and North America (the United States and Canada) have responded to meet the challenges for GCV systems through ambitious infrastructure investment programs and regulatory and legislative reforms. National communications and information networks are being introduced in these countries, attracting a continuous stream of innovative valued-added products and services. All governments have been spending billions of dollars for research and development in this field. Several challenging projects have started from September 1, 1995 in Europe within the advanced communication technologies and services (ACTS) program initiated by the European Commission.

The following section discusses the developments taking place in Europe toward the path of UMTS [34].

1.2.1 UMTS Objectives and Challenges

UMTS is a third-generation mobile communication system providing seamless personal communication services anywhere and anytime. In particular, it provides mobile broadband multimedia services. Main UMTS objectives are [35]:

- User bit rates of 144 Kbps (wide-area mobility and coverage) and up to 2 Mbps (local mobility and coverage);
- Provision of services via handheld, portable, vehicular mounted, movable and fixed terminals, in all radio environments based on single radio technology;
- High spectrum efficiency compared to the existing system;
- Speech and service quality at least comparable to current fixed network;
- Flexibility for the introduction of new services and technical capabilities;
- Radio resource flexibility to multiple networks and traffic types within frequency band.

The most important success factor for UMTS will be the existence of a mass market. A UMTS Task Force report [36] identifies that during the first 3 years of mass exploitation, 8 million UMTS terminals would be in use and the number of terminals would grow to 60 million within 10 years. The basis for the UMTS market will be the existing GSM/DCS market for speech, low and medium bit rate data up to 100 Kbps.

The current number of GSM/DCS users is over 70 million and is expected to increase manyfold by the year 2000. The GSM market will continue to grow even after UMTS introduction, and thus positioning of UMTS toward GSM is important. Most likely the first services offered by UMTS will complement the services offered by GSM/DCS. The bit rates of UMTS compared to existing and evolved second-generation systems are shown in Figure 1.5.

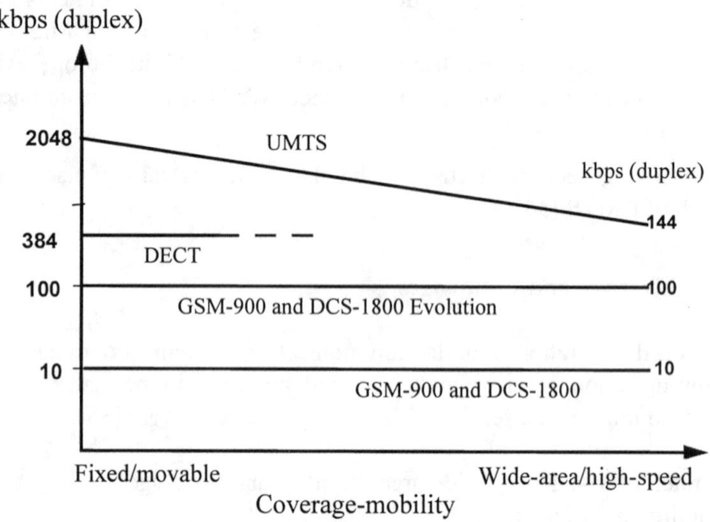

Figure 1.5 UMTS bit rates versus coverage and mobility.

To make UMTS happen, several challenges both in technical and regulatory issues must be overcome. The technical challenges include flexible design of the UMTS radio interface, effective radio resource management algorithms and protocols, and software radios covering multimode terminals. On the regulatory side, an efficient framework for UMTS spectrum licensing has to be created that will cater to both licensed and unlicensed use.

The goal of UMTS is to support a large variety of services, most of which are not known yet. UMTS air interface has to be able to cope with variable and asymmetric bit rates, up to 2 Mbps, with different quality of service requirements (bit error probability and delay) such multimedia services with bandwidth on demand. Effective packet access protocol is also essential for the UMTS air interface to handle bursty real time and non-real time data.

High- and low-rate users with different quality of service requirements will co-exist in the network, and thus, effective radio resource management has to guarantee required quality for all users. Interference management will be a key issue in achieving this. Flexible allocation of radio resources for packet-type services with unpredictable bit rates, taking into account the asymmetrical data transmission, is an essential requirement for third-generation networks. Manual network planning for a network with varying load and services would be an enormous task. Therefore, automatic network planning is an important area in UMTS development.

A UMTS objective is to cover all environments with a single interface. However, to use spectrum efficiently in different environments and for different services, the air interface has to be adaptable. Therefore, implementation of UMTS terminals have to take this adaptability into account requiring configurable terminals. In addition, utilization of the existing infrastructure will require dual-mode terminals. Development of software radios, which enable the required terminal adaptability, is one of the challenges for UMTS implementation.

To start commercial operation, several regulatory aspects such as frequency allocation need to be considered. UMTS Task Force recommended that to reduce risks in implementation of UMTS regulatory framework for services and spectrum should be defined before the end of 1997 [36]. Currently, the European Radio Committee/European Radio Office (ERC/ERO) is carrying out a frequency study for UMTS [37]. The European Telecommunication Standardization Institute (ETSI) has also contributed to this [38]. The relative need of frequencies for different services are impossible to predict exactly since they will depend on usage and penetration. Thus, more flexibility is needed in future frequency allocation policy compared to the current licensing schemes. As well, the need for additional spectrum of two times 180 MHz to create a mass market for UMTS broadband services was recognized by the UMTS Task Force [36].

1.2.2 ACTS Program

Within the European 4th Framework, ACTS has been set up. It represents the European Commission's major effort to support precompetitive research and technological development (RTD) in the context of trials in the field of telecommunications during the period of the Fourth Framework Programme of scientific research and development (1994–1998). Within ACTS several projects aiming to contribute to the UMTS system development have been set up, namely cordless business communication system (COBUCO), flexible integrated radio system technology (FIRST), future radio wideband multiple access systems (FRAMES), multimedia portable digital assistant (MULTIPORT), radio access independent broadband on wireless (RAINBOW) and software tools for the optimization of resources in mobile systems (STORMS).

1.2.3 Standardization of UMTS

The ETSI Subtechnical Committee (STC) Special Mobile Group (SMG) is responsible for definition of UMTS and coordination of UMTS standardization work. Most of the UMTS standards will be created within ETSI TC SMG and its subtechnical committees are also responsible for GSM/DCS standardization. The ETSI work program for UMTS [39] covers the UMTS standards that need to be developed. The UMTS time schedule for standardization, regulation, and implementation is given in Table 1.6.

Table 1.6
UMTS Schedule [36]

Task Name	1996	'97	'98	'99	2000	'01	'02	'03	'04	'05
GSM-900 Phase 2 + implementation		■	■	■						
Cooperative research: ACTS	■	■	■							
ETSI: Basic UMTS standards studies	■	■								
ETSI: Freezing basic parameters of UMTS standard			■							
ETSI: Basic UMTS standards			■	■						
ETSI: UMTS standards, advanced features					■	■				
Regulation: framework		■	■	■						
Regulation: license conditions			■							
Regulation: license awards					■					
System development: basic UMTS					■	■				
System development: UMTS							■	■		
Basic UMTS: planning, procurement, deployment					■	■				
Basic UMTS: operation							■	■	■	
UMTS: planning, procurement, deployment									■	■
UMTS: operation										■

Most of the UMTS ETSI technical reports (ETRs) are already completed within SMG5 [35–51]. They form the basis for standardization. These ETRs include objectives, system requirements, migration aspects, framework for network architecture, radio requirements, objectives and framework for telecommunication multimode/multimedia networks (TMN), voice band data coding aspects, security principles, and satellite issues.

The development of UMTS standards has also started. Currently, SMG1 is developing standards for service aspects [52], SMG3 for network principles and SMG5 for UMTS system concept. The UMTS system concept specifies the firm principles and priorities in accordance with the UMTS objectives and overview. It also specifies the functional elements and interfaces of UMTS. Figure 1.6 presents the environment of the UMTS system concept [53].

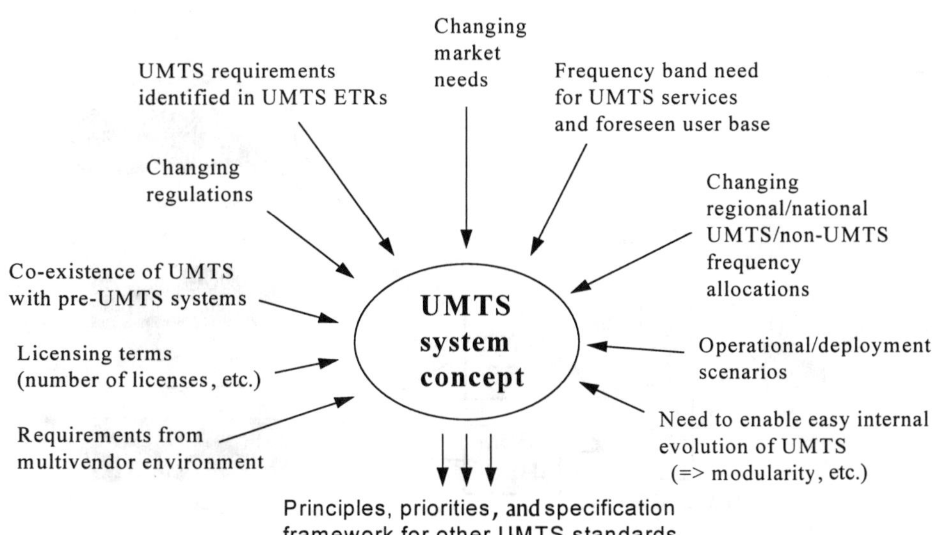

Figure 1.6 Environment for UMTS system concept.

One of the main aspects of UMTS standardization is selection of the air interface. In January 1998, the strong support behind wideband CDMA (WCDMA) led to the selection of WCDMA as the UMTS terrestrial air interface scheme for frequency division duplex (FDD) bands in ETSI. The selection of the WCDMA was also backed by the Asian and American GSM operators. For time division duplex (TDD) bands, a time division CDMA (TD-CDMA) concept was selected.

According to the global multimedia mobility (GMM) framework, the main focus of the UMTS standards will be the radio access network [54]. The generic radio access network can be connected to several core networks (e.g., evolved GSM and B/N-ISDN) for which the standards needed to implement UMTS are developed in respective standards bodies [54]. Figure 1.7 presents the idea of generic radio access network connecting to several core networks.

In ITU, development of FPLMTS (also called IMT-2000) is carried out. FPLMTS can be seen as global interworking of mobile services and especially the cornerstone of spectrum allocation for third-generation mobile radio systems. UMTS standards are more detailed than FPLMTS recommendations, covering test specifications and focusing on the European market and existing systems.

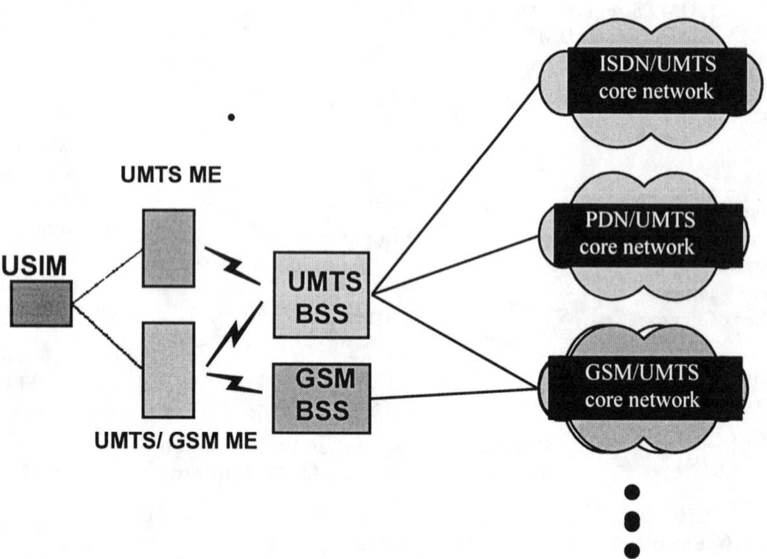

Figure 1.7 UMTS radio ACCESS and core network concept.

In addition to the actual standardization of UMTS, several other aspects need to be considered to ensure smooth implementation of UMTS networks. Therefore, UMTS

Forum has been established based on the outcome of the UMTS Task Force [36]. UMTS Forum is divided into four working groups: WG-1 Regulatory, WG-2 Spectrum, WG-3 Market Aspects, and WG-4 Technical Impact.

1.3 WIRELESS BROADBAND COMMUNICATION SYSTEMS

The theme of WBMCS is to provide its users a means of radio access to broadband services supported on customer premises networks or offered directly by public fixed networks. WBMCS will provide a mobile/movable nonwired extension to wired networks for information rates exceeding 2 Mbps with applications foreseen in wireless local area network (LAN) or mobile broadband systems. Thus, WBMCS will be a wireless extension to the broadband integrated services digital network (B-ISDN). It will be achieved with the transparent transmission of asynchronous transmission mode (ATM) cells.

Research and development of WBMCS is in progress in North America, Europe, and Japan in the microwave and millimeter-wave bands in order to accommodate the necessary bandwidth. The research in the field of WBMCS has drawn a lot of attention because of the increasing role of multimedia and computer applications in communications. There is a major thrust on three research areas: (1) microwave and millimeter-wave bands for fixed access in outdoor, public commercial networks, (2) evolution of wireless LANs for in-building systems, and (3) use of LAN technology outdoors rather than indoors. In short, WBMCS will provide novel multimedia and video mobile communication services, also related to wireless customer premises network (WCPN) and WILL.

To implement the wireless broadband communication systems, the following challenges must be considered:

- Frequency allocation and selection;
- Channel characterization;
- Application and environment recognition, including health hazard issues;
- Technology development;
- Air interface multiple access techniques;
- Protocols and networks;
- Systems development with efficient modulation, coding, and smart antenna techniques.

The spectacular growth of video, voice, and data communication via the Internet and the equally rapid pervasion of mobile telephony, justify great expectations for mobile multimedia. Research and development in the field of mobile multimedia are taking place all over the world. Before presenting our approach, we first give a concise overview of worldwide activities.

Within the European ACTS program there are four European Union funded R&D projects, namely The Magic Wand (wireless ATM network demonstration), ATM wireless access communication system (AWACS), system for advanced mobile broadband applications (SAMBA) and wireless broadband CPN/LAN for professional and residential multimedia applications (MEDIAN) [55–64].

Table 1.7 summarizes the European projects [63].

In the United States, seamless wireless network (SWAN) and broadband adaptive homing ATM architecture (BAHAMA) and two major projects in Bell Laboratories and wireless ATM network (WATMnet) are being developed in the computer and communication (C&C) research laboratories of Nippon Electric Company (NEC) in the United States [55–59].

In Japan, Communication Research Laboratory (CRL) is busy developing several R&D projects, such as a broadband mobile communication system in the super high frequency (SHF) band (from 3 to 10 GHz) with a channel bit rate up to 10 Mbps and an indoor high speed wireless LAN in SHF band with a target bit rate of up to 155 Mbps [64].

Table 1.7

Summary of European ACTS Projects

ACTS project / Parameter	WAND	AWACS	SAMBA	MEDIAN
Frequency	5 GHz	19 GHz	40 GHz	61.2 GHz
Data rate	20 Mbps	70 Mbps	2 x 41 Mbps	155 Mbps
Modulation	OFDM (orthogonal frequency division multiplexing) 16 carriers, 8 PSK (phase shift keying)	OQPSK (offset quadrature PSK), coherent detection	OQPSK	OFDM, 512 carriers, DQPSK, (differential QPSK)
Cell radius	20–50 m	50–100 m	6m × 200m 60m × 100m	10m
Radio access	TDMA/TDD (time division multiple access/ time division duplex)	TDMA/TDD	TDMA/FDD (Frequency DD)	TDMA/TDD

In the Netherlands, Delft University of Technology has been busy with a multi-disciplinary research project "Mobile Multimedia Communication" (MMC) since April 1996. The team consists of experts from the telecommunications and traffic control and information theory groups of the department of electrical engineering, the product ergonomics group of the department of industrial design engineering, and the work and organizational psychology group of the department of technology and society.

The MMC has the following objectives to achieve at 60 GHz:

- Wireless access of 155 Mbps;
- Indoor/outdoor;
- Less complex, inexpensive mobile station;
- Modified OFDM;
- Constant bit rate (CBR), variable bit rate (VBR), and adaptive bit rate (ABR) services.

1.3.1 Millimeter Waves

During recent years, millimeter waves have gained increasing interest because of bandwidth scarcity, and therefore, the study of millimeter wave communication systems has drawn the attention of many researchers [65,66]. Within Europe, cooperation in the field of scientific and technical research (COST), the promising features of the millimeter waves for communication applications are investigated in the COST 231 project dealing with evolution of land mobile radio (including personal communications) [66]. The low millimeter-wave band from 20 to 60 GHz, which is nearly unused and allows for large bandwidth applications, combines the advantages of infrared (IR) (enough free bandwidth) and ultrahigh frequency (UHF) (good coverage). Systems operating particularly in the 60 GHz frequency band can have a small reuse distance, because of oxygen absorption at the rate of 14 dB/km. However, the indoor radio channel shows adverse frequency selective multipath characteristics due to the highly reflective indoor environment, which results in severe signal dispersion and limits the maximum usable symbol rate. Another advantage is the fact that this frequency region is not in use by any other communications medium, so every channel can be allocated a large bandwidth: 100 MHz channels can be used without any bandwidth problems. A third advantage of millimeter wave technology is that antenna sizes are very small, so the equipment will be convenient. A fourth advantage is that the millimeter-wave spectrum has the potential to support broadband service access, which is especially relevant because of the advent of B-ISDN.

A major drawback of this frequency region is that the technology for transmitters and receivers has not yet been fully developed. As a consequence the hardware will be expensive in the early stages. It is worth mentioning that no definitive evidence of any hazards has been shown to date to the general public arising from prolonged exposure in fields of less than 10mW/cm^2 in millimeter-waves.

Within the European research program, the millimeter-wave spectrum has been selected for development of the mobile broadband systems (MBS) [57]. Mobile broadband system (MBS) typically addresses services above 2 Mbps. The high data rates envisaged for MBS require operation at much higher frequencies, currently estimated to be in the 60 GHz bands. Study of MBS needs a lot of investigation in terms of propagation modeling, antenna diversity, technology development, and so forth.

Chapters 13 and 16 describe in detail millimeter waves and wireless broadband multimedia communications, respectively.

1.4 PREVIEW OF THE BOOK

The book consists of 18 chapters. It covers all the necessary elements to achieve the GCV. The GCV system will provide communication services for any person anywhere in a multimedia environment by one pocket-sized unit using a personal reference number. To study the GCV system it is necessary to understand the behavior of macrocellular, microcellular, picocellular, satellitecellular, and overlaycellular systems. The propagation characteristics of these cellular systems are very hostile in nature. This book addresses related transmission system issues and problems.

To follow the mathematical concepts presented in the book, statistical radio channel models are reviewed in Chapter 2. Multipath fading, log-normal shadowing and near-far effects are discussed. Models are developed to evaluate the joint probability density function (pdf) in the presence of a number of radio signals suffering from multipath fading, log-normal shadowing, and ground wave propagation individually or combined. Level crossing rate and fade duration are also introduced. Finally, wireless channel classification and wideband characteristics are presented.

Chapter 3 reviews the well-known cellular theory by presenting its basic principle, frequency reuse, co-channel interference, carrier-to-interference ratio, hand-off mechanism and cell splitting. Next, the cellular family of macrocells, microcells, and picocells is introduced.

Chapter 4 first presents mathematical models for evaluating co-channel interference probability for a macrocellular system in the presence of Rayleigh fading, and uncorrelated lognormal shadowing. Interesting computational results are discussed. Cochannel interference probability is also evaluated analytically in the presence of correlated log-normal shadowing. Second, a mathematical model is developed to evaluate the bit error rate for a 4-quadrature amplitude modulation (QAM) and differential phase shift keying (DPSK) in the presence of co-channel interference, Gaussian and man-made noise. An analytical model is developed considering class-A, class-B, and class-C noise. Computational results are also presented which are followed by conclusions. The spectrum efficiency is defined and evaluated for both the analog and digital systems. The effect of co-channel interference, Gaussian noise, and man-made noise on the spectrum efficiency is analyzed.

Chapter 5 first presents an assessment of co-channel interference probability for a microcellular land mobile radio system by considering the desired signal as (fast) Rician fading with (slow) lognormal shadowing and co-channel interfering signals as uncorrelated (fast) Rayleigh fading superimposed over (slow) lognormal shadowing. The expression for co-channel interference probability is derived using appropriate path-loss law for microcells for four different cases: (1) Rician plus lognormal desired signal and Rayleigh plus lognormal interfering signals, (2) Rician desired signal and Rayleigh fading plus lognormal shadowing interfering signals, (3) Rician desired signal and Rayleigh interfering signal, and (4) both desired and interfering signals as Rician fading. The performance of a microcellular system is compared with that of a conventional macrocellular system.

Second, the performance of the microcellular digital mobile radio systems is investigated using DPSK modulation with co-channel interference, Gaussian, and class A man-made noise.

Chapter 6 first presents an analytical model to evaluate the co-channel interference probability using propagation measurement results of a picocellular environment at 1.9 GHz. Chapter 6 discusses propagation measurement results for the indoor wideband channel at 2.4, 4.75 and 11.5 GHz in an indoor laboratory/office environment, and bit error probability (BEP) model for the frequency selective multipath channel is developed. Expressions for the calculation of BEP in a stationary frequency selective multipath channel is derived for a coherent BPSK receiver. Numerical BEP results for a number of measured indoor channels are given.

An overview of adaptive equalization techniques is presented in Chapter 7. Simulation results are discussed for a 16-QAM system using square root raised cosine matched filtering with a decision feedback equalizer and measured impulse responses of indoor radio channels. Fast Kalman and least mean square adaptive training algorithms are investigated.

Chapter 8 reviews the CDMA systems considering direct sequence (DS), frequency hopping (FH), time hopping (TH), chirp spread spectrum, and hybrid CDMA systems. Finally, an overview of the design and properties of code sequences is presented for CDMA systems.

Chapter 9 presents an overview of contention (random) multiple access protocols in the wireless environment. First the multiple access protocols are classified and reviewed. Second, two types of random access protocols, namely slotted ALOHA, inhibit sense multiple access (ISMA) are discussed. Third, the throughput of packet radio channels is analyzed using the interference model in a radio fading environment and capture effect. Numerical results are presented, indicating the effect of propagation impairments on channel capacity. The results are of importance for mobile data networks, wireless office communications and other packet systems with contention-limited performance. Fourth, the performance of the protocols is evaluated using analytical models considering BPSK modulation. The effect of coding schemes on the performance is also investigated. Fifth, the performance of a cellular DS-CDMA system with imperfect power control and sectorization is analyzed in terms of capacity, throughput, and delay.

Chapter 10 presents the basic principles and applications of the orthogonal frequency division multiplex (OFDM) system. Analytical models are developed to evaluate the typical system performance.

The study of hybrid OFDM and DS-CDMA, known as multicarrier CDMA (MC-CDMA), has recently drawn a lot of attention. An overview of MC-CDMA and its performance analysis in a wireless environment are presented in Chapter 11.

Chapter 12 briefly presents the role of antennas in the field of wireless communication, antenna diversity, and combining techniques. Some well-known schemes such as selection diversity, maximal ratio combining, switch diversity, and polarization diversity are explained. Computational results for a DECT system are presented using mathematical models considering switched and selection diversity in a frequency-nonselective fading channel.

An overview of wireless communication systems is presented in millimeter waves in Chapter 13 considering the 60 GHz region. Outage probability and fade duration are evaluated for frequency-nonselective channel. System performance is evaluated in terms of BEP considering a frequency selective channel.

Chapter 14 describes the basic principles of dynamic channel assignment. Different channel assignment methods and DECT concept are reviewed briefly. The effect of delay time on the performance of dynamic channel assignment is investigated for a DECT system.

Chapter 15 presents an overview of worldwide air interference research activities toward the third-generation mobile communications (FPLMTS/ IMT-2000/ UMTS). Details of different CDMA, TDMA, OFDM, and hybrid air interference designs are given. Furthermore, the implications of time division duplex (TDD) operation are also discussed.

Wireless broadband multimedia communications (WBMC) are reviewed in Chapter 16. The WBMC systems are under investigation in creating a high-speed (up to 155 Mbps) local data communication link.

Chapter 17 presents a novel system using OFDM-based wireless ATM transmission for the future wireless broadband multimedia communications. First, the transmitter and receiver configuration of the proposed OFDM system are described. Then, system performance is presented and compared with the conventional OFDM transmission scheme.

Chapter 18 presents an overview of the satellite communications (SATCOM) that can be employed to provide connectivity where the terrestrial (FPLMTS/ IMT-2000/ UMTS) infrastructure is thin or lacking, as well as to enhance user mobility over large geographic regions. Special attention is given to introduce the broadband wide area networking via Internet protocol (IP) over ATM for SATCOM applications.

REFERENCES

[1] R. Prasad, *CDMA for Wireless Personal Communications*, Norwood, MA: Artech House, 1996.

[2] R. Prasad, "Research challenges in future wireless personal communications: microwave perspective," *Proc. 25th European Microwave Conference* (Keynote opening address), Bologna, Italy, pp. 4–11, September 4–7, 1995.

[3] R. Prasad, "Overview of wireless personal communications: microwave perspectives," *IEEE Comm. Mag.*, pp.104–108, Vol. 34, 1997.

[4] O.E. Dunlop, *Marconi the Man and His Wireless*, New York: The Macmillan Company, 1937.

[5] W.P. Jolly, *Marconi*, London: Constable, 1972.

[6] W.C. Jakes, Jr.(Ed.), *Microwave Mobile Communications*, John Wiley & Sons, 1974.

[7] W.C.Y. Lee *Mobile Communications Engineering*, New York: McGraw-Hill Book Company, 1982.

[8] J.E. Padgett, C. G. Gunther, and T. Hattori, "Overview of wireless personal communications," *IEEE Comm. Mag.*, Vol. 33, pp. 28–41, January 1995.

[9] K. Kinoshita, M. Kuramoto, and N. Nakajima, "Development of a TDMA digital cellular system based on Japanese standard," *Proc. 41st IEEE Vehicular Technology Conference*, pp. 642–645, 1991.

[10] H. Zhang, "Performance evaluation of the DS-CDMA transceiver for mobile communications," Ph.D. Thesis, Aalborg University, Denmark, 1996.

[11] G. Calhoun, *Digital Cellular Radio*, Norwood, MA: Artech House, 1988.

[12] B.C. Jones, "Interference modelling and cell design in micro-cellular networks," Ph.D. Thesis, Macquarie University, Australia, 1996.

[13] D.C. Cox, "Wireless network access for personal communications," *IEEE Comm. Mag.,* Vol. 30, pp. 96–115, December 1992.

[14] D.C. Cox, "Wireless personal communications: What is it?," *IEEE Personal Comm.,* Vol. 2, pp. 20–35, April 1995.

[15] D.J. Goodman, "Trends in cellular and cordless communications," *IEEE Comm. Mag,* Vol. 29, pp. 31–40, June 1991.

[16] R. Steel, "Deploying personal communication networks," *IEEE Comm. Mag,* Vol. 28, pp. 12–15, September 1990.

[17] *IEEE Comm. Mag,* special issue: *Wireless Personal Communications*, Vol. 33, January 1995.

[18] M. Mouly and M.B. Pautet, "The Evaluation of GSM," *in Mobile Communications—Advanced Systems and Components, Proc. 1994 Int. Zurich Sem. on Dig. Comm.*, (Springer-Verlag, LNCS, Vol. 783).

[19] M.R.L. Hodges, "The GSM radio interface," *Br. Telecom Tech. J.,* Vol. 8, pp. 31–43, January 1990.

[20] M. Mouly and M.B. Pautet, "Current evolution of the GSM systems," *IEEE Personal Comm.,* Vol. 2, pp. 9–19, October 1995.

[21] W.H.W. Tuttlebee (Ed.), *Cordless Telecommunications in Europe*, Springer-Verlag, 1990.

[22] W.H.W. Tuttlebee, "Cordless personal communications," *IEEE Comm. Mag.,* Vol. 30, pp. 42–62, December 1992.

[23] R. Prasad and C.K. Chatterjee, "An overview of wireless in local loop", In *Wireless Communication TDMA versus CDMA* (Ed., S.G. Glisic and P.A. Leppanen), Kluwer Academic Publishers, pp. 265–290, September 1997.

[24] D. Akerberg, "Novel radio access principles useful for third generation mobile radio systems," *Proc. Third IEEE Int. Symp. in PIMRC*, 1992, Boston, MA, September 1992.

[25] ETSI, *Digital European Cordless Telecommunications—Common Interface, Radio Equipment and Systems*, Valbonne, France.

[26] T. Hattori et al., "Personal communication—concept and architecture," *Proc. 1990 IEEE Int. Comm. Conf.* (ICC'90), pp. 1351–1357, April 1990.

[27] Research & Development Centre for Radio systems (RCR), *Personal Handy Phone Systems*, *RCR STD -28*, December 1993.

[28] K. Ogawa, K. Kohiyama, and T. Kobayashi, "Towards the personal communication era—a proposal of the radio access concept from Japan," *Int. J. Wireless Info. Networks,* Vol. 1, pp. 17–27, January 1994.

[29] S. Kato, "Developments of PCS in Japan," *Proc. IEEE Fourth Symp. on Comm. and Vehicular Technology in the Benelux*, Gent, Belgium, pp. 1–5, October 1996.

[30] K. Pahlavan and A.H. Levesque, "Wireless data communications," *Proc. IEEE,* Vol. 82, pp. 1398–1430, September 1994.

[31] P.W. Baier, P. Jung, and A. Klein, "Taking the challenge of multiple access for third generation cellular mobile radio systems—a European view," *IEEE Comm. Mag,* Vol. 34, pp. 82–89, February 1996.

[32] R. Prasad and A. Kegel, "Effects of Rician faded and log-normal shadowed signals on spectrum efficiency in micro-cellular radio," *IEEE Trans. on Vehicular Technology,* Vol. 42, pp. 274–281, August 1993.

[33] J.A. Schwarz da Silva and B.E. Fernendes, "The European research program for advanced mobile radio systems," *IEEE Personal Comm.,* Vol. 2, pp. 14–19, February 1995.

[34] R. Prasad and T. Ojanpera, "Development of UMTS in Europe," *Proc. IEEE Fourth Symp. on Comm. and Vehicular Technology in the Benelux*, Gent, Belgium, pp. 6–11, October 1996.

[35] ETSI ETR 271 *Objectives and Overview of UMTS*, February 1996.

[36] *UMTS Task Force Report,* February 1996.

[37] *ERO Report on UMTS Frequencies,* April 1996.

[38] *ETSI Technical Input to ERC/ERO Study on UMTS Frequencies,* 1996.

[39] ETSI Special Mobile Group (SMG), "Work program for the standardization of the Universal Mobile Telecommunication System (UMTS)," ETSI Doc. TRC-TR 15, June 1994.

[40] ETSI DTR/SMG-05-0102, *Vocabulary for UMTS.*

[41] ETSI ETR 291, *System Requirements for UMTS,* May 1996.

[42] ETSI DTR/SMG-05-0104, *Scenarios and Considerations for the Introduction of UMTS.*

[43] ETSI DTR/ SMG-05-0201, *Framework of Services to be Supported by UMTS.*

[44] ETSI DTR/SMG-05-0301, *Framework of Network Architecture for UMTS.*

[45] ETSI DTR/SMG-05-0401, *Overall Requirements on the Radio Interface(s) of UMTS.*

[46] ETSI DTR/SMG-05-0401, *Selection Procedures for Radio Transmission Principles for UMTS.*

[47] ETSI DTR/SMG-05-0501, *Objectives and Framework for TMN of UMTS.*

[48] ETSI DTR/SMG-05-0602, *Quality Requirements and Selection Procedure for Support of Voice Band Data Coding for UMTS.*

[49] ETSI DTR/SMG-05-0901, *Security Principles for UMTS.*

[50] ETSI DTR/SMG-05-1201, *Framework for Satellite Integration Within the UMTS.*

[51] ETSI DTR/SMG-05-1202, *Technical Characteristics, Capabilities and Limitations of Satellite System Applicable for UMTS.*

[52] ETSI ETS 050-2101, *System Concept and Reference Model for the UMTS.*

[53] ETSI ETS/SMG-05-02201, *UMTS Service Aspects; Service principles.*

[54] PAC EG5 Report, *Global Multimedia Mobility (GMM),* May, 1996.

[55] R. Prasad, "Wireless broadband communication systems," *IEEE Comm. Mag.*, Vol. 35, p. 18, January 1992.

[56] W. Honcharenko, J.P. Kruys, D.Y. Lee, and N.J. Shah, "Broadband wireless access," *IEEE Comm. Mag.,* Vol. 35, pp. 20–26, January 1997.

[57] L.M. Correia and R. Prasad, "An overview of wireless broadband communications," *IEEE Comm. Mag.,* Vol. 35, pp. 28–33, January 1997.

[58] M. Morinaga, M. Nakagawa, and R. Kohno, "New concepts and technologies for achieving highly reliable and high capacity multimedia wireless communications system," *IEEE Comm. Mag.* Vol. 35, pp. 34–40, January 1997.

[59] J.S. da Silva, B. Arroyo-Fernandez, B. Barani, J. Pereira, and D. Ikonomou, "Mobile and personal communications: ACTS and beyond," *Proc. PIMRC'97,* Helsinki, Finland, September 1997.

[60] F.D. Prisoli and R. Velt, "Design of medium access control and logical link control functions for ATM support in the MEDIAN system," *Proc. ACTS Mobile Comm. Summit'97,* pp. 734–744, Aalborg, Denmark, October 1997.

[61] R. Rheinschmitt, A. de Haz, and M. Umehinc, "AWACS MAC and LLC functionality," *Proc. ACTS Mobile Comm. Summit'97,* pp. 745–750, Aalborg, Denmark, October 1997.

[62] J. Aldid, E. Busking, T. Kleijne, R. Kopmeinen, and R. Van Nee, "Magic into reality, building the WAND modem," *Proc. ACTS Mobile Comm. Summit'97,* pp. 734–744, Aalborg, Denmark, October 1997.

[63] J. Mikkonen, C. Corrado, C. Erci, and M. Progler, "Emerging wireless broadband networks," *IEEE Comm. Mag.,* February 1988.

[64] G. Wu, Y. Hase, K. Taira, and K. Iwasak, "A wireless ATM oriented MAC protocol for-high-speed wireless LAN," *Proc. PIMRC'97,* pp. 198–203, Helsinki, Finland, September 1997.

[65] R. Prasad and L. Vandendorpe, "An overview of millimeter wave indoor wireless communication system," *Proc. 2nd Int. Conf. on Universal Personal Communications,* pp. 885–889, Ottawa, Canada 1993.

[66] E. Damosso and G. de Brito, "COST 231 achievements as a support to the development of UMTS: a look into the future," *IEEE Comm. Mag.* Vol. 34, pp. 90–96, February 1996.

Chapter 2
Radio Propagation Aspects

2.1 INTRODUCTION

Radio propagation characterization is the bread and butter of communications engineers. Without knowledge of radio propagation, a wireless system could never be developed. A radio engineer has to acquire full knowledge of the channel if he or she desires to be successful in designing a good radio communication system [1].

Therefore, knowledge of radio propagation characteristics is a prerequisite for the design of radio communication systems. A lot of measurements have been done to obtain information concerning propagation loss, spatial distribution of power when the environment is physically static, wideband and narrowband statistics concerning the random variables of received signals at a fixed location due to any movement of the surroundings, and delay spread. References [2–7] provide a good review of these measurements.

Radio channel is generally hostile (nasty) in nature. It is very difficult to predict its behavior. Therefore, radio channel is modeled in a statistical way using real propagation measurement data. In general, the signal fading in a radio environment between a transmitter and a receiver can be decomposed into a large-scale path loss component together with a medium-scale, slow-varying component having a lognormal distribution, and a small-scale fast varying component with a Rician or Rayleigh distribution depending on the presence or absence of the line of sight (LOS) situation between the transmitter and receiver [7–16]. Accordingly, a three-stage propagation model should be appropriate to describe a wireless cellular environment: (1) area mean power depending on the path loss characteristics in the range from the transmitter to the area where the receiver is located, (2) local mean power within that area, which is slow varying, can be represented by a lognormal distribution, and (3) superimposed fast-fading instantaneous power that follows a Rayleigh or Rician distribution. Figure 2.1 shows scene A multipath fading that yields fast fluctuation in the received instantaneous power. Scene B in Figure 2.1 shows the man-made structures (buildings) that cause slow variation in the received local mean power. This phenomenon is called shadowing.

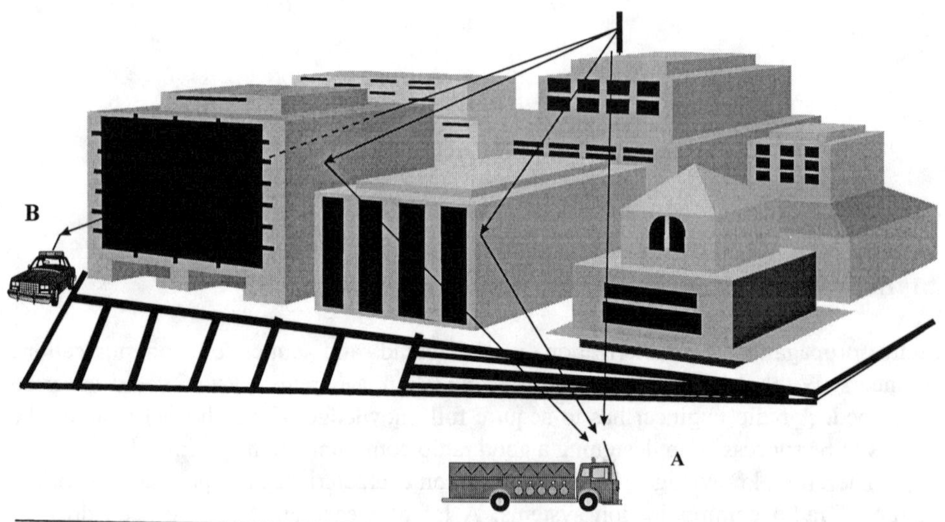

Figure 2.1 Multipath propagation (scene A) and shadowing (scene B).

The performance of a wireless communication system is strongly affected by its environment. Outdoor as well as indoor radio channels can be characterized as multiple channels with shadowing. Multiple reflections of the radio signal cause the signal to arrive at the receiver via multiple paths, which differ in amplitude, phase, and delay time. Figure 2.2 shows an impression of the multipath channel in an office environment.

This chapter is organized as follows. A very detailed account of the narrowband characteristics of fading channels is presented in Sections 2.2 to 2.6. Large-scale path loss component is discussed in Section 2.2. Sections 2.3 and 2.4 present medium-scale, slow-varying and small-scale, fast-varying components, respectively. Combined small-scale slow- and medium-scale fast-varying components are analyzed in Section 2.5. Level crossing rate and fade duration are introduced in Section 2.6.

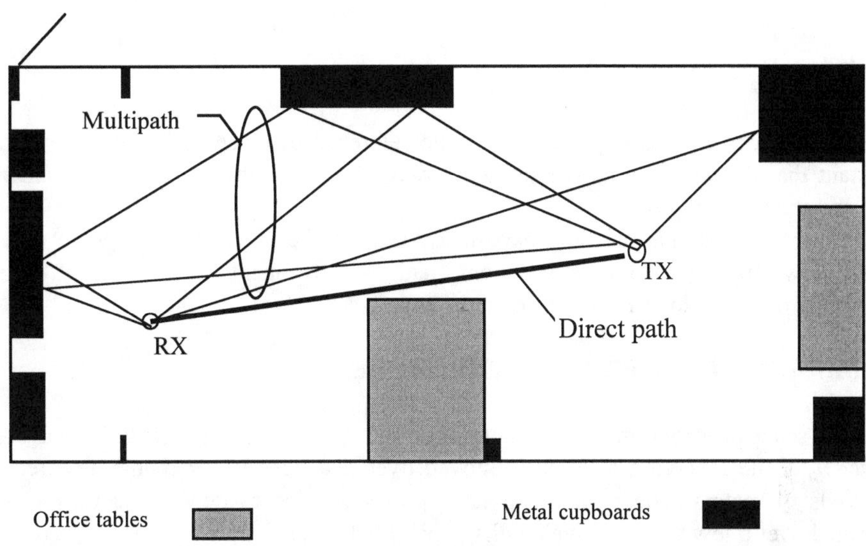

Figure 2.2 Multipath channel in an office room.

Section 2.7 describes the classification of channels. A short description of the wideband characteristics of the channel is given in Section 2.8. A detailed investigation of wideband channel is presented in Chapter 6. Finally, conclusions are given in Section 2.9.

2.2 LARGE-SCALE PROPAGATION MODEL

An extreme variation in the transmission path between the transmitter and receiver can be found, ranging from direct LOS to severely obstructed paths due to buildings, mountains, and foliage. The phenomenon of decrease in the received power with distance due to reflection, diffraction around structures, and refraction within them is known as path loss. Propagation models have been developed to determine the path loss and are known as large-scale propagation models because they characterize the received signal strength by averaging the power level over large transmitter receiver separation distances in the range of hundreds or thousands of meters. Therefore, the average power is referred to as the area mean power p_a. In general, the spatially averaged power (i.e., area mean power p_a, at a point that is distance d from the transmitter is a decreasing function of d. Usually this function is represented by a path-loss power law of the form

$$p_a = C\left(\frac{d}{d_0}\right)^{-\gamma} \tag{2.1}$$

where γ is the path-loss exponent, d_0 is the reference distance and C is an arbitrary constant that depends on the transmitted power, the transmitter, and receiver antenna gain and the wavelength.

In free space, the path-loss exponent $\gamma = 2$, so the power law obeys an inverse square law. In a wireless environment, path-loss exponent varies from 1.8 to 6 depending on the indoor and outdoor environment.

2.3 MEDIUM-SCALE PROPAGATION MODEL

Medium-scale propagation models determine the gradual changes in the local mean power p_0 if the receiving antenna is moved over distances larger than a few tens or hundreds of meters. The term local mean power is used to denote the power level averaged over a few tens of wavelengths, typically 40 λ [10,12].

Since the wireless environment continuously fluctuates due to the presence of mountains, forests, or man-made structures like buildings, the received local mean power, p_0, changes gradually. This effect is called shadowing. The received local mean power, p_0, fluctuates with a lognormal distribution about the area mean power. By lognormal it is meant that the local mean, expressed in logarithmic values such as decibels or nepers, has a normal distribution (i.e., Gaussian distribution) about the area mean power

$$f_{p_0}\{\ln(p_0)|p_a\} = \frac{1}{\sqrt{2\pi}\sigma}\exp\left[-\frac{\{\ln(p_0)-m\}^2}{2\sigma^2}\right] \tag{2.2}$$

where σ and $p_a = \exp(m)$ are the logarithmic standard deviation of the shadowing and the area mean power, respectively.
By applying the following transformation

$$f(x) = f\{\ln(x)\}\left|\frac{d\ln(x)}{dx}\right| \tag{2.3}$$

to (2.2), the local mean power, p_0, is expressed by the lognormal pdf

$$f_{p_0}\{p_0|p_a\} = \frac{1}{\sqrt{2\pi}\sigma p_0}\exp\left[-\frac{\{\ln(p_0)-m\}^2}{2\sigma^2}\right] \tag{2.4}$$

The lognormal distribution has been confirmed in a number of propagation surveys [8,17–19]. The standard deviation, σ, varies between 4 and 12 dB, depending on the severity of the shadowing.

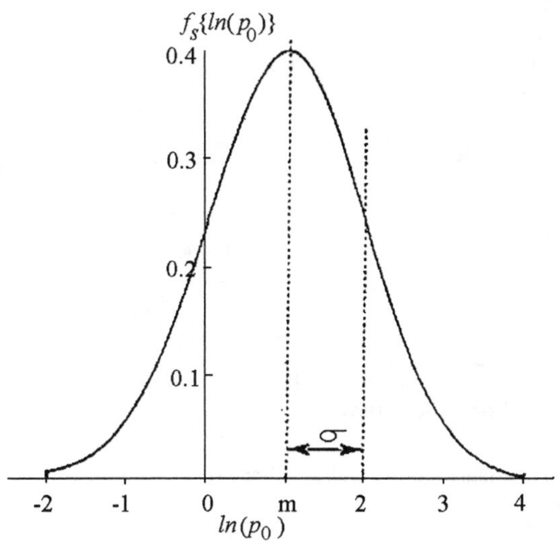

Figure 2.3 Normal distribution of $\ln(p_0)$ with $m = 1$ and $\sigma = 1$.

Figure 2.3 shows the normal distribution of $\ln(p_0)$ and the lognormal distribution of p_0 is shown in Figure 2.4.
The mean μ and variance D^2 of P_0 can be obtained as follows [20]

$$\mu = \bar{p}_0 = \text{mean} \quad of \quad p_0 = \exp(m + \frac{\sigma^2}{2}) \tag{2.5}$$

$$D^2 = \text{variance of } p_0 = [\exp(\sigma^2)-1]\exp(2m + \sigma^2) \tag{2.6}$$

The median of P_0 is

$$\hat{p}_0 = \exp(m) \tag{2.7}$$

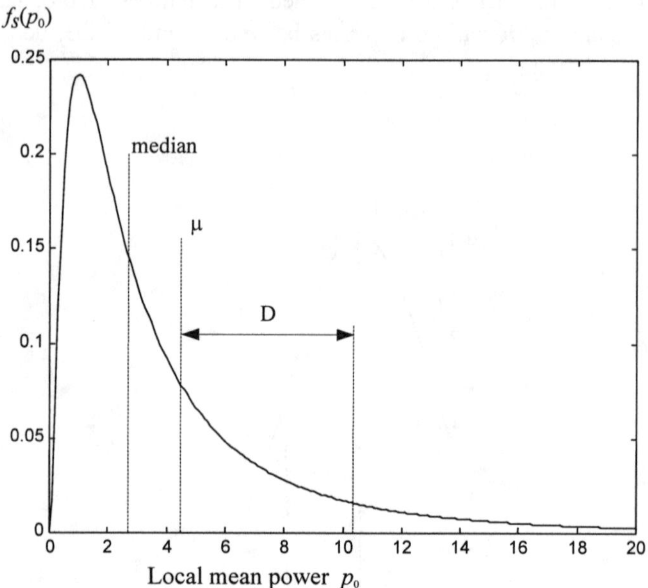

Figure 2.4 Lognormal distribution of P_0 with $m = 1$ and $\sigma = 1$.

2.3.1 Sum of Lognormal Signals

In wireless communications, simultaneous transmission of information takes place. Therefore, it is often required to find the statistics of sum of these information signals. The sum of n stochastic independent lognormal variables can be approximated by another lognormal variable [20–25]. An approximate pdf of the sum of random signals with lognormal probability distribution can be obtained using Fenton's method [20]. Unfortunately, Fenton's method is suitable for small signal standard deviation ($\sigma < 4$ dB) only, while typical values of σ lie between 4 and 12 dB in wireless channels. An improved technique for approximating the probability distributions of the sum of several random variables with such high variances, based on a method by Schwartz and Yeh [21], has been reviewed by Prasad and Arnbak [24]. Schwartz and Yeh [21] derived exact expressions for the mean and variance for sum of two lognormal variables and then used a recursive approach to obtain area mean, $p_n = \exp(m_n)$ and the variance, σ_n^2, for n variables. Therefore, the pdf for $p_{0n}\left(=\sum_{i=1}^{n} p_{0i}\right)$ is given by

$$f_{p_{0n}}(p_{0n}) = \frac{1}{\sqrt{2\pi}\sigma_n p_{0n}} \exp\left[-\frac{\{\ln(p_{0n}) - m_n\}^2}{2\sigma_n^2} \right] \tag{2.8}$$

Figure 2.5 illustrates how to calculate the area mean power and variance for n signals.

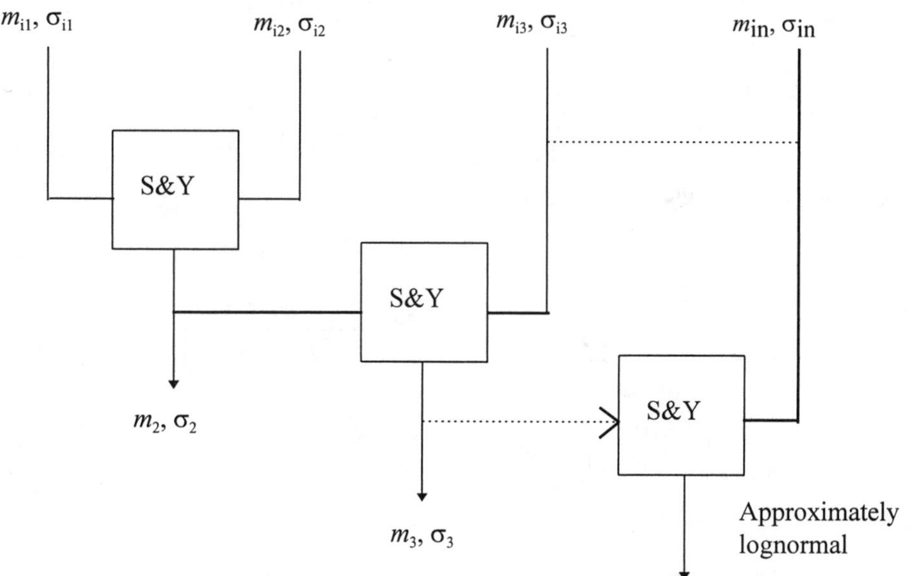

Figure 2.5 Recursive process of the Schwartz and Yeh (S & Y) r... determining mean and variance of sum of n lognormal signals.

Accordingly, the steps are shown for computing area mean power and variance for two interference variables.

Step 1: Convert m_1, m_2, σ_1, and σ_2 in natural logarithms

$$m_n \stackrel{\Delta}{=} 0.23026 m_{dB} \tag{2.9}$$

Step 2: Define

$$m_w \stackrel{\Delta}{=} m_{n2} - m_{n1} \qquad \sigma_w^2 \stackrel{\Delta}{=} \sigma_{n2}^2 + \sigma_{n1}^2 \tag{2.10}$$

Step 3: Determine

$$m_m = (m_{n1} + G_1) \qquad \sigma_m^2 = (\sigma_{n1}^2 - G_1^2 - 2\sigma_{n1}^2 G_3 + G_2) \tag{2.11}$$

where

$$G_1 \stackrel{\Delta}{=} m_w \phi\left(\frac{m_w}{\sigma_w}\right) + \frac{\sigma_w}{\sqrt{2\pi}} e^{-m_w^2/2\sigma_w^2} + \sum_{k=1}^{\infty} C_k e^{k^2\sigma^2{}_w/2}\left[e^{km_w\phi}\left(\frac{-m_w - k\sigma_w^2}{\sigma_w}\right) + T_1\right] \tag{2.12}$$

$$G_2 \stackrel{\Delta}{=} \sum_{k=1}^{\infty} b_k T_2 + \left[1 - \phi\left(-\frac{m_w}{\sigma_w}\right)\right](m_w^2 + \sigma_w^2)$$

$$+ \frac{m_{w'}\sigma_w}{\sqrt{2\pi}} e^{-m_w^2/2\sigma_w^2} + \sum_{k=1}^{\infty} b_k e^{-(k+1)m_w + (k+1)^2\sigma^2{}_{w/2}} \phi\left[\frac{m_w - \sigma_w^2(k+1)}{\sigma_w}\right]$$

$$-2\sum_{k=1}^{\infty} C_k e^{(-m_w k + k^2\sigma^2{}_{w/2})}\left[m_k \phi\left(-\frac{m_k}{\sigma_w}\right) - \frac{\sigma_w}{\sqrt{2\pi}} e^{-m_k^2/2\sigma_w^2}\right] \tag{2.13}$$

$$G_3 \stackrel{\Delta}{=} \sum_{k=0}^{\infty} (-1)^k e^{k^2\sigma^2{}_w/2} T_1 + \sum_{k=0}^{\infty} (-1)^k T_2$$

$$T_1 \stackrel{\Delta}{=} e^{-km_w} \phi\left(\frac{m_w - k\sigma_w^2}{\sigma_w}\right) \tag{2.14}$$

$$T_2 \stackrel{\Delta}{=} e^{\left[m_w(k+1) + (k+1)^2\sigma_w^2/2\right]} \phi\left[\frac{-m_w - \sigma_w^2(k+1)}{\sigma_w}\right] \tag{2.15}$$

$$b_k = \left[2(-1)^{k+1}/(k+1)\right]\sum_{j=1}^{k} j^{-1}, \quad C_k = (-1)^{k+1}/k, \quad \phi(x) = 1/2 + (1/2)erf(x/\sqrt{2}),$$

and $erf(x) = (2/\sqrt{\pi})\int_0^x \exp(-t^2)dt.$

Table 2.1 gives values for m_n and σ_n as a function of number of signals, n, with the initial value of $m_i = 0$ dB.

Table 2.1

Standard Deviation (σ_n) and Area Median Power (m_n) for $n = 1$ to 6 Interferers Subject to Lognormal Shadowing.

	$\sigma_i = 6$ dB $m_i = 0$ dB		$\sigma_i = 12$ dB $m_i = 0$ dB	
n	σ_n dB	m_n dB	σ_n dB	m_n dB
1	6.00000	0.00000	12.00000	0.00000
2	4.57930	4.57589	9.58168	7.45252
3	3.93286	6.90771	8.39718	11.20030
4	3.53877	8.43359	7.65698	13.61733
5	3.25742	9.56948	7.13423	15.37234
6	3.03997	10.47850	6.73756	16.73545

Initial value of $m_i = 0$ dB.

Fenton's method [20] makes it simple to evaluate the area-mean power and variance of n signals, and therefore this method is also reviewed. Remember that Fenton's method should be used when $\sigma < 4$ dB.

Figure 2.6 illustrates the use of Fenton's method to calculate ${\sigma_n}^2$ and m_n for n signals.

Fenton showed that the sum of n stochastic independent lognormal variables is well approximated by another lognormal variable with mean

$$m_n = \ln\left[\sum_{i=1}^{n} \exp\left(m_i + \frac{\sigma_i^2}{2} - \frac{\sigma_n^2}{2}\right)\right] \tag{2.16}$$

and logarithmic variance

$$\sigma_n^2 = \ln\left[1 - \frac{\sum_{i=1}^{n}\left\{1 - \exp(\sigma_i^2)\right\}\exp(2m_i + \sigma_i^2)}{\sum_{i=1}^{n}\exp(m_i + \frac{\sigma_i^2}{2})}\right] \tag{2.17}$$

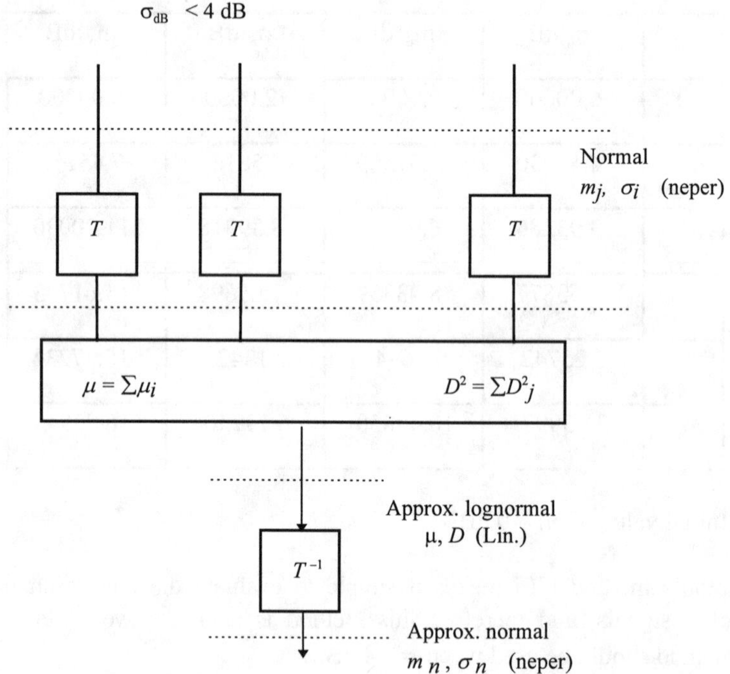

Figure 2.6 Fenton's method for determining mean and variance of sum of n lognormal signals.

2.4 SMALL-SCALE PROPAGATION MODEL

Small-scale propagation models are those that characterize the fast variation of the received signal strength over a short distance on the order of a few wavelengths or over short time durations on the order of seconds. Small-scale fast fading, also known as short-term fading or multipath fading, is due to multipath reflections of a transmitted wave by local scatterers such as houses, buildings, and man-made structures, or natural objects such as a forest surrounding a mobile unit. The pdf for the signal amplitude r of the received signal, conditional on its local mean power p_0, is given by Rayleigh density function [26–29].

$$f_r(r/p_0)\begin{cases} = \dfrac{r}{p_0}\exp(-\dfrac{r^2}{2p_0}) & (0 \le r \le \infty) \\ = 0 & (r < 0) \end{cases} \tag{2.18}$$

Rayleigh pdf is shown in Figure 2.7.

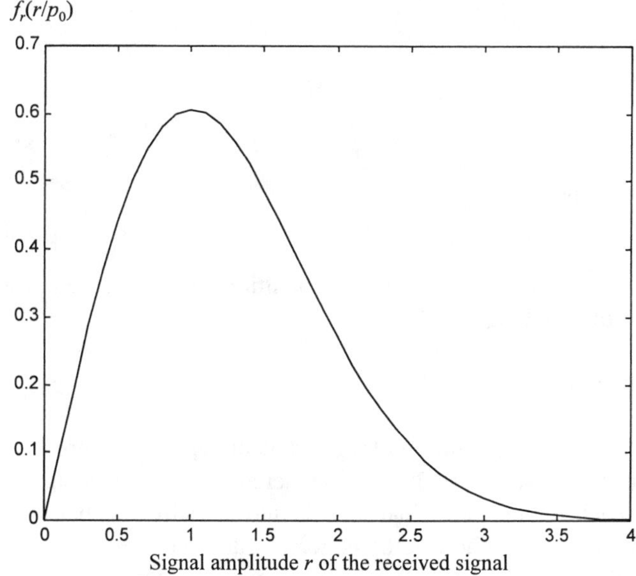

Figure 2.7 Rayleigh pdf.

The mean value is given by

$$E(r) = \int_0^\infty r f_r(r) dr = \sqrt{\frac{\pi}{2} p_0} \tag{2.19}$$

The mean square is given by

$$E(r^2) = \int_0^\infty r^2 f_r(r) dr = 2p_0 \tag{2.20}$$

The variance is given by

$$\sigma^2 = E(r^2) - [E(r)]^2 = (2 - \frac{\pi}{2}) p_0 \tag{2.21}$$

The pdf for the instantaneous power is obtained using the following transformation

$$f_y(y) = f_x(x) \left| \frac{dx}{dy} \right| \tag{2.22}$$

to (2.18) by substituting $p_i = \dfrac{r^2}{2}$

$$f_{p_i}(p_i | p_0) = \frac{1}{p_0} \exp\left(-\frac{p_i}{p_0} \right) \tag{2.23}$$

Thus, the pdf for the instantaneous power conditional to local mean power is given by the exponential distribution (2.23).

2.4.1 Sum of Exponential Signals

There are two situations of addition of signals, namely, (1) incoherent addition, and (2) coherent addition of signals [10,25,29]. Incoherent addition of n signals is required if the phases of each of the individual signals substantially fluctuates due to mutually independent modulation. Whereas the coherent addition of n signals is performed, if during the observation, the phase fluctuations caused by the modulation of signals are sufficiently small, and the carrier frequencies of the n signals are exactly equal.

In case of incoherent addition, the pdf of the sum of n incoherent signals is obtained by the n-fold convolution of the exponential distribution (2.23) with identical local mean power. Thus, the pdf of the sum of n signals is given by the gamma distribution.

$$f_{p_{in}}(p_{in}|p_0) = \frac{1}{p_0}\frac{(p_{in}|p_0)^{n-1}}{(n-1)!}\exp\left(-\frac{p_n}{p_0}\right) \tag{2.24}$$

where $p_{in} = \sum_{j=1}^{n} p_{ij}$

In the event of coherent addition, the sum of n signals behaves as a Rayleigh phasor, so the instantaneous power, p_{in}, is exponentially distributed with the local mean power levels of the individual signals.

$$f_{p_{in}}(p_{in}|p_0) = \frac{1}{np_0}\exp\left(-\frac{p_{in}}{np_0}\right) \tag{2.25}$$

2.4.2 Rician Distribution

In a situation where the distance between the transmitter and the receiver is small and the environment is static, there is a fixed spatial pattern of maxima and minima. Mostly there is a dominant stationary (nonfading) signal component, such as an LOS propagation path. In such a situation, the small-scale propagation model is given by the Rician distribution. The small-scale, fast-varying signal level r around the local mean p_0 is given by a Rician distribution where the pdf is given by

$$f_r(r|p_0')\begin{cases} = \dfrac{r}{p_0'}\exp\left[-\dfrac{r^2+s^2}{2p_0'}\right]I_0\left(\dfrac{rs}{p_0'}\right) & for(s \geq 0, r \geq 0) \\ = 0 & for(r < 0) \end{cases} \tag{2.26}$$

where $I_0(\ )$ is the modified Bessel function of the first kind and zeroth order, s is the peak value of the specular radio signal, p_0' is the average power of the signal, $p_0' = p_0/\ (K+1)$, p_0 is the total received local mean power, and K is the Rician factor that is defined as the ratio of the average specular power and the average fading power received over specular paths

$$K \underline{\Delta} \frac{s^2}{2p_0'} \tag{2.27}$$

When the direct signal does not exist (i.e., when s becomes zero ($K = 0$)), (2.26) reduces to (2.18). Also, when s is large, (2.26) is approximately Gaussian. Therefore, Rayleigh statistics are one extreme of the Rician statistics, while Gaussian are another extreme case of the Rician statistics. Therefore, the concept of a Gaussian channel to be unrealizable in wireless communications is not true. It is very possible to have a large value of Rician parameter (i.e., to have an LOS component with almost no multipath component), giving a Gaussian channel [6,30].The shape of the Rician distribution is plotted in Figure 2.8.

The pdf for the instantaneous signal power p_i is given by using $p_i = \frac{1}{2}r^2$ and transformation (2.22)

$$f_{p_i}(p_i|p_0) = \frac{1}{p_0'}\exp\left[-\frac{2p_i + s^2}{2p_0'}\right] I_0\left[\frac{\sqrt{2p_i}\,s}{p_0'}\right] \tag{2.28}$$

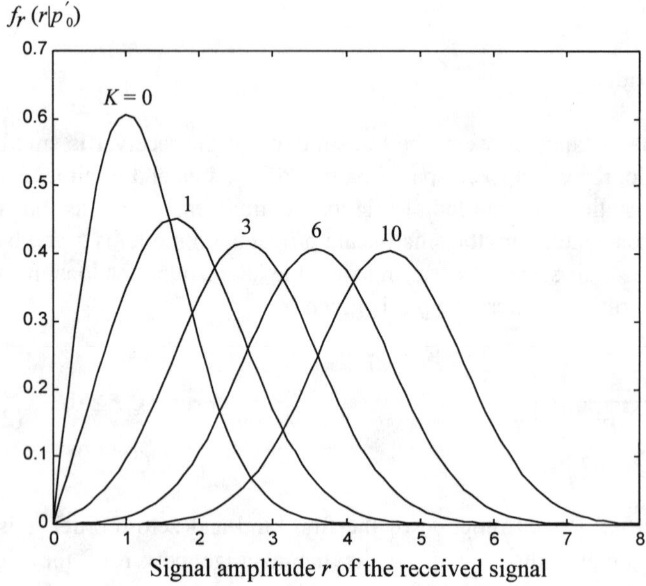

Figure 2.8 Rician pdf.

Equation (2.28) can be written as

$$f_{p_i}(p_i|p_0) = \frac{K+1}{p_0}\exp\left[-\frac{K+1}{p_0}p_i - K\right] I_0\left[2\sqrt{\frac{K^2+K}{p_0}p_i}\right] \tag{2.29}$$

The pdf for the sum of n Rician signals p_{in} with equal local mean power of each signal is contained by convolving (2.29) n times [30–32] (Appendix 2A).

$$f_{p_{in}}(p_{in}|p_0) = \left(\frac{K_{si}+1}{p_0}\right)^{\frac{n+1}{2}} \left(\frac{p_{in}}{nK_{si}}\right)^{\frac{n-1}{2}} \exp\left(-\frac{K_{si}+1}{p_0}p_{in} - nK_{si}\right) I_{n-1}\left(\sqrt{\frac{4n(K_{si}^2+K_{si})}{p_0}p_{in}}\right) \tag{2.30}$$

when K_{si} is the Rice factor of each signal.

2.4.3 Nakagami Distribution

Nakagami [31] studied the situation where amplitude and phase of the reflected signals are both randomly distributed, which leads to a realistic model of a small-scale radio channel. The pdf of the signal amplitude r is described by the Nakagami m- distribution

$$f_r(r|p_0) = \frac{2m^m r^{2m-1}}{\Gamma(m)p_0^m} \exp\left(-\frac{mr^2}{p_0}\right) \qquad m \geq \frac{1}{2} \tag{2.31}$$

where Γ(m) is the Gamma function [32], with $\Gamma(m+1) = m!$.
Figure 2.9 shows the Nakagami distribution using m as a parameter.
The Nakagami distribution is used to model the fast-multipath-fading channel due to the following [33,34]:

1. m is the fading figure associated with various propagation conditions, irrespective of anything more or less severe than the Rayleigh fading. The higher the value of m, the less the fading is. For $m = 1$, (2.31) reduces to the Rayleigh distribution (2.18). For $m = 1/2$, (2.31) becomes a one-sided Gaussian distribution and for $m \to \infty$, (2.31) becomes an impulse (no fading).

2. It can be shown that there is a direct relation between the Rice factor K and the Nakagami fading figure m, given by

$$m = \frac{(1+K)^2}{1+2K} = 0.4998528K + 0.7622159 \tag{2.32}$$

The pdf for the instantaneous power conditioned on the local mean power is given by

$$f_{p_i}(p_i|p_0) = \left(\frac{m}{p_0}\right)^m \frac{p_i^{m-1}}{\Gamma(m)} \exp\left(-\frac{mp_i}{p_0}\right) \tag{2.33}$$

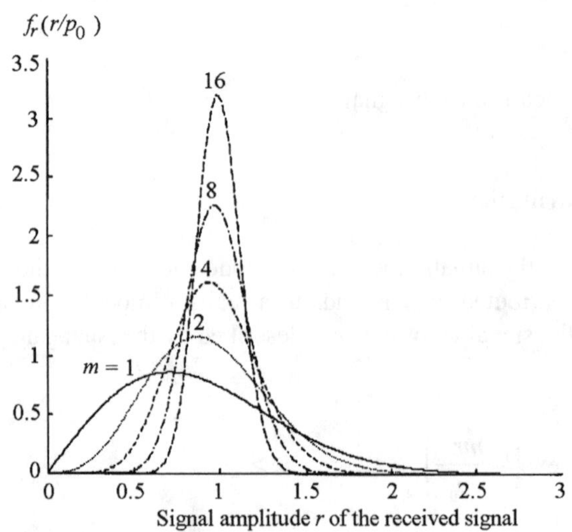

Figure 2.9 The Nakagami pdf for several values of m with $p_0 = 1$.

2.5 COMBINED RAYLEIGH FADING AND LOGNORMAL SHADOWING

The composite pdf for the instantaneous received power p_i of the ith Rayleigh faded and lognormal shadowed signal is

$$f_i(p_i) = \frac{1}{\sqrt{2\pi}\sigma_i} \int_{-\infty}^{\infty} \exp(-p_{gi}) \exp\left[-p_i \exp(-p_{gi})\right] \exp\left[-\frac{(p_{gi} - m_i)^2}{2\sigma_i^2}\right] dp_{gi} \tag{2.34a}$$

where $p_{gi} = \ln(p_{0i})$.

In the general case of combined multipath fading and shadowing, considering incoherent addition of the individual, stochastically-independent signals, the composite pdf for the (individual) ith signal power can be approximated by the lognormal pdf with the same first and second moments as those of (2.34a). The vth moment of (2.34a) with index i about the origin is

$$\alpha_v = v!\exp\left[m_i v + \frac{(v\sigma_i)^2}{2}\right] \tag{2.34b}$$

The logarithmic variance and area mean of the approximate lognormal pdf of (2.34a) with index *i* are given by

$$\sigma_i^2 = [\sigma_i^2 + 1n(2)] \tag{2.34c}$$

and $p_{ai} = \dfrac{p_{ai}}{\sqrt{2}}$ (2.34d)

Table 2.2

Standard Deviation (σ_n) and Area Median Power (m_n) for $n = 1$ to 6 Interferers Subject to Rayleigh Fading and Lognormal Shadowing.

	$\sigma_i = 6$ dB $m_i = 0$ dB		$\sigma_i = 12$ dB $m_i = 0$ dB	
n	σ_n (dB)	m_n (dB)	σ_n (dB)	m_n (dB)
1	7.00525	−1.50515	12.53290	−1.50515
2	5.41069	3.51069	10.02766	6.22217
3	4.66320	6.05705	8.79895	10.10659
4	4.20419	7.71282	8.03063	12.61139
5	3.87870	8.93309	7.48808	14.42947
6	3.62925	9.89978	7.07652	15.84395

Initial value of $m_i = 0$ dB.

The above approximation has been numerically verified. Subsequently, involving the method of Schwartz and Yeh [21], the composite pdf for the (total) sum p_{in} can then be approximated by a pure lognormal distribution with the appropriate moments.

$$f\left(p_{in}\middle|p_{an}\right) = \frac{1}{\sqrt{2\pi}\sigma_n p_n}\exp\left\{-(p_u - m_n)^2 / 2\sigma_n^2\right\} \tag{2.34e}$$

Here, $p_u \stackrel{\Delta}{=} \ln p_n$, while p_{an} (= exp m_n) and σ_n^2 are the area mean power and variance, respectively, of the equivalent lognormal variable. They are found in accordance with the procedure in Section 2.3.

Table 2.2 gives values for m_n and σ_n as a function of the number of interferers, *n*, with the initial value of $m_i = 0$ dB.

To compare the system performance in coherent and incoherent interference conditions, we write the composite pdf for *n* signals with combined Rayleigh fading and lognormal shadowing for coherent addition using (2.4), (2.8), (2.18), and (2.25):

$$f_{pin}\left(p_{in}|p_{an}\right) = \frac{1}{\sqrt{2\pi}\sigma_n}\int_0^\infty \frac{1}{p_{0n}^2}\exp\left(-p_{in}|p_{0n}\right)\exp\left\{-(\ln p_{0n} - m_n)^2 / 2\sigma_n^2\right\}dp_{0n} \tag{2.35}$$

2.6 FADING ENVELOPE STATISTICS

Investigation of fading envelope statistics is very important in the field of wireless communications, because with the help of the fading statistics, efficient error-control codes and diversity techniques can be developed. Signals in a fading environment result in a drop of the signal amplitude below the threshold; this phenomenon is called fade. Figure 2.10 illustrates the fading signal behavior. According to the signal behavior in a fading environment, fade timings are divided into fade interval (t_F) and nonfade interval (t_{NF}). The threshold is defined above the noise plus undesired (i.e., interfering) signal. The level crossing rate (LCR) and average fade duration of a fading signal are two important statistical parameters.

2.6.1 LCR

The LCR, N_R, is defined as the expected rate at which the fading signal envelope *r* crosses a specified signal level *R* in a positive-going direction. The expected rate is determined by evaluating the average number of level crossing and given by

$$N_R \Delta \int_0^\infty \dot{r} f(R,\dot{r}) d\dot{r} \tag{2.36}$$

where $f(R,\dot{r})$ is the joint pdf of r and $\dot{r}$ at r = R and $\dot{r}$ is the time derivative of $r(t)$ (i.e., dr/dt).

A Rayleigh fading signal $f(R,\dot{r})$ is given by [8]

$$f(R,\dot{r}) = \frac{R}{p_0}\exp(-R^2/2p_0)\frac{1}{\sqrt{2\pi b_2}}\exp(-\dot{r}^2/2b_2) \tag{2.37}$$

where b_2 is the second moment and expressed by

$$b_2 = 2\pi^2 f_m^2 p_0 \tag{2.38}$$

and f_m is the maximum Doppler shift given by

$$f_m = \frac{V}{\lambda} = \frac{fV}{C} \tag{2.39}$$

with V as the mobile velocity, λ as the wavelength, C as the speed of light 3.10^8 m/s and f is the frequency.

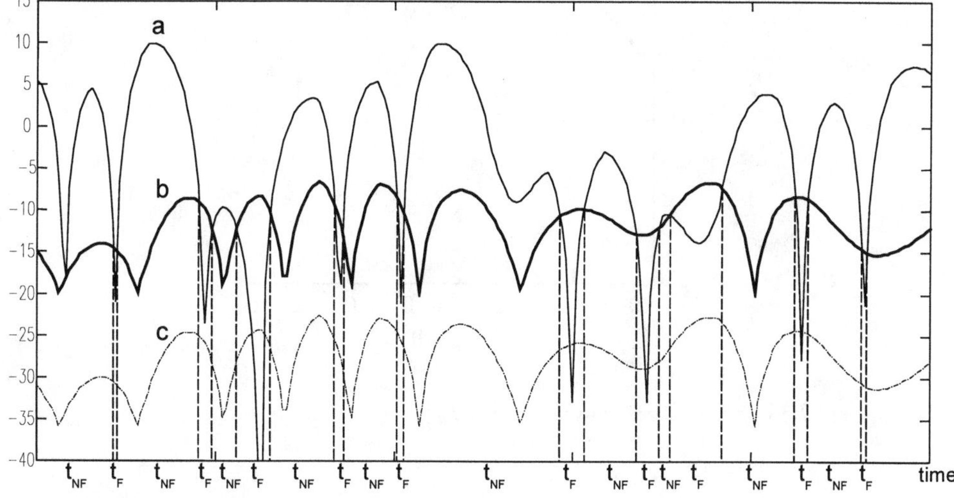

Figure 2.10 Fade t_F, and nonfade t_{NF} for (a) desired signal, (b) threshold, and (c) undesired signal.

N_R can be obtained using (2.36) and (2.37) [8]

$$N_R = \sqrt{2\pi} f_m \rho \exp(-\rho^2) \tag{2.40}$$

where $\rho = R/R_{rms}$ and R_{rms} is the local rms (root mean square) amplitude of the fading signal.

It is now evident from (2.39) and (2.40) that the LCR is a function of the mobile speed V.

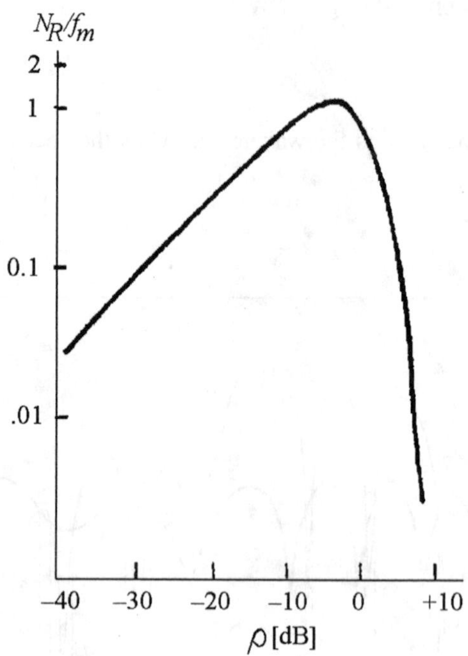

Figure 2.11 Normalized crossing rate N_R/f_m.

Figure 2.11 shows the normalized level crossing rate, N_R/f_m. It can be seen from Figure 2.11 that N_R is maximum at $\rho = -3$ dB (i.e., $\rho = \rho = 1/\sqrt{2}$). N_R decreases with both increase and decrease of the fading signal level. Since the level crossing rate is specified for a positive direction, when a low level is then set, the signal remains above this level for most of the time; a similar argument applies when a level is much greater than the RMS value selected. At f_c = 900 MHz and V = 48 Km/h, f_m =40 Hz, thus the N_R is 39 per second at $\rho = -3$ dB [13]. The fading signal experiences very deep fade only occasionally, but the shallow fade occurs more frequently.

2.6.2 Average Fade Duration

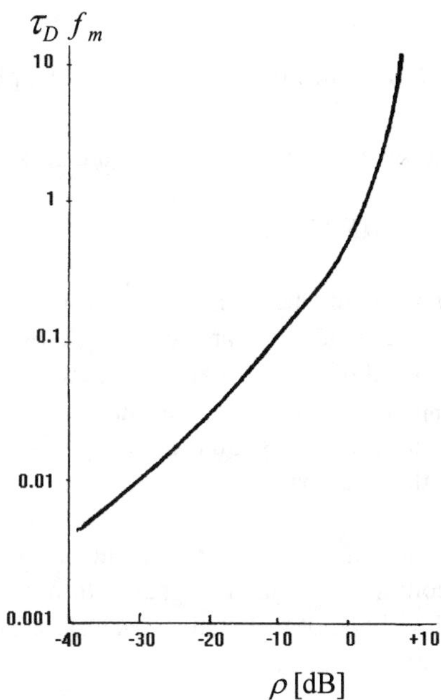

Figure 2.12 Normalized average fade duration $\tau_D f_m$.

The average fade duration, τ_D, is defined as the average duration of time for which the fading signal is below the specified signal level R,

$$\tau_D \triangleq \frac{\Pr ob(r \leq R)}{N_R} \tag{2.41}$$

For a Rayleigh fading signal (2.18) Prob ($r \leq$ R), that is, the probability that the fading signal r is less than or equal to R and is given by

$$\text{Prob}(r \leq R) = \int_0^R f(r)dr = \int_0^R \frac{r}{p_0} \exp(-\tfrac{r^2}{2p_0})dr = 1 - \exp(-\rho^2) \tag{2.42}$$

Now one obtains τ_D using (2.38), (2.40), and (2.41)

$$\tau_D = \frac{\exp(\rho^2) - 1}{\sqrt{2\pi}\rho f_m} \quad (2.43)$$

Thus, the average fade duration τ_D depends on the maximum Doppler frequency, f_m, (i.e., mobile speed).

The normalized average fade duration $\tau_{D.} f_m$ is shown in Figure 2.12.

2.7 WIRELESS CHANNEL CLASSIFICATION

The wireless channel is defined as a link between a transmitter and a receiver and classified considering the coherence bandwidth and coherence time. The coherence bandwidth is based on the delay spread of the channel and the coherence time is related to the Doppler spread. Delay spread is spreading in time and Doppler spread is spreading in frequency. The delay spread occurs if there is a significant widening of the received impulse signal due to the multipath scattering. Figure 2.13 illustrates the phenomenon of delay spread.

As shown in Figure 2.13, when the transmitter transmits an impulse signal, $s(t)$, the received impulses, $r(t)$, becomes a continuous signal with a pulse width τ if the number of scatterness in the neighborhood of the receiver increases. The delay spread is defined as the broadness received pulse width τ [9].

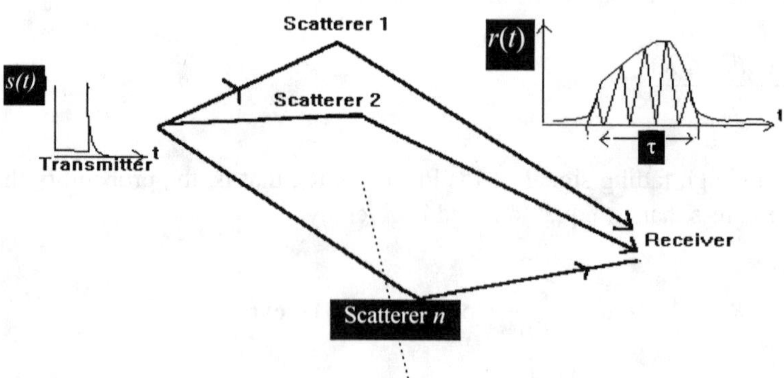

Figure 2.13 Delay spread.

The Doppler spread occurs due to the Doppler effect on the moving transmitter or receiver as shown in Figure 2.14.

The Doppler spread, f_m, is expressed by $(V/\lambda)\cos\theta$. Here, V is the velocity and λ is the wavelength. The channel classification is based on the relationship of the transmitting signal band width W with the coherence bandwidth $(\Delta f)_c$ and the signal duration or bit time, T_b, with the coherence time $(\Delta t)_c$.

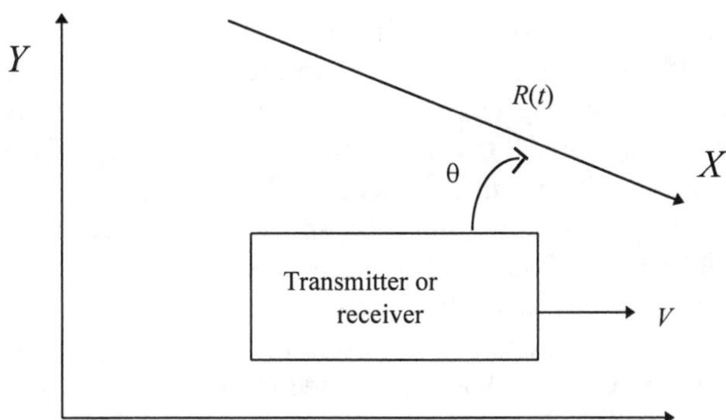

Figure 2.14 Doppler spread.

2.7.1 Coherence Bandwidth and Delay Spread

The impulse response, which can be interpreted as a power delay intensity function, yields useful information about the multipath characteristic of the channel. This power delay profile may be treated as a delay density function, where the delay is weighted by the signal level at that delay. An appropriate way to characterize the power delay profile is the rms multipath delay spread denoted by T_m, which is defined as the standard deviation (i.e., the square root of the second central moment or variance) of the power delay profile. Practically, the rms delay spread T_m is determined from the measured power delay profile by calculating the standard deviation of this profile [35,36].

$$T_m \triangleq \sqrt{E\left(\tau^2\right) - E^2(\tau)} \tag{2.44}$$

with

$$E(\tau)\Delta\frac{\sum_{l}\tau_{l}\beta_{l}^{2}}{\sum_{l}\beta_{l}^{2}} \tag{2.45}$$

and

$$E(\tau^{2})\Delta\frac{\sum_{l}\tau_{l}^{2}\beta_{l}^{2}}{\sum_{l}\beta_{l}^{2}} \tag{2.46}$$

where β_l is the received power of the first pulse.

Figure 2.15 shows an example of a measured impulse response.

Another possible way to characterize the power delay profile is by defining a maximum time delay spread, that is determined by the range of delays over which the power delay profile is essentially nonzero [37]. For measurements one can, for example, choose the maximum time delay spread to be the delay at which the power of the received pulse is 30 dB lower than the first received pulse.

Typical values for the rms time delay spread measured in an indoor environment at a 910-MHz frequency are in the range of 50 up to 250 ns [38]. In [39–41], rms delay spread values from 10 up to 20 ns are reported for a frequency band of 2.4 to 11.5 GHz. Maximum delay spreads of 300 ns have been measured in [36] and maximum delay spreads of 270 ns have been reported in [41–44] and other references for indoor communication in the frequency band of 850 MHz to 4 GHz. For a detailed overview, refer to [5]. In an outdoor environment, typical values for the rms time delay spread in urban areas are in the range of 10 to 25 μs at 892 MHz [45]. The average standard deviation and maximum value of the power delay profile at 910 MHz are 1300, 600, and 3500 ns, respectively, in an urban environment measured in New York City. Averaged typical and averaged extreme rms delay spread values at 910 MHz are in the range of 200 to 310 ns and 1960 to 2110 ns, respectively, in a suburban environment [46].

If the data bit duration is larger than the rms delay spread, then the channel introduces a negligible amount of intersymbol interference(ISI). The reciprocal of the rms time delay spread is a measure for the coherence bandwidth $(\Delta f)_C$ of the channel:

$$(\Delta f)_c \approx \frac{1}{T_m} \tag{2.47}$$

The coherence bandwidth, $(\Delta f)_C$, is the bandwidth over which the signal propagation characteristics are correlated (i.e., the width of the frequency correlation function $\varphi_C(\Delta f)$). Thus, two sinusoids with frequency separation larger than the coherence bandwidth are affected differently by the channel. If the coherence bandwidth $(\Delta f)_C$ is small compared with the bandwidth of the transmitted signal W, the channel is frequency selective. In this case, the different frequency components in the signal are subject to different gains and phase shifts. On the other hand, if the coherence bandwidth is large compared with the signal bandwidth W, the channel is frequency nonselective and all frequency components are subject to the same attenuation and phase shift.

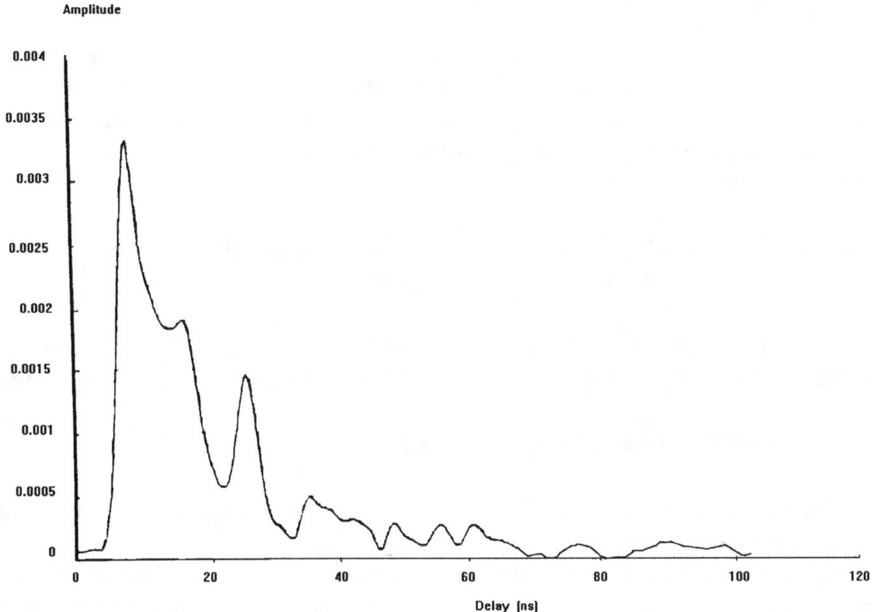

Figure 2.15 Measured power delay profile.

The frequency correlation function, $\varphi_C(\Delta f)$, and the delay profile, $\varphi_C(\tau)$, can be related by applying the Fourier transform:

$$\varphi_C(\Delta f) \xleftrightarrow{\text{Fourier}} \varphi_C(\tau) \tag{2.48}$$

It is worth mentioning here that an exact relationship between coherence bandwidth and rms delay spread does not exist.

If the coherence bandwidth, $(\Delta f)_C$, is defined as the bandwidth over which the frequency correlation function is above 0.9, then $(\Delta f)_C$ becomes approximately [7,11]

$$(\Delta f)_c \approx \frac{1}{50T_m} \tag{2.49}$$

If the correlation function is above 0.5, $(\Delta f)_C$ is approximated to

$$(\Delta f_c)_c \approx \frac{1}{5T_m} \tag{2.50}$$

Since in general spectral analysis techniques and simulation are required to determine the exact impact that time-varying multipath has on a particular transmitted signal [47–50], accurate multipath channel must be used in the design of specific modems for wireless applications [51,52].

Resuming: $W >> (\Delta f)_C \Rightarrow$ frequency selective channel
$W << (\Delta f)_C \Rightarrow$ frequency nonselective channel

The system bandwidth of a spread-spectrum system is generally larger than the coherence bandwidth, which implies that in that case the channel is frequency selective.

2.7.2 Coherence Time and Doppler Spread

To classify the time characteristics of the channel, the coherence time and Doppler spread are important parameters. The coherence time is the duration over which the channel characteristics do not change significantly. The time variations of the channel are evidenced as a Doppler spread in the frequency domain, which is determined as the width of the spectrum when a single sinusoid (constant envelope) is transmitted. Both the time correlation function $\varphi_C(\Delta t)$ and the Doppler power spectrum $\varphi_C(f)$ can be related to each other by applying the Fourier transform:

$$\varphi_C(\Delta t) \xleftarrow{\text{Fourier}} \varphi_C(f) \tag{2.51}$$

The range of values of the frequency f over which $\varphi_C(f)$ is essentially nonzero is called the Doppler spread B_d of the channel. Since $\varphi_C(f)$ is related to $\varphi_C(\Delta t)$ by the Fourier transform, the reciprocal of B_d is a measure of the coherence time $(\Delta t)_C$ of the channel; that is:

$$(\Delta t)_c \approx \frac{1}{B_d} \tag{2.52}$$

The coherence time $(\Delta t)_C$ is a measure of the width of the time correlation function. Clearly, a slow-changing channel has a large coherence time, or equivalently, a small Doppler spread. The rapidity of the fading can now be determined either from the correlation function $\varphi_C(\Delta t)$ or from the Doppler power spectrum $S_C(f)$. This implies that either the channel parameters $(\Delta t)_C$ or B_d can be used to characterize the rapidity of the fading. If the bit time T_b is large compared to the coherence time, then the channel is subject to fast fading. When selecting a bit duration that is smaller than the coherence time of the channel, the channel attenuation and phase shift are essentially fixed for the duration of at least one signaling interval. In this case the channel is slowly fading or quasi-static. Like coherence bandwidth, there is no exact relationship between coherence time and Doppler spread. If the coherence time $(\Delta t)_C$ is defined as the time over which the time correlation function is above 0.5, $(\Delta t)_C$ is approximately given by [7,50]

$$(\Delta t)_c \approx \frac{9}{16\pi B_d} \tag{2.53}$$

In the modern digital communication, a rule of thumb to define $(\Delta t)_C$ as the geometric mean of (2.52) and (2.53); is:

$$(\Delta t)_c \triangleq \sqrt{\frac{9}{16\pi B_d^2}} \tag{2.54}$$

The Doppler spread B_d, is also known as Doppler shift f_m. The Doppler spread or shift is experienced by the components of the received signal, if either the transmitter or receiver is in motion. The frequency shift is related to the spatial angle between the direction of arrival of that component and the direction of vehicular motion. If a vehicle is moving at a constant speed V along the X axis, the Doppler shift f_m of the mth plane-wave component [13] is given by

$$f_m = \frac{V}{\lambda}\cos\alpha_m \tag{2.55}$$

where α_m is the arrival angle of the mth plane-wave component relative to the direction of movement of the vehicle. It is worth noting here that waves arriving from ahead of the vehicle experience a positive Doppler shift while those arriving from behind the vehicle have a negative shift. The maximum Doppler shift occurs at $\alpha_m = 0$

$$f_m = \frac{V}{\lambda} \tag{2.56}$$

Resuming: $T_b >> (\Delta t)_c \Rightarrow$ fast-fading channel
$T_b << (\Delta t)_c \Rightarrow$ slow-fading channel

Indoor measurements [47] show that in any fixed location, temporal variations in the received signal envelope caused by movement of personnel and machinery are slow, having a maximum Doppler spread of 6.1 Hz. Table 2.3 gives an overview of the terminology that is used to characterize the time and frequency domain properties of fading.

Table 2.3

Terminology

Channel Models	$T_b << (\Delta t)_c$	$T_b >> (\Delta t)_c$	
$W<<(\Delta f)_c$	Nondispersive flat-flat fading	Time dispersive frequency-flat fading	Frequency nonselective channel
$W>>(\Delta f)_c$	Frequency dispersive time-flat fading	Doubly (time and frequency) dispersive	Frequency selective channel
	Time nonselective or slow-fading channel	Time selective or fast-fading channel	

2.8 WIDEBAND CHARACTERISTICS

The multipath nature of the channel results in dispersion of the received signal in the time-domain and frequency selectivity in the frequency domain. In Figure 2.16, the received signal power is shown as a function of frequency for a sample multipath channel.

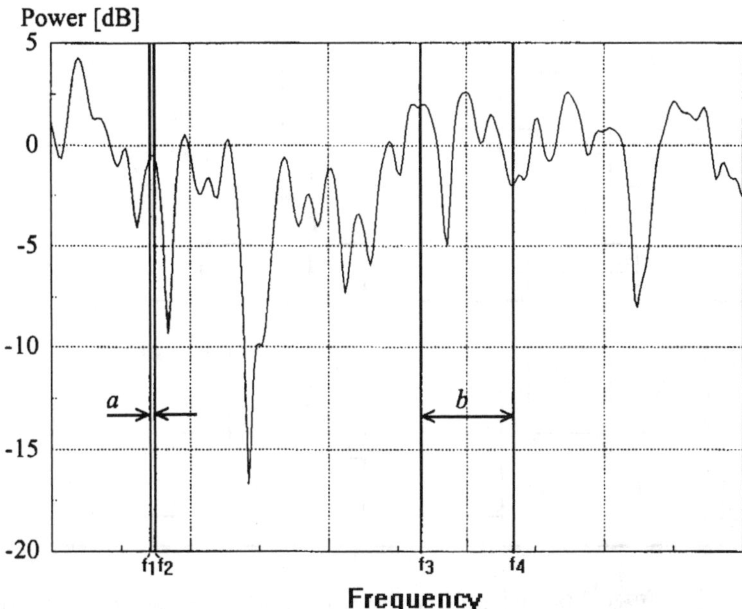

Figure 2.16 Received signal power as a function of the frequency in a multipath channel. (a) narrowband signal, (b) wideband signal.

If the signal bandwidth is smaller than the coherence bandwidth of the channel; that is, the expected bandwidth over which the transfer function remains practically constant, the signal is called narrowband. A narrowband signal suffers from amplitude (or flat) fading in a multipath channel [10], as shown in Figure 2.16(a). If the signal bandwidth is larger than the coherence bandwidth, frequency selective fading occurs as shown in Figure 2.16(b), and the signal suffers from time dispersion that causes intersymbol interference (ISI) and limits the maximum usable symbol rate [5]. So, ISI results when different symbols of the same signal interfere with each other. The distortion inflicted by ISI on the received signal is shown in Figure 2.17 for a two-path channel with impulse response $h(t) = \delta(t) + 0.5\delta(t - T_b / 2)$ with half a bit period T_b delay between the paths.

For wideband signals, maximum channel delays of many symbol periods may occur, which cause heavy distortion of the received signal. In addition, errors in the estimated carrier phase and timing instant due to the wideband nature of the signal may further degrade system performance. Several techniques can be applied to reduce or get rid of ISI in wideband signal transmission (e.g., equalization, spread-signal modulation, and antenna diversity). Some of these techniques are briefly discussed below.

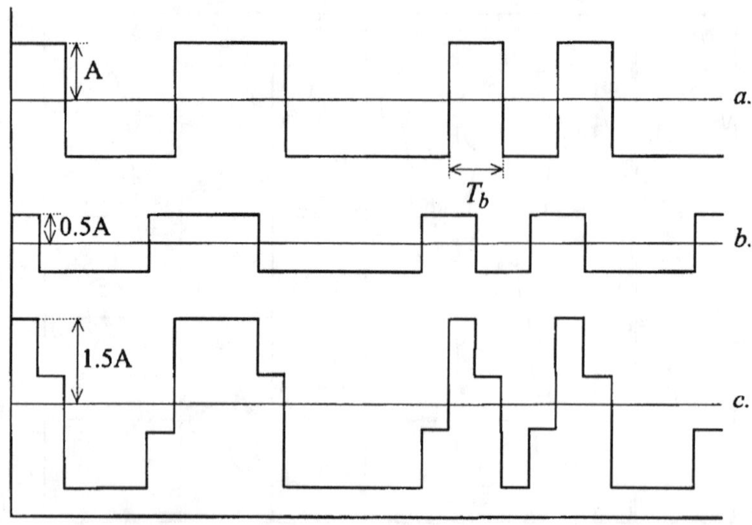

Figure 2.17 The effect of ISI: (a) the signal received via path 1, (b) the signal received via a delayed path 2, (c) the total received signal with ISI distortion.

Equalization. By using an equalizer [53], the received signal is filtered in such a way that ISI is eliminated or reduced. Ideal ISI elimination is achieved when the filter is the inverse of the channel response. Clearly, the channel must be known, or accurately estimated, to perform effective equalization. Therefore, the equalizer needs to be trained to adapt itself to the time-varying channel in wireless systems. Usually this is achieved by transmitting a training sequence. Equalization of the signal results in a decrease of ISI at the cost of a lower signal-to-noise ratio (SNR) [54]. Chapter 7 presents an overview of adaptive equalization.

Direct sequence spread spectrum (DS-SS). In DS-SS modulation [35,55] the signal is multiplied with a code that results in a signal with a much wider bandwidth than the original information-bearing signal. The ratio between the bandwidth after spreading and the bandwidth of the nonspread signal is called the processing gain. In a time-dispersive multipath channel, the spread signal replicas, which travel via different paths, are un-correlated if the path delays are more than one symbol period, T_C, apart from each other. After decorrelation in the receiver, the signal replicas from different paths are combined in a Rake receiver [54], thus all received energy is effectively used. A disadvantage of using DS-SS with high bit-rate signals is that to achieve a sufficiently high processing gain, a very large bandwidth is required. This is especially the case in an indoor environment, where the delay times between the paths are very short, in the order of 1 ns. Basic concepts of CDMA are discussed in Chapter 8.

OFDM. In OFDM [56,57], the symbols of the high bit rate signal are distributed over a large number of subcarriers. This results in a low symbol rate per carrier. The individual carrier signals, therefore, do not suffer from ISI, but still are subject to flat fading. OFDM is a promising technique for future high bit-rate applications. However, it suffers from a number of problems: a very linear amplifier in the transmitter is required to prevent signal distortion, accurate synchronization in the receiver is needed, and in the transmitter and receiver real-time discrete Fourier transform (DFT) operations have to be computed. OFDM is discussed in detail in Chapters 10 and 17.

Antenna diversity. In general, the term diversity is used to indicate techniques to enhance the transmission performance by distributing the available energy over different independent channels [54]. By combining the energies of the independent signal replicas, the probability of a bad resulting signal decreases compared with each of the individual channels. Examples of diversity techniques are frequency diversity (as used in frequency hopping (FH) spread spectrum [55]), time diversity (as used in interleaving), polarization diversity and antenna or space diversity (as used in antenna combining or selection techniques).

Antenna or space diversity is a well-known technique to suppress or repair radio channel impairments [58–60]. Signals received by different antenna elements are used for selection or combining, with the aim of improving the quality of the receiver input signal.

For microdiversity, the antenna elements are separated by a short distance in the order of a wavelength λ [61]. The configuration of antenna elements, which is called an array, becomes rather small at microwave and higher frequencies. With antenna diversity, not only a decrease in ISI is achieved but also an increase in SNR. This is not possible with equalization. Chapter 12 discusses the application of antenna diversity for wireless personal communications.

Frequency bands in the microwave and millimeter-wave region (i.e., from 5–60 GHz [62]) are candidates for high bit-rate communication systems. Advantages of these frequency bands are :

1. Availability of large bandwidths;
2. Shielding of concrete partitioning, which allows for small cell sizes [63] and reuse of the same frequency at a short distance;
3. Antenna dimensions can be small.

Millimeter-wave communications and wireless broadband multimedia communications are presented in depth in Chapters 13 and 16, respectively.

2.9 CONCLUSIONS

This chapter presents an overview of radio propagation characteristics. Since the study of a wireless channel and its application to develop a system needs a thorough understanding of the probability and random variables, some frequently used terms are revisited below [64,65].

2.9.1 Random Variable Terms

There are four types of random variables:
(1) Continuous random variables defined as a continuous distribution function, (2) discrete random variables defined as a stepped distribution function, (3) mixed-type random variables defined as a distribution function that is discontinuous but not necessarily flat between discontinuities, and (4) lattice-type random variables defined as a discontinuous random variable whose cumulative distribution function has discontinuities at regularly spaced intervals [64–66].

Moments
(a) The nth moment of a discrete random variable x is given by

$$m_n \underline{\Delta} E(x^n) = \int_{-\infty}^{\infty} x^n f(x) dx \tag{2.57}$$

where $f(x)$ is the pdf of x. Note that the first moment, m, is the same as mean.
(b) The kth central moment μ_k of a discrete random variable is given by

$$\mu_k \underline{\Delta} E\left[(x-\bar{x})^k\right] = \int_{-\infty}^{\infty} (x-\bar{x})^k f(x) dx \tag{2.58}$$

where $\bar{x}$ is the mean,
(c) The second central moment is called the variance and denoted by σ^2. The positive square root of variance is known as the standard deviation σ. The third and fourth central moments are known as the skew and kurtosis, respectively.

$$\text{(d)}\, \sigma^2 = E\left[(x-\bar{x})^2\right] = E(x^2) - [E(x)]^2 \tag{2.59}$$

Characteristic Function $\phi_X(w)$

$$\Phi_x(w) \underline{\Delta} E[\exp(jwx)] = \int_{-\infty}^{\infty} f(x)\exp(jwx)dx = F^{-1}[f(x)] \tag{2.60}$$

where F^{-1} is the inverse Fourier transform.

(a) If x and y are independent random variables and $z = x + y$, then

$$\Phi_z(w) = \Phi_x(w)\Phi_y(w) \tag{2.61}$$

$$\text{(b)}\, m_n = \left\{ j^{-n} \frac{j^n}{jw^n} \phi_x(w) \right\}_{w=0} \tag{2.62}$$

Probability Density Function

If x and y are continuous random variables, joint pdf is written as $f_{XY}(x,y)$. If x and y are statistically independent,

$$f_{xy}(x,y) = f_x(x) f_y(y) \tag{2.63}$$

$$f_x(x) = \int_{-\infty}^{\infty} f_{xy}(x,y)dy \tag{2.64}$$

$$f_y(y) = \int_{-\infty}^{\infty} f_{xy}(x,y)dx \tag{2.65}$$

The conditional pdf $f(x|y)$ and joint pdf $f(x,y)$ are related as

$$f(x|y) = \frac{f(x,y)}{f(y)} \tag{2.66}$$

If x and y are statistically independent, then

$$f(x|y) = f(x) \tag{2.67}$$

(i) Joint moment m_{kn} of order $k + n$

$$m_{kn} \underline{\Delta} E(x^k y^n) = \int_{-\infty}^{\infty}\int_{-\infty}^{\infty} x^k y^n f_{xy}(x,y)dxdy \tag{2.68}$$

(ii) Joint central moment μ_{kn}

$$\mu_{kn} \underline{\Delta} E\left[(x-\bar{x})^k (y-\bar{y})^n\right] \tag{2.69}$$

(iii) Joint central moment for discrete random variables

$$m_{kn} = E(x^k y^n) \tag{2.70}$$

Correlations

(i) Correlation $R_{xy} \underline{\Delta} E(xy) = \int_{-\infty}^{\infty}\int_{-\infty}^{\infty} xyf_{xy}(x,y)dxdy$ (2.71)

(ii) Covariance $C_{xy} \underline{\Delta} E\left[(x-\eta_x)(y-\eta_y)\right]$ $\quad |C_{xy}| \le \sigma_x\sigma_y$ (2.72)

(iii) Correlation coefficient $r_{xy} \underline{\Delta} \dfrac{C_{xy}}{\sigma_x\sigma_y}$ $\quad |r_{xy}| \le 1$ (2.73)

(iv) Uncorrelated random variables have the following features

(a) $C_{xy} = r_{xy} = 0$

(b) $E(xy) = E(x)\, E(y)$

(c) $\sigma^2_{x+y} = \sigma_x^2 + \sigma_y^2$

(d) Two statistically independent variables are also uncorrelated, but the opposite is not true.

(vi) Two variables are orthogonal if and only if $E(xy) = 0$.

Probability Densities

(i) $U = g_1(x,y)$
$V = g_2(x,y)$

$$f_{uv}(u,v) = \frac{f_{x,y}(x,y)}{|J|} \quad |(x,y) = g^{-1}(u,v) \tag{2.74}$$

where $J = det\begin{bmatrix} \frac{\delta}{\delta x} g_1 & \frac{\delta}{\delta y} g_1 \\ \frac{\delta}{\delta x} g_2 & \frac{\delta}{\delta y} g_2 \end{bmatrix}$

$$\text{(ii) } f_y(y) = f_x(y[x])\left|\frac{dy}{dx}\right| \tag{2.75}$$

where $y(x)\underline{\Delta} f^{-1}(y)$, if y = f(x)

Averages

$$\text{(i) Mean } m_n(t) = \frac{1}{N}\sum_{i=1}^{N}(t,s_i) \tag{2.76}$$

where N is the number of repetitions of the experiment, and $x(t,s_i)$ is the reduction observed in the ith repetition.

(ii) Mean or autocorrelation from the time average of a single reduction

$$\langle X(t)\rangle_T = \frac{1}{2T}\int_{-T}^{T} X(t,s)dt \tag{2.77}$$

(iii) The ergodic theorem states that if X_n is an independent, identically distributed (iid) discrete-time random process with finite mean $E[X_n] = m$, then the time average of the samples converges to the ensemble average with a probability of 1.

Autocorrelation Functions (ACF) $R_X(t_1,t_2)$

$$R_x(t_1,t_2)\underline{\Delta}E(x_1,x_2) = \int_{-\infty}^{\infty}\int_{-\infty}^{\infty} x(t_1)x(t_2)f_{x_1}f_{y_1}[x(t_1),x(t_2)]dx_1dx_2 \tag{2.78}$$

(i) ACF is nonnegative

$$\text{(ii) Power in process } x = E[x^2(t)] = R_X(t,t) = \sigma_X^2 \tag{2.79}$$

(iii) Process for which $R_X(t,t) < \propto$ for all t, is a second-order process

(iv) $R_X(t,s) = R_X(s,t)$ for all t and s (2.80)

(v) $\left|R_x(t,s)\right| \le \sqrt{R_x(t,t)R_x(s,s)}$ (2.81)

(vi) Autocovariance function (ACVF) $K_X(t_1\ t_2)$

$$K_x(t_1,t_2)\underline{\Delta}E\left[\{x(t_1)-m_x(t_1)\}\{x(t_2)-m_x(t_2)\}\right]$$
$$= R_x(t_1,t_2)-m_x(t_1)m_x(t_2) \quad (2.82)$$

(vii) $x(t)$ and $y(t)$ are uncorrelated if for all possible t and s

$$R_{XY}(t,s) = \mu_X(t)\ \mu_Y(s) \quad (2.83)$$

REFERENCES

[1] R. Prasad, "European radio propagation and subsystems research for the evolution of mobile communications," *IEEE Comm. Mag.*, Vol. 34, p. 58, February 1996.

[2] J.B. Anderson, T.S. Rappaport, and S. Yoshida, "Propagation measurements and models for wireless communications channels," *IEEE Comm. Mag.*, Vol. 33, pp. 42–49, January 1995.

[3] B.H. Fleury and P.E. Leuthold, "Radiowave propagation in mobile communications: an overview of European research," *IEEE Comm. Mag.*, Vol. 34, pp. 70–81, February 1996.

[4] D. Molkdar, "Review on radio propagation into and within buildings," *IEE Proc-H*, Vol. 138, pp. 61–73, February 1991.

[5] H. Hashemi, "The indoor propagation channel," *Proc. IEEE*, Vol. 81, pp. 943–968, July 1993.

[6] R. Steel (Ed.), *Mobile Radio Communications*, London: Pentech Press Publishers, 1992.

[7] T.S. Rappaport, *Wireless Communication Principles & Practice*, New Jersey: Prentice-Hall PTR, 1996.

[8] W.C. Jakes, Jr. (Ed.), *Microwave Mobile Communications*, New York: John Wiley & Sons, 1974.

[9] W.C.Y. Lee, *Mobile Communications Engineering*, New York: McGraw-Hill Book Co., 1982,

[10] J.-P. Linnartz, *Narrowband Land-Mobile Radio Networks*, Norwood, MA: Artech House, 1993.

[11] W.C.Y. Lee, *Mobile Cellular Telecommunications Systems*, New York: McGraw-Hill Book Co., 1989.

[12] W.C.Y. Lee, *Mobile Communications Design Fundamentals,* New York: John Wiley & Sons., Inc, 1993.

[13] R.C.V. Macrario (Ed.), *Personal & Mobile Radio Systems*, Peter Peregrinus Ltd., 1991.

[14] G.L. Stuber, *Principles of Mobile Communications*, Boston: Kluwer Academic Publishers, 1996.

[15] A. Mehrotra, *Cellular Radio Performance Engineering*, Norwood, MA: Artech House, 1994.

[16] G.C. Hess, *Land-Mobile Radio System Engineering*, Norwood, MA: Artech House, 1993.

[17] J.J. Egli, "Radio propagation above 40 Mc/s over irregular terrain," *Proc. IRE*, pp. 1383–1391, October 1957.

[18] D.M. Black and D.O. Reudink, "Some characteristics of radio propagation at 800 MHz in the Philadelphia area," *IEEE Trans. on Vehicular Technology*, Vol. 21, pp. 45–51, May 1972.

[19] T. Okumura, E. Ohmori, T. Kawano, and K. Fukuda, "Field strength and its variability in UHF and VHF land mobile services," *Rev. Elec. Comm. Lab.*,Vol. 16, p. 825, September – October 1968.

[20] L.F. Fenton, "The sum of lognormal probability distributions in scatter transmission systems," *IRE Trans. Comm. Syst.*, Vol. CS-8, pp. 57–67, March 1960.

[21] S.C. Schwartz and Y.S. Yeh, "On the distribution function and moments of power sums with lognormal components," *Bell Syst. Tech. J.*, Vol. 61, pp. 1441–1462, September 1982.

[22] N.C. Beaulieu, A.A. Abu-Dayya, and P.J. Mclane, "Comparison of methods of computing lognormal sum distributions and outages for digital wireless applications," *Proc. IEEE Int. Conf. on Comm. (ICC'94)*, New Orleans, LA, pp. 1270–1275, May 1994.

[23] J.B. Punt and D. Sparreboom, "Summing received signal powers with arbitrary probability density functions on a logarithmic scale," *Wireless Personal Comm. An Int. J.* (Kluwer Academic Publishers), Vol. 3, No. 3, pp. 215–224, 1996.

[24] R. Prasad and J.C. Arnbak, "Comments on analysis for spectrum efficiency in single cell trunked and cellular mobile radio," *IEEE Trans. on Vehicular Technology*, Vol. 37, pp. 220–222, November 1988.

[25] R. Prasad and A. Kegel, "Improved assessment of interference limits in cellular radio performance," *IEEE Trans. on Vehicular Technology*, Vol. 40, pp. 412–419, May 1991. (See also R. Prasad and A. Kegel, "Corrections to improved assessment of interference limits in cellular radio performance," *IEEE Trans. on Vehicular Technology*, Vol. 41, p. 551, November 1992.)

[26] R.H. Clarke, "A statistical theory of mobile radio reception," *Bell Syst. Tech. J.*, Vol. 47, pp. 957–1000, July 1968.

[27] M.J. Gans, "A power spectral theory of propagation in the mobile radio environment," *IEEE Trans. on Vehicular Technology*, Vol. 21, pp. 27–38, February 1972.

[28] R.C. French, "The effect of fading and shadowing on channel reuse in mobile radio," *IEEE Trans. on Vehicular Technology*, Vol. 28, pp. 171–181, August 1979.

[29] J.C. Arnbak and W.van Blitterswijl, "Capacity of slotted ALOHA in Rayleigh-fading channels," *IEEE J. Selected Areas in Comm.*, Vol. 5, pp. 261–269, February 1987.

[30] R. Prasad and C.-Y. Liu, "Throughput analysis of some mobile packet radio protocols in Rician fading channels," *IEE Proc.-I*, Vol. 139, pp. 297–302, June 1992.

[31] M Nakagami, "The m distribution; a general formula of intensity distribution of rapid fading," in *Statistical Methods in Radio Wave Propagation*, W.G. Hoffman (Ed.), New York, NY,Pergamon Press, pp. 3–36, 1960.

[32] M. Abramowitz and I.A. Stegin (Eds.), *Handbook of Mathematical Functions*, New York: Dover, 1965.

[33] U. Charash, "Reception through Nakagami fading multipath channels with random delays," *IEEE Trans. Comm.*, Vol. 27, pp. 657–670, April 1979.

[34] S. Aun Abbas and A.U. Sheikh, "A geometric theory of Nakagami fading multipath mobile radio channel with physical implementation," *Proc. IEEE 46th Veh. Technol. Conf. (VTC'96)*, Atlanta, GA, pp. 637–641, April / May 1996.

[35] R. Prasad, *CDMA for Wireless Personal Communication*, Norwood, MA: Artech House, 1996.

[36] A.M. Saleh and R.A. Valenzuela, "A statistical model for indoor multipath propagation," *IEEE J. Selected Areas in Comm.*, Vol. SAC-5, pp. 128–137, February 1987.

[37] J.C. Proakis, *Digital Communications,* 2nd Edition, New York: McGraw-Hill, 1989.

[38] R.J.C. Bulitude, "Measurement, characterization and modelling of indoor 800/900 MHz radio channels for digital communications," *IEEE Comm. Mag.*, Vol. 25, pp. 5–12, June 1987.

[39] G.J.M. Janssen and R. Prasad, "Propagation measurements in an indoor radio environment at 2.4 GHz, 4.75 GHz and 11.5 GHz," *Proc. 42nd VTS Conference Frontiers of Technology*, Denver, CO, pp. 617–620, May 10–13, 1992.

[40] G.J.M. Janssen, P.A. Stigter, and R. Prasad, "A model for BER evaluation of indoor frequency selective channels using multipath measurement results at 2.4, 3.75 and 11.5 GHz," *Proceedings 1994 International Zurich Seminar on Digital Communications*, Zurich, Switzerland, pp. 344–355, March 1994.

[41] G.J.M. Janssen, P.A. Stigter, and R. Prasad, "Wideband indoor channel measurements and BER analysis of frequency selective multipath channels at 2.4, 4.75 and 11.5 GHz," *IEEE Trans. Comm.* Vol. 44, pp.1272–1288, October 1996.

[42] D.M.J. Devarsivatham, "Time delay spread measurements of wideband radio signals within buildings," *Electron. Lett.*, Vol. 20, pp. 950–951, 1984.

[43] D.M.J. Devarsivatham, "Time delay spread and signal level measurements of 850 MHz radio waves in building environments," *IEEE Trans. Antennas and Propagation*, Vol. AP-34, pp. 1300–1305, November 1986.

[44] D.M.J. Devarsivathm, C. Banarjee, M.J. Krain, and D.A. Rappaport, "Multi-frequency radiowave propagation measurements in the portable radio-environment," *Proc. ICC'90*, Vol. 4, pp. 335.1.1–335.1.7, April 1990.

[45] T.S. Rappaport, S.Y. Seidel, and R. Singh, "900 MHz multipath projection measurements for U.S. digital cellular radio telephone," *IEEE Trans. on Vehicular Technology*, pp. 132–139, May 1990.

[46] D.C. Cox, "Delay Doppler characteristics of multipath delay spread and average excess delay for 910 MHz urban mobile radio paths," *IEEE Trans. Antennas and Propagation*, Vol. 20, pp. 625–635, September 1972.

[47] S.J. Howard and K. Pahlavan, "Performance of a DFE modem evaluated from measured indoor radio multipath profiles," *Proc. ICC'90*, Vol. 4, pp. 335.2.1–335.2.5, April 1990.

[48] J. Chuang, "The effects of time delay spread on portable communications channels with digital modulation," *IEEE J. Selected Areas in Comm.*, Vol. 5, pp, 879–889, June 1987.

[49] V. Fung, T.S. Rappaport, and B. Thoma, "Bit error simulation for π/4 DQPSK mobile radio communication using two-ray and measurement-based impulse response models," *IEEE J. Selected Areas in Comm.*, Vol. 11, pp. 393–405, April 1993

[50] R. Steel (Ed.), *Mobile Radio Communications*, IEEE Press, 1994.

[51] T.S. Rappaport, S.Y. Seidel and R. Singh, "Statistical channel impulse response models for factory and open plan building radio communication system design," *IEEE Trans. Comm.*, Vol. 39, pp. 794–806, May 1991.

[52] B.D. Woerner, J.H. Reed and T.S. Rappaport, "Simulation issues for future wireless modems," *IEEE Comm. Mag.*, Vol. 32, pp. 19–35, July 1994.

[53] S.U.H. Qureshi, "Adaptive equalization," *Proc. IEEE*, Vol. 73, No. 9, pp. 1349– 1387, February 1985.

[54] J.G. Proakis, *Digital Communications, Third Edition*, New York: McGraw-Hill, Inc., 1995.

[55] R.A. Scholtz, "The origins of spread spectrum communications," *IEEE Trans. Comm.*, Vol. 30, No. 5, pp. 822–854, May 1982.

[56] J.A.C. Bingham, "Multicarrier modulation for data transmission : an idea whose time has come," *IEEE Comm. Mag.*, pp. 5–14, May 1990.

[57] L.J. Cimini Jr., "Analysis and simulation of a digital mobile channel using orthogonal frequency division multiplexing," *IEEE Trans. Comm.,* Vol. 33, No. 7, pp. 665–675, July 1985.

[58] J.H. Winters, "The impact of antenna diversity on the capacity of wireless communications systems," *IEEE Trans. Comm.,* Vol. 42, No. 2/3/4, pp. 1740–1751, February/March/April 1994.

[59] L.C. Godara, "Applications of antenna arrays to mobile communication, part I : performance improvement, feasibility, and system considerations," *Proc. IEEE, Vol. 85, No.* 7, pp. 1031–1060, July 1997.

[60] L.C. Godara, "Applications of antenna arrays to mobile communication, part II : beam-forming, and direction-of-arrival considerations," *Proc. IEEE,* Vol. 85, No. 7, pp. 1195–1245, August 1997.

[61] A.A.Abu-Dayya and N.C. Beaulieu, "Micro- and Macro-diversity NCFSK(DPSK) on shadowed Nakagami-fading channels," *IEEE Trans. Comm.,* Vol. 42, No. 9, pp. 2693–2702, September 1994.

[62] L.M. Cooreia and R. Prasad, "An overview of wireless broadband communications," *IEEE Comm. Mag.,* Vol. 35, No. 1, pp. 28–33, January 1997.

[63] S.Y. Seidel and T.S. Rappaport, "Path loss prediction in multi-floored buildings at 914 MHz, Electron. Lett., pp. 1384–1387, Vol. 27, No.15, July 1991.

[64] C.B. Rorabaugh, "*Communications Formulas & Algorithms for System Analysis & Design,*" New York: McGraw-Hill, Inc., 1990.

[65] A. Leon-Garcia, "*Probability and Random Processes for Electrical Engineering,*" Reeding, MA: Addison-Wesley Publishing Co., 1989.

[66] L. Kuipers and R. Timman, *Handbook der Wiskunde I*, Amsterdam: Scheltema & Hollema (in Dutch), 1966.

APPENDIX 2A

Assuming equal Rice factor, K_{si} and local mean power, p_0, for each signal, the pdf of the instantaneous power, p_i, of a signal is in the form of (2.29) with the simple substitutions: $p \rightarrow p_i$, and $K \rightarrow K_i$ and can be written as

$$f(p_i) = C_i e^{-A_i p_i} I_0(B_i \sqrt{p_i}) \tag{2A.1}$$

$$C_i = \frac{K_{si}+1}{p_0} e^{-K_{si}}, \quad A_i = \frac{K_{si}+1}{p_0}, \quad B_i = \sqrt{\frac{4K_{si}(K_{si}+1)}{p_0}}$$

The pdf of n interferers is found by convolving (2A.1) n times, assuming they are independent. In the Laplace domain, convolution becomes a multiplication.

The Laplace transform of zero-order modified Bessel function is given by [66, p. 747]

$$I_0(2\sqrt{p}) \overset{L}{\leftrightarrow} \frac{e^{\frac{1}{s}}}{s} \tag{2A.2}$$

With the help of standard Laplace transforms, the Laplace transform for the pdf of one interferer is found as:

$$C_i\, e^{-A_i p_i} I_0\left(B_i \sqrt{p_i}\right) \overset{L}{\leftrightarrow} C_i \frac{u e^{\frac{1}{u(s+A_i)}}}{u(s+A_i)} \quad \text{with} \quad u = \frac{4}{B_i^2} \tag{2A.3}$$

n times multiplication of (2A.3) is given by

$$C_i^n \frac{1}{(s+A_i)^n} e^{\frac{1}{d(s+A_i)}} \quad \text{with} \quad d = \frac{u}{n} \tag{2A.4}$$

which transforms back into

$$C_i^n e^{-A_i p_n} \int_0^{p_n} \int_0^{y_{n-1}} \int_0^{y_{n-2}} --- \int_0^{y_1} I_0(B_i \sqrt{n x_1}) dx_1 dx_2 --- dx_n \tag{2A.5}$$

The integral of (2A.4) is of the form [32, p. 484]

$$\int_0^y x^{\frac{v-1}{2}} I_{v-1}(B_i \sqrt{nx}) dx = \frac{2}{B_i \sqrt{n}} Y^{\frac{v}{2}} I_v(B_i \sqrt{ny}) \tag{2A.6}$$

When we take $v = 1$, we get the inner integral of (2A.5),

$$\int_0^{y_1} I_0(B_i \sqrt{n x_1}) dx_1 = \frac{2}{B_i \sqrt{B_i n}} \left[Y_1^{\frac{1}{2}} I_1(B_i \sqrt{n Y_1}) \right] \tag{2A.7}$$

The part between the brackets on the right of (2A.7) is again of the form (2A.6). Putting this back into (2A.6) the result will be 1 order higher between the brackets with a multiplication by a constant. This results in a closed form for the multiple integral of (2A.5)

$$\left[\frac{2\sqrt{p_n}}{B_i\sqrt{n}}\right]^{n-1} I_{n-1}(B_i\sqrt{np_n}) \tag{2A.8}$$

where p_n = the total interference power.

The pdf of n interferers is then given by:

$$f_{p_n}(p_n) = C_i^n e^{-A_i p_n}\left[\frac{2\sqrt{p_n}}{B_i\sqrt{n}}\right]^{n-1} I_{n-1}(B_i\sqrt{np_n}) \tag{2A.9}$$

Substituting the A_i, B_i, and C_i back into (2A.9) results in (2.30).

Chapter 3
Cellular Concepts

3.1 INTRODUCTION

In the beginning, mobile systems were developed much like radio or television broadcasting (i.e., a large area was covered by installing a single, high-power transmitter on a tower situated at the highest point in the area [1,2]). A single high-power transmitter mobile radio system gave good coverage with a small number of simultaneous conversations depending on the number of channels N_C. The $(N_C + 1)$ caller was blocked. Figure 3.1 shows the early mobile systems.

To increase the number of simultaneous conversations, a large area can be divided into a large number of small areas, N_a. Each small area is called a "cell". To cover a cell, a single low-power transmitter is required. If every cell uses the same frequency that was available for a large area and available frequency is divided into same number of channel N_C, then instead of N_C simultaneous conversations for a large area there would be N_C simultaneous conversation for each cell. Thus, now there can be $N_a\ N_C$ simultaneous conversations in the entire large area as compared with only N_C. Figure 3.2 illustrates the idea of dividing a large area into cells.

But the idea of using the same frequency in all the cells does not work because of the interference between mobile terminals operating on the same channel in adjacent cells. Therefore, the same frequency cannot be used in each cell and it is necessary to skip a few cells before the same frequency is used. Cellular concept is illustrated in Figure 3.3.

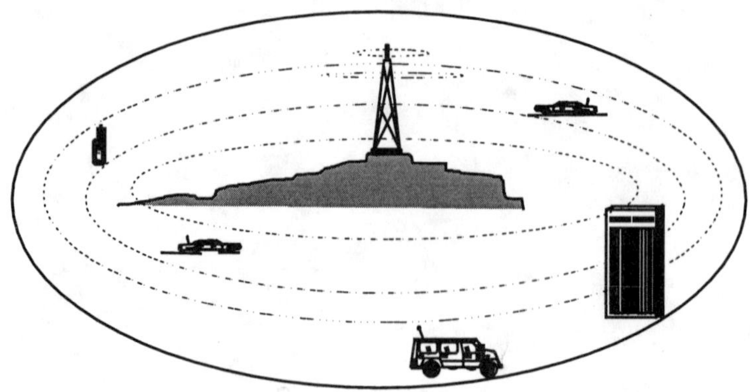

Figure 3.1 Early mobile systems.

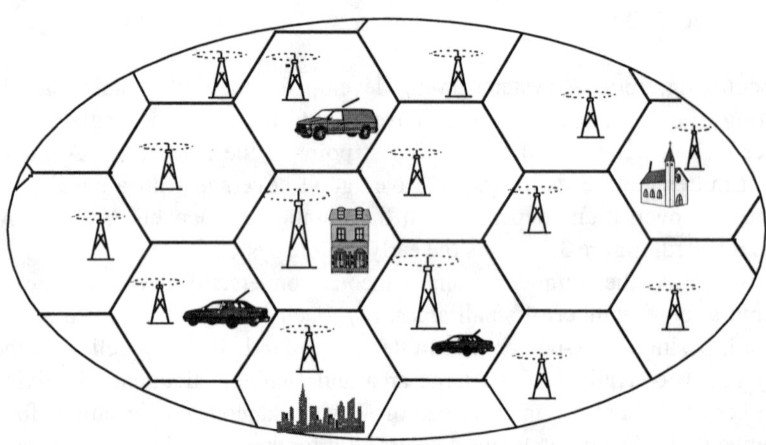

Figure 3.2 Division of a large area in cells.

The cellular concept is a wireless system designed by dividing a large area into several small cells, replacing a single, high-power transmitter in a large area with a single, low-power transmitter in each cell, and reusing the frequency of a cell to another cell after skipping several cells. Thus, the limited bandwidth is reused in distant cells, causing a virtually infinite multiplication of the available frequency. To efficiently utilize the available frequency, the design of cellular systems depends on five basic elements [3]:

1. Frequency reuse;
2. Co-channel interference;
3. Carrier-to-interference ratio;
4. Handover / Handoff mechanism;
5. Cell splitting.

This chapter briefly introduces these basic elements. In addition, different types of cellular systems are reviewed.

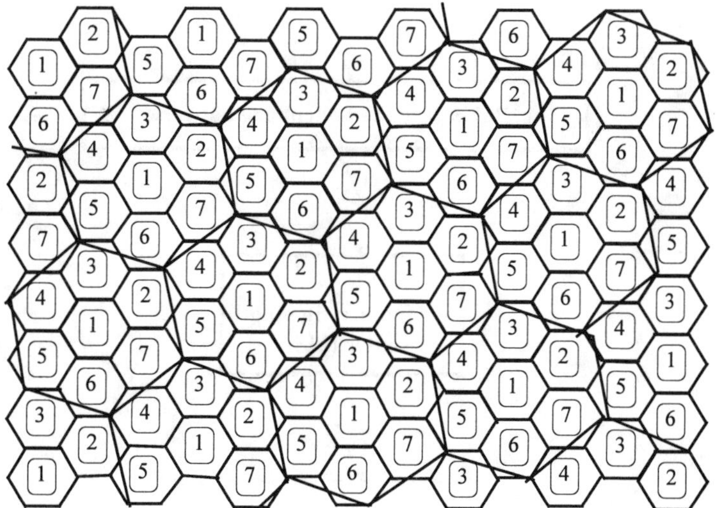

Figure 3.3 Cellular concept.

3.2 FREQUENCY REUSE

The cellular structure was introduced due to capacity problems of mobile communication systems. In a cellular radio system the area covered by the mobile radio system is divided into cells. In theory, the cells are considered hexagonal, but in practice they are less regular in shape. Figure 3.3 shows the (theoretical) layout of the cellular radio system. Each cell contains a base station, which is connected to the mobile switching center (MSC). This MSC is connected to the fixed telecommunication system, namely the telephone system public switched telephone network (PSTN). MSC serves as the central coordinator and controller for the cellular radio system and as the interface between mobile and PSTN. The cellular radio user in a car or train or on the street picks up a handset, dials a number, and immediately can talk to the person called.

Behind this simple act is an enormous array of technology. Figure 3.4 illustrates the cellular radio system structure. Each cell is assigned a part of the available frequency spectrum. Cellular radio system offers the possibility of using the same part of the frequency spectrum more than once. This is called frequency reuse. Cells with identical channel frequencies (i.e., the same part of the frequency spectrum) are called co-channel cells. The co-channel cells have to be sufficiently separated to avoid interference. The distance between these co-channel cells is ensured by creating a cluster of cells. A cluster is defined as a group of cells that are assigned different channel frequencies. The total frequency spectrum is then divided over the cells within a cluster.

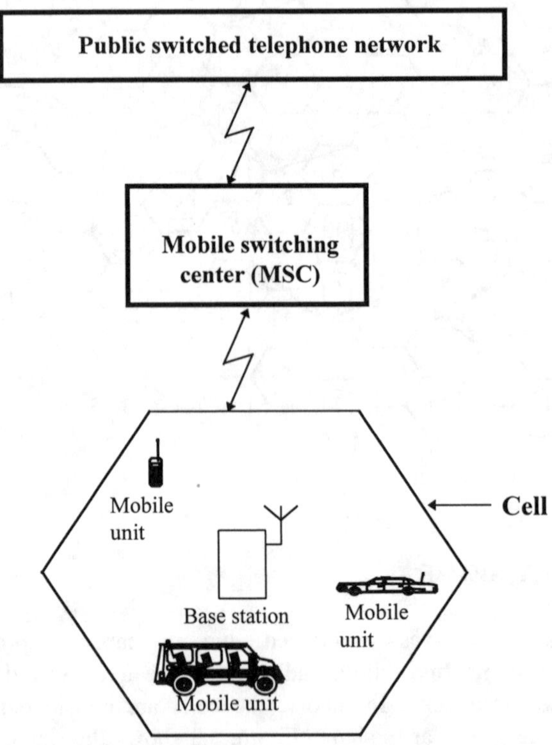

Figure 3.4 Cellular radio system structure.

The number of cells within a cluster is defined as the cluster size C. Figure 3.3 shows an example of the cellular network with cluster of size 7. In Figure 3.3, cells with identical numbers make use of the same part of the frequency spectrum [1–6]. The total number of channels N_{tc} in a cellular radio system is given by

$$N_{tc} = N_r N_c C \tag{3.1}$$

where N_r is the number of times a cluster is replicated within the system, N_c is the number of channel in a cell, and C is the cluster size.

It is not possible to choose an arbitrary value of the cluster size. The cluster size is determined by

$$C = i^2 + ij + j^2 \tag{3.2}$$

where i and j are non-negative integers. So there can be only selected values of C.

$$C = 3,4,7,9,12,13,... \tag{3.3}$$

The cluster size can be chosen and determines the amount of frequency reuse within a certain area. An important design parameter, denoting the amount of frequency reuse in a certain area is called the (normalized) reuse distance. The normalized reuse distance, R_u, is defined as the ratio of the reuse distance D between the centers of the nearest co-channel cells and the cell radius R.

$$R_u \triangleq \frac{D}{R} \tag{3.4}$$

This is shown in Figure 3.5.

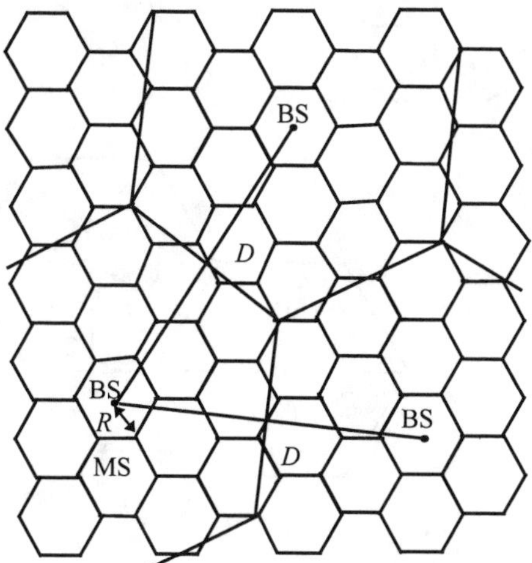

Figure 3.5 Normalized reuse distance.

The following steps should be followed to find the nearest co-channel cells:

1. Walk i cells along any chain of hexagons;
2. Take a turn of 60 degrees counterclockwise and walk j cells.

Figure 3.6 illustrates the method of locating co-channel cells for $C = 19$ ($i = 3, j = 2$). The relation between reuse distance R_u and cluster size C can be derived using Figures 3.7 and 3.8.

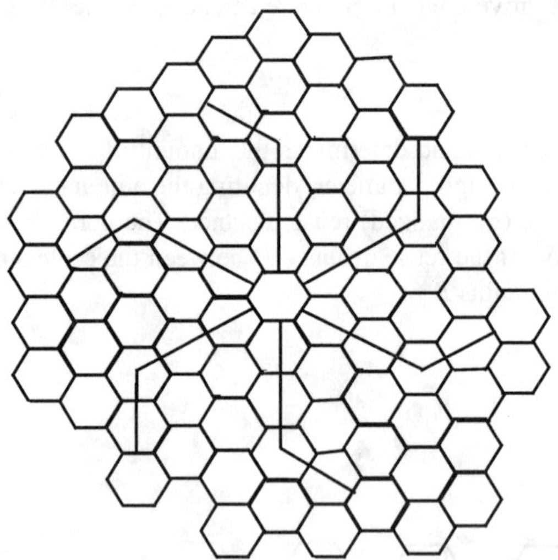

Figure 3.6 Locating co-channel cells.

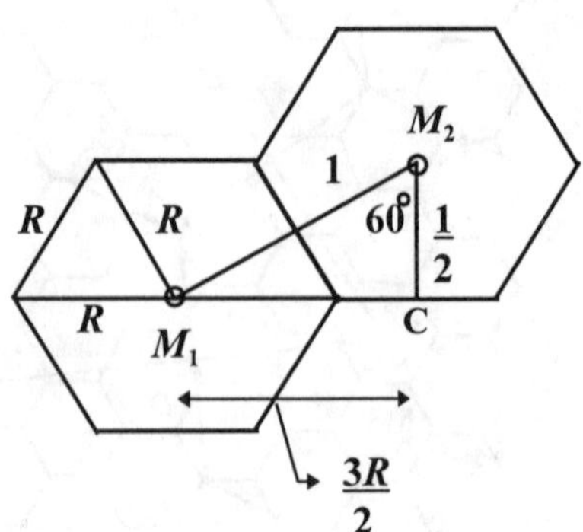

Figure 3.7 Cluster size length.

Using triangle M_1M_2C of Figure 3.7, R can be written as

$$R = \frac{1}{\sqrt{3}} \tag{3.5}$$

Now using (3.4) and (3.5), R_u is given by

$$R_u = \frac{D}{R} = \sqrt{3}D \tag{3.6}$$

Using Figure 3.8, C and D can be related to

$$C = D^2 = i^2 + j^2 - 2ij\cos(120^0)$$

$$= i^2 + ij + j^2 \tag{3.7}$$

Thus relationship between R_u and C is obtained using (3.6) and (3.7)

$$R_u = \sqrt{3C} \tag{3.8}$$

Therefore,

$$C = \frac{R_u^2}{3} \tag{3.9}$$

3.3 CO-CHANNEL INTERFERENCE

Figure 3.9 illustrates the co-channel interference in a cellular system.
Co-channel interference is one of the major elements of a cellular radio system. Equation (3.9) shows that with the increase in the cluster size C, the reuse distance also increases. An advantage of large clusters is the fact that the interference from co-channel cells decreases because the distance between the co-channel cells also increases with the increase in cluster size. On the other hand, the part of the frequency spectrum assigned to a cell also becomes small because the total available frequency spectrum has to be distributed over a large number of cells. Thus, use of the available frequency spectrum becomes less efficient. This efficiency, which is also an important parameter of a cellular radio system, is called the spectrum efficiency.

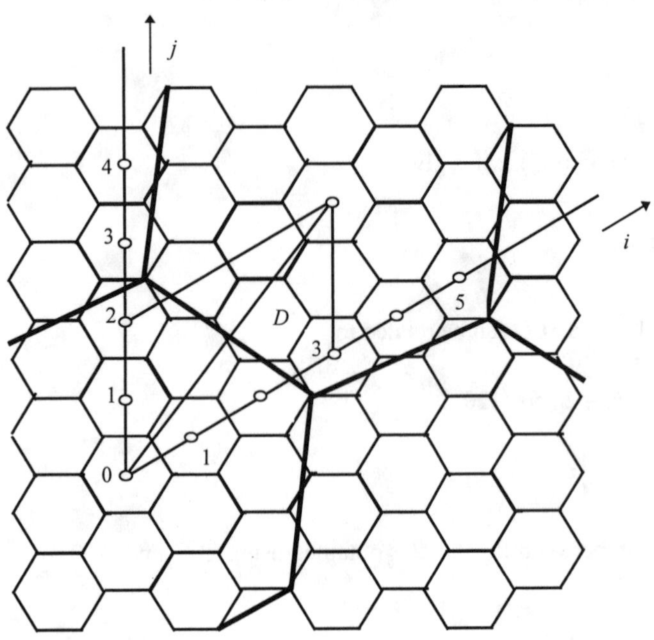

Figure 3.8 Cluster size.

Spectrum efficiency, E_S, for a circuit switched cellular network, is defined as the carried traffic per cell, A_C, divided by the product of bandwidth per channel, W, number of channels per cell, N_C, and cluster size C cells, given a unit cell area of S [7]. Thus

$$E_s \triangleq \frac{A_c}{N_c WCS} \quad \text{erlang/MHz/km}^2 \tag{3.10}$$

Figure 3.10 shows plots for the cluster size and reuse distance determined using (3.9) and for spectrum efficiency versus reuse distance obtained using (3.10) taking $W = 25$ KHz, $A_C = 5$ erlang, $N_C = 10$, and $S = 1$ km^2.

We can infer from Figure 3.10 that the spectrum efficiency can be increased by accepting an increase in the co-channel interference (i.e., decrease in reuse distance and cluster size). Thus, a trade-off must obtained between co-channel interference and spectrum efficiency.

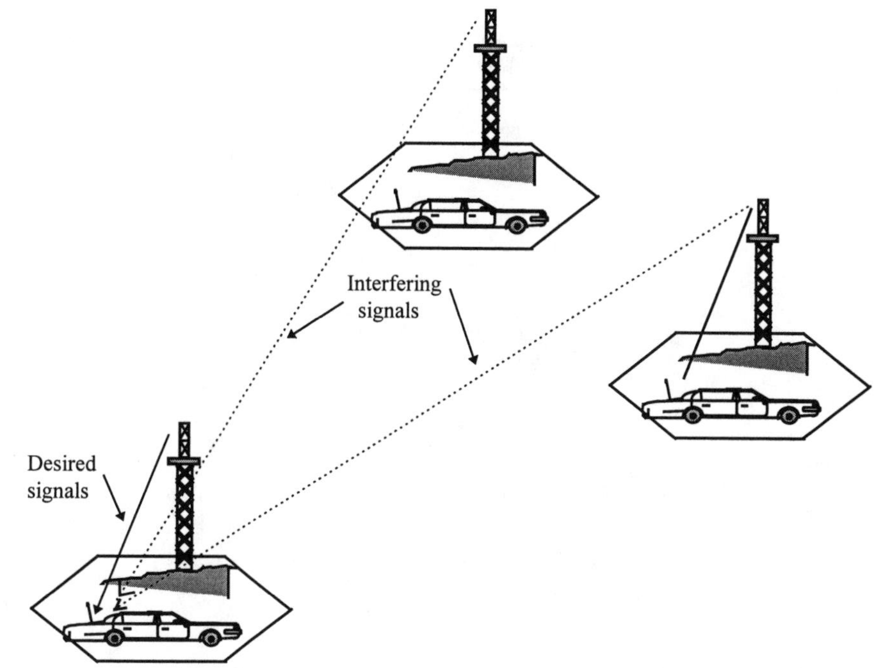

Figure 3.9 Co-channel interference.

3.4 CARRIER-TO-INTERFERENCE RATIO

The carrier-to-interference ratio (CIR) at the desired mobile unit is given by

$$CIR = \frac{P_c}{\sum_{i=1}^{n} P_i} \tag{3.11}$$

where P_c is the carrier power, P_i is the interference power, and n is the number of co-channel interfering cells. In a mobile radio environment, the received power is given by

$$P_r \propto r^{-\gamma} \tag{3.12}$$

where γ is the path-loss exponent and r is the distance between transmitter and receiver. For the worst case (when the mobile is at the cell corner), CIR is given by using (3.11) and (3.12)

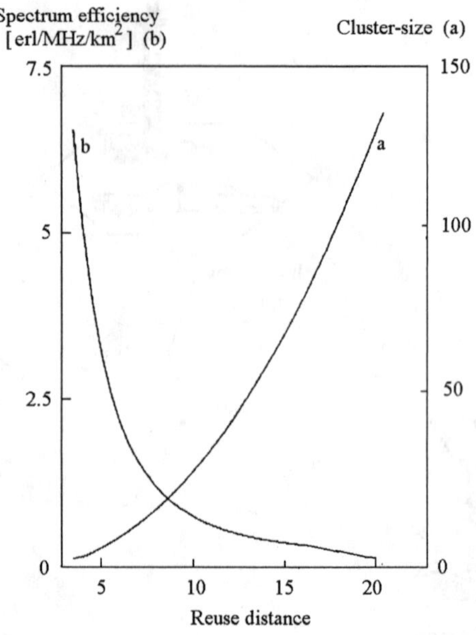

Figure 3.10 Influence of normalized reuse distance (R_u) on (a) cluster size (C) and (b) spectrum efficiency (E_s) for A_c = 5 erlang, W = 25 kHz, S = 1 km² and N_c = 10.

$$CIR = \frac{R^{-\gamma}}{\sum_{i=1}^{n} D_i^{-\gamma}} = \frac{1}{\sum_{i=1}^{n} (R_{ui})^{-\gamma}} \tag{3.13}$$

where D_i and R_{ui} are the reuse and normalized reuse distance of ith co-channel interfering cell.

In a hexagonal cellular system, there are six co-channel interfering cells in the first tier. The maximum number of co-channel cells, n, in the first tier can be shown as [3]

$$n = \frac{2\pi D}{D} \approx 6 \tag{3.14}$$

CIR can be derived for a 7-cell cluster using Figure 3.11 [2] and neglecting co-channel interference due to second and higher tiers

$$CIR = \frac{R^{-\gamma}}{2(D-R)^{-\gamma} + (D-R/2)^{-\gamma} + (D+R/2)^{-\gamma} + (D+R)^{-\gamma} + D^{-\gamma}} \tag{3.15}$$

Equation (3.15) is written in terms of normalized reuse distance R_u

$$CIR = \frac{1}{2(R_u-1)^{-\gamma} + (R_u-1/2)^{-\gamma} + (R_u+1/2)^{-\gamma} + (R_u+1)^{-\gamma} + R_u^{-\gamma}} \tag{3.16}$$

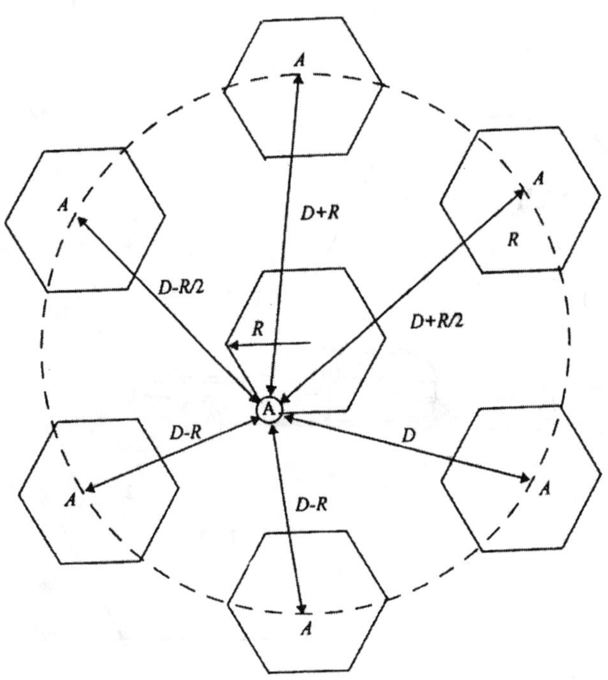

Figure 3.11 CIR evaluation due to first tier.

Assuming R_{ui} are same or $R_u > 1$, using (3.16) R_u is expressed as

$$R_u = (6\text{CIR})^{1/\gamma} \tag{3.17}$$

The value of CIR depends on the required system performance and the specified value of γ that is based on the terrain environment. For a given value of CIR and γ, R_u is determined using (3.15). Once R_u is determined, the cluster size is evaluated using (3.9) and the reuse distance found by deciding a radius R in (3.4).

A detailed investigation of co-channel interference is presented in Chapters 4, 5, and 6.

3.5 HANDOVER/HANDOFF MECHANISM

Handover, also known as handoff, is a process to switch an ongoing call from one cell to the adjacent cell as a mobile user approaches the cell boundary [1–6].

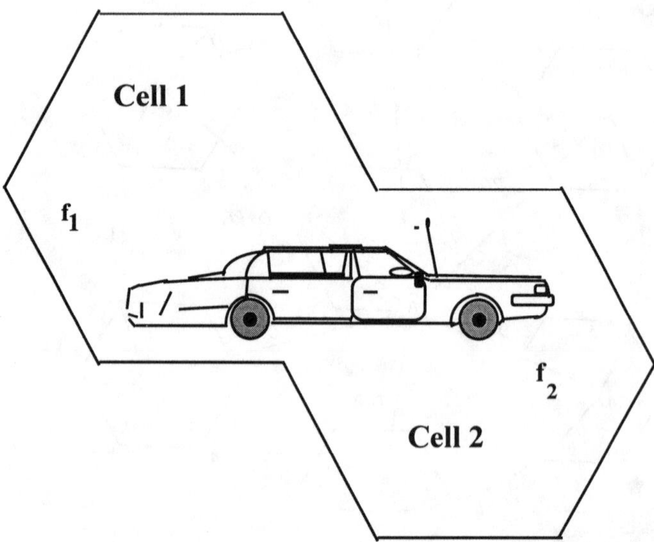

Figure 3.12 Handover/handoff mechanism.

Figure 3.12 shows that as the user moves from cell 1 to cell 2, the channel frequencies will be automatically changed from the set f_1 to the set f_2. Handover is an automatic process, if the signal strength falls below a threshold level. It is not noticed by the user because it happens very quickly within 200 to 300 ms.

The need for a handover may be caused by radio, operation and management (O&M), or by traffic.

Radio causes the majority of handover requests. Parameters involved are low signal level or high error rate. This can be caused by a mobile moving out of a cell or signal blocking by objects.

O&M-generated handovers are rare. They evolve from maintenance of equipment, equipment failure, and channel rearrangement.

Handovers due to unevenly distributed traffic may cause some mobiles at the border of a cell to be handed over to an adjacent cell. The handover procedure should meet several quality criteria. First, the handover rate should be kept low, as each handover causes the network overhead to facilitate the handing over of a mobile station (MS) to another base station (BS). This is possible by blocking unnecessary handovers, which occur often at the border of a cell, where the received signals of some BSs are about equal. Furthermore, the handover blocking probability should also be kept low. A handover is blocked when a mobile is handed over to a BS with no more free channels, and the call is dropped.

This section discusses the basic performance aspects of handover. The performance metrics used to evaluate handover algorithms are [8]:

- Handover blocking probability: the probability that a handover attempt is blocked. It should be kept low, since an unsuccessful handover attempt usually results in a dropped call. A typical probability is 0.05.
- Call blocking probability: the probability that a new call attempt is blocked. This may be higher than the handover blocking probability, since the damage caused by a blocked call is less than that by a dropped call, because the former was not yet set up. The caller is able to retry.
- Handover probability: the probability that, while communicating with a particular cell, an ongoing call requires a handover before the call terminates. This metric is translated into the average number of handovers per call.
- Call dropping probability: the probability that a call terminates due to handover failure. This metric is derived directly from the handover blocking probability and the handover probability.
- Rate of handover: the average number of handovers per unit time. Combined with the average call duration, it allows the calculation of the average number of handovers per call, and thus the handover probability. It should be kept low, because handovers cause network overhead.
- Probability of an unnecessary handover: the probability that a handover is stimulated by a particular handover algorithm when the existing radio link is still adequate. It should be kept low as well, because a decrease of unnecessary handovers obviously lowers the handover rate.
- Duration of interruption: the length of time during a handover for which the mobile terminal is in communication with neither base station. This metric is heavily dependent on the particular network topology and the duration of the handover procedure.
- Delay (distance): the distance by which the mobile moves from the point where the handover should occur up to the point where it actually does occur. This parameter should also be kept low, because mobiles drifting too far into adjacent cells can cause interference with stations further away.

In addition to the previously mentioned performance metrics, the ping-pong effects (the continuous handing over of the mobile between base stations) are often investigated.

A handover is performed in three stages. The MS continuously gathers information on the received signal level of the BS it is connected with, and of all other BSs it can detect. This information is then averaged to filter out fast-fading effects. The averaged data are then passed on to the decision algorithm, which decides if it will request a handover to another station. When it decides to do so, handover is executed by both the old BS and the MS, resulting in a connection to the new BS.

As stated before, the received signal level suffers from fading effects. To prevent handover resulting from temporary fluctuations in received signal level, the measurements must be averaged. This is done by an averaging window whose length determines the number of samples to be averaged. Longer averaging lengths give more reliable handover decisions, but also result in longer handover delays. Detailed studies were done to determine the averaging window shape, that is to determine whether recent measurements should be treated as more reliable than older ones. The averaging window is used to trade off between handover rate and handover delay.

Published reports on the subject of handover study have presented several initiation algorithms. Zhang and Holtzmann [9] have developed an analytical model of a handover algorithm based on both absolute and relative measurements. According to this algorithm, a mobile is handed over to a new BS only if the current signal level drops below a threshold, and the target BS is stronger than the current one by a given hysteresis margin. They have proved that this algorithm outplays the other decision algorithms.

Given the hysteresis margin *h*, handover in Figure 3.13 will occur at point A if the threshold is either *T*1 or *T*2, and will occur at point B if the threshold is *T*3.

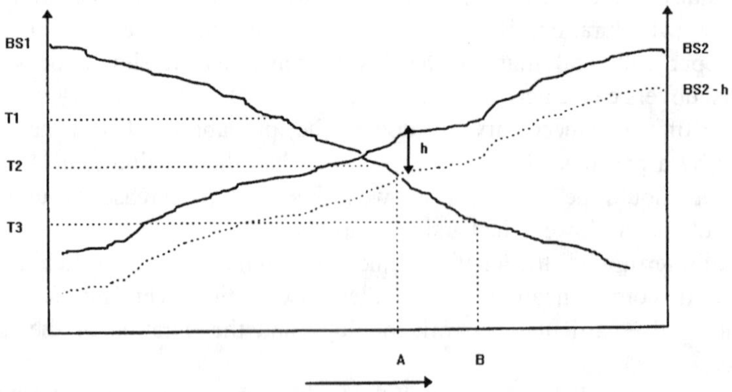

Figure 3.13 Received signal strengths from BS1 and BS2 versus distance.

Handover can be executed in two ways: forward and backward.

When using the backward handover process, the mobile sends to the network a command message containing the identity of the target BS. The resource acquisition at the target BS is then performed by the network, and the target BS will forward the allocated channel information or the negative acknowledgment message to the MS via the "old" signaling link; that is, via the network and the current BS. The network forwards the handover command message to the target BS and achieves all the operations required to establish a connection at the target BS side. Only after the new link has been set up, is the old one released.

The forward handover approach process makes use of another approach. In this case the mobile transmits an access packet on the access channel of the target BS. The access acknowledgment is transmitted to the mobile on the target BSs downlink channel. The mobile sends, when the access has succeeded, its identity and the old BS's identity in order to be connected with the network.

Both methods can be used in the case of a soft handover, where communication with both the original and the target BS is possible. In the case of a suddenly dropped connection (hard handover), it is not possible to communicate with the old BS any more, and thus only the forward handover process can be used to save the call.

A highway LAN requires backward handover process, because a temporary obstruction of the LOS path with the new BS might result in a dropped call if the link with the old BS is released before the conclusion of the handover procedure.

Reference [10] presents in depth the handover mechanism in GSM.

3.6 CELL SPLITTING

In principle, a cellular system can provide services for an unlimited number of users. However once a system is installed, it can only provide to a certain fixed number of users. As soon as the number of users increases and approaches the maximum that can be served, some technique must be developed to accommodate the increasing number of users. There are various techniques to enhance the capacity of a cellular system. One technique is cell splitting, a mechanism by which cells are split into smaller cells, each having the same number of channels as the original large cells [1–6].

Suppose cell splitting is achieved in such a way that the radius of the new cell is half of the original cell. To cover the complete area of the original cell, four new cells are needed. Thus, the new traffic capacity per unit area will be four times larger than the original traffic capacity per unit area.

Figure 3.14 illustrates cell splitting [6].

Cell splitting provides an increase in capacity without the need for extra bandwidth, although the increase in capacity (i.e., spectrum efficiency) is achieved at the cost of additional infrastructure. However, the additional cost is offset by the increase of users' subscription revenue.

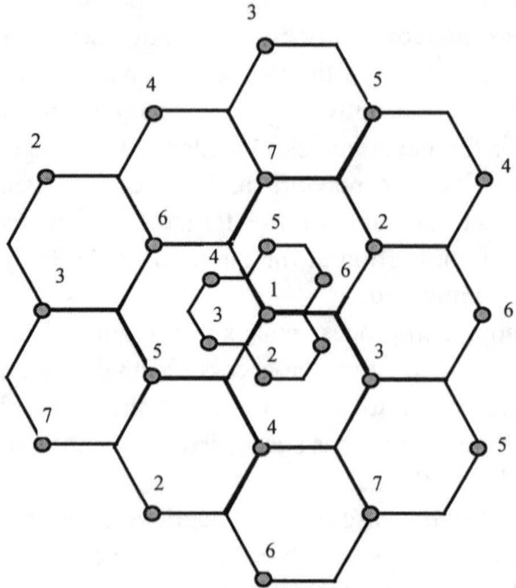

Figure 3.14 Cell splitting.

3.7 TYPES OF CELLULAR NETWORKS

Based on the radius of the cells, there are three types of cellular networks:

1. Macrocells;
2. Microcells;
3. Picocells.

3.7.1 Macrocellular Radio Networks

Macrocells are mainly used to cover large areas with low traffic densities. These cells have radii between 1 and 10 km. A distinction between large macrocells and small macro- cells should be made.

Large macrocells have radii between 5 and 10 km. They are used for rural areas. To ensure good radio coverage, base stations have to be installed in a smart way. If directional antennas are used, they must be positioned with care. Macrocellular radio nets need good planning, based on a combination of field measurements and theoretical modeling based on these measurements. Also, reliable predictions concerning traffic demands must be done to allocate the correct number of channels to a cell.

Small cells have radii between 1 and 5 km. These cells are used if the traffic density in large cells is so high that it will cause blocking of calls. They thus provide large cells with extra capacity (cell splitting). Small cells demand the same planning requirements as do large cells. Planning small cells is more difficult since traffic predictions for relatively small areas cannot be easily done. This is because the offered traffic per square kilometer is dynamic and will fluctuate more rapidly as the area over which the traffic is offered, become smaller. The path-loss exponent for macrocells lies between 2 and 5, depending on the critical mobile radio environment. The signals undergo multipath Rayleigh fading and lognormal shadowing. The standard deviation of lognormal shadowing signal lies between 4 and 12 dB. Typical rms delay spread is 8 μs.

3.7.2 Microcellular Radio Networks

Microcellular radio networks are used in areas with high traffic density, like (sub-) urban areas. The cells have radii between 200m and 1 km. For such small cells, it is hard to predict traffic densities and area coverage. Models for such parameters prove to be quite unreliable in practice. This is because the shape of the cell is time dynamic (i.e., the shape changes from time to time) due to propagation characteristics.

We can distinguish one- and two-dimensional microcells. One-dimensional microcells are placed in a chainlike manner along main highways with high traffic densities. Two-dimensional microcells usually cover one or two house blocks. Antennas are placed at street lamp elevations. Surrounding buildings block signals propagating to adjacent co-channel cells. This improves the ability to reuse frequencies, since co-channel interference is reduced drastically by the shadowing effect caused by the infrastructure. Microcells follow a dual path-loss law. The value of the "turning point" depends on the type of environment and the position of the transmitting antenna. The signal undergoes Rician fading and lognormal shadowing. Typical rms delay spread is 2 μs.

3.7.3 Picocellular Radio Networks

Picocells or indoor cells have cell radii between 10 and 200m. For indoor applications, cells have three-dimensional structures. Fixed cluster sizes, fixed channel allocations, and prediction of traffic densities are difficult for indoor applications. Today, picocellular radio systems are used for wireless office communications.

Path-loss exponent varies from 1.2 to 6.8. Signals in picocells are always Rician faded. The Rician parameter lies between 6.8 and 11 dB. Typical values of rms delay spread lie between 50 and 300 μs.

Table 3.1 summarizes the three cellular systems.

Table 3.1

Different Cell Characteristics

Cell Type	Size	Transmission Power	Antenna Height/ Location	Path-loss Exponent	Signal Characteristics	RMS Delay Spread	Use
Macro cell	2–20 km diameter	0.6–10W	> 30m, top of tall building	2–5	Rayleigh fading and lognormal shadowing	< 8 μs	Large area coverage—reduce infrastructure cost
Micro cell	0.4–2 km diameter	< 20 mW	< 10m street lamp elevation	Dual path- loss law	Rician fading and lognormal shadowing	< 2 μs	Urban area coverage
Pico cell	20–400 m diameter	On the order of a few milliwatts	Ceiling/ top of book shelf	1.2–6.8	Rician fading	50–300 ns	Mainly for indoor areas—areas with high terminal density

REFERENCES

[1] G. Calhuon, *Digital Cellular Radio*, Norwood, MA: Artech House, 1988.

[2] T.S. Rappaport, *Wireless Communication Principle and Practice*, New Jersey: Prentice Hall, 1996.

[3] W.C.Y. Lee, *Mobile Cellular Telecommunication Systems*, New York: McGraw Hill Book Co., 1989.

[4] W.C.Y. Lee, *Mobile Communications Design Fundamentals*, New York: John Wiley & Sons, Inc., 1993.

[5] W.C. Jakes, Jr. (Ed.), *Microwave Mobile Communications*, New York: John Wiley & Sons, Inc., 1974.

[6] V.H. MacDonald, "The Cellular Concept," *Bell Syst. Tech. J.*, Vol. 58, pp. 15–42, January 1979.

[7] R. Prasad and A. Kegel, "Improved assessment of interference limits in cellular radio performance," *IEEE Trans. on Vehicular Technology*, Vol. 40, pp. 412–419, May 1991.

[8] G.P. Pollini, " Trends in Handover Design," *IEEE Comm. Mag.* Vol. 34, pp. 82–90, March 1996.

[9] N. Zhang and J.M. Holtzmann, "Analysis of handoff algorithms using both absolute and relative measurements," *IEEE Trans. on Vehicular Technology*, Vol. 45, No. 1, pp. 174–179, February 1996.

[10] M. Mouly and M.-B. Pautet, *The GSM System for Mobile Communications*, M. Mouly and M.-B. Pautet Publishing Company, 1991.

Chapter 4
Macrocellular Systems

4.1 INTRODUCTION

A system designer always considers the influence of co-channel interference and Gaussian and impulsive noise on the system performance in a mobile radio environment. Improved assessment of co-channel interference in a macrocellular system to evaluate the system performance has enjoyed a high priority in mobile radio research even before the beginning of the cellular era. The reason for such extensive investigations is that there were no classifications of a cellular system. Generally, all the cell sizes were in the order of 2 to 20 km in diameter.

To design any cellular system, it is imperative to understand and analyze the effect of co-channel interference probability (CIP) caused by a number of co-channel interferers [1–23]. A realistic evaluation of CIP helps establish the optimal frequency reuse distance between the co-channel cells for a given choice of a cellular radio system. Evaluation of CIP requires a realistic model of the wave propagation in a fading and shadowing environment. A mobile radio signal envelope in a macrocell is modeled as a Rayleigh fading signal superimposed as a shadowing signal. Rayleigh fading is caused by a multipath propagation, while blocking by the local horizon is responsible for shadowing. The resulting random fluctuations in the signal amplitude is described approximately by a lognormal pdf averaged over long periods, and a Rayleigh pdf for much shorter periods. Rayleigh fading of different signals is uncorrelated; however, in a shadowing environment, the mean levels of the desired and interfering signals might be correlated. Section 4.2 presents the analysis of CIP in a Rayleigh fading, lognormal shadowing, and combined fading and shadowing environment. Effect of correlated lognormal shadowing is investigated in Section 4.3. A mathematical model is developed in Section 4.4 to study the performance of a macrocellular network with co-channel interference, Gaussian and impulsive noise for two modulation techniques, namely 4-QAM and DPSK. The impulsive noise is also known as non-Gaussian noise, man-made noise, or electromagnetic interference. The large contribution of man-made noise to the total interference on the mobile radio systems, especially in urban areas, motivated researchers to include the impulsive noise together with the co-channel interference and Gaussian noise in the study of the performance of a cellular radio system. Middleton [24] has classified the impulsive noise in three categories; namely (1) class A (narrowband), (2) class B (broadband), and (3) class C (combination of class A and class B).

Spectrum efficiency (E_S), expressed in erlang/MHz/km^2, is defined as the ratio of carried traffic per cell and the product of bandwidth, number of cells per cluster, and cell area. Thus, E_S is an important parameter for assessing the frequency requirements of cellular radio mobile systems. Sections 4.2, 4.3, and 4.4 discuss the effect of cluster size, protection ratio, and carried traffic on the spectrum efficiency.

4.2 CO-CHANNEL INTERFERENCE PROBABILITY

The cellular telephony concept is described in numerous papers (e.g., [25]). To investigate the reuse distance and spectrum efficiency of a cellular system, it is necessary to determine CIP [1–23] defined as

$$F(\mathrm{CI}) \stackrel{\Delta}{=} \sum_n F(\mathrm{CI}|n) F_n(n) \tag{4.1}$$

Here, $F_n(n)$ is the probability of n co-channel interferers being active. $F(\mathrm{CI}|n)$ is the corresponding conditional CIP,

$$F(\mathrm{CI}|n) \stackrel{\Delta}{=} \mathrm{prob}\{p_d / p_n < \alpha\} \tag{4.2}$$

where p_d is the instantaneous power of the desired signal, p_n is the joint interference power from n active channels, and α is the specified co-channel protection ratio given by $[p_d / p_n]_{\min}$.

The (fast) amplitude fluctuation of mobile radio signals is described by the Rayleigh density function. The pdf for the signal amplitude r_i of the ith interferer, conditional on its local mean power p_{0i}, is given by

$$f_{ri}(r_i | p_{0i}) = \frac{r_i}{p_{0i}} \exp(-r_i^2 / 2p_{0i}) \tag{4.3}$$

The corresponding pdf for the instantaneous power p_i is

$$f_{pi}(p_i|p_{0i}) = \frac{1}{p_{0i}} \exp\left(-\frac{p_i}{p_{0i}}\right) \quad (4.4)$$

The local mean power p_{0i} is itself a (more slowly varying) stochastic variable with the lognormal pdf

$$f_{p_{0i}}(p_{0i}) = \frac{1}{\sqrt{(2\pi)}\sigma_i p_{0i}} \exp\left\{-(\ln p_{0i} - m_i)^2 / 2\sigma_i^2\right\} \quad (4.5)$$

Here, σ_i^2 and $\xi_i = \exp(m_i)$ are the logarithmic variance and the area mean power of the interfering signal, respectively.

The amplitude, r_d, instantaneous power, p_d, and local mean, p_{0d}, of the desired signal are also described by density functions of the forms (4.3) to (4.5), respectively, with the following substitutions:

$r_i \to r_d$
$p_i \to p_d$
$p_{0i} \to p_{0d}$
$\sigma_i \to \sigma_d$
$m_i \to m_d$
$\xi_i \to \xi_d$ (4.6)

The formulation of conditional CIP for three different propagations conditions is described as follows.

4.2.1 Rayleigh Fading Only

In the event of incoherent cumulation with equal local mean power i.e., ($p_{0i} = p_0$), the pdf for the joint interference power p_n is obtained by convolving (4.4) n times and given by Gamma distribution [26, 27]

$$f_{p_n}(p_n) = \frac{1}{p_0} \frac{(p_n / p_0)^{n-1}}{(n-1)!} \exp\left(-\frac{p_n}{p_0}\right) \quad (4.7)$$

The conditional co-channel interference is derived using (4.2), (4.4), (4.6), and (4.7):

$$F(\mathrm{CI}|n) = \int_0^\infty dp_n \int_0^{\alpha p_n} \frac{1}{p_0 p_{0d}} \frac{(p_n / p_0)^{n-1}}{(n-1)!} \exp\left(-\frac{p_d}{p_{0d}} - \frac{p_n}{p_0}\right) dp_d \quad (4.8)$$

After integrating, (4.8) becomes

$$F(\mathrm{CI}|n) = 1 - \left(\frac{1}{\alpha p_0 / p_{0d} + 1}\right)^n \tag{4.9}$$

4.2.2 Lognormal Shadowing Only

The sum of n stochastic-independent lognormal variables is well approximated by another lognormal variable [28,29]. Schwartz and Yeh [29] and others [11] derived exact expressions for the mean and variance for the sum of two lognormal variables, and then used a recursive approach to approximate the mean, ξ_u, and the variance, σ_u^2, for n variables. Therefore, the pdf for $p_{0u}(=\sum_{i=1}^{n} p_{0i})$ is of the form (4.5) with the following substitutions:

$$\begin{aligned} p_{0i} &\rightarrow p_{0u} \\ \sigma_i &\rightarrow \sigma_u \\ m_i &\rightarrow m_u \\ \xi_i &\rightarrow \xi_u \end{aligned} \tag{4.10}$$

Substituting (4.5), (4.6) and (4.10) in (4.2) yields the conditional CIP

$$\begin{aligned} F(\mathrm{CI}|n) = \int_0^\infty dp_{0u} \int_0^{\alpha p_{0u}} & \frac{1}{2\pi\sigma_d\sigma_u p_{0d} p_{0u}} \exp\{-(\ln p_{0u} - m_u)^2 / 2\sigma_u^2\} \\ & \exp\{-(\ln p_{0d} - m_d)^2 / 2\sigma_d^2\} dp_{0d} \end{aligned} \tag{4.11}$$

Equation (4.11) simplifies to

$$F(\mathrm{CI}|n) = \frac{1}{\sqrt{(2\pi)}} \int_x^\infty \exp(-t^2/2) dt \tag{4.12}$$

where $x \stackrel{\Delta}{=} [\ln(\xi_d / \alpha\xi_u)] / (\sigma_d^2 + \sigma_u^2)^{\frac{1}{2}}$ (4.13)

Table 2.1 gives values for m_u and σ_u as a function of number of interferers, n, with the initial value of $m_i = 0$ dB.

4.2.3 Rayleigh Fading Plus Lognormal Shadowing

In the general case of combined multipath fading and shadowing, considering incoherent addition of the individual, stochastically-independent signals [27], the composite pdf for the (individual) ith interfering signal power can be approximated by the lognormal pdf with logarithmic variance $[\sigma_i^2 + \ln(2)]$ and area-median power $\xi_i / \sqrt{2}$ [30]. According to the suggestion of Schwartz and Yeh [27], the composite pdf for the (total) interference sum p_n can then be approximated by a pure lognormal distribution with the appropriate moments

$$f_n(p_n) = \frac{1}{\sqrt{2\pi}\sigma_n p_n} \exp\{-(p_u - m_n)^2 / 2\sigma_n^2\} \tag{4.14}$$

Here, $p_u \stackrel{\Delta}{=} \ln p_n$, while $\xi_n(= \exp m_n)$ and σ_n^2 are the area-median (interference) power and variance, respectively, of the equivalent lognormal variable. They are found in accordance with the procedure in [11,29]. Hence, using (4.2) through (4.6) and (4.14), conditional CIP is given as

$$F(\mathrm{CI}|n) = \int_0^\infty dp_{0d} \int_0^\infty dp_n \int_0^{\alpha p_n} \frac{1}{2\pi\sigma_d\sigma_n p_{0d}^2 p_n} \exp\left(-\frac{p_d}{p_{0d}}\right) \exp\left[\{-(\ln p_n - m_n)^2 / 2\sigma_n^2\}\right.$$

$$\left. + \{-(\ln p_{0d} - m_d)^2 / 2\sigma_d^2\}\right] dp_d \tag{4.15}$$

After repeated integration, (4.15) becomes

$$F(\mathrm{CI}|n) = \frac{1}{\sqrt{\pi}} \int_{-\infty}^{\infty} \exp(-w^2) fw dw \tag{4.16}$$

where

$$f(w) = 1 - \exp\left[-\exp\{m_n + \alpha_0 - m_d - [2(\sigma_d^2 + \sigma_n^2)]^{1/2} w\}\right] \tag{4.17}$$

$$\alpha_0 \stackrel{\Delta}{=} \ln \alpha \tag{4.18}$$

Table 2.2 gives values for m_n and σ_n as a function of the number of interferes, n, with the initial value of $m_i = 0$ dB.

To compare the system performance in coherent and incoherent interference conditions, conditional CIP for the coherent case is also derived. To this end, we write the composite pdf for n interfering signals with combined Rayleigh fading and lognormal shadowing for coherent addition using (4.3) to (4.6)

$$f_{fs}(p_n) = \frac{1}{\sqrt{2\pi}\sigma_u} \int_0^{\infty} \frac{1}{p_{0u}^2} \exp(-p_n / p_{0u}) \exp\{-(\ln p_{0u} - m_u)^2 / 2\sigma_u^2\} dp_{0u} \tag{4.19}$$

The composite pdf for the desired signal with Rayleigh fading and lognormal shadowing is again of the form (4.19), with the following substitutions:

$$\begin{aligned} p_n &\rightarrow p_d \\ p_{0u} &\rightarrow p_{0d} \\ \sigma_u &\rightarrow \sigma_d \\ m_u &\rightarrow m_d \end{aligned} \tag{4.20}$$

The conditional CIP is obtained using (4.2), (4.19), and (4.20)

$$F(\mathrm{CI}|n) = \frac{1}{2\pi\sigma_d\sigma_u} \int_0^{\infty} dp_{0d} \int_0^{\infty} dp_{0u} \int_0^{\infty} dp_n \int_0^{\alpha p_n} \frac{1}{p_{0u}^2 p_{0d}^2} \exp\left\{-\left(\frac{p_d}{p_{0d}} + \frac{p_n}{p_{0u}}\right)\right\}$$

$$\exp\left[\{-(\ln p_{0u} - m_u)^2 / 2\sigma_u^2\} + \{-(\ln p_{0d} - m_d)^2 / 2\sigma_d^2\}\right] dp_d \tag{4.21}$$

Equation (4.21) reduces to

$$F(\mathrm{CI}|n) = \frac{1}{\sqrt{\pi}} \int_{-\infty}^{\infty} \left[\{\exp(-u^2)\} / \left\{1 + (\xi_d / \alpha\xi_u) \exp\left[-\{2(\sigma_d^2 + \sigma_u^2)\}^{1/2} u\right]\right\}\right] du \tag{4.22}$$

It is assumed throughout that all co-channel interferers are statistically independent and identically distributed. If, moreover, only interfering signals from the nearest neighboring six co-channel cells are considered, and both the blocking probability B and the number of channels n_c are the same in all cells, $F_n(n)$ *is* written as [7]

$$F_n(n) = \binom{6}{n} B^{n/n_c} (1 - B^{1/n_c})^{6-n} \qquad n = 0, 1, \ldots, 6 \tag{4.23}$$

$F(n)$ *is* also expressed in terms of carried traffic a_c per channel [8,9]

$$F_n(n) = \binom{6}{n} a_c^n (1 - a_c)^{6-n} \tag{4.24}$$

where $a_C = A_C/n_C$, and A_C is the carried traffic per cell given by A(1-B), A is the offered traffic per cell in erlang and B is the blocking probability defined as the probability that the system is fully occupied. Generally, (4.24) is a more accepted representation in traffic analysis [31].

4.2.4 Reuse Distance

The (normalized) channel reuse distance, R_u, is defined as the ratio of the distance D between the centers of the nearest neighboring co-channel cells and the cell radius R, that is,

$$R_u \stackrel{\Delta}{=} \frac{D}{R} \tag{4.25}$$

The signal-to-interference ratio at a receiving base station located at the cell center is

$$\frac{\xi_d}{\xi_u} = \frac{R^{-\gamma}}{D^{-\gamma}\left[\exp\sum_{i=1}^{n} G_{ij}\right]} \tag{4.26}$$

where γ is the ground-wave propagation path-loss slope (we take $\gamma = 4$), n is the number of co-channel interfering cells (here $n = 6$), and G_{ij} is a correction to the area-median power dependent on n and given by (2.12) - (2.15). Using (4.25) and (4.26), R_u is written as

$$R_u = \left[\frac{\xi_d}{\xi_u}\exp\left(\sum_{i=1}^{n} G_{ij}\right)\right]^{1/4} \tag{4.27}$$

Now, CIP is computed as a function of R_u, for the three different propagation conditions, namely (1) Rayleigh fading only, (2) lognormal shadowing only, and (3) Rayleigh fading plus lognormal shadowing only.

The results are shown in Figure 4.1 for $\sigma_i = \sigma_d = \sigma = 6$ and 12 dB. It is confirmed from Figure 4.1 that a large reuse distance is essential to avoid excessive values of co-channel interference. We also observe that shadowing is the predominant factor in determining the reuse distance and CIP, in particular for large σ.

Typical results for CIPs in the event of pure lognormal shadowing with $\sigma = 6$ and 12 dB, $\alpha = 8$ dB, $n_C = 10$, and $B = 0.02$, are shown in Figure 4.2, for both Fenton's classical method [28], Schwartz and Yeh's technique [29], and [11]. Table 4.1 shows that Schwartz and Yeh's method results in up to 89% higher CIPs for $R_u = 15$.

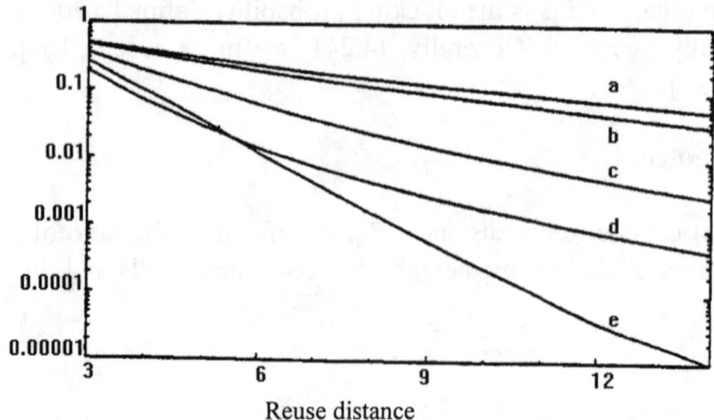

Figure 4.1 CIP versus normalized reuse distance for protection ratio α=8 dB, blocking probability B = 0.02, carried traffic per cell A_c = 5 erlang and n_c = 10 channels. Incoherent cumulation of interferers for (a) combined Rayleigh fading and shadowing (σ = 12 dB), (b) shadowing only (σ = 12 dB), (c) combined Rayleigh fading and shadowing (σ = 6 dB), (d) Rayleigh fading only (σ = 0), and (e) shadowing only (σ = 6 dB).

Table 4.1

Percentage Difference in CIP for Lognormal Shadowing Environment Only, Comparing Schwartz and Yeh's Technique With Fenton's Method for α = 8 dB, n_c = 10, and β = 0.02

R_u	σ_i = 6 dB (%)	σ_i =12 dB (%)
3	4	25
6	9	47
9	12	64
12	14	78
15	15	89

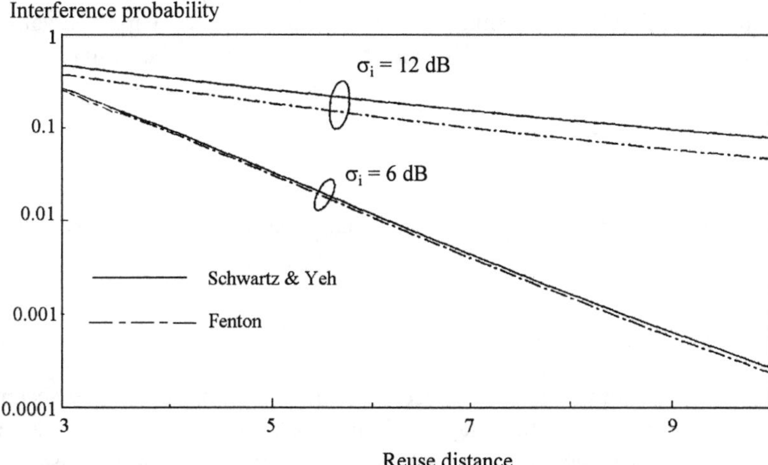

Figure 4.2 CIP versus normalized reuse distance for $\alpha = 8$ dB, $A_c = 5$ erlang, $B = 0.02$ and $n_c = 10$ for shadowing only ($\sigma = 6$ and 12 dB), calculated using Fenton's and Schwartz and Yeh's techniques.

4.2.5 Spectrum Efficiency

Spectrum efficiency, E_s, is defined as the carried traffic per cell, A_c, divided by the product of bandwidth per channel, W, number of channels per cell, n_c, and cluster size C cells, given a unit cell area of S. Thus

$$E_s \stackrel{\Delta}{=} \frac{A^c}{n_c WCS} \text{erlang / MHz / km}^2 \tag{4.28}$$

Carried traffic is obtained by

$$A_c \stackrel{\Delta}{=} A(1-B) \tag{4.29}$$

where A is the offered traffic per cell in erlang. The blocking probability B is determined using the erlang-B formula [31,32]:

$$B = \frac{A^n c}{n_c! \sum_{n=0}^{n_c} \left(\frac{A^n}{n!} \right)} \tag{4.30}$$

The cells are assumed to form a cluster of size C, located around a reference cell and repeated around each of its co-channel cells. However, the exact shape of a valid cluster need not be precisely specified [25]. The cluster size is taken on the form (3.2)

$$C = i^2 + ij + j^2 \qquad i, j \geq 0 \tag{4.31}$$

with integers i and j.

The reuse distance and the number of cells per cluster are related by [25,33], (3.8)

$$R_u = (3C)^{1/2} \tag{4.32}$$

It can be inferred from Figure 4.3 that spectrum efficiency can increase by accepting an increase in CIP. The cluster size plays an important role in determining the optimum spectrum efficiency for meeting a certain system requirement— a decrease in cluster size (and hence reuse distance) increases the spectrum efficiency. Referring to Figure 4.1, this is equivalent to accepting a higher CIP.

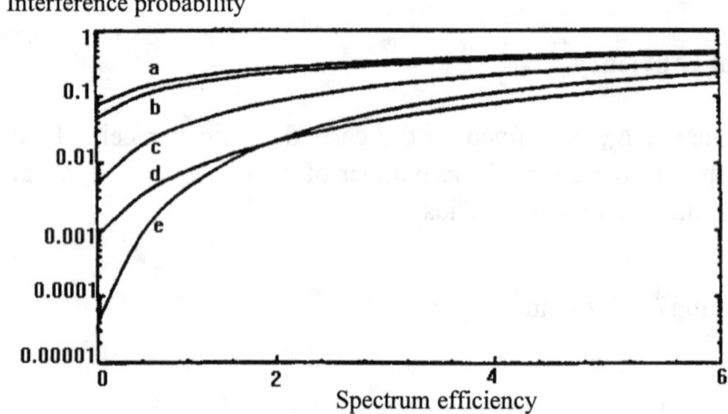

Figure 4.3 CIP versus spectrum efficiency (E_s). Channel bandwidth $W = 25$ kHz; unit cell area $S = 1$ km^2; other parameters as in Figure 4.1.

To ascertain the difference between coherent and incoherent cumulation of interference, Table 4.2 compares the corresponding CIPs for $\sigma = 6$ and 12 dB, and different reuse distances. While coherent conditions do allow smaller interference probability, Table 4.2 shows that the difference is rather small (below 10% in the cases considered), and decreases at larger reuse distances. Therefore, in the following numerical examples, we confine ourselves to incoherent cumulation interferers.

Table 4.2

CIP for Coherent and Incoherent Cumulation
of Co-channel Interferers With σ = 6 and 12 dB as a Function of Reuse Distance

	σ_i = 6dB		σ_i = 12dB	
R_u	F(CI) Incoherent	F(CI) Coherent	F(CI) Incoherent	F(CI) Coherent
3	0.35867	0.32317	0.50531	0.47402
4	0.19453	0.17840	0.38644	0.36220
5	0.10791	0.10107	0.30083	0.28155
6	0.06236	0.05954	0.23971	0.22285
7	0.03767	0.03655	0.19187	0.17910
8	0.02374	0.02333	0.15501	0.14574
9	0.01555	0.01543	0.12743	0.11995

Figures 4.4 and 4.5 show interference probability as a function of spectrum efficiency, with protection ratio (Figure 4.4) and the carried traffic (Figure 4.5) as a parameter. Clearly, the spectrum efficiency can be increased by tolerating a lower value of protection ratio (Figure 4.4). We see from Figure 4.5 that a higher blocking probability results in higher spectrum efficiency for a fixed value of interference.

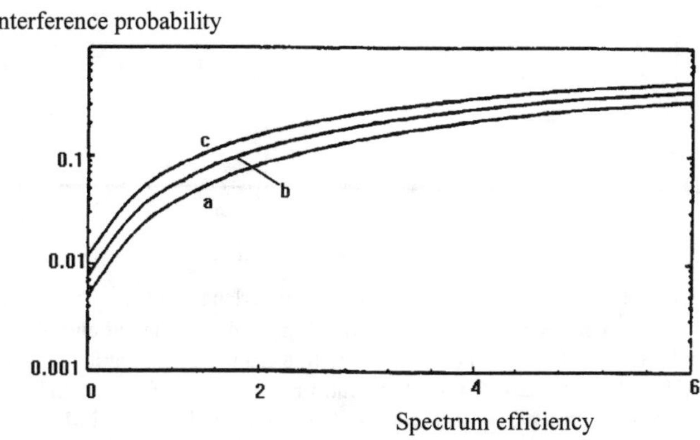

Figure 4.4 CIP versus spectrum efficiency for combined Rayleigh fading and shadowing (σ = 6 dB) using incoherent cumulation of interference with A_c = 5 erlang, B = 0.02, n_c = 10, S = 1 km^2, W = 25 kHz, and (a) α = 8 dB, (b) α = 10 dB, and (c) α = 12 dB.

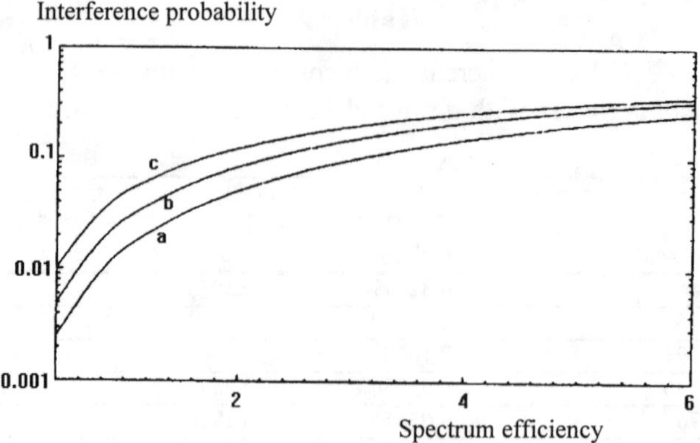

Figure 4.5 CIP versus spectrum efficiency for combined Rayleigh fading and shadowing using incoherent cumulation of interference with $n_c = 10$, $S = 1$ km^2, $\sigma = 6$ dB, $W = 25$ kHz, $\alpha = 8$ dB, and (a) $a_c = 0.99$ erlang/channel ($B = 0.8$), (b) $a_c = 0.5$ erlang/channel ($B = 0.02$), and (c) $a_c = 0.23$ erlang/channel ($B = 0.0001$).

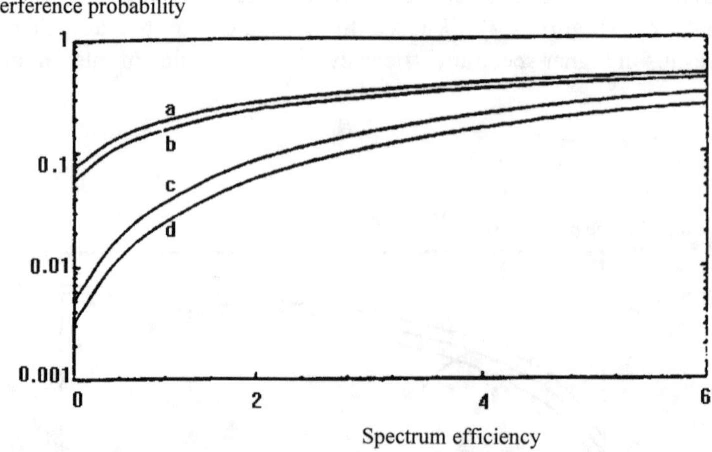

Figure 4.6 CIP versus spectrum efficiency for $a_c = 0.5$ erlang/channel, $B = 0.02$, $S = 1$ km^2, and $n_c = 10$ for combined fading and shadowing using incoherent cumulation of interference with (a) 25-kHz digital modulation with 8-dB protection ratio and $\sigma = 12$ dB, (b) 12.5-kHz FM with 12-dB protection ratio and $\sigma = 12$ dB, (c) 25-kHz digital modulation with 8-dB protection ratio and $\sigma = 6$ dB, and (d) 12.5-kHz FM with 12-dB protection ratio and $\sigma = 6$ dB.

Figures 4.6 to 4.8 compare the spectral efficiencies of the different modulation methods of interest in cellular telephony shown in Table 4.3. The protection ratio (α) is defined as the minimum value of the wanted (carrier)-to-unwanted (interference) signal ratio at the receiver input such that a specified reception quality of the wanted signal is achieved at the receiver output [32].

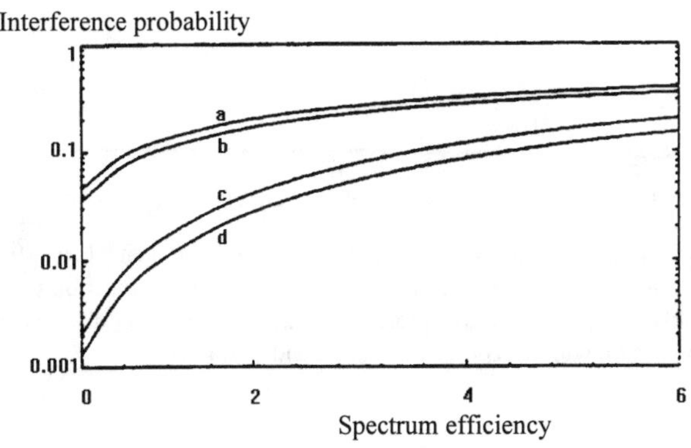

Figure 4.7 CIP versus spectrum efficiency for a_c = 0.5 erlang/channel, B = 0.02 and n_c = 10 for combined fading and shadowing using incoherent cumulation of interference with (a) 5-kHz SSB with α = 18-dB and σ = 12 dB, (b) 5-kHz SSB with α = 16 dB and σ = 12 dB, (c) 5-kHz SSB with α = 18 dB and σ = 6 dB, and (d) 5-kHz SSB with α = 16 dB and σ = 6 dB.

Table 4.3

Channel Spacing and Protection Ratio for Different Modulation Methods

Modulation Method	Channel Spacing	α (dB) Protection Ratio	References
Analog FM	12.5 kHz	12	[2], [35]
Analog FM	30 kHz	18	[25], [34], [36], [37]
Digital modulation	25 kHz	8	[36]
Single side band (SSB)	5 kHz	16 and 18	[2], [34], [35]

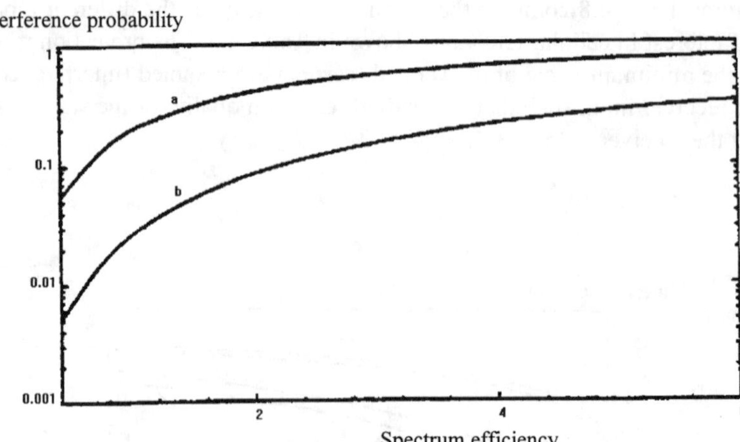

Figure 4.8 CIP versus spectrum efficiency for $a_c = 0.5$ erlang/channel, $B = 0.02$ and $n_c = 10$ for combined fading and shadowing using incoherent cumulation of interference with (a) 30-kHz FM with 18-dB protection ratio and σ = 6 dB, and (b) 25-kHz digital modulation with 8-dB protection ratio and σ = 6 dB.

For analog FM systems, as used for the U.S. AMPS system and the U.K. TACS, satisfactory operation requires approximately α = 18 dB [25, 34, 36–38]. The channel spacing in AMPS is 30 kHz, whereas in TACS a closer channel spacing of 25 kHz has been adopted. The NMT-900 system even uses a 12.5-kHz channel spacing. For a digitally modulated system, we have used 8 dB and a channel spacing of 25 kHz (a European standard) for the purpose of computations. For the sake of completeness, the spectrum efficiency is also evaluated (Figure 4.7) for SSB. An SSB system with a 5-kHz channel spacing suffers similar subjective effects of co-channel interference as a 25-kHz FM system under real mobile radio field conditions [34]. In both cases, a signal-to-interference ratio of the order of 16 dB is necessary to achieve a fair quality of reception. Therefore, the efficiency of SSB systems is evaluated for α = 16 and 18 dB with a 5-kHz channel spacing.

The computational results (Figures 4.6 to 4.8) show that SSB offers the highest efficiency among the modulation techniques considered in this section. Our efficiency ranking of these methods agree with [2] but also include for 30-kHz spaced FM (AMPS) and digital GSM.

4.2.6 System Bandwidth

System bandwidth represents the total bandwidth required to serve a cluster. Thus, it can be expressed as the product of the number of channels per cluster and the bandwidth per channel; that is,

$$S_w \overset{\Delta}{=} n_c WC = \frac{A_c}{SE_s} \text{MHz} \tag{4.33}$$

Thus, as distinct from the spectrum efficiency, the system bandwidth is an absolute measure of the frequency resources required to meet a certain traffic demand in a given area. The cluster size *C* and the modulation method influence the system complexity, and hence, the economic resources required to realize the cellular network and mobile terminals. A judicious trade-off between system options can therefore be based on S_W, *C*, and the modulation parameters.

To illustrate the typical differences between cellular system options, Table 4.4 shows spectrum efficiency, system bandwidth, and cluster size for the two FM systems, one digital modulation system (DM) and two SSB systems dealt with in Figures 4.6 to 4.8. The cluster size *C* is approximated by the nearest valid number, according to (4.31). Table 4.4 clearly reveals the relative merits of the different systems, in terms of frequency demands and cellular system complexity.

Table 4.4

Spectrum efficiency E_S, System Bandwidth S_W and Cluster Size *C* for Different Types of Modulation Techniques with $n_C = 10$, B = 0.02, $a_C = 0.5$ Erlang/Channel, and $F(\text{CI}|n) = 10^{-1}$ in a Combined Fading and Shadowing Environment With $\sigma = 6$ dB

Type of Modulation	E_s (erlang/ MHz/ km^2)	S_w (MHz)	*C*
$\mathbf{FM_{30}}$	0.6	8.4	28
$\mathbf{DM_{25}}$	2.2	2.25	9
$\mathbf{FM_{12.5}}$	2.8	2	16
$\mathbf{SSB_{18}}$	3.6	1.4	28
$\mathbf{SSB_{16}}$	4.5	1.25	25

4.2.7 Conclusions

CIPs for mobile radio systems exposed to realistic propagation impairments were calculated using Fenton's classical method and compared with our values, obtained by Schwartz and Yeh's technique. Our results indicate significantly higher interference probabilities, up to 89% more than suggested by the classical method. This indicates that earlier results for co-channel interference calculations in cellular radio systems based on Fenton's method are optimistic. The use of Schwartz and Yeh's technique is recommended for mobile scenarios in which the variance of the shadowing lies between 6 and 12 dB.

Earlier studies of CIP in a cellular mobile network were carried out assuming coherent cumulation of interference. Now, results are also available for the more realistic assumption of incoherent cumulation of interfering signals with independent Rayleigh fading, lognormal shadowing, and UHF groundwave path loss. Computed results show that coherent cumulation of interferers leads to somewhat lower values of CIP; that is, a slightly higher spectrum efficiency than with incoherent cumulation. However, for large reuse distances the differences become minute.

We presented a detailed assessment of the spectrum efficiency and system bandwidth of general cellular systems. It is confirmed that the spectrum efficiency is higher for lower protection ratio, and that a trade-off has to be made between cluster size and CIP to achieve the desired spectrum efficiency. In particular, it cannot be assumed that higher carried traffic (higher blocking probability) always leads to better spectral efficiencies, because of the associated increase in cluster size required to maintain a given interference probability.

In practical terms, a design trade-off will more often be between the overall system bandwidth (absolute bandwidth required to provide a certain service) and the associated network complexity (in terms of cluster size, modulation type, etc.). The approach described in this section would assist planners and designers in obtaining a more realistic assessment of different system options, especially in the many cases where the shadowing of mobiles cannot be assumed to be moderate.

4.3 CORRELATED SHADOWING SIGNALS

Shadowing of radio signals by buildings and hills leads to gradual changes in the local mean signal level, which can be represented by lognormal pdf

$$f_{P_{0d}}(p_{0d}) = \left[\frac{1}{\sqrt{2\pi}\sigma_d p_{0d}}\right] \exp\left\{-\left[\frac{1}{(2\sigma_d^2}\right]\left[\ln\left(\frac{p_{0d}}{\xi_d}\right)\right]^2\right\} \tag{4.34}$$

where σ_d, $p0d$, and ξ_d are the standard deviation, local mean power, and area mean power of the desired signal, respectively.

The local mean, p_{0i}, of the interfering signal is also described by the density function of the form (4.34) with σ_i, and ξ_i, respectively, the standard deviation and area mean of the interferer. Since some of the signals are shadowed by the same obstacles, the local means of the signals in the same cell are partially correlated [2, 5, 8]. Assuming a correlation between p_{0d} and p_{0i}, the joint pdf of p_{0d} p_{0i} is defined as [39]

$$f_\Lambda(\Lambda) = \int_0^\infty f(p_{0d}, p_{0i}) |\delta(p_{0d}, p_{0i}) / \delta(\Lambda, w)| dw \tag{4.35}$$

where $\Lambda = p_{0d} / p_{0i}$, $\quad w = p_{0i}$, $\quad |(\delta p_{0d}, p_{0i}) / \delta(\Lambda, w)| = w$

is Jacobian and $f(p_{0d}, p_{0i})$ is the joint density function [33]:

$$f(p_{0d}, p_{0i}) = \left[\frac{1}{\left(2\pi\sigma_d\sigma_i p_{0d} p_{0i}\sqrt{1-\rho^2}\right)}\right]$$

$$\exp\left\{-\left[\frac{1}{\left(2(1-\rho^2)\right)}\right]\left[u_d^2 - 2\rho u_d u_i - u_i^2\right]\right\} \tag{4.36}$$

where $u_d = (1/\sigma_d)\ln(p_{0d}/\xi_d)$ and $u_i = (1/\sigma_i)\ln(p_{0i}/\xi_i)$

The conditional CIP for shadowing only, $F_s(\text{CI}|n)$, is obtained using (4.2) and (4.36):

$$F_s(\text{CI}|n) = \int_0^\alpha \frac{1}{2\sqrt{\pi}\sigma_{\text{eff}}\Lambda} \exp\left\{-[\ln(\Lambda/\Lambda_m)/2\sigma_{\text{eff}}]^2\right\} d\Lambda$$

$$= \int_{p_0/(\sqrt{2}\sigma_{\text{eff}})}^\infty \left(\frac{1}{\sqrt{2\pi}}\right) \exp\{-u^2/2\} du \tag{4.37}$$

where $\Lambda_m = \xi_d / \xi_i$, $\quad p_0 = \ln\{\xi_d / \xi_i \alpha\}$, and σ_{eff} is the effective standard deviation given as

$$\sigma_{\text{eff}} = \left[(\sigma_d^2 + \sigma_i^2 - 2\rho\sigma_d\sigma_i)/2\right]^{1/2} \tag{4.38}$$

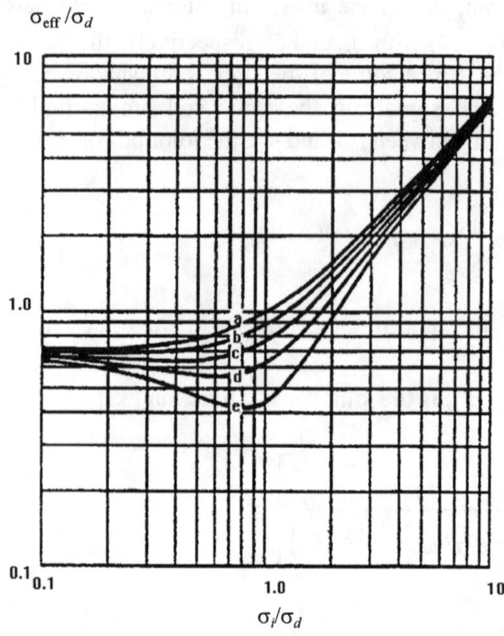

Figure 4.9 Curves for determining σ_{eff} for given values of σ_d and σ_i and ρ as a parameter. (a): ρ = 0, (b): ρ = 0.2, (c): ρ = 0.4, (d): ρ = 0.6, and (e): ρ = 0.8.

Equation (4.37) is the correct expression, which should replace (6) of [8]. An obvious difference is observed in the expression of σ_{eff}. The expression of σ_{eff} of [8] should therefore be replaced by (4.38). A correct graph for determining σ_{eff} for given values of σ_d or σ_i or both, and ρ as a parameter using (σ) is shown in Figure 4.9. Observe that σ_{eff} of [8]is correct for $\rho = 0$ and $\sigma_d = \sigma_i$ for $\rho = 0$. For $\sigma_i = 0$, σ_{eff} of [8] reduces to $\sigma_d(1-\rho)^{1/2}$, which is evidently incorrect. Note that (4.37) reduces to (17) of [2] for the uncorrelated case ($\rho = 0$) and $\sigma_i = \sigma_d$.

Figure 4.10 shows the variation of co-channel interference with p_0 for various values of the correlation coefficient, $\sigma = \sigma_d = \sigma_i = 6$ and 12 dB. We observe from Figure 4.10 that CIP becomes smaller as the signal become more correlated for all values of the p_0. This is explained intuitively by the fact that, as the signals become more correlated, the fluctuations in P_{0d} and P_{0i} would affect the probability that $p_{0d}/p_{0i} < \alpha$ to a lesser extent. Figure 4.10 also confirms that CIP increases with the increase in the shadowing effect (i.e., σ) for equal correlation coefficient. For unequal ρ, it is not always true, as Figure 4.10 shows that $F(\mathrm{CI}|\mathrm{n})$ is lower for σ = 12 dB with correlated signals $\rho = 0.8$ than $F(\mathrm{CI}|\mathrm{n})$ for σ = 6 dB with uncorrelated signals $\rho = 0$ for a given value of p_0.

In mobile radio systems, the received signal envelope fluctuates rapidly due to multipath propagation and wave interference. These fluctuations are described by Rayleigh statistics, that is, the signal envelope has Rayleigh pdf conditional to the local mean level that fluctuates typically 4 to 12 dB due to shadowing. The envelopes of signals, suffering Rayleigh fading, are uncorrelated while their local means are partially correlated as a result of shadowing [2].

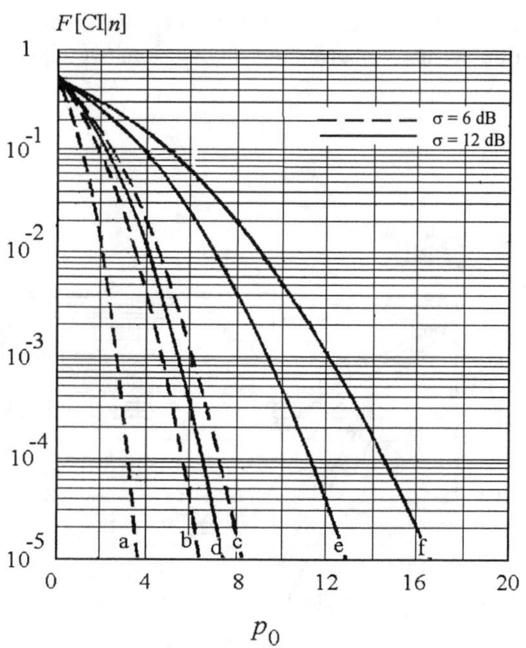

Figure 4.10 Conditional CIP $F(\text{CI}|n)$ with shadowing only for correlation coefficient and standard deviation as parameters. $\sigma = 6$ dB, (a): $\rho = 0.8$, (b): $\rho = 0.4$, (c): $\rho = 0$. $\sigma = 12$ dB, (d): $\rho = 0.8$, (e): $\rho = 0.4$, and (f): $\rho = 0$.

In case of combined Rayleigh fading and lognormal shadowing, conditional CIP, $F_{fs}(\text{CI}|n)$, is derived using (4.2), (4.34), (4.35) and considering the joint pdfs of p_{0d}/p_{0i} and p_d/p_i as independent occurrence. Here p_d and p_i are the instantaneous power of the desired and interference signals, respectively, with their pdfs as exponential distributions:

$$F_{fs}(\mathrm{CI}|n) = \left(\frac{1}{\sqrt{\pi}}\right) \int_{-\infty}^{\infty} \exp\{-v^2\} / \left[1 + \exp\{p - 2\sigma_{\text{eff}} v\}\right] \; dv \tag{4.39}$$

Again, (7) of [8] should be correctly used with σ_{eff} given by (4.38). Note that (4.39) reduced to (12) of [2] for the special case of $\rho = 0$ and $\sigma_i = \sigma_d$. The above integral has been evaluated numerically and is plotted in Figure 4.11 as a function of p_0 for $\sigma = \sigma_d = \sigma_i = 6$ and 12 dB, $\alpha = 8$ dB and with correlation coefficient as a parameter. As in Figure 4.10, which deals only with shadowing, the increasing correlation between the shadowed signals decreases CIP. However, the presence of Rayleigh fading may be observed to increase the probability of co-channel interference compared with the absence of fading, due to the uncorrelated fluctuations of signal envelopes in the presence of fading.

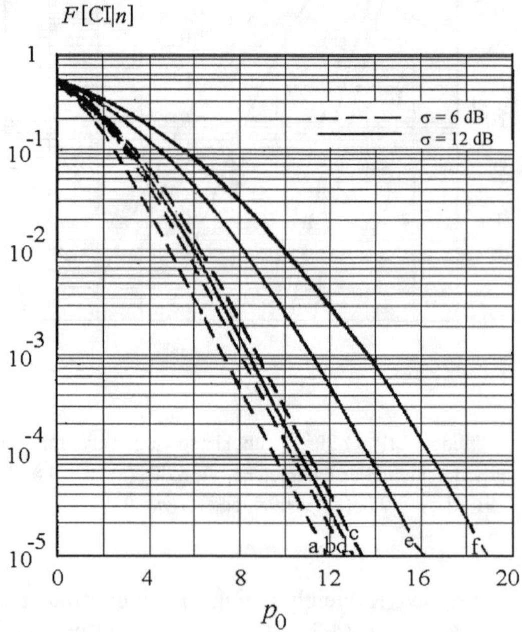

Figure 4.11 Conditional CIP $F(\mathrm{CI}|n)$ with fading and shadowing for correlation coefficient and standard deviation as parameters. $\sigma = 6$ dB. (a) $\rho = 0.8$, (b) $\rho = 0.4$, (c) $\rho = 0$. $\sigma = 12$ dB, (d) $\rho = 0.8$, (e) $\rho = 0.4$, and (f) $\rho = 0$.

4.3.1 Sectorized Cell Layouts

A sectorized cell layout is defined as a pattern of ($N \times S$) sectors with N and S the number of cell sites per cluster and number of sectors per cell site, respectively. (7 × 6), (7 × 3), (4 × 6), (4 × 3) sectors are examples among numerous sectorized cell layouts.

Directional antennas are used to obtain the desired sectorized cell layout. It means that three 120-deg and six 60-deg beam directional antennas are needed to obtain (7 × 3) and (7 × 6) sectors, respectively [33, 37, 40]. As with an omnidirectional antenna system, a directional antenna system (i.e., sectorized cell) offers two main advantages:

(1) reduces the number of co-channel interferers, and (2) increases correlated signals. The sectorized cell layouts yield reduced CIP, thereby enhancing gain in spectrum utilization.

(7 × 6) and (4 × 6) sectorized cell layouts are shown in Figure 4.12(a) and 4.12(b), respectively as the worst-case design [37].

In both sectorized cells, the number of interferers is reduced to one. The (normalized) reuse distance for (7 × 6) sectors is given by

$$R_u = (\xi_d / \xi_i)^{1/\gamma} - 0.7 \tag{4.40}$$

R_u for (4 × 6) sectors is written as

$$R_u = (\xi_d / \xi_i)^{1/\gamma} - 1 \tag{4.41}$$

Computational results are obtained for shadowing only using (4.37), (4.40) ($R_u = 4.6$), and (4.41) ($R_u = 3.46$) and shown in Figure 4.13 as conditional CIP versus protection ratio.

Figure 4.13 shows that for a particular value of protection ratio (7 × 6) sectorized cell layout causes lower conditional CIP than that due to (4 × 6) sectors.

In the case of omnidirectional antenna systems, the lower and upper bounds on CIP is estimated due to one and six interferers, respectively. For the worst-case condition [37], R_u is written as

$$R_u = 1 + (n\xi_d / \xi_i)^{1/\gamma} \qquad n = 1, 2, \ldots, 6 \tag{4.42}$$

The lower bound ($n = 1$) of CIP is presented in Figure 4.14 for shadowing only (4.37) ($R_u = 3.46$, 4.6, and 6).

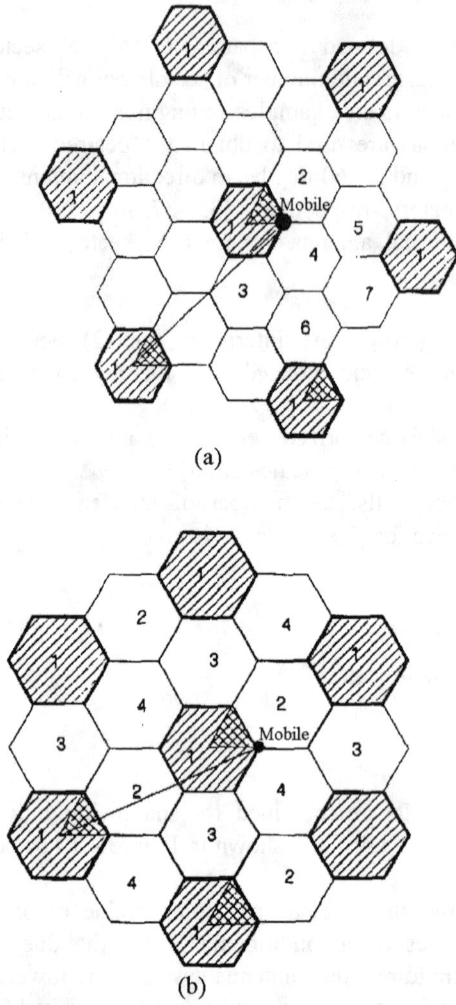

Figure 4.12 Determination of C|I in a directional antenna system for worst case in a 60-deg directional antenna system: (a) (7 × 6) sectorized cell pattern, and (b) (4 × 6) sectorized cell pattern.

Comparing Figures 4.13 and 4.14 shows that the use of sectorized cell layouts leads to considerable reduction in the conditional CIP. In fact, conditional CIP due to sectorized cell layouts is much reduced, compared with the upper bound ($n = 6$) of conditional CIP due to the omnidirectional antenna system not presented here. To have same level of conditional CIP as in the case of sectorized cell patterns, a higher value of reuse distance is required for omnidirectional cell patterns. Thus, a low value of reuse distance is sufficient for the sectorized cells, and hence it gives a higher correlation between signals.

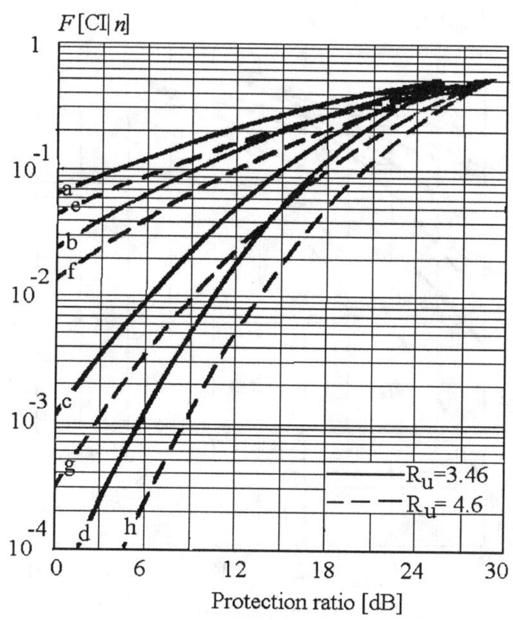

Figure 4.13 Conditional CIP $F(\text{CI}|n)$ versus protection ratio for shadowing only using sectorized cell layouts for correlation coefficients and reuse distances as parameters. $\sigma = 12$ dB: (a) $\rho = 0$, $R_u = 3.46$; (b) $\rho = 0.4$, $R_u = 3.46$; (e) $\rho = 0$, $R_u = 4.6$; (f) $\rho = 0.4$, $R_u = 4.6$. $\sigma = 6$ dB: (c) $\rho = 0$, $R_u = 3.46$; (d) $\rho = 0.4$, $R_u = 3.46$; (g) $\rho = 0$, $R_u = 4.6$; (h) $\rho = 0.4$, $R_u = 4.6$.

The influence of protection ratio α on conditional CIP is shown in Figures 4.13 and 4.14. Figures 4.13 and 4.14 show that conditional CIP increases with increase in the protection ratio. For the future Pan-European digitally modulated system GSM, α is approximately 9.5 dB and $\alpha = 18$ dB for the U.K. TACS with a channel spacing of 25 kHz (a European standard) [36]. Thus, using Figures 4.13 and 4.14, it is confirmed that conditional CIP for GSM will be less than that for TACS for a given set of parameters.

Figure 4.14 shows that $R_u = 4.6$ is not sufficient for $\alpha = 9.5$ dB and 18 dB to maintain $F(\text{CI}) = 10^{-1}$. Therefore, a higher value of R_u (i.e., larger cluster size) is required for omnidirectional cell pattern (e.g., $R_u = 6$ is sufficient, Figure 4.14) whereas in the case of sectorized pattern, even $R_u = 3.46$ is adequate to maintain $F(\mathrm{CI}) = 10^{-1}$ for $\sigma = 6$ dB (Figure 4.14).

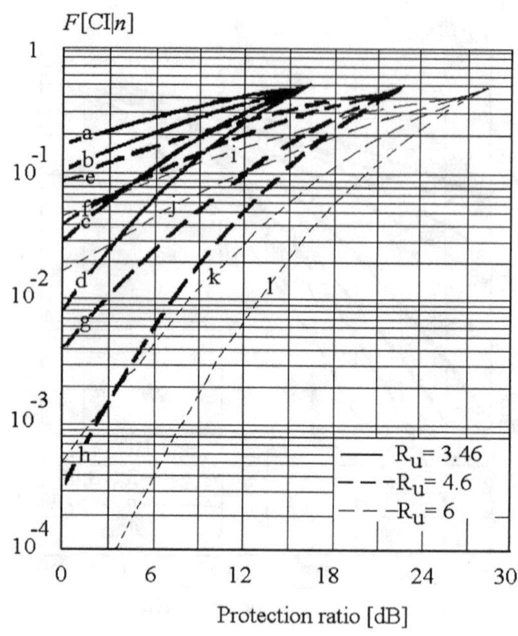

Figure 4.14 Conditional CIP $F(\mathrm{CI}|n)$ versus protection ratio for shadowing only using omnidirectional cell layouts for correlation coefficients and reuse distances as parameters. $\sigma = 12$ dB: (a) $\rho = 0$, $R_u = 3.46$; (b) $\rho = 0.4$, $R_u = 3.46$; (e) $\rho = 0$, $R_u = 4.6$; (f) $\rho = 0.4$, $R_u = 4.6$; (i) $\rho = 0$, $R_u = 6$; (j) $\rho = 0.4$, $R_u = 6$. $\sigma = 6$ dB: (c) $\rho = 0$, $R_u = 3.46$; (d) $\rho = 0.4$, $R_u = 3.46$; (g) $\rho = 0$, $R_u = 4.6$; (h) $\rho = 0.4$, $R_u = 4.6$; (k) $\rho = 0$, $R_u = 6$; (l) $\rho = 0.4$, $R_u = 6$.

4.3.2 Conclusions

An expression for the effective standard deviation was derived in terms of the standard deviation of the desired and interference signals and correlation factor. A graph was also provided to obtain the effective standard deviation, which shows that the correlation effect is equivalent to offering lower standard deviation with uncorrelated shadowing.

This section also dealt with the effect of the correlation between signals on CIP in mobile radio systems. The multipath effects resulting in uncorrelated Rayleigh fading were considered. The influence of correlation on the normalized reuse distance was also investigated.

The results show that the correlation between the signals reduces conditional CIP and its effect becomes more pronounced for higher variances of the shadowed signals, which leads to higher conditional CIPs.

Again, we confirm that the presence of fading with shadowing causes considerable increase in conditional CIP. Further, the correlation between the signals allow the use of shorter reuse distances compared with the uncorrelated case.

Conditional CIP is evaluated as a function of protection ratio for (7 × 6) and (4 × 6) sectorized cell layouts with R_u = 4.6 and 3.46, respectively. A comparative analysis is presented between sectorized and omnidirectional cell layouts.

Omnidirectional cell layouts cause the highest level of conditional CIP as compared with (7 × 6) and (4 × 6) sectors. Computational results confirm that the digital Pan-European GSM system using Minimum Shift Keying (MSK) modulation is superior to the U.K. TACS in terms of the level of conditional CIP. An omnidirectional antenna system requires higher cluster size as compared with the directional antenna to maintain acceptable conditional co-channel interference. The influence of two or more correlated lognormal interferers is not considered in this section.

4.4 CO-CHANNEL INTERFERENCE, THERMAL, NARROWBAND AND BROADBAND MAN-MADE NOISE

Many researchers have studied the performance of mobile radio networks with the typical constraints of a cellular system, such as interference. But until now no theoretical investigations have been conducted to study the effects of narrowband and broadband man-made noise in combination with co-channel interference on mobile radio systems. It is known that (non-Gaussian) man-made noise is a primary source of errors in the data transmission over telephone networks. For mobile radio, especially in urban areas, we expect a large contribution of non-Gaussian automotive noise to the system. It would be very useful to study the effect of generalized noise on cellular system performance to provide insight for system designers.

Noise in general, and especially thermal noise in circuits, is a well-known source of disturbance in communication systems. For the analysis of communication systems, unwanted noise is often taken into account by modeling the effect as a Gaussian distributed random variable; this can be quite accurate due to the central limit theorem. It is also assumed that the noise is white (i.e., it has a flat power spectrum over the frequency band of interest). However, communication systems are seldom interfered with by Gaussian noise alone. Man-made noise has become a problem of great concern in the telecommunications field, particularly in the face of limited available bandwidth.

Man-made noise is basically "impulsive;" it manifests itself in structured bursts, with a high probability of large interference levels, unlike the thermal noise processes in systems. To guard against unacceptable performance, the true characteristics of the noise must be taken into account.

There are several names for non-Gaussian noise. We mention the most common names here: man-made noise, electromagnetic interference, generalized noise, burst noise, and impulsive noise. A noise burst is called impulsive if it causes a response in the front end of the receiver whose shape is identical to the impulsive response of the front end, and only the magnitude and phase may differ. Thus, an essential requirement of a burst noise is that it is white noise, at least across the passband of interest of the receiver. To avoid confusion, we may think of man-made noise in general as pulsed noise with different waveforms for different types of noise sources. Depending on the frequency contents, or sharpness, of the pulsed noise we can divide the noise sources into two groups; the noise bandwidth (Δf_n) is either (1) smaller, or (2) larger than the bandwidth of the front end of the receiver (Δf_r). Of course the bandwidth of the front end of a receiver is controlled by the smallest bandwidth in the circuit before demodulation. Following Middleton [24] we can define the first group $(\Delta f_n < \Delta f_r)$ as class A noise and the second group $(\Delta f_n > \Delta f_r)$ as class B noise. A combination of class A and class B is defined as class C noise.

Numerous research papers report on the analysis of CIP (e.g., [14]) and BEP (e.g., [41]) separately in a fading channel. CIP is evaluated due to the Gaussian noise in [15], and BEP is analyzed in [42] and [43] due to impulsive noise and an additive mixture of Gaussian and impulsive noise, respectively. A number of papers consider Middleton's class A noise for the evaluation of BEP (e.g., [44, 45]). Middelton's class B model is used in [46, 47]. It is mentioned in [44–50] that locally and asymptotically optimum receivers are also analyzed. Those types of receivers are currently not widely used, especially in cellular radio systems.

In this section, the performance of narrowband digital cellular radio systems for two modulation techniques, namely 4-QAM and DPSK is investigated, considering standard receivers optimized against (white) Gaussian noise. The effects of co-channel interference, Gaussian (background) noise, and man-made noise on the performance of cellular radio systems in a Rayleigh fading environment are studied. The performance parameters are BEP and the spectrum efficiency (E_S).

4.4.1 Performance Model

The model for performance analysis is developed using two modulation techniques, binary 4-QAM and DPSK. 4-QAM has the advantage of two-times higher modulation speed and better performance in Gaussian noise compared with DPSK. From the point of efficient use of the frequency spectrum, 4-QAM is preferred. On the other hand, DPSK is more suited to a hostile fading environment, since no absolute carrier phase reference is needed (only nearly constant carrier phase during two successive bit intervals), and so DPSK allows for simpler receiver hardware.

First, the relevant performance parameters for a cellular mobile radio system are defined. To evaluate the BEP in the mobile system, the model is explained with decision variables for 4-QAM and DPSK modulation schemes. The co-channel interference and Rayleigh fading models are described, and finally the conditional probability of bit error is defined.

Performance Parameters

Our main goal is to calculate BEP for the system under consideration, followed by calculation of the spectrum efficiency (E_S). To determine the spectrum efficiency, we use the definition (4.28):

$$E_s = \frac{A_c}{n_c W C S_u} \qquad \text{erlang / MHz / km}^2 \tag{4.43}$$

where A_C = carried traffic per cell in erlang, n_C = number of channels per cell, W = bandwidth per channel in megahertz, C = cluster size, and S_U = cell area in square kilometers.

An important parameter, the reuse distance, R_U, is defined as $R_u \overset{\Delta}{=} D / R$, where D is the distance between two nearest co-channel cells, and R is the radius of the cell. When defining a hexagonal cellular structure, the cluster size C is the number of cells that occurs in the available bandwidth. This basic structure, based on C cells, is repeated until a certain area is covered. R_U is related to the cluster size C as follows:

$$R_u = \sqrt{3C} \tag{4.44}$$

where $C = i^2 + j^2 + ij$, with $i, j = 0, 1, 2,\ldots$

In [51] it is found that

$$\frac{P_d}{P_u} = \frac{(R_u - 1)^{\gamma}}{n} \tag{4.45}$$

Here P_d and P_U are, respectively, the desired and undesired (co-channel) local mean power, γ is the path-loss exponent, and n is the number of active co-channel cells. Equation (4.45) is a worst-case expression considering the downlink (from base station to mobile), with UHF groundwave propagation loss ($\gamma = 4$ here) and assuming the mobile to be at the cell border. BEP, P_C, due to interference from the six nearest co-channels, can be written as [51]

$$P_c = \sum_{n=0}^{6} P(e|n) F_n(n) \tag{4.46}$$

F_n is the probability of n co-channels being active, where $F_n(n)$ is given by (4.24):

$$F_n(n) = \binom{6}{n} a_c^n (1 - a_c)^{6-n} \tag{4.47}$$

a_C is the carried traffic per channel, n is the number of active co-channel cells, and $P(e|n)$ is the conditional BEP.

Decision Variables for 4-QAM and DPSK

To recover the transmitted bits, a decision must be made in the receiver. For each receiver the decision variable depends on its structure. Given a modulation technique, there exist more or less standard structures for optimum reception (minimum error) in white Gaussian noise. Figure 4.15(a) is a block diagram of an optimum coherent receiver for 4-QAM.

The block diagram of a DPSK receiver is shown in Figure 4.15(b). The received signal(s) in a 4-QAM receiver can be viewed as two phase shift keying (PSK) signals [48], so for the bit error analysis of a 4-QAM receiver we use the decision variable, D, given by:

$$D = A_0 + n_i \tag{4.48}$$

where A_0 represents the received bit and n_i is the in-phase component of the noise.

The Gaussian noise samples are assumed to be independent from bit to bit. In practice, a large number of noise sources contributes to the burst noise, so a Poisson process with independent noise burst is assumed. The narrowband noise burst is also independent in the front end of the receiver. Even in the case of broadband impulsive noise with a practical (raised cosine) filter in the front end of the receiver the noise pulse overlap is small, so the non-Gaussian noise is also assumed to be uncorrelated from bit to bit.

Using binary DPSK modulation with 0 and π as possible phase shifts and the carrier frequency as an integer multiple of the data bit rate, the decision variable, D, is [49, 50].

$$D = A_0^2 + A_0 n_{i,k} + A_0 n_{i,k-1} + n_{i,k} n_{i,k-1} + n_{q,k} n_{q,k-1} \tag{4.49}$$

Here, $n_{i,k}$ and $n_{q,k}$ are the samples of the in-phase and quadrature noise component, respectively, at time interval *k*.

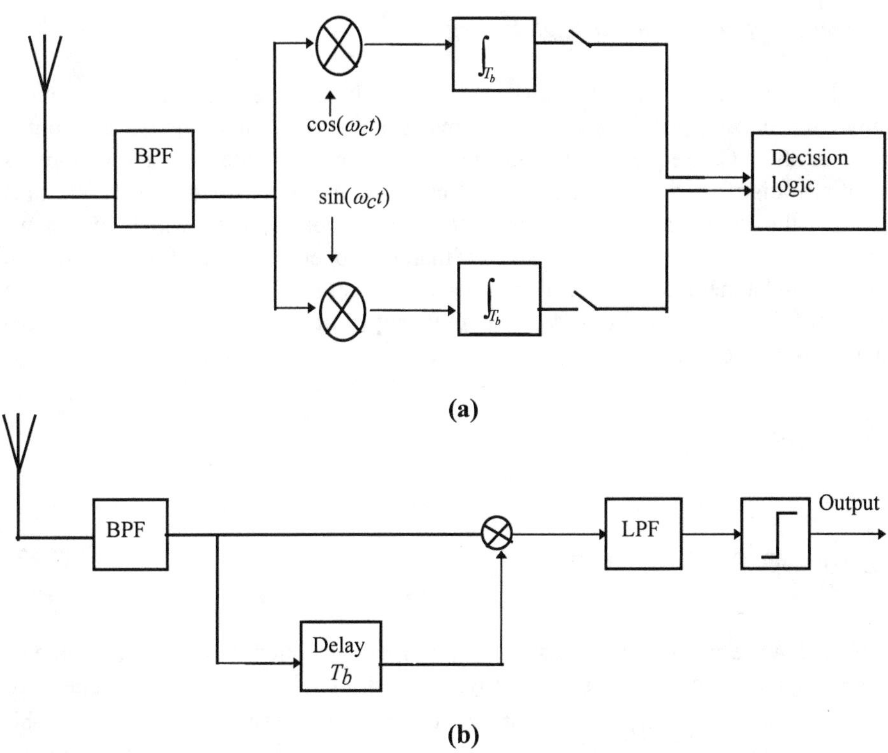

Figure 4.15 Block diagrams of digital receivers: (a) 4-QAM receiver, and (b) DPSK receiver.

BEP with symmetric noise/interference probability density functions is given by:

$$P_{e,4\text{-QAM}} = \Pr\{D < 0 | a_n = 1\}$$

$$= \Pr\{D > 0 | a_n = -1\} \tag{4.50}$$

$$P_{e,\text{DPSK}} = \Pr\{D < 0 | a_n a_{n-1} = 1\}$$

$$= \Pr\{D > 0 | a_n a_{n-1} = -1\} \tag{4.51}$$

where a_n and a_{n-1} are the data bits at two consecutive sampling times.

Co-channel Interference and Rayleigh Fading

It is well known that mobile radio signals suffer from fading due to multipath reception. The envelope of a signal in cellular systems is modeled as a Rayleigh distributed random variable. Consequently, the instantaneous power of the signal envelope is exponentially distributed. Frequency nonselective fading is considered here. We also assume that the fading is constant during the observation interval, which means we assume the bit duration to be (much) smaller than the coherence time of the channel. In the case of co-channel interference from n co-channel cells with a coherent receiver, the reception of the in-phase and quadrature component (u_1, u_2) of the total undesired (co-channel) signal becomes:

$$u_1 = \sum_{j=1}^{n} \alpha_j a_j \cos\phi_j + n_i \qquad a_j = \pm 1 \tag{4.52}$$

$$u_2 = \sum_{j=1}^{n} \alpha_j a_j \sin\phi_j + n_q \qquad a_j = \pm 1 \tag{4.53}$$

Here, α_j, and ϕ_j are random variables due to multipath reception. Since the envelope of a received signal in a cellular system is Rayleigh distributed, u_1 and u_2 are independent Gaussian variables, and ϕ is uniformly distributed. Assuming from bit-to-bit uncorrelated received (fading) signals, the co-channel interference contributes to the Gaussian noise for the long-term average BEP. Notice that in the case of slow fading for DPSK this is an approximation. With mean power P_{0j} for each interferer and mean power σ_g^2 for the Gaussian noise the long term average Gaussian noise and the long term average Gaussian power σ^2 is written as

$$\sigma^2 = \sigma_g^2 + \sum_{j=1}^{n} P_{0j} \tag{4.54}$$

Using (4.45), the signal-to-(total) interference ratio is [52]:

$$S = \frac{P_d}{\sigma_g^2 + n(R_u - 1)^{-4} P_d + \sigma_p^2} = \frac{\text{SNR}}{1 + n(R_u - 1)^{-4}\text{SNR}} \tag{4.55}$$

where SNR is the received SNR at the front end of the receiver in the absence of co-channel interference, which is defined as $P_d/(\sigma_g^2 + \sigma_p^2)$, and σ_p^2 is the average impulsive noise power.

Conditional BEP

The conditional BEP $P(e|n)$, is evaluated by averaging the BEP $P(e|\lambda,n)$, given a constant received power λ during 1 bit period, over the Rayleigh fading as follows:

$$P(e|n) = \int_0^\infty p_\lambda(\lambda) P(e|\lambda,n) d\lambda \tag{4.56}$$

where λ is defined as the ratio of the instantaneous signal power, λ_1, and the instantaneous (total) interference power, λ_2, that is,

$$\lambda = \frac{\lambda_1}{\lambda_2} \tag{4.57}$$

and $p_\lambda(\lambda)$ is the pdf of the instantaneous signal-to-interference ratio in a Rayleigh fading environment with average signal-to-interference ratio S.

$$p_\lambda(\lambda) = \frac{1}{S}\exp\left(-\frac{\lambda}{S}\right) \tag{4.58}$$

4.4.2 Co-channel Interference Plus Narrowband Class A Noise

In [52], BEP for 4-QAM is given as (Appendix 4A)

$$P_{e,4\text{-QAM}} = \sum_{n=0}^{6}\left[\frac{1}{2} - \frac{e^{-A_\alpha}}{2}\sum_{m=0}^{\infty}\frac{A_\alpha^m}{m!}\sqrt{\frac{S(n)}{S(n)+\sigma_m^2(n)}}\right]F_n(n) \tag{4.59}$$

where A_α is the index which represents the number of noise burst per unit time interval. BEP for DPSK is [52]

$$P_{e,\text{DPSK}} = \sum_{n=0}^{6}\left[\frac{e^{-A_\alpha}}{2}\sum_{m=0}^{\infty}\frac{A_\alpha^m}{m!}\frac{\sigma_m^2(n)}{\sigma_m^2(n)+S(n)}\right]F_n(n) \tag{4.60}$$

BEP for 4-QAM and DPSK modulation is compared in Figures 4.16 and 4.17. Figure 4.16 gives the probability of bit error for 4-QAM and DPSK as a function of the reuse distance with $A_\alpha = 0.1$, $\Gamma' = 0.01$, $a_c = 0.5$, and a fixed SNR =20 dB.

The probability of bit error as a function of SNR is presented in Figure 4.17 for the same values of parameters, except $\Gamma' = 10^{-3}$, and a fixed $R_u = 4$. We observe that in both cases 4-QAM performs better than the performance due to DPSK.

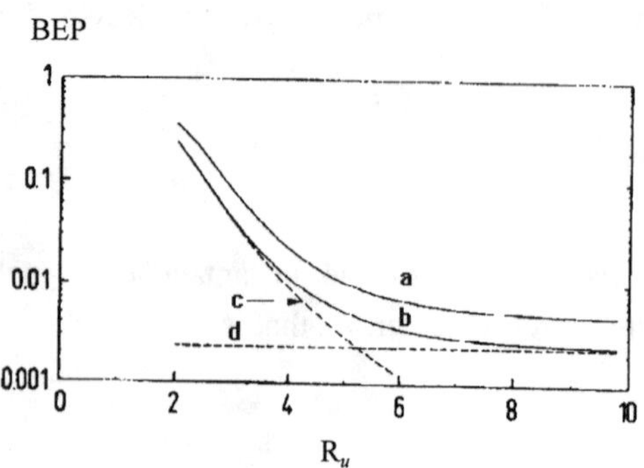

Figure 4.16 Probability of bit error as a function of reuse distance R_u (class A). $A_\alpha = 0.1$, $\Gamma' = 0.01$, SNR = 20 dB and $a_c = 0.5$. (a) DPSK modulation, (b) 4-QAM modulation, (c) co-channel interference only, and (d) noise floor.

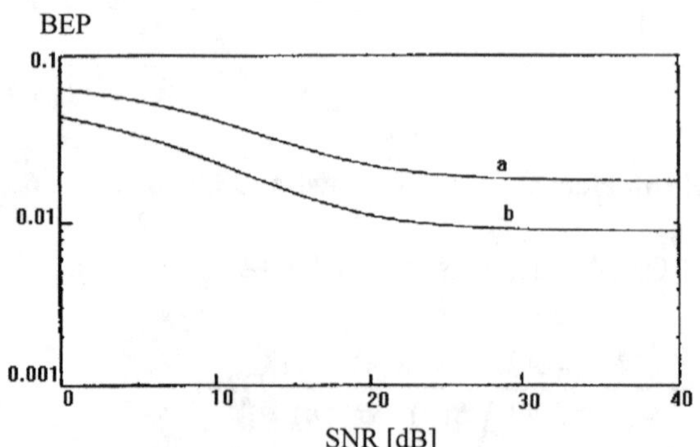

Figure 4.17 Probability of bit error as a function of SNR (class A). $A_\alpha = 0.01$, $\Gamma' = 10^{-3}$, $R_u = 4$, and $a_c = 0.5$. (a) DPSK modulation, and (b) 4-QAM modulation.

A lower limit for the probability of bit error is observed from Figures 4.16 and 4.17. The lower limit as a function of R_u is easily understood; by increasing the reuse distance the co-channel interference decreases, but a fixed amount of thermal and man-made noise remains, and beyond a certain value R_u has no effect on the error probability. For fixed R_u, increasing SNR also gives rise to a limit. Since we assume a symmetric cellular system, increasing the SNR also results in a higher transmitted power in all co-channel cells, and thus also results in more interference; after a certain value for SNR, the noise power is negligible and P_e is determined by the (constant) ratio of p_{0d} and p_{0u}. This limit, given a fixed R_u, is calculated for 4-QAM and DPSK, respectively, as

$$P_{e,\text{lim},4\text{-QAM}} = \sum_{n=0}^{6}\left[\frac{1}{2} - \frac{1}{2}\sqrt{\frac{1}{1+n(R_u-1)^{-4}}}\right]F_n(n) \tag{4.61}$$

$$P_{e,\text{lim,DPSK}} = \sum_{n=1}^{6}\frac{1}{2}\frac{F_n(n)}{1+\frac{1}{n}(R_u-1)^4} \tag{4.62}$$

Other interesting results are discussed and explained in [52].

4.4.3 Co-channel Interference Plus Broadband Class B Noise

Class B noise is of special interest to cellular mobile radio systems, since automotive ignition noise (class B) dominates the interference at frequencies greater than 100 MHz in urban areas [53]. A complete model for class B noise is reported in [24] with 2 pdfs and 6 basic (unknown) parameters, which makes its application very complicated. To avoid complexity in the analysis, a simplified pdf of the normalized instantaneous interference plus Gaussian background noise amplitude z is used here ((4.57) [47])

$$p_z(z) = \frac{e^{-\frac{z^2}{\Omega}}}{\pi\sqrt{\Omega}}\sum_{m=0}^{\infty}\frac{(-1)^m}{m!}A_\alpha^m\Gamma\left(\frac{m\alpha+1}{2}\right)\ {}_1F_1\left(-\frac{m\alpha}{2};\frac{1}{2};\frac{z^2}{\Omega}\right) \qquad |z| \le \infty \tag{4.63}$$

Here, ${}_1F_1(.)$ is the confluent hypergeometric function, $\Gamma(.)$ is the gamma function, α is a propagation parameter, A_α is the effective overlap index, and Ω is a normalizing parameter. Ω is defined as $2/E^2$, where E is the envelope rms power [24]. The corresponding pdf for the noise envelope is found in [47]. Since the second moment of the model is infinite, a saturation level for the noise envelope must be assumed to calculate Ω. When $\Omega = 1$, the process is normalized to the Gaussian energy of the noise [46]. Depending on the impulsiveness, the impulsive noise also contains a certain Gaussian part (e.g., [24]). When the impulsive noise is moderately highly impulsive, the Gaussian energy is determined mainly by the thermal background noise. Then BEP with co-channel interference plus class B noise is compared to the case with co-channel interference and thermal noise only to see the effect of broadband impulsive noise explicitly.

When $\alpha = 1$, the model (4.63) reduces to [47]

$$p_z(z) = \frac{2A_\alpha \sqrt{\Omega}}{\pi(4z^2 + \Omega A_\alpha^2)} \tag{4.64}$$

Again, as with class A noise it is assumed that the co-channel interference contributes to the Gaussian noise, and noise bursts are independent from bit to bit. This is not strictly true, but even in the case of broadband impulsive noise with a practical filter in the front end of the receiver (to avoid ISI), the noise pulse overlap will be small.

Given a constant amplitude level of the desired signal during the observation interval (1-, or 2-bit times), A_0, the conditional probability of bit error, $P(e|n)$, is obtained by averaging over the fading statistics, that is,

$$P(e|n) = \int_0^\infty \frac{A_0}{S'} \exp\left\{-A_0^2 / (2S')\right\} P(e|A_0, n) dA_0 \tag{4.65}$$

Here , S' is the average power of A_0.

With asymmetric noise distribution and assuming a "mark" was sent with 4-QAM and two marks were sent during two consecutive bit times with DPSK, $P(e|A_0, n)$ is calculated:

$$P(e|A_0, n) = \int_{-\infty}^{0} p(D) dD \tag{4.66}$$

where D (function of A_0, noise, and interference) is the decision variable for 4-QAM, or DPSK, given in (4.48) and (4.49), respectively. Defining $A_0 = \frac{\lambda_1'}{\lambda_2'}$ where λ_1' and λ_2' are the instantaneous signal amplitude, and instantaneous (total) interference amplitude, respectively, the pdf of the decision variable for 4-QAM is found

$$p(D)_{4\text{-QAM}} = p_z(D - A_0) \tag{4.67}$$

The integral in (4.66) for 4-QAM is solved for $\alpha = 1$ as

$$P(e|A_0, n) = \left[\xi - \frac{1}{\pi}\arctan\left(\frac{2A_0}{\sqrt{\Omega}A_\alpha}\right)\right] \tag{4.68}$$

where $\xi = 1/2$ when $\Omega = 1$, and $\xi = \pi^{-1}\arctan(2a_{0,\text{sat}} / \sqrt{\Omega}A_\alpha$ for $\Omega \neq 1$ with $a_{0,\text{sat}}$ the amplitude saturation level due to the normalization. For $\alpha \neq 1$ (4.66) requires numerical integration.

The probability of bit error for 4-QAM is found using (4.65) to (4.67), (4.46) and (4.47), where S' becomes S (defined in (4.55)). The pdf of the decision variable for DPSK is

$$p(D)_{\text{DPSK}} = \int_{-\infty}^{\infty}\int_{-\infty}^{\infty}\int_{-\infty}^{\infty} p_z\left(\frac{D\ \delta_3\delta_4}{\delta_2} - A_0\right)p_z(\delta_2 - A_0)p_z(\delta_3)p_z(\delta_4)\frac{1}{|S_2|}d\delta_2 d\delta_3 d\delta_4 \tag{4.69}$$

The integral in (4.66) cannot be expressed in terms of analytical functions for DPSK. Numerical integration is applied to solve (4.66). Following the same procedure as for 4-QAM, BEP for DPSK is found using (4.65), (4.66), (4.69), (4.46), and (4.47).

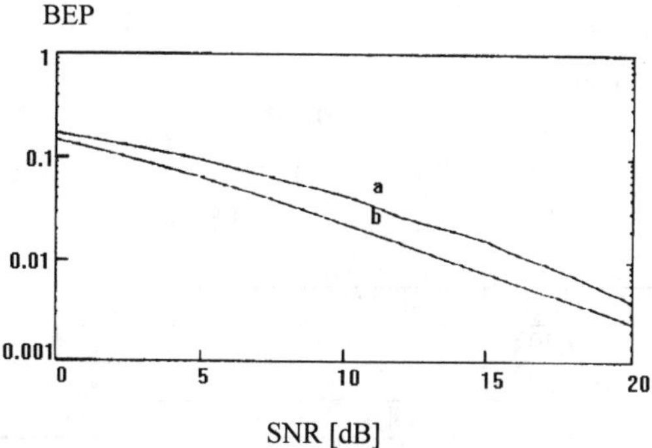

Figure 4.18 Probability of bit error as a function of SNR with 4-QAM (class B). (a) Broadband noise, $A_\alpha = 1.0$, $\Omega = 1$ and $\alpha = 1$, and (b) Gaussian noise only.

In Figure 4.18, we compare P_e as a function of the SNR for Rayleigh fading 4-QAM in the case of Gaussian noise only; we use $\Omega = 1$, $A_\alpha = 1.0$, and $\alpha = 1.0$. For the Gaussian case, $P(e|n)$ is given by:

$$P(e|n) = \frac{1}{2}\left(1 - \sqrt{\frac{S}{S+1}}\right) \tag{4.70}$$

Here S is the Signal-to-interference ratio given by (4.55), where SNR becomes P_{0d} / σ_g^2, because the impulsive noise is absent. In most practical cases, A_α seldom is larger than 1.0, and P_e decreases with decreasing (effective) impulsive index (Figure 4.19(a)) as we will see. We can conclude that the performance degrades with maximum 3 dB when broadband impulsive noise is present. Here, we have the situation normalized to the Gaussian energy; compared with Gaussian noise, class B noise (and impulsive noise in general) is characterized by more significant probability of high noise levels, and higher noise levels in general, although not always present (e.g., Figure 1 in [52]).

When an impulsive class B noise pulse is present, it is therefore more sure to cause a (bit) error. This is how we can understand class B noise gives a higher BEP than Gaussian noise. P_e is depicted in Table 4.5 as a function of the reuse distance with $A_\alpha = 1.0$, $\alpha = 1.0$, SNR =10 dB, and $a_c = 0.5$, for 4-QAM and DPSK. We see that 4-QAM performs better than DPSK, with a maximum of 3.4 times improvement in P_e over DPSK at $R_u = 2$ (i.e., small reuse distance), where the Gaussian component of the noise has a large influence. In Figure 4.19, P_e is presented as a function of the reuse distance for 4-QAM with the same parameters as in Table 4.5, and A_α here is 1.0 and 0.1.

Table 4.5

Probability of Bit Error With Class B Noise for 4-QAM and DPSK With Ω=1, α=1, A_α= 1.0 and SNR = 10 dB

R_u	P_e, DPSK	P_e, 4-QAM	Improvement in P_e Using 4-QAM
2	$(2.9)10^{-2}$	$(8.5)10^{-2}$	3.4
3	$(1.1)10^{-2}$	$(4.1)10^{-2}$	2.2
4	$(6.8)10^{-2}$	$(4.4)10^{-2}$	1.5
6	$(4.7)10^{-2}$	$(4.1)10^{-2}$	1.4
8	$(4.4)10^{-2}$	$(4.1)10^{-2}$	1.3
10	$(4.4)10^{-2}$	$(4.1)10^{-2}$	1.3

In Figure 4.19(b), we illustrate the case with the same parameters as in Figure 4.19(a) for $\alpha = 1.0$ and $\alpha = 0.5$. P_e decreases with decreasing A_α or α. In Figure 4.19(c), the results for the noise normalized to the total amount of noise power are given in terms of BEP versus reuse distance with $A_\alpha = 1.0$, $\alpha = 1.0$, SNR = 10 dB and $a_c = 0.5$, for

4-QAM with $\Omega = 1$ (normalized to Gaussian noise power), and $\Omega = 10^{-3}$: assumed saturation level 50 dB (automotive noise, data from [54]) above the Gaussian level. The results are compared with the case where the total amount of noise is assumed white Gaussian. When the results of Figure 4.19(c) are compared with the situation where the total amount of noise is assumed Gaussian, as with class A, we observe that again performance improves when part of the noise is impulsive.

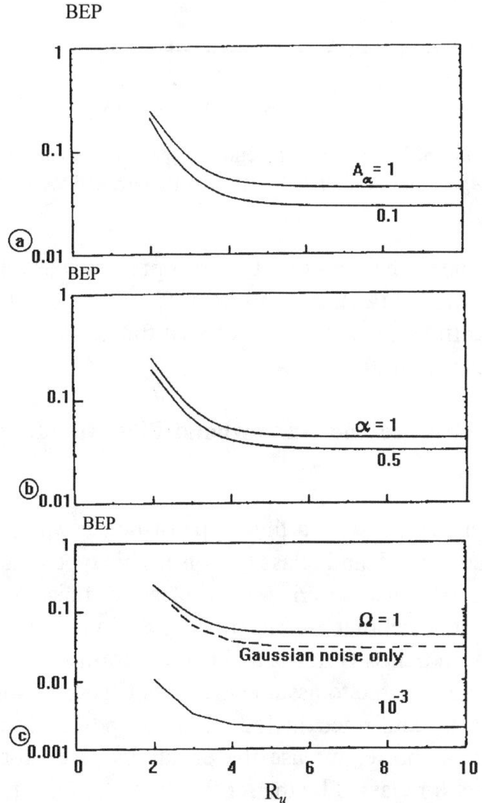

Figure 4.19 Probability of bit error as a function of reuse distance R_u with 4-QAM (class B). SNR = 10 dB and $a_c = 0.5$. (a) $\alpha = 1$, $\Omega = 1$, (b) $A_\alpha = 1$, $\Omega = 1$, and (c) $A_\alpha = 1$, $\alpha = 1$.

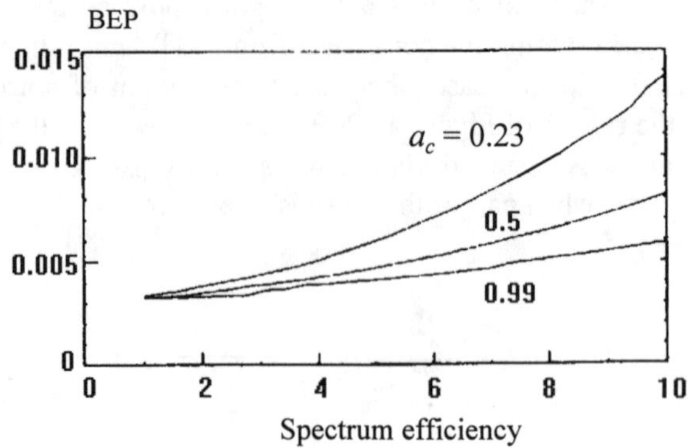

Figure 4.20 Probability of bit error as a function of spectrum efficiency with 4-QAM (class B). $\Omega = 1$, $A_\alpha = 1$, $\alpha = 1$, SNR = 10 dB, $W = 10$ kHz, and $S = 1$ km^2.

In Figure 4.20, the probability of bit error is presented as a function of spectrum efficiency for class B with 4-QAM, $W = 10$ kHz, $S_u = 1$ km^2, SNR =20 dB, $A_\alpha = 1.0$, $\alpha = 1.0$, and $\Omega = 1$ for different values of the carried traffic. Again, spectrum efficiency increases with higher carried traffic.

4.4.4 Co-channel Interference and Narrowband Plus Broadband Impulsive Class C Noise

To complete our model, we consider a third type of noise, which is a combination of narrowband (class A) and broadband (class B) noise. This type of noise is called class C noise [24]. In this section we describe how class C noise is included with the performance analysis of a cellular mobile radio system. Since class C noise is a combination of class A and class B noise, and class A form is incorporated in the class B statistical form, it is reasonable to assume the class C statistics will be similar to the class B statistics. This is confirmed in [52]. For the pdf or cumulative distribution function (cdf) of class C noise, we use the equations given for class B noise, and replace the parameter set for class B by the set for class C [52], namely:

$$P_c = \left\{ A_A, A_B; \Gamma_{AB}; \sigma_{pA}^2, \sigma_{pB}^2 \middle| \alpha, b_{1\alpha}, \Omega_c \right. \tag{4.71}$$

where $A_A + A_B = A_{AB}$, $\Gamma'_{AB} = \sigma_{AB}^2 / \Omega_{AB}$, $\sigma_{AB}^2 = \sigma_{gA}^2 + \sigma_{gB}^2$, $\Omega_{AB} = \sigma_{pA}^2 + \sigma_{pB}^2$, b_{ix} is an implicit parameter defined in [24] and $\Omega \rightarrow \Omega_c$. Here the parameters A, Γ', σ_g^2, and σ_p^2 are labeled according to the noise process they belong to. With these changes, the procedure for performance analysis including class C noise is obtained following the procedure for class B.

When we assume that the ratio of mean and rms impulsive noise power is the same in both (narrowband and broadband) noise processes, it is derived from [24, 52] that for $\Omega = 1$, and $\alpha = 1$:

$$A_{\alpha,c} \sim C_1 \sqrt{\frac{A_{AB}}{1+\Gamma'_{AB}}} \tag{4.72}$$

Here, C_1 is a constant number (>1). In Figure 4.21, the probability of bit error with class C noise is depicted under the assumptions that the Gaussian part of the class C noise is determined only by thermal noise (highly, or moderately impulsive broadband noise), the impulsive broadband noise power is twice the narrowband power, then $\Gamma'_{AB} = \frac{1}{3}\Gamma'_B$, and $A_{\alpha,C} \sim A_{\alpha,B}$ according to (4.72). The remaining parameters are : $\Omega =$ 1, $\alpha = 1$, $A_A = 0.1$, $A_B = 10^{-3}$, $A_{\alpha,B} = 1$, $\Gamma'_B = 10^{-4}$, SNR = 20 dB, and $a_C = 0.5$. Figure 4.21 shows that in the class C environment, system performance is poorer than with only class B noise present.

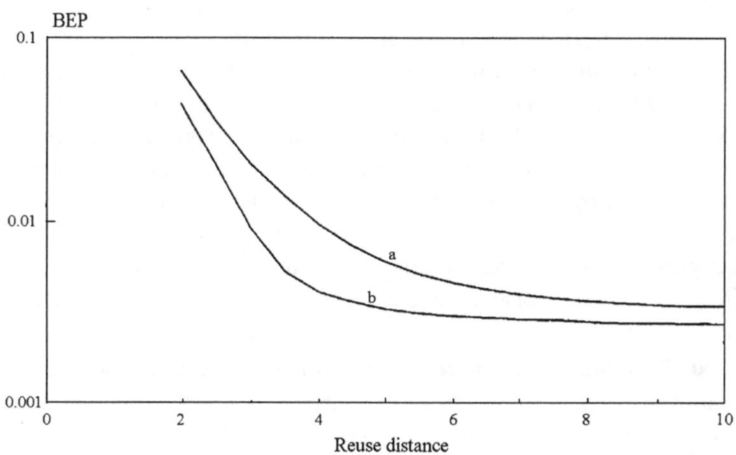

Figure 4.21 Probability of bit error as a function of reuse distance R_u with 4-QAM (class B). $\Omega =$ 1, $A_A = 0.1$, $A_B = 10^{-3}$, $A_{\alpha,B} = 0.1$, $\Gamma'_B = 10^{-4}$, $\alpha = 1$, SNR = 20 dB and $a_C = 0.5$. (a) class C noise, and (b) class B noise.

This combination noise process causes poorer performance; one can see from the parameter set and (4.72), with $\Gamma' << 1$ and A_A, $A_B < 1$ (as usual), the effective impulsive index for class C is always larger than the impulsive index of either one of the individual class A or class B processes, resulting in higher probability of error.

4.4.5 Conclusions

A model was developed that is used for performance analysis of a cellular radio system. The model includes co-channel interference, thermal (Gaussian) noise, narrowband impulsive noise, broadband impulsive noise, or a combination of both, and Rayleigh fading. In this section, a cellular system with 4-QAM and DPSK modulation was considered. For DPSK, the probability of bit error was evaluated for an optimum receiver in additive Gaussian white noise, with arbitrary noise or interference distribution(s).

For a system where the noise is (partially) narrowband-impulsive, we found that the performance was better than for a system with the same amount of purely white Gaussian noise. When the overlap index $A < 1$, and with decreasing overlap index, the performance improves. When the narrowband impulsive part of the noise increases (Γ' decreases), the probability of bit error decreases. This improvement due to a decreasing Γ' is significant for an overlap index $A << 1$ at low SNR (< 10 dB) [52].

As with Gaussian noise, 4-QAM performs better than DPSK when a combination of Gaussian and non-Gaussian noise is present.

In the case of broadband impulsive noise, the probability of bit error increases compared with a system with only Gaussian noise. The degradation has a maximum of approximately 3 dB for 4-QAM modulation considering a (practical maximum) typical effective overlap index of 1. When the density of the broadband impulsive noise increases, represented by an increasing effective overlap index, the performance of a system degrades. The probability of bit error decreases with an increasing path loss, as one would expect with less interference due to the increased path loss.

With respect to the noise parameters, the same trend is seen with class B noise as with class A noise. The performance for $\Omega = 10^{-3}$ is better than when total noise is assumed to be Gaussian. The improvement is not known a priori, since P_e is strongly dependent on the normalization parameter Ω and the saturation level, which are determined by the particular noise process under consideration.

Spectrum efficiency increases with decreasing overlap index, with decreasing noise power ratio, Γ' and with higher carried traffic. Considering class C noise, we conclude that the combination of narrowband and broadband noise results in a higher probability of bit error than with class B noise alone.

REFERENCES

[1] F. Hansen and F. Meno, "Mobile fading—Rayleigh and lognormal superimposed," *IEEE Trans. on Vehicular Technology*, Vol. VT-26, pp. 332–335, November 1977.

[2] R.C. French, "The effect of fading and shadowing on co-channel reuse in mobile radio," *IEEE Trans. on Vehicular Technology*, Vol. VT-28, pp. 171–181, August 1979.

[3] W. Gosling, "A simple mathematical model of co-channel and adjacent channel interference in land mobile radio," *IEEE Trans. on Vehicular Technology*, Vol. VT-29, pp. 361–364, November 1980.

[4] L. Jasinski, "Statistical analysis of communications range and reliability in the presence of interference," *IEEE Trans. on Vehicular Technology*, Vol. VT-30, pp. 123–130, August 1981.

[5] M Hata, K. Kinoshita and K. Hirade, "Radio link design of cellular land mobile communication systems," *IEEE Trans. on Vehicular Technology*, Vol. VT-31, pp. 25–31, January 1982.

[6] D.C. Cox, "Co-channel interference considerations in frequency reuse small-coverage-area radio systems," *IEEE Trans. Comm.*, Vol. COM-30, pp. 135–142, January 1982.

[7] R. Muammar and S.C. Gupta, "Co-channel interference in high capacity mobile radio systems," *IEEE Trans. Comm.*, Vol. COM-30, pp. 1973–1978, August 1982.

[8] K. Daikoku and H. Ohdate, "Optimal channel reuse in cellular land mobile radio systems," *IEEE Trans. on Vehicular Technology*, Vol. VT-32, pp. 217–224, August 1983.

[9] Y. Nagata and Y. Akaiwa, "Analysis for spectrum efficiency in single cell trunked and cellular mobile radio," *IEEE Trans. on Vehicular Technology*, Vol. VT-35, pp. 100–113, August 1987.

[10] J. Stjernvall, "Calculation of capacity and co-channel interference in a cellular system," in *Proc. Nordic Seminar on Digital Land Mobile Radiocommunication*, Espoo, Finland, pp. 209–217, February 1988.

[11] R. Prasad and J.C. Arnbak, "Comments on analysis for spectrum efficiency in single cell trunked and cellular mobile radio," *IEEE Trans. on Vehicular Technology*, Vol. 36, pp. 220–222, November 1988.

[12] R. Prasad, A. Kegel, and J.C. Arnbak, "Analysis of system performance of high capacity mobile radio," *Proc. IEEE 39th Veh. Technology Conf.*, San Francisco, CA, pp. 306–310, May 1989.

[13] A. Safak, R. Prasad, and J.P.M.G. Linnartz, "Outage probability analysis for mobile radio systems due to correlated lognormal interferers," *Proc. 1990 Bulkent Int. Conf. on New Trends in Communication, Control, and Signal Processing*, Bulkent University, Ankara, Turkey, pp. 518–524, July 1990.

[14] R. Prasad and A. Kegel, "Improved assessment of interference limits in cellular radio performance," *IEEE Trans. on Vehicular Technology*, Vol. 40, pp. 412–419, May 1991. (See also R. Prasad and A. Kegel, "Correction to improved assessment of interference limits in cellular radio performance," *IEEE Trans. on Vehicular Technology*, Vol. 41, p. 551, November 1992.)

[15] A. Kegel, W. Hollemans, and R. Prasad, "Performance analysis of interference and noise limited cellular land mobile radio," *Proc. IEEE 41st Vehicular Technology Conf.*, St. Louis, MO, pp. 817–821, May 1991.

[16] R. Prasad, W. Hollemans, and A. Kegel, "Performance analysis of a digital mobile radio cellular system in presence of co-channel interference, thermal and impulsive noise," *IEEE 1991 Int. Symp. on Personal , Indoor and Mobile Radio Communications (PIMRC'91)*, London, pp. 154–159, September 1991.

[17] A. Safak and R. Prasad, "Effects of correlated shadowing signals on channel reuse in mobile radio systems," *IEEE Trans. on Vehicular Technology*, Vol. 40, pp. 708–713, November 1991.

[18] J.P.M.G. Linnartz, "Effects of fading and interference in narrowband land-mobile networks," Ph.D. Thesis, Delft University of Technology, December 1991.

[19] J.P.M.G. Linnartz, "Exact analysis of the outage probability in multiple-user mobile radio," *IEEE Trans. Comm.*, Vol. 40, pp. 20–23, January 1992.

[20] A. Safak and R. Prasad, "Multiple correlated lognormal and interferers in mobile cellular radio systems," *Electron. Lett.*, Vol. 28, pp. 1319–1321, July 1992.

[21] J.P. Linnartz, *Narrowband Land-Mobile Radio Networks*, Norwood, MA: Artech House, 1993.

[22] W. Hollemans, A. Kegel, and R. Prasad, "Performance of digital mobile cellular radio networks in presence of Rayleigh fading co-channel interference, thermal, narrow and broadband man-made noise," *Archiv für Electronik ünd is betragungstechnik* (*AEÜ*), Vol. 48, pp. 84–92, March/April 1994.

[23] A. Safak, "Effects of correlated interferers in multi-user radio systems," Ph.D. Thesis, Delft University of Technology, The Netherlands, June 1994.

[24] D. Middleton, "Statistical-physical models of electromagnetic interference," *IEEE Trans.,* Vol. EMC-19, pp. 106–127, 1977.

[25] V.H. MacDonald, "The cellular concept," *Bell Syst. Tech. J.* Vol. 58, pp. 15–41, January 1979.

[26] P. Beckman, *Probability in Communication Engineering,* New York: Harcourt, Brace and World, 1967.

[27] J.C. Arnbak and W. van Blitterswijk, "Capacity of slotted ALOHA in Rayleigh fading channels," *IEEE J. Selected Areas in Comm.*, Vol. SAC-5, pp. 261–269, February 1987.

[28] L.F. Fenton, "The sum of lognormal probability distributions in scatter transmission systems," *IRE Trans. Comm. Syst.*, Vol. CS-8, pp. 57–67, March 1960.

[29] S.C. Schwartz and Y.S. Yeh, "On the distribution function and moments of power sums with lognormal components," *Bell Syst. Tech. J.*, Vol. 61, pp. 1441–1462, September 1982.

[30] R. Prasad, "Throughput analysis of nonpersistent inhibit sense multiple access in multipath fading and shadowing channels," *European Trans. on Telecommunications and Related Technologies*, Vol. 2, pp. 313–317, May–June 1991.

[31] J.A. Elldin and G. Lind, *Elementary Telephone Traffic Theory*, Stockholm, Sweden: Telefonaktiebolaget LM ERICSSON, 1967.

[32] M Schwartz, *Telecommunication Networks Protocols, Modelling and Analysis*, Reading, MA: Addison-Wesley, 1987.

[33] W.C.Y. Lee, "Elements of cellular mobile radio systems," *IEEE Trans. on Vehicular Technology*, Vol. VT-35, pp. 48–56, May 1986.

[34] Recommendations and Reports of the CCIR, Vol. VIII-I, 1986.

[35] W. Gosling, "Protection ratio and economy of spectrum use in land mobile radio," *Proc. Inst. Elect. Eng.*, Vol. 127, pt. F, pp. 174–178, June 1980.

[36] J. Walker, *Mobile Information Systems*, Norwood, MA: Artech House, 1990.

[37] W.C.Y. Lee, *Mobile Cellular Telecommunication Systems,* New York: McGraw-Hill, 1989.

[38] C.J. Hughes and M.S. Appleby, "Definition of a cellular mobile radio system," *Proc. Inst. Elect. Eng.,* Vol. 132, F, pp. 416–424, August 1985.

[39] A. Papoulis, "*Probability, Random Variables and Stochastic Processes,*" NewYork: McGraw-Hill, 1965.

[40] S. Heeralall and C.J. Hughes, "High capacity cellular patterns for land mobile radio systems using directional antennas," *Proc. Inst. Elec. Eng.*, Vol. 136, pt. I, no. 1, February 1989.

[41] R. Izzo and L. Paura, "Error probability for fading CPSK signals in Gaussian and impulsive atmospheric noise environments," *IEEE Trans.*, Vol. AES-17, pp. 719–722, 1981.

[42] S.A. Kosmopoulos, P.T. Mathiopoulos, and M.D. Gouta, "Fourier-Bessel error performance analysis and evaluation of M-ary QAM schemes in an impulsive noise environment," *IEEE Trans.*, Vol. COM-39, pp. 398–404, 1991.

[43] L. Izzo, L. Panico, and L.Paura, "Error rates for fading NCFSK signals in an additive mixture of impulsive and Gaussian noise," *IEEE Trans.*, Vol. COM-30, pp. 2434–2438, 1982.

[44] D. Middleton, "Optimum reception in an impulsive interference environment—part I, coherent detection," *IEEE Trans.*, Vol. COM-25, pp. 910–922, 1977.

[45] A.M. Maras, H.D. Davidson, and A.G.J. Holt, "Weak-signal DPSK detection in narrowband impulsive noise," *IEEE Trans.*, Vol. COM-33, pp. 1008–1011, 1985.

[46] A.D. Spaulding, "Optimum threshold signal detection in broadband impulsive noise employing both time and spatial sampling," *IEEE Trans.*, Vol. COM-29, pp. 147–150, 1981.

[47] A.D. Spaulding, "Locally optimum and suboptimum detector performance in a non-Gaussian interference environment," *IEEE Trans.*, Vol. COM-33, pp. 509–517, 1985.
[48] J.G. Proakis, *Digital Communications*, New York: McGraw-Hill, 1989.
[49] J.H. Park, Jr., "On binary DPSK detection," *IEEE Trans.*, Vol. COM-26, pp. 484–486, 1978.
[50] D. Middleton, "Canonical non-Gaussian noise models: their implications for measurements and for prediction of receiver performance," *IEEE Trans.*, Vol. EMC-21, pp. 209–219, 1979.
[51] R. Prasad, W. Hollemans, and A. Kegel, "BEP of a digital mobile radio cellular system in presence of co-channel interference, thermal and impulsive noise," *European Trans. Telecommunications and Related Technologies,* Vol. 3, pp. 237–241, May – June 1992.
[52] D. Middleton, "Procedures for determining the parameters of the first-order canonical models of class A and class B electromagnetic interference," *IEEE Trans.*, Vol. EMC-21, pp. 190–208, 1979.
[53] E.N. Skomal, *Man-Made Radio Noise*, Van Nostrand Reinhold, 1978.
[54] R.A. Shepherd, "Measurements of amplitude probability distributions and power of automobile ignition noise at HF," *IEEE Trans.*, Vol. VT-23, pp. 72–82, 1974.

Appendix 4A

4A.1. *Basic Relations*

BEP in a cellular system is defined as

$$P_e \stackrel{\Delta}{=} \sum_n P(e|n) F_n(n) \tag{4A.1}$$

Here, $F_n(n)$ is the probability of n co-channel interferers being active, and $P(e|n)$ is the corresponding conditional BEP. Assuming all co-channel interferers are statistically independent and identically distributed, and considering only interfering signals from the nearest neighboring 6 co-channel cells, $F_n(n)$ is written as

$$F_n(n) = \binom{6}{n} a_c^n (1 - a_c)^{6-n} \tag{4A.2}$$

where $a_c = A_c / n_c$ with A_c as the carried traffic per cell and n_c is the number of channels that is the same in all cells.

$$A_c \stackrel{\Delta}{=} A(1 - B) \tag{4A.3}$$

where A is the offered traffic per cell in erlang and B is the blocking probability determined using the erlang-B formula

$$B = \frac{A^{n_c}}{n_c!\sum_{n=0}^{n_c}\left(\frac{A^n}{n!}\right)} \tag{4A.4}$$

Now, the concept of reuse distance is introduced to study the effect of co-channel interference power on BEP. The (normalized) co-channel reuse distance R_u, is defined as the ratio of the distance D between the centers of the nearest neighboring co-channel cells and cell radius R, that is,

$$R_u \stackrel{\Delta}{=} \frac{D}{R} \tag{4A.5}$$

The reuse distance R_u is related to the cluster size C as follows

$$R_u = (3C)^{1/2} \tag{4A.6}$$

where $C = i^2 + j^2 + ij$ with $i, j = 0, 1, 2, \ldots$

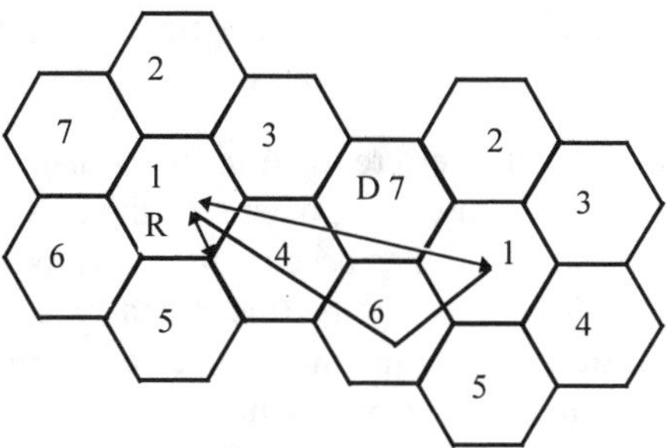

Figure 4A.1 A 7-cell cellular system with $i = 2$ and $j = 1$.

Since BEP is derived as a function of R, (4A.6) enables us to relate BEP to spectrum efficiency. A typical cell structure for 7 cells is shown in Figure 4A.1.

Considering the downlink (from base station to a mobile terminal) in the worst-case situation, with the mobile unit at the cell border, one finds with n active co-channels

$$\frac{P_d}{P_u} = \frac{P_t R^{-\gamma}}{nP_t(D-R)^{-\gamma}} \tag{4A.7}$$

where P_t is the transmitted power from a base station, P_d and P_u are respectively the desired local mean power and the sum of the individual undesired (co-channel) local mean power, and γ is the UHF groundwave propagation-loss law ($\gamma = 4$ is considered here). Using (4A.5), (4A.7) reduces to

$$\frac{P_d}{P_u} = \frac{(R_u - 1)^{\gamma}}{n} \tag{4A.8}$$

4A.2 Probability density function

The pdf for the amplitude Z of class A noise is given by [42]

$$f_z(z) = \exp(-A_\alpha)\sum_{m=0}^{\infty}\frac{A_\alpha^m \exp(-Z^2/2\sigma_m^2)}{m!\sqrt{2\pi\sigma_m^2}} \tag{4A.9}$$

where A_α is the impulsive index that represents the number of noise burst per unit time interval, $\sigma_m^2 = \dfrac{(m/A+\Gamma)}{1+\Gamma}$, and Γ is the ratio of mean Gaussian noise power (σ_g^2) and mean impulsive noise power (σ_p^2).

In the case of co-channel interference from n co-channel cells with a coherent receiver, the in-phase and quadrature components of the total undesired (co-channel) signal are independent Gaussian variables because the envelope of a received signal in a cellular system is Rayleigh distributed. In the (theoretical) receiver, we assume that the data bits are uncorrelated. In that case, the interference contributes to the Gaussian part of the noise. Consequently, in the presence of the co-channel interference, the pdf for the amplitude Z of class A noise is of the form (4A.9) with the following substitution:

$$\Gamma \rightarrow \Gamma_1 \tag{4A.10}$$

Here, Γ_1 is defined as the ratio of the sum of total co-channel interference local mean power P_u and Gaussian noise power to the mean impulsive noise power (σ_p^2), that is,

$$\Gamma_1 \stackrel{\Delta}{=} \frac{\sigma_g^2 + P_u}{\sigma_p^2} \tag{4A.11}$$

Using (4A.8), (4A.11) is written as

$$\Gamma_1 = \Gamma + n(R_u - 1)^{-4}(\Gamma + 1)E_s / N_0 \tag{4A.12}$$

where E_S / N_0 is SNR given by

$$E_s / N_0 = P_d / 2(\sigma_g^2 + \sigma_p^2) \tag{4A.13}$$

The (fast) amplitude fluctuation of mobile radio signals is described by the Rayleigh density function. The corresponding pdf for the instantaneous power p is

$$f_p(p) = \frac{1}{p}\exp(-p / P_d) \tag{4A.14}$$

The pdf $f_\lambda(\lambda)$ for λ (ratio of instantaneous desired power and mean co-channel interference plus mean noise power) can also be described by a density function of the form (4A.14) with the following substitution:

$$p \to \lambda \qquad P_d \to S \tag{4A.15}$$

Here S is defined as the ratio of the mean desired power (P_d) and mean co-channel interference (P_u) plus mean noise power ($\sigma_g^2 + \sigma_p^2$), given by

$$S \stackrel{\Delta}{=} \frac{P_d}{\sigma_g^2 + \sigma_p^2 + P_u} \tag{4A.16}$$

For both DPSK and 4-QAM, using (4A.8) and (4A.16), S is written as

$$S = \frac{E_s / N_0}{1 + n(R_u - 1)^{-4} E_s / N_0} \tag{4A.17}$$

4A.3 *BEP*

For slow (frequency nonselective) fading, conditional BEP $P(e|n)$ is determined by averaging BEP, given a constant received power λ during 1-(4-QAM) or 2-(DPSK) bit intervals, over the fading statistics.

$$P(e|n) = \int_0^\infty f_\lambda(\lambda) P(e|\lambda, n) \; d\lambda \tag{4A.18}$$

The conditional BEP for DPSK modulation is obtained referring [45] and (4A.7) through (4A.18).

$$P_{\mathrm{DPSK}}(e|n) = \int_0^\infty \frac{1}{S} \sum_{m=0}^{\infty} \frac{A_\alpha^m}{m!} \frac{1}{2} \exp\left(- A_\alpha - \frac{\lambda}{\sigma_m^2} - \frac{\lambda}{S} \right) d\lambda \tag{4A.19}$$

The BEP is derived using (4A.1) to (4A.19) and written as

$$P_{e,\mathrm{DPSK}} = \sum_{n=0}^{6} \left[\frac{1}{2} \exp(-A_\alpha) \sum_{m=0}^{\infty} \frac{A_\alpha^m}{m!} \frac{\sigma_m^2}{\sigma_m^2 + S} \right] F_n(n) \tag{4A.20}$$

Similarly, the conditional probability for 4-QAM is obtained referring [46] and (4A.7) to (4A.18):

$$P_{4\text{-QAM}}(e|n) = \int_0^\infty \frac{1}{2S} \exp\left(-A_\alpha - \frac{\lambda}{S} \right) \sum_{m=0}^{\infty} \frac{A_\alpha^m}{m!} \, erfc\left(\frac{\lambda}{\sigma_m^2} \right)^{1/2} d\lambda \tag{4A.21}$$

Finally, using (4A.1) to (4A.17) and (4A.20) we obtain BEP for 4-QAM

$$P_{e,4\text{-QAM}} = \sum_{n=0}^{6} \frac{1}{2} \left[1 - \exp(-A_\alpha) \sum_{m=0}^{\infty} \frac{A_\alpha^m}{m!} \left(\frac{S}{S + \sigma_m^2} \right)^{1/2} \right] F_n(n) \tag{4A.22}$$

4A.4 *Spectrum Efficiency*

Spectrum efficiency, E_S, is defined as the carried traffic per cell, A_C, divided by the product of bandwidth per channel, W number of channels per cell, n_C, and cluster size C cells, given a unit cell area of S_u. Thus

$$E_s \stackrel{\Delta}{=} \frac{A_c}{n_c W C S_u} [\mathrm{erlang / MHz / km^2}] \tag{4A.23}$$

Chapter 5
Microcellular Systems

5.1 INTRODUCTION

Macrocellular systems were studied extensively in Chapter 4. Microcells need special attention due to their propagation characteristics [1–17]. The radio signal propagation in a microcell differs from that in a macrocell in two ways: (1) an LOS path may exist between the two antennas, and (2) the path-loss law follows dual path-loss characteristics. Microcells cover high density areas with 0.4 to 2 km diameter with antennas at street lamp elevation and radiating power less than 20 mW.

This chapter consists of three sections. Section 5.2 presents a mathematical model to analyze the co-channel interference probability assuming the desired signal within a cell is Rician faded with lognormal shadowed. The path-loss law is characterized by a dual path-loss model assuming the first slope to be "a" (approximately 2), the second slope to be "d_S" (approximately 4), and the turning point "g" (between 100m to 200m). The second slope d_S (= $a + b$) is because of an additional attenuation rate coefficient "b" (approximately 2) for distance greater than g. CIP is also evaluated for two different channel situations for the purpose of comparison, namely desired signal as Rician faded and interfering signals as Rayleigh distributed, and both desired and interfering signals as Rician distributed.

In Section 5.3, a model is developed for performance analysis of a microcellular digital mobile radio system with Rayleigh-faded co-channel interference, Gaussian noise, and narrowband impulsive noise (Middleton's class A noise) using the DPSK modulation technique. The desired signal is assumed to be Rician faded. The performance is measured in terms of BEP and spectrum efficiency. The influence of reuse distance, that is, cluster size, traffic intensity, blocking probability, impulsive index, Rician parameter, and turning point of the dual path-loss law characteristic of microcells on the performance parameters are investigated in detail.

A brief description of the propagation model is repeated in each section so readers will have a clear understanding of each section independently.

5.2 CIP

An assessment of spectrum efficiency for a microcellular land mobile radio system is presented by considering the desired signal as (fast) Rician fading with (slow) lognormal shadowing, and co-channel interfering signals as uncorrelated (fast) Rayleigh fading superimposed over (slow) lognormal shadowing. Spectrum efficiency is defined in terms of reuse distance (i.e., cluster size, traffic intensity, bandwidth of the system, and area of a cell) by considering CIP. The expression for co-channel interference probability is derived using the appropriate path-loss law for microcells for four different cases: (1) Rician plus lognormal desired signal and Rayleigh plus lognormal interfering signals, (2) Rician desired signal and Rayleigh fading plus lognormal shadowing interfering signals, (3) Rician desired signal and Rayleigh interfering signals, and (4) both desired and interfering signals as Rician fading. Finally, the performance of a microcellular system is compared with that of a conventional macrocellular system.

5.2.1 Propagation Model

In the microcellular environment, the desired signal level r around the local mean is given by a Rician distribution, where the pdf is given by

$$f_r(r|p_{0d}) = \frac{r}{p'_{0d}}\exp\left[-\frac{r^2+s^2}{2p'_{0d}}\right]I_0\left(\frac{rs}{p'_{0d}}\right)$$

$$R_d = \frac{s^2}{2p'_{0d}} \qquad\qquad p_{0d} = (R_d+1)p'_{0d}$$

$$0 \le r \le \infty \qquad s \ge 0 \tag{5.1}$$

where $I_0(\ )$ is the modified Bessel function of the first kind and zeroth order, s is the peak value of the specular radio signal, p'_{0d} is the average power of the scattered signal, R_d is the Rician factor that depends on the ratio of the signal power from the dominant signal path relative to that of the scattered signal, and p_{0d} is the local mean power.

The pdf for the instantaneous signal power p_d is given by using $p_d = (1/2)r^2$

$$f_{pd}(p_d | p_{0d}) = \frac{1}{p'_{0d}} \exp\left(-\frac{2p_d + s^2}{2p'_{0d}}\right) I_0\left(\frac{\sqrt{2p_d}\, s}{p'_{0d}}\right) \tag{5.2}$$

Here $p'_{0d} = p_{0d} / (R_d + 1)$ (5.3)

The slow-varying local mean p_{0d} is given by a lognormal pdf:

$$f_{p_{0d}}(p_{0d}) = \frac{1}{\sqrt{2\pi}\sigma_d p_{0d}} \exp\left\{-(\ln p_{0d} - m_d)^2 / 2\sigma_d^2\right\} \tag{5.4}$$

Here, σ_d^2 and $\xi_d = \exp(m_d)$ are the logarithmic variance and the area mean power of the desired signal, respectively.

The conditional pdf for the instantaneous power of a co-channel interference p_i is

$$f_{p_i}(p_i | p_{0i}) = \frac{1}{p_{0i}} \exp\left(-\frac{p_i}{p_{0i}}\right) \tag{5.5}$$

where p_{0i} is the local mean of the interfering signal. Equation (5.5) is obtained because the co-channel interferer is a Rayleigh faded signal.

The pdf for the p_{0i} is described by a density function of the form (5.4) with the following substitutions:

$$\begin{aligned} p_{0d} &\rightarrow p_{0i} \\ \sigma_d &\rightarrow \sigma_i \\ m_d &\rightarrow m_i \\ \xi_d &\rightarrow \xi_i \end{aligned} \tag{5.6}$$

Using a dual path-loss model, the received signal power is given by [13]

$$p_r = Cp_t\left[\frac{1}{d^a\left(1 + \frac{d}{g}\right)^b}\right] \tag{5.7}$$

where C is a constant, p_r is the average received power, p_t is the average transmitter power, d is the distance between the base station and mobile unit, g is the turning point of the path-loss curve, a is the basic attenuation rate for short distances (approximately

2), and b is the additional attenuation rate coefficient (approximately 2) for distances greater than 100m to 200m.

5.2.2 Analytical Model

CIP is defined as (4.1)

$$F(\mathrm{CI}) \stackrel{\Delta}{=} \sum_{n} F(\mathrm{CI}|n) F_n(n) \tag{5.8}$$

where $F_n(n)$ is the probability of n co-channel interferers being active and $F(\mathrm{CI}|n)$ is the corresponding conditional CIP

$$F(\mathrm{CI}|n) \stackrel{\Delta}{=} \text{Prob}\ \{p_d / p_n < \alpha\} \tag{5.9}$$

where p_n is the composite interference power from n active channels, and α is the specified protection ratio.

The formulation of conditional CIP for four different propagation conditions is described below.

A. Rician Plus Lognormal Desired Signal and Rayleigh Plus Lognormal Interfering Signals

A three-stage propagation model should be appropriate to describe a microcellular environment: (1) area mean depending on the path-loss characteristics in the range from the transmitter to the area where the receiver is located, (2) local mean without that area, which is slow varying, is well represented by a lognormal distribution, and (3) superimposed fast-fading instantaneous power that follows a Rician distribution for the desired signal and a Rayleigh distribution for the interfering signal. The coexistence of Rician and lognormal conditions has also been found in the land mobile satellite channel [18,19].

The composite pdf for instantaneous power of the desired signal is given by

$$f_{p_d}(p_d) = \int_0^{\infty} \frac{(R_d+1)}{\sqrt{2\pi}\sigma_d p_{0d}^2} \exp\left[-\frac{(R_d+1)(2p_d+s^2)}{2p_{0d}} - \frac{(\ln p_{0d} - m_d)^2}{2\sigma_d^2} \right]$$

$$I_0\left(\frac{\sqrt{2p_d}\, s(R_d+1)}{p_{0d}} \right) dp_{0d} \tag{5.10}$$

It is worth mentioning here that s is also a function of p_{0d} and is given by $\left[2R_d p_{0d}/(R_d+1)\right]^{1/2}$. In the present analysis, s, the LOS component is considered a function of distance d, and its variability is included in the system performance calculation by using Rician factor R_d. The Rician factor varies significantly as the distance between the base station and mobile terminal changes [11]. Since in the following analysis interference probability is evaluated for the worst-case situation, computational results are shown only for typical values of Rician factors.

Developing a statistical model of the Rician factor and using it to make a detailed study of a microcellular system would be an interesting subject for future study.

The composite pdf for n interfering signals with combined Rayleigh fading and lognormal shadowing [20] is given by using (5.5) and (5.6):

$$f_{p_n}(p_n)=\frac{1}{\sqrt{2\pi}\sigma_n}\int_0^{\infty}\frac{1}{p_{0n}^2}\exp\left[-\frac{(\ln p_{0n}-m_n)^2}{2\sigma_n^2}-\frac{p_n}{p_{0n}}\right]dp_{0n} \tag{5.11}$$

where $p_{0n}=\sum_{i=1}^{n}p_{0i}$, σ_n^2 and $\xi_n=\exp(m_n)$ are the logarithmic and area mean of a lognormal variable, which is approximately equivalent to the sum of n independent lognormal interferers. The parameters m_n and σ_n^2 are determined using Schwartz and Yeh's method [21,22].

Using (5.10) and (5.11), conditional interference probability is given by

$$F(CI|n)=\int_0^{\infty}dp_{0n}\int_0^{\infty}\frac{W}{2\pi\sigma_d\sigma_n p_{0d}p_{0n}}$$

$$\exp\left[-\frac{(\ln p_{0d}-m_d)^2}{2\sigma_d^2}-\frac{(\ln p_{0n}-m_n)^2}{2\sigma_n^2}-\frac{R_d W p_{0d}}{\alpha p_{0n}(R_d+1)}\right]dp_{0d} \tag{5.12}$$

where $W=\dfrac{\alpha}{\frac{p_{0d}}{p_{0n}(R_d+1)}+\alpha}$ (5.13)

In (5.12), $m_d=\ln\xi_d$ is obtained using the following relation for the worst case of desired SNR

$$\xi_d=\xi_i\frac{R_u^a(G+R_u)^b}{(G+1)^b} \tag{5.14}$$

Here, R_u is the (normalized) reuse distance, which is defined as the ratio of the distance D between the centers of the nearest neighboring co-channel cells and the cell radius R:

$$R_u \stackrel{\Delta}{=} \frac{D}{R} \tag{5.15}$$

and $G = g/R$.

In deriving (5.14), it is assumed that the transmit power p_t is the same for desired and interfering mobiles. Equation (5.14) has two limiting cases:

(1) At distances significantly less than the turning point (i.e., $G > R_u$), (5.14) becomes

$$\frac{\xi_d}{\xi_i} = R_u^a \tag{5.16}$$

(2) At distances greater than the turning point (i.e., $G < 1$), (5.14) reduces to

$$\frac{\xi_d}{\xi_i} = R_u^{a+b} \tag{5.17}$$

Equations (5.14) through (5.17) are based on the propagation model reported in [13], and accordingly, the turning point g lies between 100m and 200m.

CIP is computed using (5.8), (5.12), (5.13), and $F_n(n)$. It is assumed that all co-channel interferers are statistically independent and identically distributed. Further, it is assumed that only interfering signals from the nearest neighboring six co-channel cells are considered, the blocking probability B is the same in all cells, and the channel is uniformly loaded. Accordingly, $F_n(n)$ is given as

$$F_n(n) = \binom{6}{n} a_c^n (1 - a_c)^{6-n} \tag{5.18}$$

where $a_c = A_c / n_c$, n_c is the number of channels per cell, and A_c is the carried traffic per cell defined as

$$A_c \stackrel{\Delta}{=} A(1 - B) \tag{5.19}$$

Here, A is the offered traffic per cell in erlang, and B is the blocking probability that is determined using erlang B formula [23,24]:

$$B = \frac{A^{n_c}}{n_c! \sum_{n=0}^{n_c} \left(\frac{A^n}{n!} \right)} \tag{5.20}$$

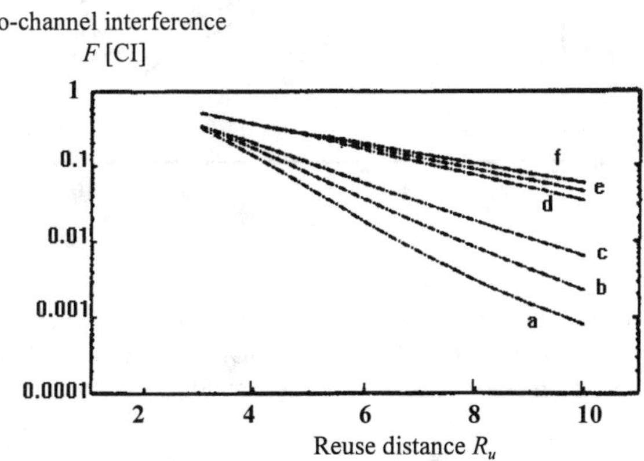

Figure 5.1 CIP versus normalized reuse distance for shadowed Rician desired signal and shadowed Rayleigh interfering signals with $A_c = 5$ erlang, $n_c = 10$, $G = 0.67$, $\alpha = 8$ dB, $R_d = 7$ dB, $B = 0.02$, $n = 6$: (a) $\sigma_d = 0$ dB, $\sigma_i = 6$ dB; (b) $\sigma_d = 4$ dB, $\sigma_i = 6$ dB; (c) $\sigma_d = 6$ dB, $\sigma_i = 6$ dB; (d) $\sigma_d = 0$ dB, $\sigma_i = 12$ dB; (e) $\sigma_d = 4$ dB, $\sigma_i = 12$ dB; and (f) $\sigma_d = 6$ dB, $\sigma_i = 12$ dB.

Using (5.8) to (5.20), CIP is calculated for six interferers and shown in Figures 5.1 and 5.2. Figure 5.1 shows CIP F(CI) versus reuse distance R_u for $A_c = 5$ erlang, $n_c = 10$, $G = 0.67$, $\alpha = 8$ dB, and $R_d = 7$ dB with the spread of desired and interfering signals as parameters. Figure 5.1 shows that in the interfering signals with a spread of 6 dB, F(CI) increases with the increase of spread desired signal by a higher value than in the case where the interfering signal has a spread of 12 dB.

CIP versus distance as shown in Figure 5.2 for Rician factor R_d as a parameter, $G = 0.67$, $\alpha = 8$ dB, $A_c = 5$ erlang, $\sigma_d = 4$ dB, $\sigma_i = 6$ dB, and $n_c = 10$. $R_d = 0$ represents the conditions for a macrocellular environment. The computational results for Figure 5.2 are obtained using (5.8), (5.12), and (5.18) through (5.20) and six interferers. Figure 5.2 shows that for a given value of reuse distance, F(CI) is lower for a microcellular system than for a macrocellular system, and as R_u increases F(CI) decreases. Therefore, to maintain a specific value of F(CI), a much smaller value of the reuse distance is adequate for a microcellular system as compared with the conventional macrocellular system.

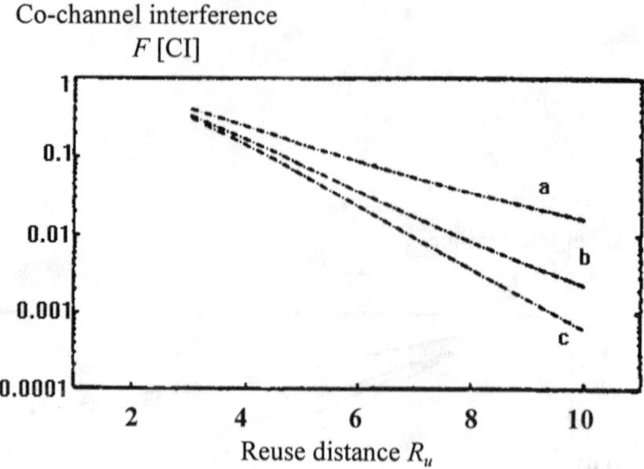

Figure 5.2 CIP versus normalized reuse distance for shadowed Rician desired signal and shadowed Rayleigh interfering signals with $A_c = 5$ erlang, $n_c = 10$, $G = 0.67$, $\alpha = 8$ dB, $B = 0.02$, $n = 6$: $\sigma_d = 4$ dB, $\sigma_i = 6$ dB; $n = 6$: (a) $R_d = 0$, (b) $R_d = 7$ dB, and (c) $R_d = 12$ dB.

B. *Rician Desired Signal and Rayleigh Plus Lognormal Interfering Signals*

Conditional CIP for a Rician desired signal and Rayleigh plus lognormal interfering signals is obtained from (5.12) after removing the shadowed condition on the local-mean power p_{0d} of the desired signal, (i.e., p_{0d} is a constant):

$$F(CI|n) = \int_0^\infty \frac{W}{\sqrt{2\pi}\sigma_n p_{0n}} \exp\left[-\frac{(\ln p_{0n} \quad m_n)^2}{2\sigma_n^2} - \frac{R_d W p_{0d}}{\alpha p_{0n}(R_d + 1)}\right] dp_{0n} \tag{5.21}$$

C. *Rician Desired Signal and Rayleigh Interfering Signals*

In the case of a Rician desired signal and Rayleigh interfering signals, conditional CIP is obtained from (5.12). However, instead of using (5.12), a simple, direct, and exact method for the evaluation of $F(\mathrm{CI}|n)$ for this case is given below.

In the case of n Rayleigh independent interfering signals, the pdf for the joint interference power p_n is obtained by convolving (5.5) n times and given by a gamma distribution:

$$f_{p_n}(p_n) = \frac{1}{p_{0i}} \frac{(p_n / p_{0i})^{n-1}}{(n-1)!} \exp\left(-\frac{p_n}{p_{0i}} \right) \tag{5.22}$$

Equation (5.22) is obtained by assuming that each interferer has equal local mean power p_{0i} and $p_n = \sum_{i=1}^{n} p_i$

Using (5.5), (5.9), and (5.22), conditional interference probability is given by

$$F(\mathrm{CI}|n) = W \exp\left[-R_d W \frac{p'_{0d}}{\alpha p_{0i}} \right] \sum_{k=0}^{n-1} \left(W \frac{p'_{0d}}{\alpha p_{0i}} \right)^k L_k(-kW) \tag{5.23}$$

where L_k = Laguerre polynomial [25,26] and W is obtained from (5.13) after substituting p_{0i} for p_{0n}.

CIP is computed using (5.8), (5.18) to (5.20) and (5.23) for α = 8 dB, A_c = 5 erlang, R_d = 7 dB, G = 0.67, and n_c = 10 for n = 1, 6 as a function of R_u and is shown in Figure 5.3. For n = 1, (5.23) reduces to

$$F(\mathrm{CI}|1) = W \exp\left[-R_d W \frac{p'_{0d}}{\alpha p_{0i}} \right] \tag{5.24}$$

For the purpose of comparison, computational results for one co-channel interferer with Rician pdf are also included. $F(\mathrm{CI}|n)$ for a single Rician co-channel interferer is given by

$$F_r(\mathrm{CI}|1) = \frac{W}{\alpha} \left\{ 1 - Q\left(\left[\frac{2R_d W}{\alpha} \right]^{\frac{1}{2}}, \left[2R_d W \right]^{\frac{1}{2}} \right) \right\} + WQ\left(\left[2R_d W \right]^{\frac{1}{2}}, \left[\frac{2R_d W}{\alpha} \right]^{\frac{1}{2}} \right) \tag{5.25}$$

Here, $Q(a,b)$ is a Marcum's Q-function given by

$$Q(a,b) = \int_b^\infty x \exp[-\frac{x^2 + a^2}{2}] I_0(ax) dx \tag{5.25a}$$

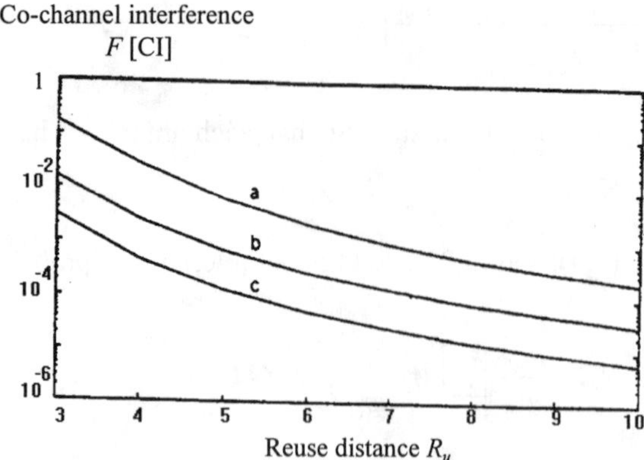

Figure 5.3 CIP versus normalized reuse distance for Rician faded desired signal with protection ratio α = 8 dB, blocking probability B = 0.02, carried traffic per cell A_c = 5 erlang, n_c = 10 channels per cell, Rician factor R_d = 7 dB and g/R = 0.67: (a) six Rayleigh co-channel interferers, (b) one Rayleigh co-channel interferer, and (c) one Rician co-channel interferer.

Figure 5.3 shows that CIP is always minimum for one Rician interferer.

D. Desired and Interfering Signals as Rician pdf

Finally, conditional CIP is derived for a special case in which desired and interfering signals are Rician pdf. If the pdf for the interfering signals is also Rician, the pdf for the instantaneous power is of the same form as (5.2) with the simple substitutions of changing index d to i.

In a microcellular system, generally interfering signals are Rayleigh faded because they travel a longer distance to arrive at the base station. But in the case of a very small microcell size, the distance between the centers of co-channel cells may not be large, and, therefore, the interfering signals may be modeled by Rician fading. For example, if R = 200m, and R_u = 3, then D = 600m. For such cases, assuming the pdf for interfering signals as Rician distributed is justified.

Assuming n-Rician independent interfering signals, the pdf for the joint interference p_n is given by

$$f_{p_n}(p_n) = \frac{1}{p'_{0i}}\left(\frac{2p_n}{p}\right)^{\frac{n-1}{2}} \exp\left(-\frac{2p_n + p}{2p'_{0i}}\right) I_{n-1}\left(\frac{\sqrt{2p_n p}}{p'_{0i}}\right) \tag{5.26}$$

where $p = \sum_{i=1}^{n} s_i^2$ and $p_n = \sum_{i=1}^{n} p_i$

conditional CIP is obtained using (5.2), (5.9), and (5.26), assuming $R_d = R_i$:

$$F(\mathrm{CI}|n) = 1 - \left(\frac{1}{\sqrt{2nR_d}}\right)^{n-1} \int_0^\infty u^n \exp\left(-\frac{u^2 + 2nR_d}{2}\right) I_{n-1}\left(u\sqrt{2nR_d}\right)$$

$$Q\left(\sqrt{2R_d}, u(\alpha p_{0i} / p_{0d})^{\frac{1}{2}}\right) du \tag{5.27}$$

where $Q(x,y)$ is a Marcums Q-function.

5.2.3 Conclusions

A mathematical model was developed to evaluate CIP in a microcellular radio system having Rician faded and lognormal shadowed desired signal, and Rayleigh faded and lognormal shadowed interfering signals by considering a dual path-loss model with a turning point. CIP thus obtained is used to study the effects of carried traffic, protection ratio, shadowing spread, Rician factor, and turning point ($G = g/R$) on spectrum efficiency. The value of the turning point varies between 100m and 200m . As a special case, performance is also evaluated for Rician faded desired and interfering signals.

Computational results show that slow fading (shadowing) causes increase in CIP as compared with the case in which interfering signals are Rayleigh faded only. Further, by introducing shadowing in both the desired and interfering signals, it is found that, with the same value of spread for the desired signal and a different spread of the interfering signal, the increase in CIP is less profound with a higher spread of interfering shadowing.

In the case of Rician faded interferers, performance of a microcellular system improves as compared with the systems having Rayleigh faded interferers; for $G < 1$ and for $G \geq 1$ the situation gets reversed; Rayleigh faded interferers yield superior performance.

This study will be useful for designers and planners in developing any type of microcellular system (e.g., streets, highways, forests, and rural areas).

5.3 CO-CHANNEL INTERFERENCE, NATURAL, AND MAN-MADE NOISE ENVIRONMENT

Most studies consider interfering noise as Gaussian. But a realistic electromagnetic (EM) environment contains also man-made noise that is partly impulsive. It would be interesting to analyze the effect of such noise on system performance to provide a more reliable tool for the systems designer. In this section, the effects of co-channel interference, Gaussian, and man-made noise on a microcellular network are investigated. The modulation scheme considered is DPSK. The performance criteria studied are BEP and spectrum efficiency.

5.3.1 Man-Made Noise

During the past two decades, man-made electromagnetic interference (or noise) has become a problem of increasing concern to the telecommunication community.

The deteriorating effects of man-made noise on system performance is now generally recognized. The sources of man-made noise are numerous: incidental radiation from electrical devices of all sorts, out-of-band modulation products from radio communication systems, automotive ignition systems, electric power lines, and so on.

This man-made contribution to the EM environment is basically impulsive, it has a highly structured form, characterized by significant probabilities of interference levels.

In a metropolitan area, man-made interference can be present in the 30-Hz to 7-GHz radio spectrum range. Above 100 MHz, the spectrum is dominated by man-made noise with automotive ignition noise, and several man-made sources form a mixture. At lower frequencies, the impulse widths and amplitudes of the impulsive emission patterns are greater, but the average occurrence rate of the largest pulses may increase with the frequency [27].

These highly non-Gaussian random processes can have severe degrading effects on system performance, particularly on most conventional systems, which are designed for optimal or near optimal performance against normal noise.

Therefore, it is important that the true EM environment (natural plus man-made noise) is modeled by a physical statistical model, providing a better basis for system design and comparison. A lot of research was done on the development of statistical-physical models. The model proposed by Middleton [28], including both man-made noise and Gaussian noise, and dividing EM interference into three classes, is used here and therefore briefly reviewed.

Three classes of EM interference (including both man-made and Gaussian noise) are identified in Middleton's theory [28]:

- *Class A*. The spectral bandwidth of the noise entering the receiver is comparable to or less than the bandwidth of the receiving system. Transient effects are ignorable.
- *Class B*. The bandwidth of the noise is greater than the bandwidth of the receiving system, (i.e., the noise pulses produce transients in the receiver).
- *Class C*. A linear sum of class A and class B.

Middleton's class A interference model is governed basically by three parameters (A,Γ,Ω), where [29]: A = the impulsive index, also known as overlap or unstructure index, which has been defined as the average number of radiation 'events' per second times the mean duration of a typical source emission. The smaller A, the more the interference is structured (in time). When $A \to \infty$ the noise is Gaussian. $\Gamma = \sigma_G^2/\sigma_p^2$ = the Gaussian factor = the ratio of average intensity of the Gaussian component. σ_p^2 = the mean intensity of the non-Gaussian (or impulsive) noise component of the interference. Ω is defined as $2/E^2$, where E is the envelope rms power.

From [30] we find that the pdf for the instantaneous amplitude yields

$$P_{\text{ampl}}(X) \cong e^{-A} \sum_{m=0}^{\infty} \frac{A^m}{m!\sqrt{2\pi\sigma_m^2}} \exp\left(-\frac{x^2}{2\sigma_m^2}\right) \tag{5.28}$$

where $$\sigma_m^2 \equiv \frac{\frac{m}{A}+\Gamma}{1+\Gamma} \tag{5.29}$$

Note that $P_{\text{ampl}}(X)$ is in fact a weighted sum of Gaussian distributions with increasing variance. The distribution function of the envelope is also obtained from [30]:

$$P_1(E \geq E_0) \cong e^{-A} \sum_{m=0}^{\infty} \frac{A^m}{m!} \exp\left(-\frac{E_0^2}{\sigma_m^2}\right) \tag{5.30}$$

This density function is a weighted sum of Rayleigh distributions with increasing variance.

5.3.2 Microcellular Mobile Radio Systems

Measured propagation results indicate that the received signal envelope is Rician distributed [14,15]. Then the pdf for the ratio of instantaneous signal power to the sum of co-channel interference, Gaussian, and impulsive noise is written as

$$f(\rho) = \frac{1}{\sigma^2} \exp\left(-\frac{2\rho+S^2}{2\sigma^2}\right) I_0\left[\frac{\sqrt{2\rho S}}{\sigma^2}\right]$$

where $\sigma^2 = \sigma_a^2 / (P_u + \sigma_g^2 + \sigma_p^2)$,

$\rho = \rho_a / (P_u + \sigma_g^2 + \sigma_p^2)$,

$$S^2 = S_a^2 / (P_u + \sigma_g^2 + \sigma_p^2) \tag{5.31}$$

Where S_a denotes the LOS signal, σ_a^2 is the average scattered power, P_u is the total mean co-channel interference power, σ_g^2 is the mean power for the Gaussian noise, σ_p^2 is the mean power for impulsive noise, and $I_0[]$ is the modified Bessel function of the first kind and zero order.

The absence of an LOS component in the co-channel interference is a realistic assumption because of the relatively large distances, so we assume that the co-channel interferer is Rayleigh faded and its power is exponentially distributed. For the *i*th interferer, the pdf of the interference power, p_i, is expressed as

$$f(p_i) = \frac{1}{P_{0i}} \exp\left(-\frac{p_i}{P_{0i}}\right) \qquad 0 \le p_i < \infty \tag{5.32}$$

where P_{0i} is the local mean power of the *i*th interferer.

Another aspect of microcell propagation is the path-loss characteristic. In microcells, the logarithmic attenuation slope is modeled by two straight lines, one representing the (inverse square) slope between the base site and so-called turning point, and the other (inverse fourth power) between the turning point and infinity [13,17]. The turning point has to be experimentally obtained; 100m to 200m is a typical value for an urban environment. This path-loss characteristic yields [13,17,31]

$$P_0 = P_t C \frac{g_b g_m}{d^a \left(1 + \frac{d}{g}\right)^b} \tag{5.33}$$

where C is a constant, P_0 is the average received power, P_t is the average emitter power, g_b and g_m are the antenna gains, d is the distance between the base site and receiver, a is the basic attenuation rate for short distances (approximately 2), b is the additional attenuation rate for distances larger than 100m to 200m and is approximately 2 and g is the turning point. It is expected that b and g depend on the environment (types of buildings, amount of street traffic, etc.).

5.3.3 DPSK Modulation Performance

In this section, a DPSK system is described. The advantage of DPSK over many other modulation types is that in DPSK no absolute phase reference is needed; only nearby constant carrier phase during two-bit successive intervals is required. That property of DPSK makes it appropriate for a fading environment.

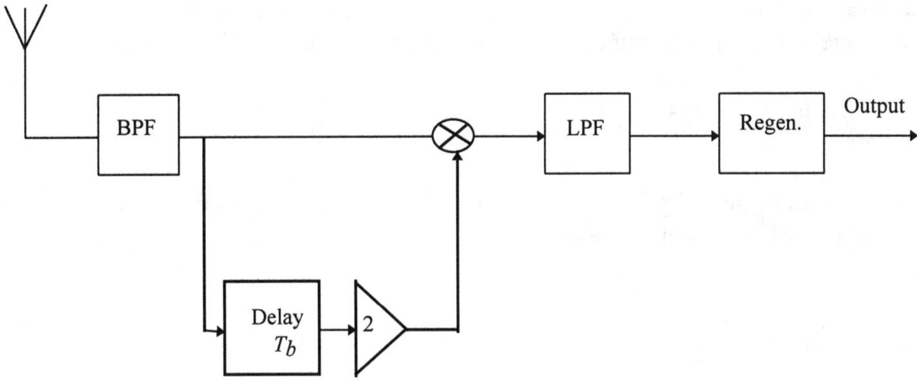

Figure 5.4 The suboptimum DPSK receiver.

Given a differentially encoded PSK signal, there are several demodulation methods [32]. In practice, this demodulation is often done by delaying the received signal by the bit interval T and using the delayed signal to multiply the received signal. This is visualized in Figure 5.4. Then the decision variable is the projection of the received signal on the phase shifted delayed signal. When using binary DPSK with 0 and π as possible phase shifts and the carrier frequency is an integer multiple of the signaling rate, this decision variable is [32]

$$D_v = 2\,\mathrm{Re}\left[V_n V_{n-1}^*\right] = \left[V_n V_{n-1}^* + V_n^* V_{n-1}\right] \tag{5.34}$$

where $V_n = A_{0,n} e^{j\phi} + N_n$

$$N_n = \left[n_i \cos(\theta_n) + i n_q \sin(\theta_n)\right] e^{j\phi} \tag{5.35}$$

Here, V_n is the received signal in the nth interval and N_n the noise component. θ_n is the noise phase and ϕ is the carrier phase, which is presumed to be constant over at least $2T$. The inphase and quadrature noise components are denoted as n_i and n_q. Without loss of generality, we assume that $A_{0,n} = A_{0,n-1}$ which is allowed under the condition that $A_{0,n}$ is likely to be +1. Then D_V yields [31,33]

$$D_v = A_0^2 + A_0 n_{i,n} + A_0 n_{i,n-1} + n_{i,n} n_{i,n-1} + n_{q,n} n_{q,n-1} \tag{5.36}$$

Because the modulation phases are 0 and π, the decision level is zero. This is an advantage in Rayleigh fading environments, since the level itself is not sensitive to that fading. Now BEP is written as

$$P_{\text{bit error}} = \Pr\left\{D_v < 0 \middle| A_0^2 = 1\right\} \tag{5.37}$$

For Gaussian noise, the pdf of D_v contains a quadratic Gaussian random variable and is derived in [28]. The final expression yields

$$P_{\text{bit error, Gaussian noise}} = \frac{1}{2}\exp(-\rho_b) \tag{5.38}$$

where ρ_b is the SNR, and the noise is assumed to be independent from bit interval to interval.

However, we are interested in the BEP in Middleton's class A noise. Deriving that probability from the decision variable as defined in (5.36) is extremely complicated because of the noise product terms. Therefore, an approximation is presented here. This estimated performance measure is proposed by Marras et al. [34]. It is based on the conditional error probability for the noncoherrent correlation (i.e., suboptimum) FSK receiver given by Spaulding and Middleton [35, (51b)]. Using the relation of DPSK to FSK in white Gaussian noise, given by Viterbi [36], Marras et al. derive an approximation that yields [34]

$$P_e(\rho) = \frac{1}{2}\exp(-A)\sum_{m=0}^{\infty}\frac{A^m}{m!}\exp\left(-\frac{\rho}{\sigma_m^2}\right) \tag{5.39}$$

where ρ is the normalized power (so the SNR) and A and σ_m are the parameters defined in the previous section. Equation (5.39) is the main tool for the performance analysis.

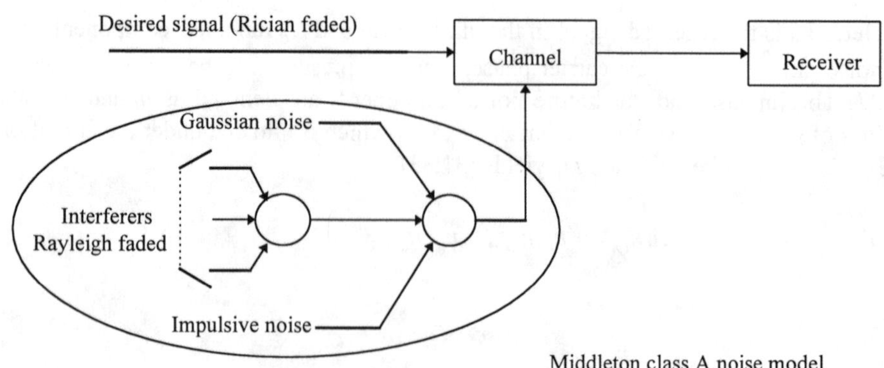

Figure 5.5 Diagram of the additive channel model.

Another issue in our model description is the inclusion of the Rayleigh faded interferers. We assume an additive model as shown in Figure 5.5.

All co-channel signals are independent and uncorrelated. The phases are assumed to be uniformed randomly distributed from bit to bit, but constant during one bit period. Since we are considering bandpass signals here, we write the inphase and quadrature components of the sum $z(t)$, u_i and u_q, respectively

$$u_i = \sum_{j=1}^{n} \alpha_j a_j \cos\phi_j$$
$$u_q = \sum_{j=1}^{n} \beta_j a_j \sin\phi_j \qquad a_j = \pm 1 \tag{5.40}$$

In (5.40), α_j and β_j are Rayleigh distributed. Multiplying these variables with the randomly phased cosine or sine terms gives a Gaussian variable as a product. In this event, the joint interference power is added to the Gaussian noise. This assumption has been applied in several papers [37–39]. With mean power p_{0j} of the jth interferer and mean power σ_g^2 for the Gaussian noise, we write for the new long-term average Gaussian power σ'^2_j

$$\sigma'^2_g = \sigma_g^2 + \sum_{j=1}^{n} P_{0j} = \sigma_g^2 + P_u \tag{5.41}$$

Now the co-channel interference can be included in the Gaussian noise portion — it has been implemented as an addition to the Middleton noise parameter Γ just by substituting σ_g^2 by σ'^2_j.

The described model is used to analyze the microcellular network. The reciprocal of the mean received signal power to noise plus interference power at a receiving mobile station located at the microcell border, γ, is written as

$$\gamma \stackrel{\Delta}{=} \frac{\sigma_g^2 + \sigma_p^2 + P_u}{(1+K)\sigma^2} = \frac{n}{(R_u - 1)^2}\left(\frac{G+1}{G+(R_u-1)}\right)^2 + \frac{1}{S_n} \tag{5.42}$$

where $R_u \left(\stackrel{\Delta}{=} D/R\right)$ is the (normalized) reuse distance, defined as the ratio of the distance between the centers of the nearest neighboring co-channel cells (D) to the cell radius R, as visualized in Figure 5.6, n is the number of active co-channel interferers, K is the Rician factor ($= S^2/2\sigma^2$), S_n is the total received signal to mean Gaussian noise power ratio, and $G(=g/R)$ is defined as the normalized turning point being the ratio of the path-loss turning point g and the cell radius R.

The conditional BEP $P(e|n)$ is obtained by integrating the product of $f(\rho)$ and $P_{e(\rho)}$ from zero to infinity:

$$P(e|n) = \frac{1}{2\sigma^2}\exp\left(-A - \frac{S^2}{2\sigma^2}\right)\sum_{m=0}^{\infty}\frac{A^m}{m!}\int_0^{\infty}\exp\left\{-\rho\left(\frac{1}{\sigma^2}+\frac{1}{\sigma_m^2}\right)\right\}I_0\left[\frac{\sqrt{2\rho S}}{\sigma^2}\right]d\rho \qquad (5.43)$$

Equation (5.43) simplifies to

$$P_n(e|n) = \frac{1}{2}\exp(-A-K)\sum_{m=0}^{\infty}\frac{A^m}{m!}\alpha\exp(K\alpha)$$

where $\alpha \stackrel{\Delta}{=} \dfrac{\sigma_m^2}{\sigma_m^2 + \frac{1}{(1+K)\gamma}}$ (5.44)

and K is the Rice factor.

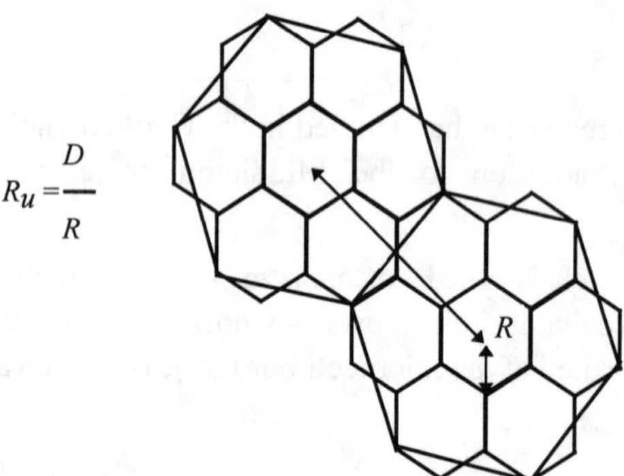

Figure 5.6 Configuration of a cellular system and definition of reuse distance.

Finally, BEP is evaluated by using

$$P_e \stackrel{\Delta}{=} \sum_n P_n(e|n)F_n(n) \qquad (5.45)$$

where $F_n(n)$ is the probability of active n co-channel cells. If we adopt the commonly assumed hexagonal cell structure, $F_n(n)$ is expressed in terms of carried traffic a_c per channel

$$F_n(n) = \binom{6}{n} a_c^n (1 - a_c)^{6-n} \tag{5.46}$$

Additionally, to BEP, another performance measure, the spectrum efficiency E_s, is defined as

$$E_s \overset{\Delta}{=} \frac{a_c}{WCS_c} \qquad \text{[erlang/MHz/km}^2\text{]} \tag{5.47}$$

where a_c is again the carried traffic per channel in erlang, W the channel bandwidth in megahertz, C the cluster size, and S_c the cell area in square kilometers. Spectrum efficiency is evaluated using (5.47) and the relation between reuse distance and the number of cells per cluster; that is

$$R_u = (3C)^{\frac{1}{2}} \tag{5.48}$$

In the model described above, setting the Rice factor K and normalized turning point G to zero corresponds to a macrocellular system, where the desired signal is assumed to be Rayleigh faded and the path-loss function is of the second order. In this case, (5.44) reduces to

$$P_{K=0}(e|n) = \frac{1}{2}\exp(-A)\sum_{m=0}^{\infty}\frac{A^m}{m!}\frac{\sigma_m^2}{\sigma^2 + \sigma_m^2} \tag{5.49}$$

Another special condition is the hypothetical case where all base station transmitter powers increase to infinity. Then BEP is completely determined by the interfering co-channel cells and (5.44) reduces to

$$P(e|n) = \frac{1}{2}\frac{1+K}{\gamma+1+K}\exp\left(\frac{-K\gamma}{\gamma+1+K}\right) \tag{5.50}$$

Another boundary is the situation where the co-channel interference is absent and BEP is completely determined by Gaussian and impulsive noise. If we again consider (5.44), we see that in this case

$$P(e|n) = \frac{1}{2}\exp(-A-K)\sum_{m=0}^{\infty}\frac{A^m}{m!}\frac{\sigma_m^2}{S_n+\sigma_m^2}\exp\left(\frac{\sigma_m^2}{S_n+\sigma_m^2}K\right) \tag{5.51}$$

If transmitter power decreases to zero, the conditional bit error probability increases to 1/2, giving an overall error P_e of 1/2 at the receiver.

BEP is numerically evaluated using (5.39) through (5.46) and the results are found in Figures 5.7 to Figure 5.9. Figure 5.7 shows the BEP P_e against reuse distance R_u for a carrier-to-noise ratio of 10 dB, carried traffic per channel a_c = 0.5 erlang, Rice factor K = 7 dB, normalized turning point G = 0.2, and Γ =0.001 for several values of A. Parameter A denotes the character of the noise; individual Gaussian and impulsive noise powers are invariant for A. Figure 5.7 shows that decreasing A improves the performance. Therefore, we conclude that the more structured (in time) the noise is, the less impact it has on BEP.

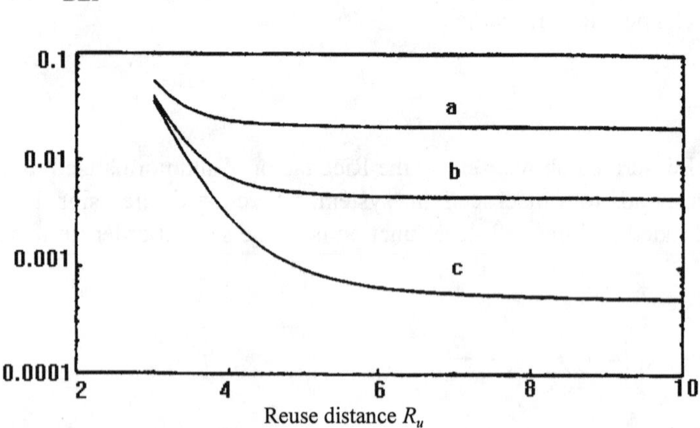

Figure 5.7 BEP P_e against reuse distance R_u for a carrier-to-noise ratio of 10 dB, a_c = 0.5 erlang, Rice factor K = 7 dB, G = 0.2, and Γ = 0.001 for several parameter A values, (a) A = 0.1, (b) A = 0.01, and (c) A = 0.001.

Figure 5.8 depicts BEP P_e against reuse distance R_u for a carrier-to-noise ratio of 10 dB, a_c = 0.5 erlang, A = 0.001, and Γ = 0.001 with Rice factor K and normalized turning point G as parameters. Here, bit error decreases with increasing K and decreasing G values. Since for k = 0, (5.33) denotes the pdf of a Rayleigh faded signal, this can represent the macrocellular case, if G is also taken to be zero.

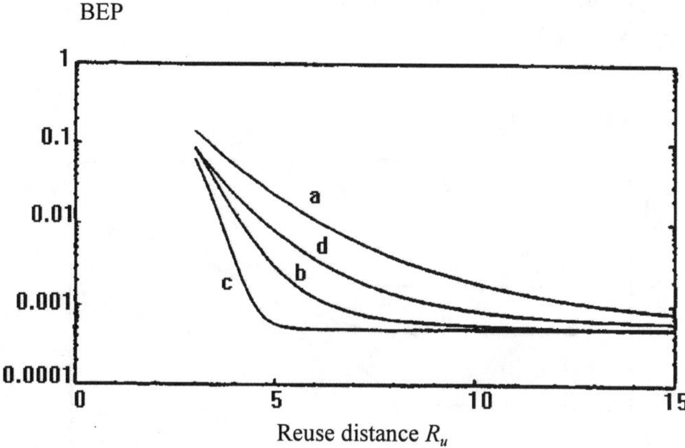

Figure 5.8 BEP P_e against reuse distance R_u for a carrier-to-noise ratio of 10 dB, a_c = 0.5 erlang, A = 0.001, Γ =0.001 with Rice factor K, and normalized turning point G as parameters. (a) K = 0, G =2, (b) K = 7 dB, G =2, (c) K = 12 dB, G =2, and (d) K = 0, G =0.2.

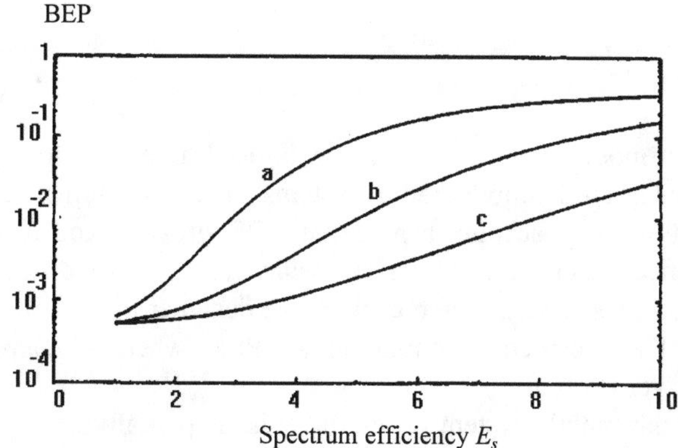

Figure 5.9 BEP P_e against spectrum efficiency E_s for a carrier-to-noise ratio of 10 dB, A = 0.001, G = 0.2, Γ =0.001, W = 0.025 MHz, and K = 7 dB with carried channel traffic a_c as a parameter. (a) a_c = 0.23 erlang, (b) a_c = 0.5 erlang, and (c) a_c = 0.99 erlang.

In Figure 5.9, BEP P_e against spectrum efficiency E_s for a carrier-to-noise ratio of 10 dB, A = 0.001, G = 0.2, Γ = 0.001, and bandwidth W = 0.025 MHz, Rice factor K = 7 dB is given, with carried channel traffic a_c as a parameter. Here, the spectrum efficiency increases with carried traffic a_c for a given BEP. BEP P_e against spectrum efficiency E_s for several A values is also shown in Figure 5.10, but for a Rician factor K

of 12 dB and $G = 0.2$, a carrier-to-noise ratio of 10 dB, $a_c = 0.5$ erlang, $\Gamma = 0.001$, and channel bandwidth $W = 0.025$ MHz.

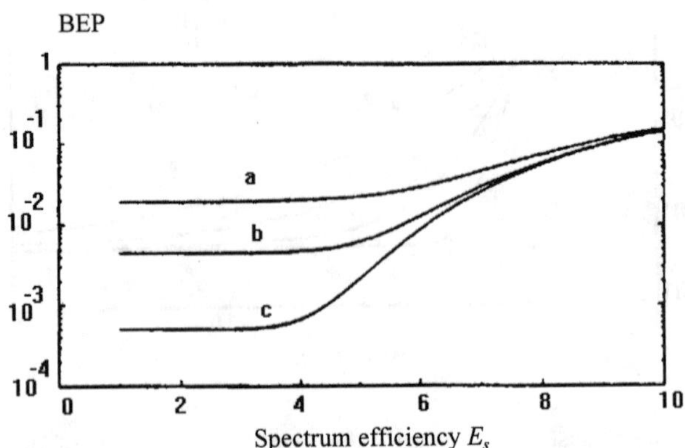

Figure 5.10 BEP P_e against spectrum efficiency E_s . Here $G = 0.2$, $K = 12$ dB, carrier-to-noise ratio is 10 dB, $a_c = 0.5$ erlang, $\Gamma = 0.001$, and $W = 0.025$ MHz. Parameter A values are (a) $A = 0.1$, (b) $A = 0.01$, and (c) $A = 0.001$.

5.3.4 Conclusions

In this section, a model was developed that is used as an aid for system design of digital microcellular land mobile radio systems. The model includes Rician faded desired signal, taking into account a potential LOS situation and Rayleigh faded co-channel interference. Also included are Gaussian noise, narrowband impulsive noise, and a realistic path-loss model, which enables the designer to investigate small base to mobile distance. The considered modulation is DPSK, where an approximation of the suboptimum receiver is used.

For a cellular mobile system where the noise is partially narrowband-impulsive, we found that performance is better compared with a system with purely Gaussian noise, given equal noise power in both cases. Another conclusion we draw is that given an amount of impulsive noise, a smaller impulsive index A yields an improvement in performance. So, large-amplitude, short-duration pulses degrade the system less than small-amplitude, large-duration pulses. Once the impulsive index is above a certain threshold, the impulsive noise power is irrelevant; only the frequency of impulsive events times their mean duration determines the error performance.

Finally, in this section we presented a tool for microcellular radio system planning. The system performance is evaluated from two view points: BEP and spectrum efficiency. From these perspectives, the model allows us to study the effects of the varying parameters (e.g., cell size or transmitter power) on the performance.

REFERENCES

[1] R. Prasad, A. Kegel, and J. Olsthoorn, "Spectrum efficiency analysis for microcellular mobile radio systems," *Electron. Lett.*, Vol. 27, pp. 423–425, February 1991.

[2] R. Prasad and A. Kegel, "Spectrum efficiency of microcellular systems," *Proc. IEEE 41st Vehicular Technology Conf.* , St. Louis, MO., pp. 357–361, May 1991.

[3] A. Kegel, H.J. Wesselman, and R. Prasad, "Bit error probability for fading DPSK signals in microcellular land mobile radio systems," *Electron. Lett.*, Vol. 27, pp. 1647–1648, August 1991.

[4] R. Prasad, A. Kegel, and O.L. Van Linden, "Performance evaluation of microcellular systems with shadowed Rician / Rayleigh faded multiple co-channel interference," *Proc. IEEE 42nd Vehicular Technology Conf.* Colorado, pp. 427–430, May 1992.

[5] R. Prasad, A Kegel, and A. de Vos, "Performance of microcellular mobile radio in a co-channel interference, natural and man-made noise environment," *IEEE Trans. on Vehicular Technology*, pp. 33–40, February 1993.

[6] R. Prasad and A. Kegel, "Effects of Rician faded and lognormal shadowed signals on spectrum efficiency," *IEEE Trans. on Vehicular Technology*, Vol. 42, pp. 274–281, August 1993.

[7] J.-P.M.G. Linnartz, A.J. 'T Jong, and R. Prasad, "Effect of coding in digital microcellular personal communication systems with co-channel interference, fading, shadowing and noise," *IEEE J. Selected Areas in Commun.*, Vol. 11, pp. 901–910, August 1993.

[8] E. Wesselman, R. Prasad, and A. Kegel, "Bit error probability of MDPSK in a Rician faded microcellular radio network with diversity reception," *Archuv für Electronik und Ubertragungstechnik (AEU)*, Vol. 48, pp. 19–27, January/February 1994.

[9] Y.-D. Yao and A.U.H. Sheikh, "Outage probability analysis for microcell mobile radio systems with co-channel interferers in Rician/Rayleigh fading environment," *Electron. Lett.*, Vol. 26, pp. 864–866, June 1990.

[10] R.H. Muammar, "Co-channel interference in microcellular mobile radio system," *Proc. 41st IEEE Vehicular Technology Conf.*, St. Louis, MO, pp. 198–203, May 1991.

[11] S.T.S. Chia, "Radiowave propagation and handover criteria for microcells," *Br. Telecom. Tech. J.*, Vol. 8, pp. 50–61, October 1990.

[12] E. Green, "Radio link design for microcellular systems," *Br. Telecom. Tech. J.*, Vol. 8, pp. 85–96, January 1990.

[13] H. Harley, "Short distance attenuation measurements at 900 MHz and 1.8 GHz using low antenna heights for microcells," *IEEE J. Selected Areas in Comm.*, Vol. 7, pp. 5–10, January 1989.

[14] R.J.C. Bultitude and G.K. Bedal, "Propagation characteristics on microcellular urban mobile radio channels at 910 MHz," *IEEE J. Selected Areas in Comm.*, Vol. 7, pp. 31–39, January 1989.

[15] S.T.S. Chia, R. Steel, E. Green, and A. Baran, "Propagation and bit error ratio measurements for a microcellular system," *J. Inst. Elect. Radio Eng.*, Vol. 57, No. 6 (Supplement), pp. S255–266, November/December 1987.

[16] R. Davies, A. Simpson, and J.P. McGreechan, "Propagation measurements at 1.7 GHz for microcellular urban communications," *Electron. Lett.*, Vol. 26, pp. 1053–1055, July 1990.

[17] R.L. Pickholtz, L.B. Milstein, D.L. Schilling, M Kullback, D. Fishman, and W.H. Biederman, "Field tests designed to demonstrate increased spectral efficiency for personal communications," *Proc. IEEE Globecom'91,* Phoenix, AZ, pp. 878–882, 1991.

[18] C. Loo, "Digital transmission through a land-mobile satellite channel," *IEEE Trans. Comm.*, Vol. 38, pp. 693–697, May 1990.

[19] D.J.R. van Nee, H.S. Misser, and R. Prasad, "Direct-sequence spread spectrum in a shadowed Rician fading land-mobile satellite channel ," *IEEE J. Selected Areas in Comm.*, Vol. 10, pp. 350–357, February 1992.

[20] R. Prasad and A. Kegel, "Improved assessment of interference limits in cellular radio performance," *IEEE Trans. on Vehicular Technology*, Vol. 40, pp. 412–419, May 1991.

[21] R. Prasad and J.C. Arnbak, "Comments on analysis for spectrum efficiency in single cell trunked and cellular mobile radio," *IEEE Trans. on Vehicular Technology*, Vol. 37, pp. 220–222, November 1988.

[22] S.C. Schwartz and Y.S. Yeh, "On the distribution function and moments of power sums with lognormal components," *Bell Syst. Tech. J.*, Vol. 61, pp. 1441–1462, September 1982.

[23] J.A. Elldin and G. Lind, *Elementary Telephone Traffic Theory*. Stockholm, Sweden: Telefonaktiebolaget LM ERICSSON, 1967.

[24] M Schwartz, *Telecommunication Networks Protocols, Modelling and Analysis,* Reading, MA: Addison-Wesley, 1987.

[25] M Abramowitz and I.A. Stegun (Eds.), *Handbook of Mathematical Functions and Formulas, Graphs and Mathematical Tables*, New York: Dover, 1972.

[26] H.L. van Trees, *Detection, Estimation, and Modulation Theory*. New York: John Wiley & Sons, 1971.

[27] E.N. Skomal, *Manmade Radio Noise*, Princeton, NJ: Van Nostrand, 1978.

[28] D. Middleton, "Statistical-physical models for urban radio noise environments—part I: foundations," *IEEE Trans. Electromagn. Compat.*, Vol. EMC-14, pp. 38–56, May 1972.

[29] D. Middleton, "Canonical non-Gaussian noise models: their implications for measurement and for predetection of receiver performance," *IEEE Trans. Electromagn. Compat.*, Vol. EMC-21, pp. 209–220, August 1979.

[30] D. Middleton, “Statistical-physical models for electromagnetic interference,” *IEEE Trans. Electromagn. Compat.*, Vol. EMC-19, pp. 106–127, August 1977.

[31] W.C. Jakes, Jr.(Ed.), *Microwave Mobile Communications*, New York: John Wiley & Sons, 1974.

[32] J.G. Proakis, *Digital Communications, 2nd Ed.* New York: McGraw Hill, 1983.

[33] J.H. Park, Jr., “On binary DPSK detection,” *IEEE Trans. Comm.*, Vol. COM-26, April 1978.

[34] A.M. Maras, H.D. Davidson, and A.G.J. Holt, “Weak-signal DPSK detection in narrow-band impulsive noise,” *IEEE Trans. Comm.*, Vol. COM-33, pp. 1008–1011, September 1985.

[35] A.D. Spaulding and D. Middleton, “Optimum reception in an impulsive interference environment — part II: incoherent detection,” *IEEE Trans. Comm.*, Vol. COM-25, pp. 924–934, September 1977.

[36] A.J. Viterbi, *Principles of Coherent Communications*, New York: McGraw-Hill, 1965.

[37] R. Prasad, A. Kegel, and W. Hollemans, “Performance analysis of a digital mobile radio cellular system in presence of co-channel interference, thermal and impulsive noise,” *Proc. Int. Symp. Personal, Indoor & Mobile Radio Communications at King's College*, London, pp. 154–159, September 1991.

[38] J.P.M.G. Linnartz, H. Goossen, and R. Hekmat, “Comments on slotted Aloha radio networks with PSK modulation in Rayleigh fading channels,” *Electron. Lett.*, Vol. 26, No. 9, April 1990.

[39] R. Prasad, W. Hollemans, and A. Kegel, “Bit error probability of digital mobile radio cellular system in presence of co-channel interference, thermal and impulsive noise,” *European Trans. Telecommunications and Related Technologies*, Vol. 3, No. 3, pp.237–241, May/June 1992.

Chapter 6
Picocellular Systems

6.1 INTRODUCTION

Picocellular systems are especially suited for indoor wireless communications. Picocells cover 20m to 400m diameter with antennas placed on top of a bookshelf and radiating power in the order of a few milliwatts. Interest in indoor wireless communications (IWC) has grown rapidly in recent years because of its advantages over cable networks, such as mobility of users, elimination of casting, flexibility of changing, and creating various new communication services. IWC offers highly attractive means of combined voice and data transmission in business environments. Therefore, intensive studies of the IWC environment have drawn the attention of many researchers for both propagation measurements and theoretical analysis [1–16]. Whether an IWC system functions using dynamic channel allocation (DCA) or fixed cellular principles, it is essential to evaluate the outage probability for a given CIR.

Section 6.2 investigates the behavior of the outage probability in terms of CIR using propagation measurement results of the 23-story Electrical Engineering Department Building of the Delft University of Technology, The Netherlands. The propagation measurements were conducted in the rooms on the 19th and 20th floors of the building where the Telecommunications and Traffic-Control Systems Laboratory is situated. The rooms are located on both sides of a 65m long, 2m wide, straight corridor. Each room is 4m wide, 6m long, and 2.80m high. The partitions between the rooms are made of plasterboard, rockwool, and steel frames. The wall that separates the room from the corridor is made of brick. The floor is made of 50cm reinforced concrete.

Above the door at a height of 2.10m a glass window runs up to the ceiling. On both ends of the corridor an open stairway leads to the upper and lower floors.

In indoor communication, special propagation problems arise due to the highly reflective and shadowing environment, which are related to the problems also present in outdoor mobile communication. In the indoor channel, which depends heavily on the type of building (materials, dimensions, etc.), radio signals propagate via multiple paths that differ in amplitude, phase, and delay time. Therefore, the received information signals are distorted by time dispersion and amplitude fading. These anomalies limit the maximum channel symbol rate significantly when no special precautions are taken (e.g., adaptive equalization or antenna diversity). Recently, radio propagation measurements in the indoor radio channel were reported by several researchers [1–8]. These propagation measurements were conducted in different types of buildings at frequencies of 900 MHz, and 1.7, 4.0, and 5.8 GHz.

In Section 6.3, the results of coherent wideband propagation measurements, which have been conducted simultaneously at 2.4, 4.75, and 11.5 GHz in wideband indoor channels, are presented. The effect of frequency on path loss, root mean square (rms) delay spread and coherence bandwidth is evaluated. In this chapter, we use the terms wideband and narrowband to indicate whether the occupied bandwidth for measurements or information transmission is larger or smaller, respectively, than the coherence bandwidth of the channel.

A theoretical model for evaluation of BEP of the stationary frequency selective indoor channel is developed for the coherent binary phase shift keying (BPSK) receiver. This model is used to compute BEP results for data rates up to 50 Mbps, using measured complex channel impulse responses, at 2.4, 4.75, and 11.5 GHz.

6.2 PROPAGATION MEASUREMENTS AT 1.9 GHz FOR WIRELESS COMMUNICATIONS

This section has three objectives. First, the propagation measurement results are presented and the statistical parameters are obtained using these results. Second, an analytical model is developed to evaluate the outage probability for the desired Rician signal and uncorrelated interfering signals with different Rician factor and equal local mean power. An outage probability expression is also derived for zero Rician factor and different local mean power for each interferer. The outage probability is analytically evaluated using the extracted Rician factor from the propagation measurements. Finally, an illustration is presented to design a cellular structure for the 19th floor.

6.2.1 Propagation Measurements

This section first discusses the measurement procedure, then determines the propagation parameters from a typical propagation result and finally extracts the propagation parameters to study the indoor system. Propagation results were obtained for the 19th and 20th floors. Results for both floors displayed the same general trend. This section specifically makes use of the results for the 19th floor.

A. Measurement Procedure

The measurement equipment was developed by the radio communication and electromagnetic compatibility (EMC) department of PTT Research, Leidschendam, The Netherlands, to plan IWC systems. The equipment consists of a fixed transmitter and a mobile receiver, which operate at a continuous wave frequency of 1.9 GHz. Both transmitter and receiver are equipped with a $\lambda/4$ dipole antenna. The antennas of the transmitter and the receiver are positioned, at a height of 2.5m and 1.80m respectively.

The receiver is operated by an acquisition program that runs on a laptop PC. Before a measurement is done, the transmitted power is gauged. The difference in the transmitted and received signal power is digitized and stored in the laptop's memory. The difference is taken because the transmitter is unstable over a long period, which would make a comparison between different measurement results incorrect. The acquisition program measures the signal every traveled centimeter when it receives an interrupt from the wheel attached to the antenna pole. Localization of the measurement data is done with the help of a graphics tablet that gets the attention of the acquisition program via an interrupt. When for instance a corner is turned, an index is given. The position of the antenna pole relative to the reference points on the graphics tablet will be stored together with the index number and the corresponding measured signal. The location of the measurement data between two such indexes can be determined provided the measurement route is a straight line between the index points. When the PC is not engaged with the interrupts of the wheel or the graphics tablet, the acquisition program writes the measurement results to the screen to inform the user. The system makes it possible to measure the received power within 1 dB accuracy over a range of −110 to 5 dBm.

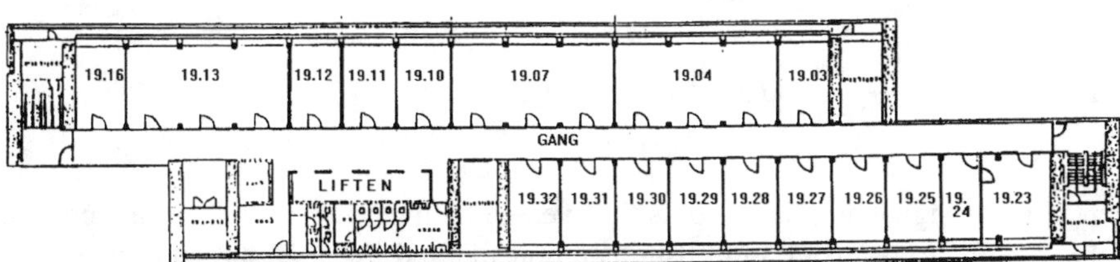

Figure 6.1 Diagram of the 19th floor.

Figure 6.1 shows the 19th floor. The transmitter is positioned in room 19.28 about 1m from the common wall with room 19.29 and 3m from the window. This is the only position used during the measurements due to time limitations. Data are obtained by randomly driving the receiver around in a room over a distance of about 20m. The large rooms are split up into three separate fictional rooms to simplify the illustration of a cellular structure in Section 6.2.4. Figure 6.2 illustrates that the large room ABCD is divided in to three separate fictional rooms; namely, AEFD, EGHF, and GBCH. During the measurements, the receiver was driven within each fictional room as shown in Figure 6.2.

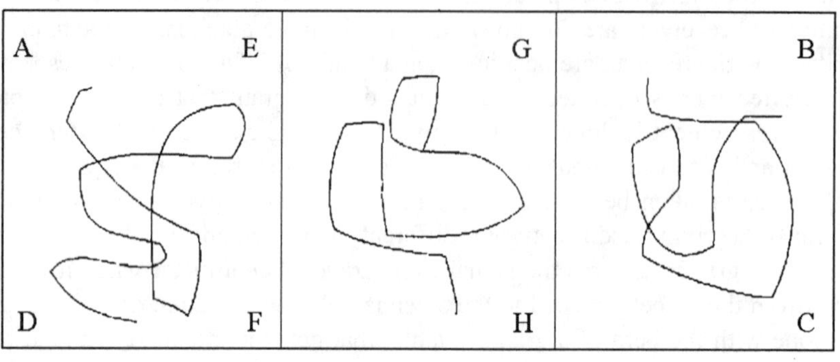

Figure 6.2 The measurement method for the large rooms.

B. Propagation Parameters

Figure 6.3 is a typical propagation result attained from room 19.23, which shows the received instantaneous power versus the traveled distance. The fast variation of the instantaneous power is called multipath fading. The fast-varying amplitude is described by the Rician distribution

$$f(A) = \frac{A}{\sigma^2}\exp\left(-\frac{A^2+s^2}{2\sigma^2}\right)I_0\left(\frac{A_s}{\sigma^2}\right) \qquad 0 \le A < \infty \tag{6.1}$$

where A is the signal amplitude, σ^2 is the average fading power, s is the peak amplitude of the dominant received multipath component, and $I_n(.)$ is the modified Bessel function of the first kind and nth order.

From (6.1), the pdf of the fast-varying instantaneous power, $p = A^2/2$, is found as

$$f(p) = \frac{K+1}{p_0}\exp\left(-\frac{K+1}{p_0}p - K\right)I_0\left(2\sqrt{\frac{K^2+K}{p_0}p}\right) \qquad 0 \le p < \infty \tag{6.2}$$

where $p_0 = (s^2 + 2\sigma^2)/2$ is the local mean power, and K the Rice factor, which is defined as the ratio of the average power of the dominant multipath component and the average fading power received over the nondominant paths

$$K = \frac{s^2}{2\sigma^2} \tag{6.3}$$

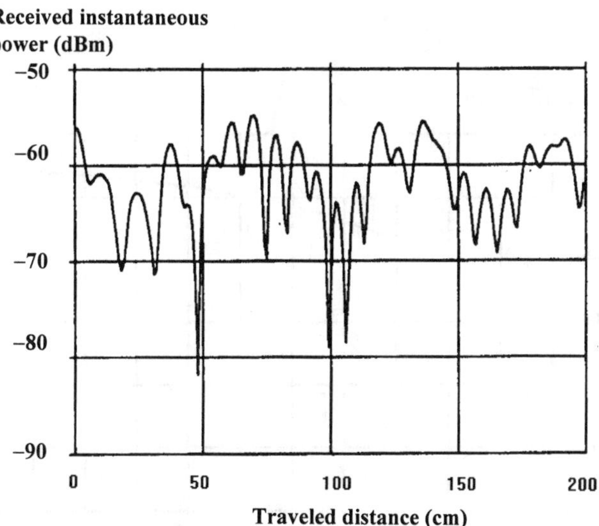

Figure 6.3 The received instantaneous signal power versus the traveled distance.

Equation (6.2) shows that the modeling parameters are K and p_0. These parameters are to be extracted from the received signal. The local mean power is extracted by averaging the instantaneous signal power over 1m (100 samples). In [17] it is shown that this results in a 95% reliability interval for the local mean power.

The Rice factor is extracted from the received signal using the following expression (Appendix 6A):

$$\frac{E[A]}{\sqrt{E[A^2]}} = \sqrt{\frac{\pi}{4(K+1)}} \exp\left(-\frac{K}{2}\right)\left[(K+1)I_0\left(\frac{K}{2}\right) + KI_1\left(\frac{K}{2}\right)\right] \tag{6.4}$$

where $E[A]$ is the average amplitude and $E[A^2]$ the average of the squared amplitude. Figure 6.4 shows (6.4) graphically.

The K value that is generated by (6.4) is checked by plotting the cdf obtained using the measured data, and the cdf obtained using (6.2) and (6.4). The Kolmogorov-Smirnov test is used for the best fit test [18]. This test generates a reliability interval around the cdf of the received signal power normalized on its local mean power. Plots using many different measurements showed that (6.4) is a reliable tool in the determination of K. An example of a typical plot is shown in Figure 6.5. $K = 0$ is used as a reference. It equals the exponential distribution and therefore becomes a straight line in a lognormal scaled graph. The cdf obtained using (6.2) and (6.4) corresponds to $K = 5$. When $K \geq 0$, a clear deviation occurs from the exponential distribution at the lower probabilities.

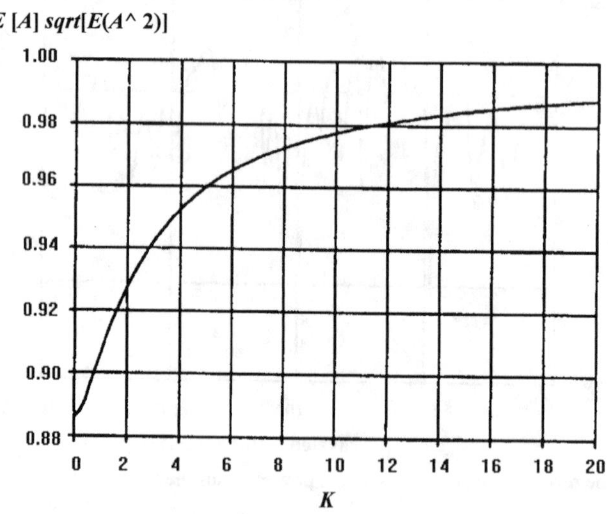

Figure 6.4 The relationship between the statistics of the received amplitude and the Rice factor.

C. Extracted Results

In this section, the measured propagation parameters are presented. Some remarks are made to explain the resulting numerical values. In Table 6.1, the measurement results of the 19th floor are displayed. In Figure 6.6, a more useful overview is given. The remarks are concerned with the behavior of the Rice factor. Figure 6.6 shows that $K \neq 0$ in areas where no LOS is present. The corridor acts as a sort of waveguide that results in a dominant multipath component in the rooms not in sight of the transmitter. The Rice factor varies between 0 and 2.4 for rooms more than one room distant from the transmitter. However, in the transmitter room and its two neighbors, K varies between 0 and 6.6. The high K value in the transmitter room is to be expected because of the LOS in that room.

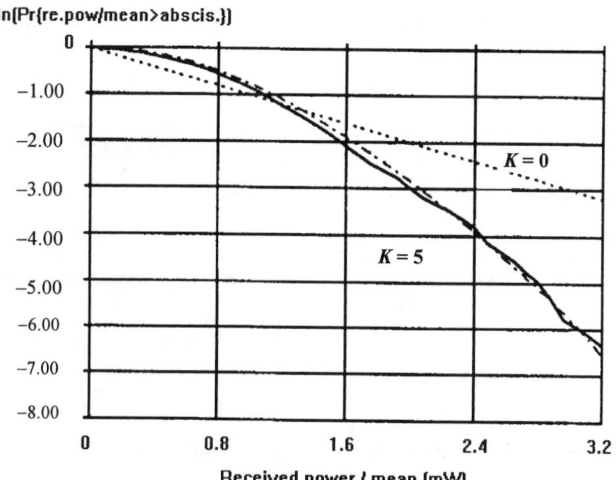

Figure 6.5 A typical example of the CDF plot due to the measured data and (6.2) and (6.4). . . . correspond to $K = 0$; - - - corresponds to distribution via (6.2) and (6.4); and — corresponds to distribution that results from dividing the number of equal samples by the total sample size.

The large difference between the K value found in room 19.29 of 6.6, and the zero K value in room 19.27, is not expected. An explanation could be that the transmitter is positioned closer to room 19.29, which causes a large component through the wall bordering room 19.29 and a scattered one through the wall bordering room 19.26.

The measured local mean power is used for the computational results in Section 6.2.2 and for the evaluation of the picocellular system in Section 6.2.4.

6.2.2 Formulation of CIP

The cellular concept is described in Chapter 3. To investigate the reuse distance of a cellular system, it is necessary to determine CIP [19,20], defined in Chapters 4 and 5.

$$F(\mathrm{CI}) \stackrel{\Delta}{=} \sum_{n} F(\mathrm{CI}|n) F_n(n) \tag{6.5}$$

Here, $F_n(n)$ is the probability of n co-channel interferers being active, $F(\mathrm{CI}|n)$ is the corresponding conditional CIP,

$$F(\mathrm{CI}|n) \stackrel{\Delta}{=} \mathrm{prob}\left\{\frac{p_d}{p_n} < \alpha\right\} \tag{6.6}$$

where p_d is the instantaneous power of the desired signal, p_n is the joint interference power from n active channels, and α is the specified co-channel protection ratio.

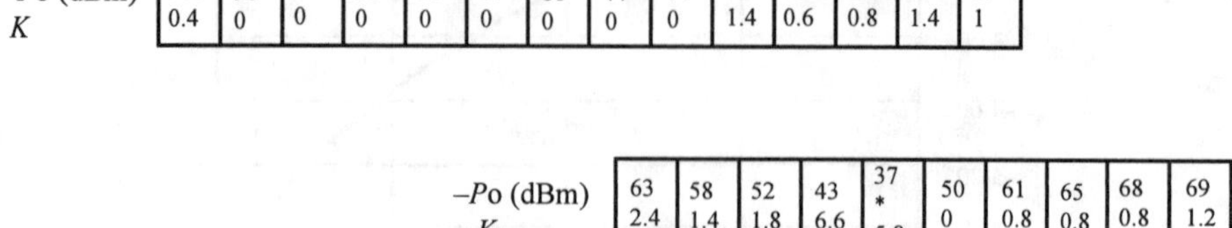

–Po (dBm)	100	98	97	95	91	87	81	77	74	72	68	66	57	67
K	0.4	0	0	0	0	0	0	0	0	1.4	0.6	0.8	1.4	1

–Po (dBm)	63	58	52	43	37 *	50	61	65	68	69
K	2.4	1.4	1.8	6.6	5.8	0	0.8	0.8	0.8	1.2

Figure 6.6 An overview of the measured propagation parameters. K and Po, on the 19th floor. * = Transmitter.

The measurement results show that the fast amplitude fluctuations of a mobile radio signal are described by a Rician density function with a Rice factor varying between 0 and 6.6. This means that the desired and interfering signals are described by a Rician density function. The pdf for the joint interference power p_n is obtained by convolving (6.2) n times (2.30), (Appendix 2A)

$$f_{p_n}(p_n) = \left(\frac{K_i+1}{p_{0i}}\right)^{\frac{n+1}{2}} \left(\frac{p_n}{nK_i}\right)^{\frac{n-1}{2}} \exp\left(-\frac{K_i+1}{p_{0i}}p_n - nK_i\right) I_{n-1}\left(\sqrt{\frac{4n(K_i^2+K_i)}{p_{0i}}p_n}\right) \tag{6.7}$$

where p_{0i} is the local mean power of each interfering signal, K_i is the Rice factor of each interfering signal and n is the number of co-channel interferers.

Table 6.1

Measured Propagation Parameters of the 19th Floor

Room Number 19th floor	Rice Factor K	$-p_0$ (dBm)	Standard Deviation of p_0(dB)
03	1	66.2	3.0
04	1.4	56.5	5.7
05	0.8	66.0	2.3
06	0.6	68.1	1.4
07	1.4	72.3	2.1
08	0	74.2	1.1
09	0	76.8	1.2
10	0.8	81.3	1.6
11	1	86.4	1.8
12	0.6	90.8	1.6
13	0	95.2	1.0
14	0	96.0	1.6
15	0	98.0	1.1
16	0.4	99.5	1.9
23	1.2	69.4	2.1
24	0.8	66.7	1.5
25	0.8	65.0	1.5
26	0.8	61.3	1.5
27	0	50.0	1.5
28	5.8	36.3	2.9
29	6.6	43.2	2.6
30	1.8	51.6	2.6
31	1.4	58.0	4.3
32	2.4	63.5	3.4

Conditional CIP is derived using (6.2), (6.5) to (6.7)

$$F(\mathrm{CI}|n) = 1 - (2nK_i)^{\frac{1-n}{2}} e^{-nK_i} \int_0^\infty Q\left(\sqrt{2K_d}, \sqrt{\frac{K_d+1}{K_i+1}\frac{\alpha p_{0i}}{p_{0d}}x^2}\right) I_{n-1}\left(\sqrt{2nK_i x^2}\right)dx \qquad (6.8)$$

where K_d is the Rice factor of the desired signal, p_{0d} is the local mean power of the desired signal, and $Q(a,b)$ is Marcum's Q-function [21]. A heavy restriction on the applicability of (6.8) in practice is the assumption that the local mean power is equal for each interfering signal. Table 6.1 shows that this is not the case in a real environment.

When K_i is taken as zero, it is possible to calculate conditional CIP with a different local mean power p_{0i} for each interferer (Appendix 6B)

$$F(\text{CI}|n) = \sum_{i=1}^{n} \frac{1}{1+\frac{\text{SI}_i}{(K_d+1)\alpha}} \exp\left[-\frac{\frac{K_d}{(K_d+1)\alpha}}{\frac{(K_d+1)\alpha}{SI_i}+1}\right] \prod_{j=1 \wedge j\neq i}^{n} \frac{1}{1-\frac{P_{0j}}{P_{0i}}} \tag{6.9}$$

where SI_i is p_{0d} / p_{0i}, p_{0d} and p_{0i} are the local mean power of the desired and of the ith interfering signals, respectively.

In Section 6.2.3 it is shown that, given some practical system requirements and the measured propagation parameters, (6.9) does not significantly differ from (6.8). The simplicity of (6.9) and its flexibility concerning p_{0i} promises new possibilities for the evaluation of an IWC system, which uses dynamic channel allocation.

6.2.3 Computational Results

This section shows that the Rice factor of the interfering signals has no significant influence on the reuse distance in the measured environment. First, it is necessary to specify the system requirements; namely, the maximum allowed CIP and the protection ratio:

- $F(\text{CI} \mid n) \leq 10\%$
- $\alpha = 10$ dB [22]

$F(\text{CI} \mid n)$ is evaluated using the ratio of desired and total interfering local mean power Si_t. A base station is placed in every room because this is the most efficient cell size concerning the spectrum efficiency [23]. This means that the propagation parameters of the desired and interfering signals have the following values or bounds:

- $K_d = 5.8$, $P_{0d} = -37$ dBm
- $0 \leq K_i \leq 6.6$, $-100 \leq P_{0i} \leq -43$ dBm

Figure 6.7 depicts $F(\text{CI} \mid n)$ versus SI_t for $K_d = 5.8$, $n = 1$, $K_i = 0$ and 6.6. When $F(\text{CI} \mid n) \approx 0.01$, the largest difference in allowed SI_t occurs; namely, 2 dB in favor of $K_i = 0$. This may be unacceptable, however; Figure 6.7 shows that the minimum allowed $\text{SI}_t = 14$ dB and the combination of $K_i = 6.6$ and $\text{SI}_t \geq 14$ dB does not occur in the measured environment as seen in Figure 6.6. This means that the bounds on the interfering signal's propagation parameters alter to

- $0 \leq K_i \leq 2.4$, $-100 \leq P_{0i} \leq -52$ dBm

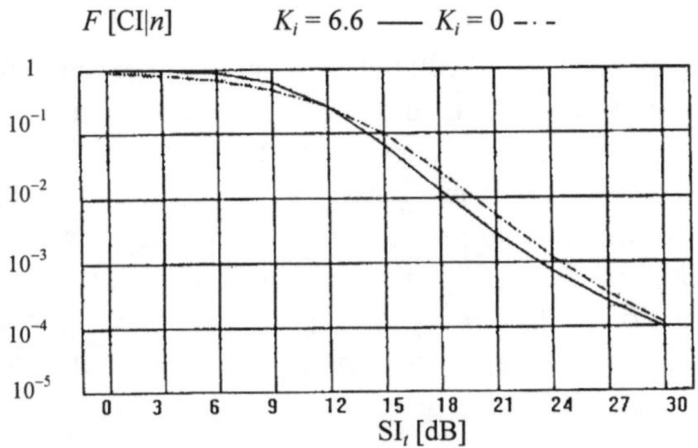

Figure 6.7 Conditional CIP $F(\mathrm{CI}|n)$ versus the ratio of the desired and the total interfering local mean power SI_t ($n = 1$, $\alpha = 10$ dB, $K_d = 5.8$ $K_i = 0$ and 6.6).

Figure 6.8 depicts $F(\mathrm{CI}|n)$ versus SI_t for $K_i = 0$ and 2.4. There is no significant difference in the allowed SI_t for $K_i = 0$ and 2.4. This means that the Rice factor of the interfering signals can be taken to zero without loss of accuracy in the determination of the reuse distance. Equation (6.9) is therefore used in the measured environment given the specified system requirements.

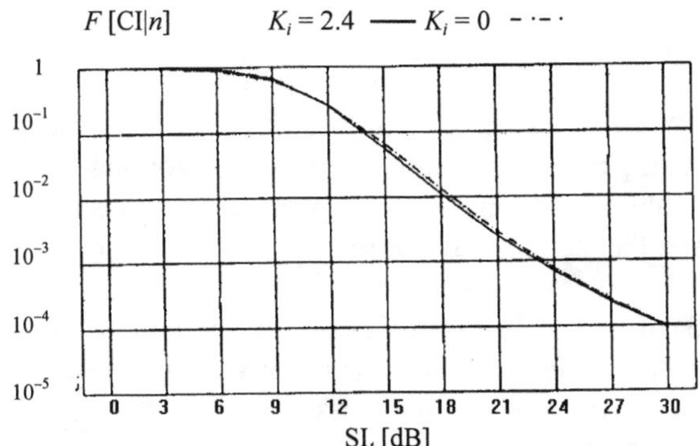

Figure 6.8 Conditional CIP $F(\mathrm{CI}|n)$ versus the ratio of the desired and the total interfering local mean power SI_t ($n = 4$, $\alpha = 10$ dB, $K_d = 5.8$, $K_i = 0$, and 2.4).

6.2.4 An Illustration to Evaluate a Picocell

An IWC system, which uses a fixed cellular structure, has been determined for the 19th floor of the building. The objective is to have as many co-channel users as possible on the floor. This means that the reuse distance must be very small. When path loss of the desired signal is kept low, co-channel users are allowed to be close to each other because the reuse distance is dependent on SI_t. For this reason, we chose a cell size of one room (see also [22]). The propagation results of Table 6.1 and the Rayleigh interferers approximation (6.9) are used. Figure 6.9 shows the resulting cellular structure for the 19th floor. The structure covers the whole floor with eight clusters of three cells with a maximum CIP of 1.46%.

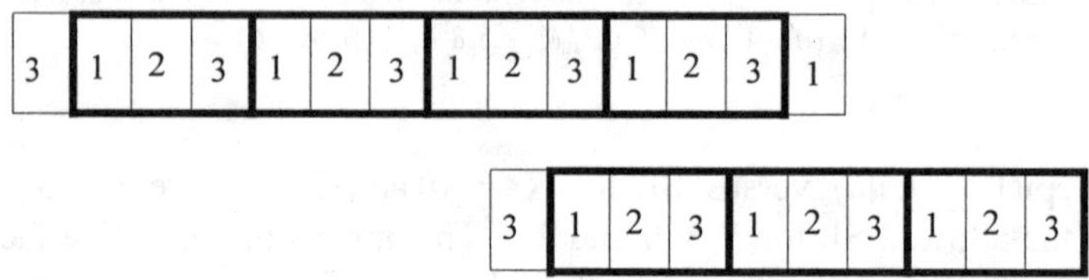

Figure 6.9 The resulting cellular structure for the 19th floor.

6.2.5 Conclusions

This section presented measurement results conducted in an office environment. The measured propagation parameters; namely, the Rice factor and the local mean power, are presented in tables. An analytical expression was derived from which the relationship between the statistics of the received signal and the Rice factor becomes clear.

The Rice factor is not necessarily dependent on an LOS to the transmitter. An explanation is that dominant multipath components are generated by waveguiding of the corridor. The numerical values of the measured Rice factor vary between 0 and 6.6. High Rice factors are measured only in rooms close to the transmitter. More than one room away from the transmitter, the Rice factors vary between 0 and 2.4.

An analytical expression was found for the CIP for Rician desired signal and n Rayleigh interfering signals. Each interfering signal introduces its own local mean power to the expression. This possibility offers a degree of freedom not present in earlier calculations. The Rayleigh approximation is made because the Rice factor of the interfering signals does not influence the used distance significantly, given the measured propagation parameters and the specified system requirements. The expression promises to be useful in capacity calculations for dynamic channel allocation (DCA) IWC systems because of its simplicity and extra degree of freedom.

Finally, an evaluation of a picocellular IWC system is done for the 19th floor. When the fixed transmitter is placed in a room, the smallest cluster size will be three with a one-room cell area. The small reuse distance justifies the use of a cellular system in an indoor environment.

6.3 WIDEBAND INDOOR CHANNEL MEASUREMENTS AT 2.4, 4.75, and 11.5 GHz

The content of this section is organized as follows. In Section 6.3.1, the indoor multipath channel model is briefly reviewed, and important channel parameters are discussed. Propagation measurement results for the indoor wideband channel at 2.4, 4.75, and 11.5 GHz of an indoor laboratory/office environment are presented in Section 6.3.2. The measurements were conducted at the TNO Physics and Electronics Laboratory, The Hague. The measurement principle, measurement system, and indoor environment are described in detail. The measurement results and values for the path-loss law, rms delay spread (τ_{rms}), and coherence bandwidth (B_{coh}) are given for three different situations, LOS, obstructed direct path (OBS), and the effect of people present in the room. A summary of these results is presented in [9]. In Section 6.3.3, an analytical BEP model for the stationary frequency selective multipath channel is derived for the coherent BPSK receiver. In this model, the measured channel impulse response is used. Numerical BEP results for a number of measured indoor channels are given. Finally, conclusions are given in Section 6.3.4.

6.3.1 The Indoor Radio Channel

The indoor radio channel is an adverse communication channel that causes undesired frequency selective fading and dispersion of the information signal. The baseband complex channel impulse response is modeled as [4]

$$h(t) = \sum_{k=0}^{n} \beta_k \exp(j\theta_k)\delta(t - \tau_k) \tag{6.10}$$

Here, k is the path index, β_k is the path gain, θ_k is the phase shift and τ_k is the time delay of the kth path. $\delta(.)$ is the Dirac delta function.

The total number of paths is $N + 1$. Because the absolute delay of the channel is not important, the first arriving path is taken as time reference by setting $\tau_0 = 0$. Another way to represent the channel response is in the frequency domain, by taking the Fourier transform $H(f) = F\{h(t)\}$. In the case of changing environment due to transmitter, receiver, or people movement, the parameters β_k, θ_k, and τ_k are functions of time. However, in this section the channel is assumed to be quasi-static. This assumption is reasonable when the transmitter and receiver have a fixed position. To determine the characteristic of wideband channels, a wideband modulated sounding signal is used (e.g., a narrowpulse). For a pulselike signal, the transmitted sounding signal is given as $s(t) = p(t)\cos(\omega_C t + \phi)$ where ω_C indicates the carrier frequency, ϕ the carrier phase, and $p(t)$ is pulse signal. For $\phi = 0$, the received signal is given by

$$r(t) = \Re\{s(t) * h(t)\} = \sum_{k=0}^{N} \beta_k p(t - \tau_k) \cos\left(\omega_c(t - \tau_k) + \theta_k\right) \tag{6.11}$$

The pulse width or the measurement bandwidth determines the time resolution that can be achieved. The channel parameters; namely, power delay profile (PDP), path loss as a function of distance, coherence bandwidth, and delay spread τ_{rms}, are defined as follows.

A. Power Delay Profile

The PDP $P(t)$ gives the time distribution of the received signal power from a transmitted δ-pulse, and is defined as

$$P(t) \stackrel{\wedge}{=} h(t)h^*(t) = |h(t)|^2 = \sum_{k=1}^{N} \beta_k^2 \delta(t - \tau_k) \tag{6.12}$$

In practical measurements, the transmitted pulses $p(t)$ have finite width. When the pulse width of $p(t)$ is less than the delay time differences between the paths, the PDP is given by

$$P(t) = |r(t)| = \sum_{k=1}^{N} \beta_k^2 |p(t - \tau_k)|^2 \tag{6.13}$$

B. Path Loss

Different models have been presented for the path loss in an indoor environment. The following simple model is frequently used to describe path loss [5,6]

$$Loss = S_0 + 10a\log_{10}(d / d_0) \tag{6.14}$$

where a is the power loss exponent of the environment and d [m] is the distance between the transmit and receive antennas. S_0 is the free-space loss of a path of d_0 meter. Often, d_0 = 1m is chosen. When this parameter is measured with the antenna in the far field region, the antenna gains are also taken into account in the model.

C. RMS Delay Spread

The rms delay spread, τ_{rms}, is a measure for the amount of signal dispersion [28]. τ_{rms} is defined as

$$\tau_{\text{rms}} = \sqrt{\int_{-\infty}^{\infty} (t - \tau_m)^2 P_{\text{norm}}(t) dt} \tag{6.15}$$

where τ_m is the mean excess delay time defined as $\tau_m = \int_{-\infty}^{\infty} t P_{\text{norm}}(t) dt$, and $P_{\text{norm}}(t)$ is the normalized PDP: $P_{\text{norm}}(t) = P(t) / P_{\text{tot}}$, with $P_{\text{tot}} = \int_{-\infty}^{\infty} P(t) dt$.

In the literature, τ_{rms} is used to give a rough indication of the maximum data rate that can be reliably supported by the channel, when no special precautions are taken, such as equalization or diversity techniques. The estimate of the maximum reliable data rate R_{max} is given by [29]

$$R_{\text{max}} = 1 / 4\tau_{\text{rms}} \tag{6.16}$$

D. Coherence Bandwidth

The coherence bandwidth, B_{coh}, is the statistical average bandwidth of the radio channel, over which signal propagation characteristics are correlated. The definition of the coherence bandwidth B_{coh} is based on the complex autocorrelation function $|R(\Delta f)|$ of the frequency response $H(f)$. $R(\Delta f)$ is defined as [7]

$$R(\Delta f) \stackrel{\wedge}{=} \int_{-\infty}^{\infty} H(f) H^*(f + \Delta f) df \tag{6.17}$$

B_{coh} is now defined as the value of Δf where $|R(\Delta f)|$ has decreased with 3 dB. $B_{\text{coh}} = 1/\alpha\tau_{\text{rms}}$ [31], [32], where α is constant. In case the PDP is exponentially distributed, $\alpha = 2\pi$ [32]. However, in practical situations, the value of α is not a constant; it depends on the impulse response. From results presented in [7,8,29], values of α between 5 and 7 were found.

6.3.2 Propagation Measurements

This section describes the measurement principle, setup, locations, and results.

A. Measurement Principle

Instead of transmitting a narrowpulse width signal in the time domain, the measurement technique applied here is a wideband coherent frequency response measurement, which consists of coherent amplitude/phase measurements at 801 equidistant frequency points in a selected frequency band. Thus, complex samples of *H*(*f*) are taken in the frequency domain. The sampled frequency response of the multipath radio channel is converted into the impulse response in the time domain by taking the inverse Fourier transform.

The important measurement parameters are as follows.

(1) The measurement bandwidth over which the samples are taken, $BW_{meas} = f_{max} - f_{min}$. BW_{meas} determines the time resolution, τ_{res}, that is achieved in the time domain. The relation between BW_{meas} and τ_{res} is given by

$$\tau_{res} \approx \frac{1}{BW_{meas}} \tag{6.18}$$

(2) The frequency distance between successive sample points Δf_{meas} determines the unambiguity in range and delay time. The unambiguous range of the measurements is the range [*m*] or delay [*s*] over which the received energy can be unambiguously related to a certain transmitted pulse. For larger distances, aliasing due to under sampling in the frequency domain will occur. The unambiguous range R_{unamb} is given by

$$R_{unamb} = \frac{c}{\Delta f_{samp}} [m] \tag{6.19}$$

where *c* is the speed of light. The unambiguous time τ_{unamb} is given by

$$\tau_{unamb} = \frac{1}{\Delta f_{samp}} [s] \tag{6.20}$$

(3) Influence of the processing window. Before transformation of the measured frequency response to the impulse response in the time domain, the samples are weighted in order to suppress undesired sidelobes at the cost of a slightly decreased resolution. For this weighting, the Hanning window function was chosen, which is a good compromise between main peak width and sidelobe suppression. The Hanning window causes peak width widening by a factor 1.5. The parameters applying to the measurements performed at the frequencies 2.4, 4.75, and 11.5 GHz are given in Table 6.2.

Table 6.2

Measurement Parameters

Frequency (GHz)	Bw_{meas} (GHz)	τ_{res} (ns)	τ_{unamb} (μs)	R_{unamb} (m)
2.4	0.5	3	1.6	480
4.75	0.5	3	1.6	480
11.5	1.0	1.5	0.8	240

B. Measurement Setup

The measurements were performed using an HP 8510C network analyzer. The measurement system is depicted in Figure 6.10. Biconical antennas were selected because of their large bandwidth and constant impedance over a large frequency band. These antennas have an omnidirectional radiation pattern in azimuth and vertical polarization. Different antennas were dimensioned for each frequency band with equal gains of 2.5 dB, and equal beamwidth in the vertical plane of about 100 deg. Sucoflex 106A-type cables were used to connect the antennas, amplifiers, and HP8511A test set. The measurement system was dimensioned in such a way that the SNR at the test set was not less than 20 dB. The distance between transmit and receive antennas was maximum 30m. The maximum acceptable path loss was 123, 139, and 123 dB at 2.4, 4.75, and 11.5 GHz, respectively.

Before the measurements were carried out, the system was calibrated carefully at all three frequencies. Calibration was performed by disconnecting the antennas and connecting the transmitter output and receiver input by a calibrated attenuator at that point. The calibration results were used to compensate for the influence of phase and amplitude variations caused by the cables, amplifiers, and measurement equipment.

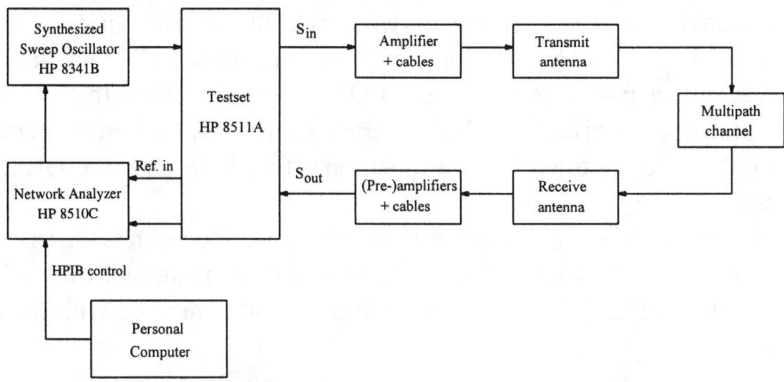

Figure 6.10 Measurement system setup for wideband indoor measurements.

The calibrated channel measurement results now contain the influence of the channel as well as both transmit and receive antennas. The influence of the antenna gain variations over the frequency band are so small that they are neglected. Experiments showed that movement of cable after calibration had a negligible effect on the measurements.

Table 6.3

Dimensions of the Rooms Used for Measurements

Location	Length (m)	Width (m)	Height (m)
LR	5.4	5.0	3.5
OR1, 2, 3	3.6	5.0	3.5
CR	27	10	3.5
HW	35	3.0	3.5

LR; laboratory room, OR; office room, CR; conference room, HW; hallway.

C. Description of Environment and Measurements

The measurements were performed in a laboratory/office environment on the top floor of a three-story building. The walls in this building are made of brick/stone and plasterboard, and the floors are made of concrete. The glass in the windows in the rooms is not metal coated. The measurements were performed at four types of locations: one laboratory room (indicated as LR), three office rooms (indicated as OR1, OR2, and OR3), one conference room (indicated as CR), and in a hallway (indicated as HW).The locations of the rooms (except for CR) are shown in Figure 6.11. The dimensions of the rooms are given in Table 6.3.

The rooms OR1 and OR2 are separated by a plasterboard wall, and OR2 and OR3 are separated by a brick wall. The CR room is subdivided by an RF permeable wall. In all rooms one side consists of windows. In the LR, measurements were performed on ten receiver (RX) positions, with one transmitter (TR) position as shown in Figure 6.12; seven positions had direct LOS, three had OBS. In OR1, OR2, and OR3, transmitter and receiver were in different rooms, and all measurements were OBS. In the CR, there were ten receiver positions (five LOS and five OBS) with one transmitter position.

In HW, 22 positions as shown in Figure 6.11 were chosen with 1m separation, with the transmitter/receiver located in OR1, so nearly all positions were OBS. In LR the influence of people in the room was investigated and compared with the results for the empty room.

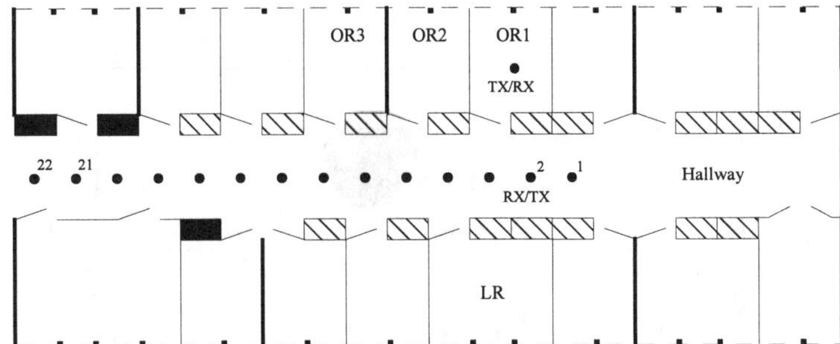

Figure 6.11 Schematic overview of the measurement locations.

At every measurement position, except for the positions in HW, a cluster of measurements was conducted at all three frequencies. A cluster consists of six positions located on a circle with 12.5-cm diameter. The minimum distance between two positions on the circle is $\lambda/2$ at 2.4 GHz, λ at 4.75 GHz, and 2λ at 11.5 GHz. The position of a cluster was fixed for all three frequencies, so that the results could be compared. The receiver antenna height was 1.5m. The transmitter antenna height was 1.5 and 3m. During the time required to measure the channel, special precautions were taken to keep the channel as static as possible.

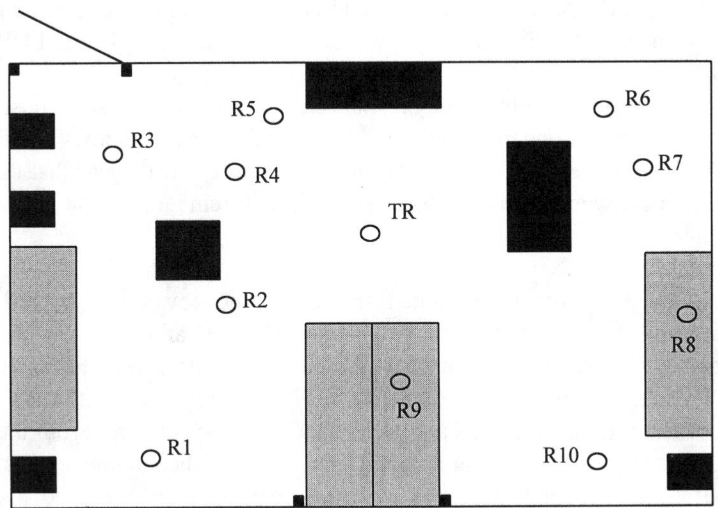

Figure 6.12 Schematic overview of the measurement positions in the LR.

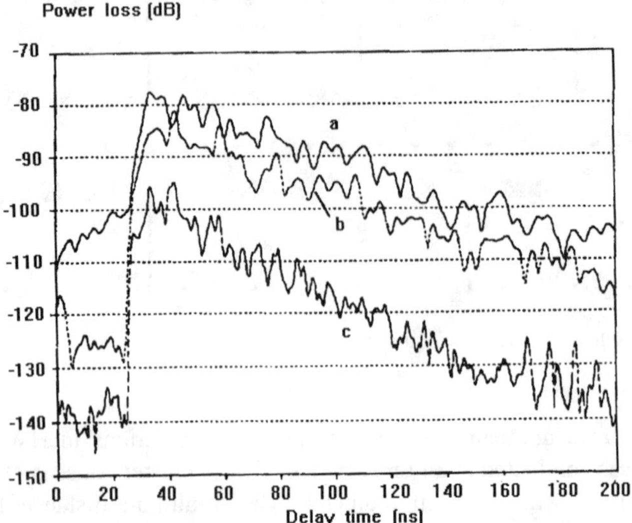

Figure 6.13 PDPs in an OBS situation of a channel between OR1 and OR3, with (a) 2.4 GHz, (b) 4.75 GHz, and (c) 11.5 GHz.

D. Measurement Results

In Figure 6.13, the measured PDPs are given for all three frequencies at the same position for an OBS path between OR1 and OR3. For short delays, some paths can be matched at different frequencies; however for longer delays the PDPs are quite different.

In Figure 6.14, the frequency response at 2.4 GHz for the same position is given that shows a highly frequency-selective channel. Figure 6.15 shows the PDP and the average PDP of a cluster in the LR for an LOS channel. It is clear that the PDPs of a cluster do not show great differences when the measurement position is changed over a few wavelengths.

Path loss. The path loss is computed as the average power loss over the measured frequency band (0.5 GHz bandwidth at 2.4 and 4.75 GHz and 1 GHz at 11.5 GHz). The path loss as a function of distance at the three frequencies is treated separately for LOS and OBS situations. In Figures 6.16 and 6.17, the path-loss results are shown as a function of the distance for LOS and OBS channels, respectively. From the results, the path loss is determined by fitting a straight line through the attenuation-distance scatter plot. The simple model that comes with this regression is described by a modified version of (6.14).

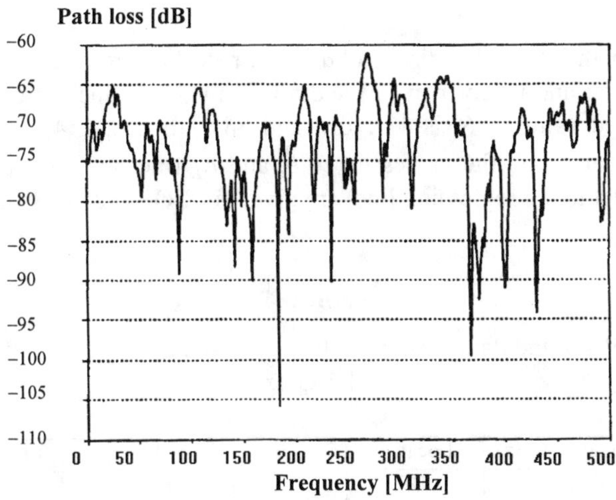

Figure 6.14 Frequency response at 2.4 GHz for the same channel between OR1 and OR2 as shown in Figure 6.13.

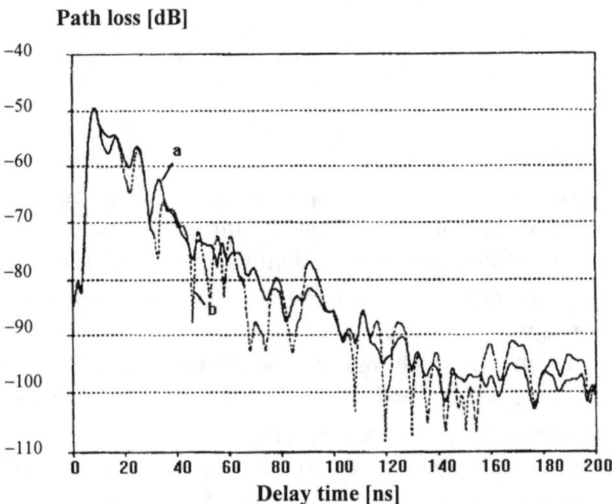

Figure 6.15 PDP for an LOS channel (receiver position 1) in LR, with (a) averaged PDP over six cluster positions, and (b) one specific PDP of the cluster.

$$\text{Loss} = S_0 + 10a \log_{10}(d) + b \tag{6.21}$$

where S_0 is the path loss at 1m distance, a is the path-loss exponent, d is the distance between transmitter and receiver in (m) and b is the value of the regression line at d = 1m. The measured value of S_0 is 43.1 dB, and 56.6 dB at 2.4, 4.75, and 11.5 GHz, respectively. In Table 6.4, the values for a, b, and σ_L (the rms error of the measured values with respect to the regression line) are given for LOS and OBS situations.

Table 6.4

Calculated Results of the Path-Loss Variables a, b, and the rms Error at 2.4, 4.75, and 11.5 GHz

Frequency (GHz)	Location	a	b [dB]	σ_L [dB]
2.4	LOS	1.86	– 1.6	1.6
	OBS	3.33	– 5.4	3.6
4.75	LOS	1.98	– 1.2	2.0
	OBS	3.75	– 6.5	4.1
11.5	LOS	1.94	– 1.7	2.3
	OBS	4.46	– 8.9	5.0

For LOS situations, the path-loss exponent value is very close to the expected value for free-space propagation, a = 2, at all three frequencies. This was to be expected, because the channel response is dominated by the direct LOS path. The values for b are slightly negative, but close to 0 dB. The rms error σ_L is small (<2.5 dB); but increases with frequency.

For OBS situations, the path-loss exponent value a increases with increasing frequency. The calculated values for b also are increasingly negative with increasing frequency. The rms error σ_L also increases with frequency.

From these results, we conclude that the simple model used here performs well in LOS situations, but is not accurate for OBS paths. More complicated models, which take into account attenuation jumps caused by walls and other obstructions, are proposed in [5,6].

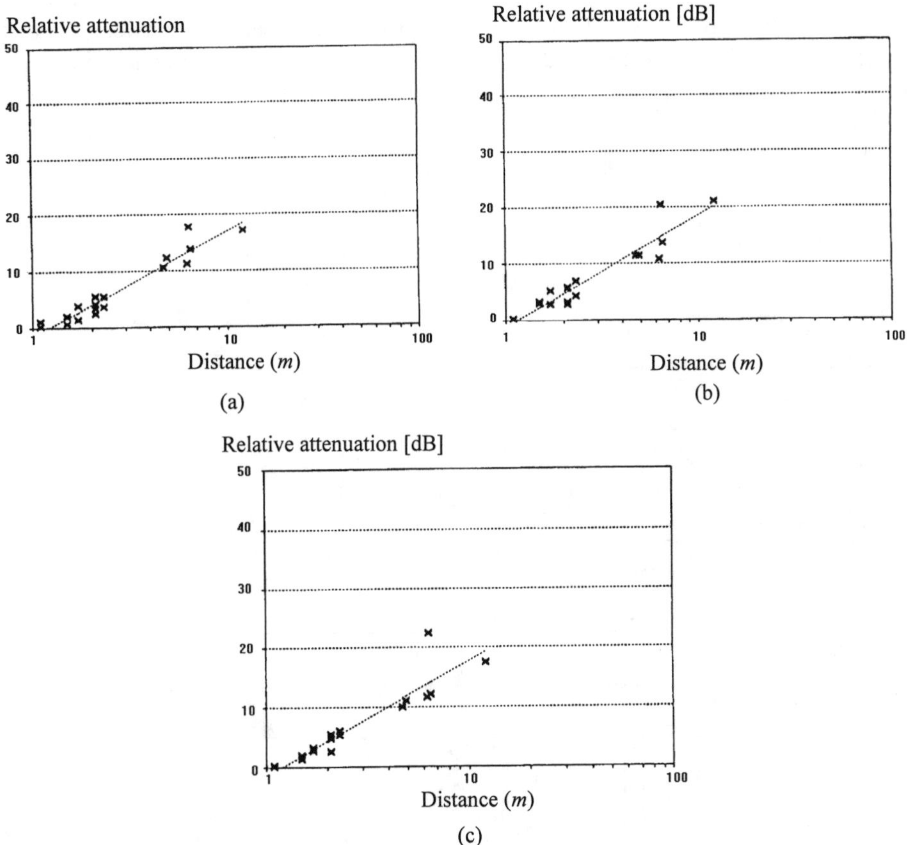

Figure 6.16 Attenuation as a function of the distance for LOS at (a) 2.4 GHz, (b) 4.75 GHz, and (c) 11.5 GHz.

Influence of people. In the LR, measurements were conducted to investigate the influence of people located at different positions in the room on the transmission channel. The measurements were performed for one single radio channel (transmitter-receiver setup TRR10 in Figure 6.12) and 10 different people configurations (five LOS and five OBS; when people obstruct the direct path). For every frequency, the peoples' locations were also different. Two transmitter antenna heights were used: 1.5 and 3m. The path-loss results with people are compared with the results for the empty room. The excess attenuation that was found is shown in Table 6.5.

These results show that the presence of people in the LOS case only slightly influences the received power level (less than 1 dB). Obstruction of the direct path has a clear negative effect. For the transmitter antenna at 1.5m, the excess attenuation is 4 to 5.5 dB; for the antenna at 3m, the excess attenuation was found to be less: 1.5 to 3.5

dB. In both situations, LOS and OBS, no clear difference in behavior between the three frequencies was found. One should be careful about drawing general conclusions from these results since they are derived from a very limited number of measurements at only one location.

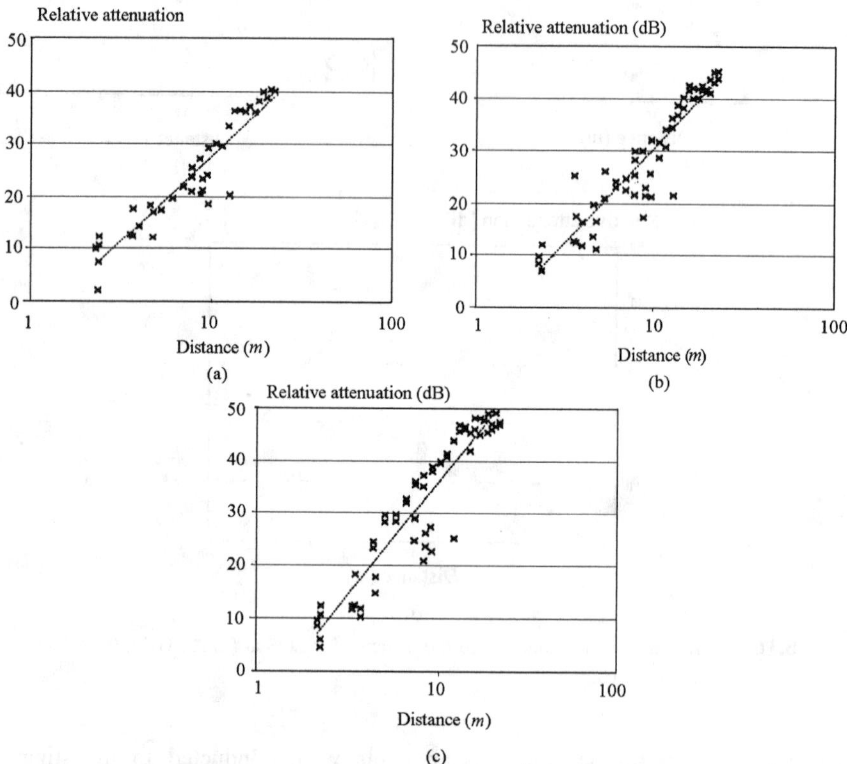

Figure 6.17 Attenuation as a function of the distance for OBS at (a) 2.4 GHz, (b) 4.75 GHz, and (c) 11.5 GHz.

The rms delay spread. The rms delay spread τ_{rms}, as defined in (6.15) is a measure for the dispersion of the radio channel, and can therefore be used to estimate the maximum usable symbol rate. For all measured channel responses, τ_{rms} was calculated. The average values and standard deviation of τ_{rms} at 2.4, 4.75, and 11.5 GHz are given for different locations and for two transmitter antenna heights in Tables 6.6 and 6.7, for the LOS and OBS situations, respectively.

Table 6.5

Excess Attenuation and Variables for the LR with People Present, Compared with the Empty LR at 2.4, 4.75, and 11.5 GHz

Frequency (GHz)	Location	Antennas Height (m)	Average Excess Attenuation (dB)	σ (dB)
2.4	LOS	1.5	− 0.33	0.3
		3.0	0.24	0.3
	OBS	1.5	4.84	1.8
		3.0	2.73	2.3
4.75	LOS	1.5	0.64	1.1
		3.0	− 0.3	0.8
	OBS	1.5	5.35	3.7
		3.0	1.46	1.4
11.5	LOS	1.5	0.19	0.6
		3.0	0.28	0.8
	OBS	1.5	4.33	1.5
		3.0	3.59	0.6

Table 6.6

Rms Delay Spread (Mean and Standard Deviation) for LOS Channels in Different Types of Rooms and for Transmitter Antenna Heights of 1.5m and 3m

Location	2.4 GHz				4.75 GHZ				11.5 GHz			
	$\bar{\tau}_{\mathrm{rms}}(ns)$		$\sigma_{\tau_{\mathrm{rms}}}(ns)$		$\bar{\tau}_{\mathrm{rms}}(ns)$		$\sigma_{\tau_{\mathrm{rms}}}(ns)$		$\bar{\tau}_{\mathrm{rms}}(ns)$		$\sigma_{\tau_{\mathrm{rms}}}(ns)$	
Antenna height (m)	1.5	3	1.5	3	1.5	3	1.5	3	1.5	3	1.5	3
LR	8.0	6.3	2.2	1.4	8.5	8.5	1.4	2.4	6.2	4.6	0.9	1.3
LR people	6.4	8.3	1.6	0.5	6.9	11.8	1.2	1.0	5.1	5.6	0.4	0.5
CR	-	14.9	-	2.9	-	18.0	-	5.2	-	12.8	-	3.7
HW	-	5.4	-	0.9	-	10.2	-	2.6	-	4.3	-	1.4

From these results we conclude that the average $\bar{\tau}_{rms}$ is minimum at 11.5 GHz for all LOS as well as in most OBS channels. The values of $\bar{\tau}_{\mathrm{rms}}$ at 11.5 GHz are on the average 30% less than at 2.4 and 4.75 GHz. In LOS channels, the height of the transmitter antenna only slightly affects $\bar{\tau}_{rms}$. However, reducing the antenna height in OBS channels yields an increase of both $\bar{\tau}_{rms}$ and σ_{rms}. In the larger rooms (e.g., the

CR and the HW), large τ_{rms} values are found due to large distances between reflective walls and objects. As well, the variation $\sigma_{\tau_{rms}}$ is large at these locations.

However, this observation is not always true for LOS channels. When transmitter and receiver are in adjacent rooms (OR1 $\rightarrow$ OR3), τ_{rms} is less compared with the situation when there is an empty room in between (OR1$\rightarrow$ OR3). The presence of people in the LR do not show a significant effect on τ_{rms}, even when the direct path is obstructed.

Table 6.7

Rms Delay Spread (Mean and Standard Deviation) for OBS Channels in Different Types of Rooms and for Transmitter Antenna Heights of 1.5m and 3m

Location	2.4 GHz				4.75 GHz				11.5 GHz			
	$\bar{\tau}_{rms}(ns)$		$\sigma_{\tau_{rms}}(ns)$		$\bar{\tau}_{rms}(ns)$		$\sigma_{\tau_{rms}}(ns)$		$\bar{\tau}_{rms}(ns)$		$\sigma_{\tau_{rms}}(ns)$	
Antenna height (m)	1.5m	3m	1.5m	3m	1.5m	3m	1.5m	3m	1.5m	3m	1.5m	3m
LR	16.4	12.4	6.2	2.5	12.4	11.6	2.5	2.3	6.9	6.5	2.1	1.[illegible]
LR people	10.8	8.8	3.1	1.1	10.6	11.1	1.1	1.2	5.4	5.4	0.9	1.[illegible]
OR1$\rightarrow$ OR2	19.8	12.1	1.4	1.1	21.2	12.6	2.9	1.1	14.4	9.7	1.1	1.[illegible]
OR1$\rightarrow$ OR3	23.1	19.2	1.7	3.0	21.5	19.3	1.3	1.9	13.6	12.4	1.8	1.[illegible]
CR	-	21.1	-	5.5	-	23.6	-	4.3	-	20.1	-	2.[illegible]
HW	-	21.6	-	16.0	-	15.2	-	4.4	-	18.6	-	10[illegible]

Antennas with wide opening angles were used during the measurements. This leads to relatively low τ_{rms} values compared to cases with high directive omnidirectional antennas [30]. The variation of τ_{rms} within a cluster is very small, usually less than 10% of the average value. In Figure 6.18, the variations in τ_{rms} are shown for the different locations by means of the cumulative distribution functions of τ_{rms} at the three frequencies. In general, the values for τ_{rms} are less than 30 ns.

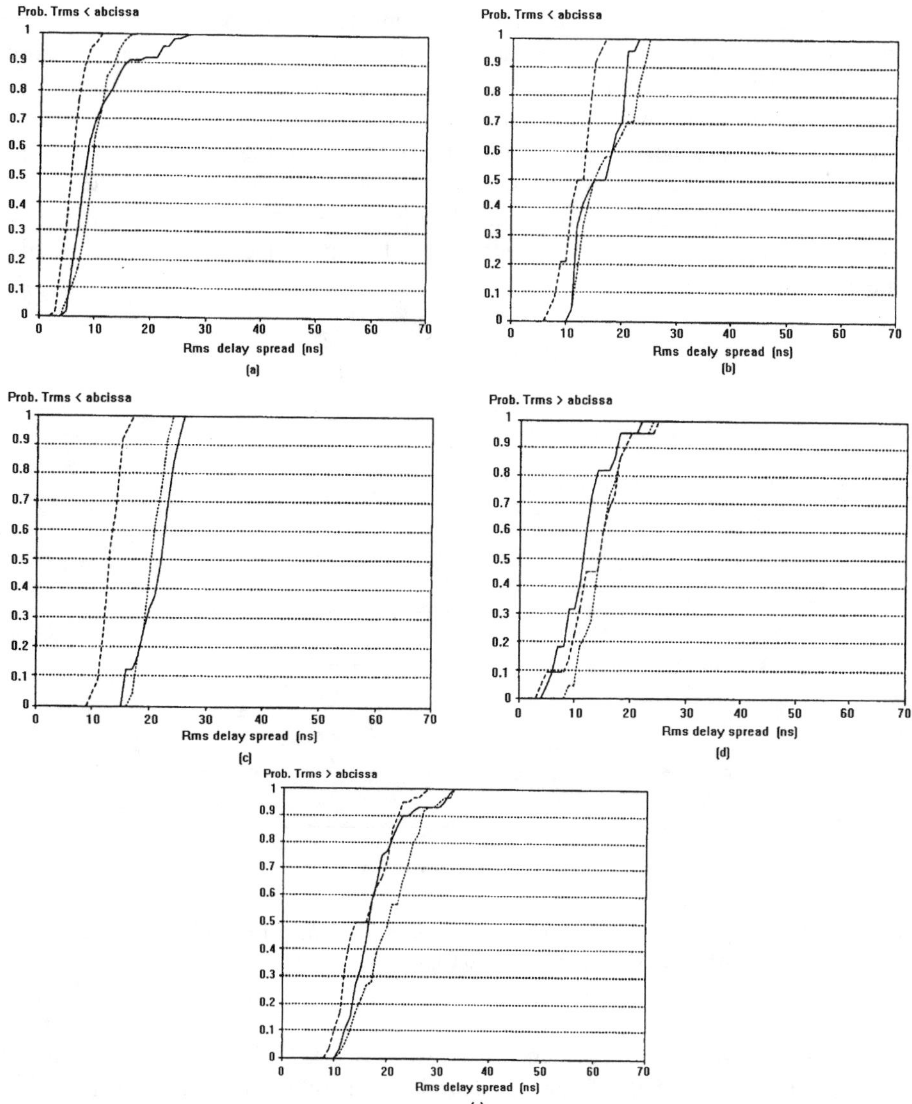

Figure 6.18 The cdf of τ_{rms} with transmitter and receiver in (a) the LR, (b) adjacent rooms OR1↔OR2, (c) OR1↔OR3, (d) the CR, and (e) with transmitter in the HW and receiver in OR1 where the frequency band is indicated by: —: 2.4 GHz; ....: 4.75 GHz: ----: 11.5 GHz.

This however, is not true for the measurements done in the HW. In Figures 6.19 and 6.20, the results are given for τ_{rms} as a function of the distance between transmitter and receiver when the transmitter is in OR1 and the receiver is in the HW and vice versa. These results show that the variation of τ_{rms} in the HW is large. τ_{rms} seems to increase with the distance between transmitter and receiver. Excessively large values in the order of 70 ns were observed on some positions at about the middle of the hallway. The measurement with the receiver in the HW show larger τ_{rms} than with the transmitter placed in the HW. This illustrates clearly with the environment of the receiver has a large impact on τ_{rms}.

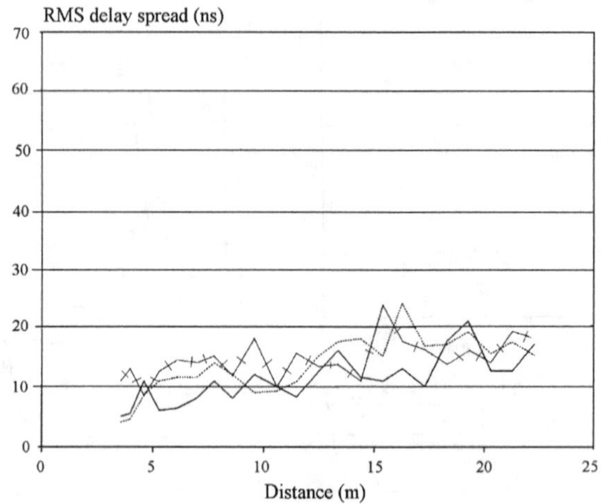

Figure 6.19 τ_{rms} versus distance with transmitter in the HW and receiver in OR1, at: — : 2.4 GHz; ┼ : 4.75 GHz; ----: 11.5 GHz.

Coherence bandwidth. The coherence bandwidth B_{coh} is defined as in (6.17). It was found that the shape of the frequency correlation function strongly depends on the characteristics of the multipath channel. The B_{coh} results show a large variance in most of the situations also within a cluster. In many LOS situations, the correlation level did not drop below the –3 dB level for a bandwidth of over 250 MHz. This effect was seen for about 30% of the measurements in the OR at the three frequencies. In the CR, this percentage was much less (about 6%). A remarkable observation is that for the lower frequencies, about 40% of the clusters had at least one position with B_{coh} > 250 MHz. At 11.5 GHz, this was even 65%.

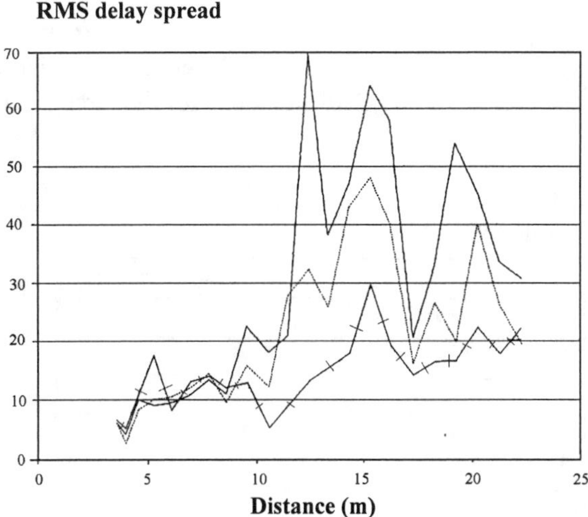

Figure 6.20 τ_{rms} versus distance with transmitter in OR1 and receiver in the HW, at:—: 2.4 GHz; +: 4.75 GHz; ----: 11.5 GHz.

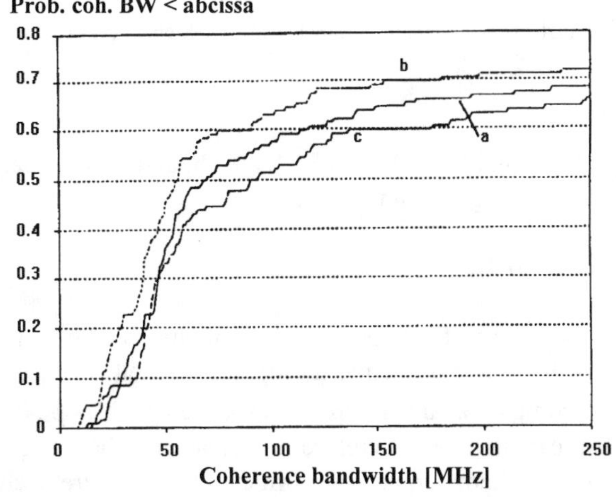

Figure 6.21 The cdf of the coherence bandwidth for LOS channels: (a) 2.4 GHz, (b) 4.75 GHz, and (c) 11.5 GHz.

The cumulative distribution functions for B_{coh} are calculated from all measured LOS and OBS channels, and are shown in Figures 6.21 and 6.22. We find that for LOS situations, more than 27% of the positions have B_{coh} of over 250 MHz. For OBS situations this percentage is much less (5%). The relation $\alpha = 1\ (\tau_{rms}\ B_{coh})$ was evaluated for the measured results. In Figure 6.23, the cumulative distribution functions of α are given separately for LOS and OBS situations for the three frequencies.

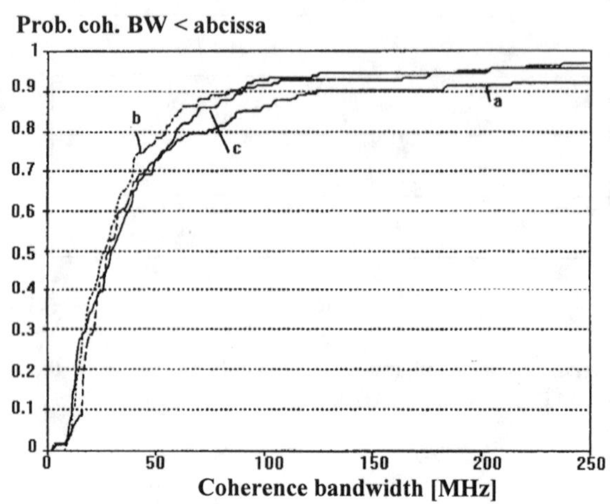

Figure 6.22 The cdf of the coherence bandwidth for OBS channels (a) 2.4 GHz, (b) 4.75 GHz, and (c) 11.5 GHz.

The median values of α found for LOS and OBS are about 4 or 5, respectively, which is in agreement with the results given in [31,32].

6.3.3 Analytical Model for BEP Evaluation

This section describes the analytical approach to evaluate BEP of a BPSK modulated signal in a stationary frequency-selective multipath channel, using a coherent BPSK receiver [33,34]. The complex channel impulse response variables β_k, θ_k, and τ_k are used in the model for BEP calculations. Therefore, the measured channel impulse response, which has a continuous character, is transformed to the discrete channel response variables β_k, θ_k, and τ_k, which are the amplitude, phase, and time delay at the local maxima of the response. These maxima are identified as discrete paths.

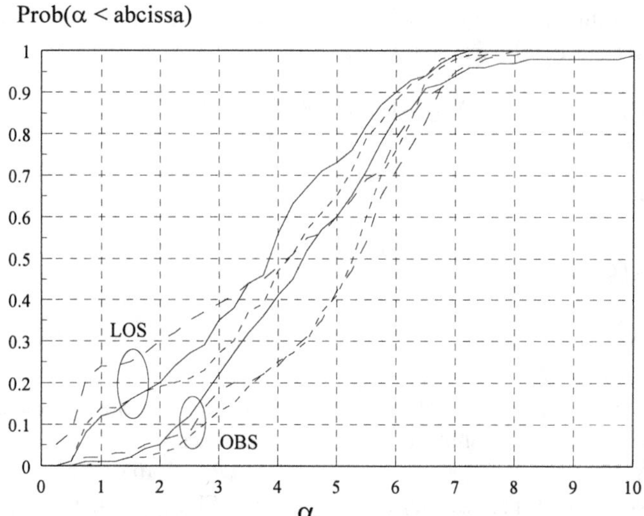

Figure 6.23 The cdf of $\alpha = 1/(\tau_{rms}.B_{coh})$ for LOS and OBS channels —: 2.4 GHz; ....: 4.75 GHz; ----: 11.5 GHz.

A. Signal Model

The complex equivalent baseband representation of a BPSK modulated signal is written as

$$\bar{s}_B(t) = \sqrt{P}d(t)\exp(j\phi) \tag{6.22}$$

where P denotes signal power and ϕ indicates carrier phase. The overbar denotes a complex signal. The real modulated signal is given by

$$s(t) = \Re\left\{\sqrt{2}\bar{s}_B(t)\exp(j\omega_c t)\right\} = \sqrt{2P}d(t)\cos(\omega_c t + \phi) \tag{6.23}$$

where ω_C is carrier frequency and $\Re\{z\}$ denotes the real part of the complex value z. Without loss of generality we assume that $\phi = 0$, $d(t)$ is a random data signal that consists of rectangular pulses with duration T, and is given by

$$d(t) = \sum_{k=0}^{\infty} a_k \prod(t - kT) \tag{6.24}$$

where $a_k \in \{-1, +1\}$ with equal probability of occurrence. When $s(t)$ reaches the receiver via a multipath channel with impulse response $h(t)$ as given by (6.10), the equivalent baseband received signal $r_B(t)$ is

$$r_B(t) = \Re\{\bar{s}_B(t) * h(t)\} = \sqrt{P}\sum_{k=0}^{N}\beta_k d(t-\tau_k)\cos\theta_k \tag{6.25}$$

and the real received signal is

$$r(t) = \Re\left\{\sqrt{2}\bar{r}_B(t)\exp(j\omega_c t)\right\} \tag{6.26}$$

B. BPSK Receiver Model

In this model, a conventional coherent BPSK receiver matched to the additive white Gaussian noise (AWGN) channel is used. The reference carrier is recovered from the received signal by means of a squaring loop [35]. A block diagram of the receiver and the squaring loop carrier recovery circuit is shown in Figure 6.24. The bandwidth of the phase locked loop (PLL) of the carrier recovery circuit is assumed to be very small compared with the bit rate, so that phase-jitter due to AWGN and ISI, caused by multipath signals, is neglected. The recovered carrier phase for the multipath channel is determined from

$$\bar{r}_B^2(t) = P\sum_{i=0}^{N}\sum_{j=0}^{N}\beta_i\beta_j d(t-\tau_j)\exp\left(j(\theta_i+\theta_j)\right) \tag{6.27}$$

which is the complex baseband equivalent signal of the signal component at $2f_c$ after squaring of the real signal $r(t)$. The PLL locks on this component at $2f_c$, whereas the low-pass signal is filtered out. In the ideal case that the PLL bandwidth $\to$ 0, the recovered phase is given by

$$\phi_c = \frac{1}{2}\arg\left\{E\left[\bar{r}_B^2(t)\right]\right\} + n\pi$$

$$= \frac{1}{2}\arg\left\{E\left[P\sum_{i=0}^{N}\sum_{j=0}^{N}\beta_i\beta_j d(t-\tau_i)d(t-\tau_j)\exp(j(\theta_i+\theta_j))\right]\right\} + n\pi$$

$$= \frac{1}{2}\arg\left\{\sum_{i=0}^{N}\sum_{j=0}^{N}\beta_i\beta_j R_d\left(\tau_i-\tau_j\right)\exp\left(j(\theta_i+\theta_j)\right)\right\} + n\pi \tag{6.28}$$

where $R_d(\tau_i - \tau_j)$ is the ACF of the data signal d(t) and its value indicates the fraction of the bit time that two paths carry the same bit during a bit interval, with

$$R_d(\tau_i - \tau_j) = 1 - \frac{|\tau_i - \tau_j|}{T} \qquad \text{if } |\tau_i - \tau_j| \leq T$$

$$= 0 \qquad \text{if } |\tau_i - \tau_j| > T \qquad (6.29)$$

The uncertainty of $n\pi$ is caused by the phase ambiguity of the squaring loop circuit. This ambiguity requires the data signal to be differentially encoded.

(a)

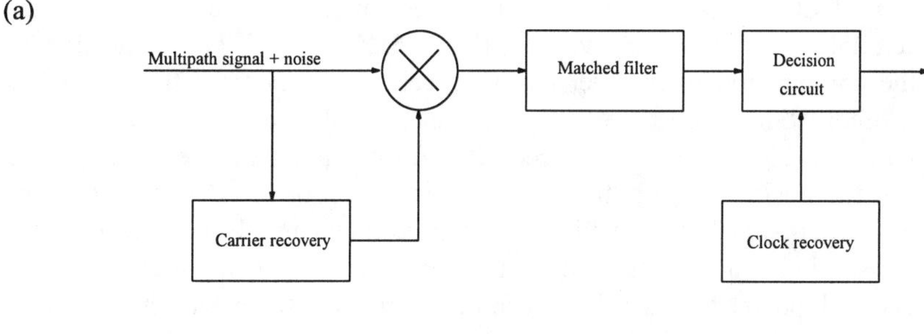

(b)

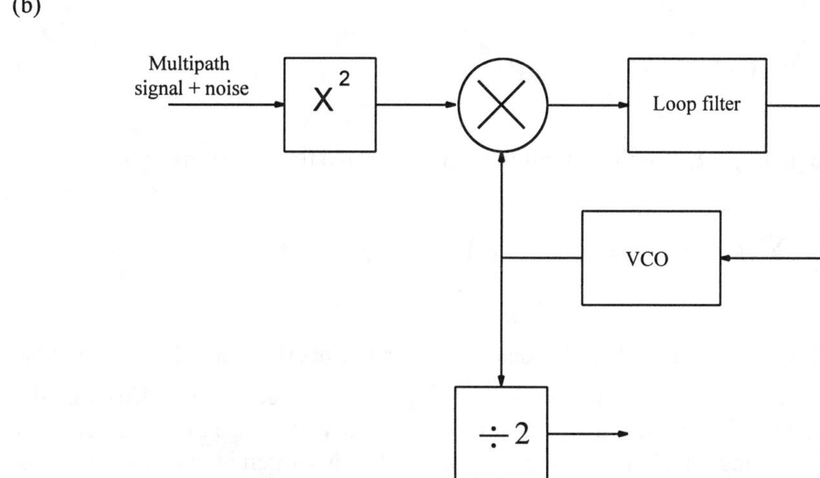

Figure 6.24 Block diagram of (a) the coherent BPSK receiver and (b) the squaring loop carrier recovery circuit.

C. BEP for a Stationary Frequency Selective Multipath Channel

The input signal of the matched filter is the baseband component that results after multiplication of the received signal $r(t)$ with the recovered reference carrier

$$r_{\text{dem}}(t) = r(t)\cos\left(\omega_c t + \phi_c\right) = \sqrt{P}\sum_{k=0}^{N}\beta_k d\left(t-\tau_k\right)\cos\left(\theta_k - \phi_c\right) \tag{6.30}$$

This signal is the sum of the contributions of all $N + 1$ paths. The maximum number of bits that are active in the channel during one bit interval T is $L + 2$ with $L = \left\lceil\left(\tau_N - \tau_0\right)/T\right\rceil = \left\lceil \tau_N / T\right\rceil$, where $\lceil x \rceil$ indicates the integer value larger or equal to x.

Because the energy received during a bit time results from $L + 2$ bits, it is necessary for the BEP calculation to select a desired bit from the active bits in the channel, as the bit detected by the receiver (see Figure 6.24). The BEP of the desired bit depends on the combination of receiver timing, received energy from that bit in the corresponding integration interval, and the amount of ISI energy in that interval. Different criteria for selection of the desired bit and bit timing, can be used: (1) the lowest BEP for the desired bit, (2) the largest contribution of the desired bit to the matched filter output signal, (3) maximum signal-to-intersymbol interference ratio (SIIR) for the desired bit, and (4) maximum total energy at the matched filter output.

The received power for a single bit can be determined by transmitting a test bit over the multipath channel, as depicted in Figure 6.25. Let the test bit be defined as $p_{test}(t) \overset{\wedge}{=} U(t) - U(t-T)$; then the equivalent received complex baseband signal is

$$r_{\text{B-test}}(t) = \Re\left\{p_{\text{test}}(t) * h(t)\right\} = \sum_{k=0}^{N}\beta_k p_{\text{test}}(t-\tau_k)\cos\theta_k \tag{6.31}$$

The output signal $s_{\text{test}}(\tau_s)$ for the test bit for a given sample time τ_s, is given by

$$s_{\text{test}}(\tau_s) = \int_{\tau_s - T}^{\tau_s}\left[\sum_{k=0}^{N}\beta_k p_{\text{test}}\left(t-\tau_k\right)\cos\left(\theta_k - \phi_c\right)\right]dt \tag{6.32}$$

τ_s is the sample time referred to the delay time of the paths in the channel impulse response $h(t)$, $\tau_s \in [T, \tau_N + T]$. The sign of s_{test} is determined by the phases of the contributing paths. When a real data signal is transmitted, a sequence of bits is active. If we denote the desired bit a_d, the sequence of active bits is written in vector notation as

$\boldsymbol{a} = (a_{-1}, a_0, a_1, \ldots, a_{d-1}, a_d, a_{d+1}, \ldots, a_L)$

with $a_j \in \{-1, +1\}$.

The position of a_d in $\boldsymbol{a}$ depends on $h(t)$ and the selection criterion used to determine τ_s. The index d is equal to $d = \lceil \tau_s / T \rceil$. In the integration interval $[\tau_s - T, \tau_s]$ in general, two bits, a_{i-1} and $a_i \in \boldsymbol{a}$, are active in the kth path, where the index i is given by $i = \lceil \tau_k / T \rceil$. The contribution of path k to the output signal of the matched filter is now given by

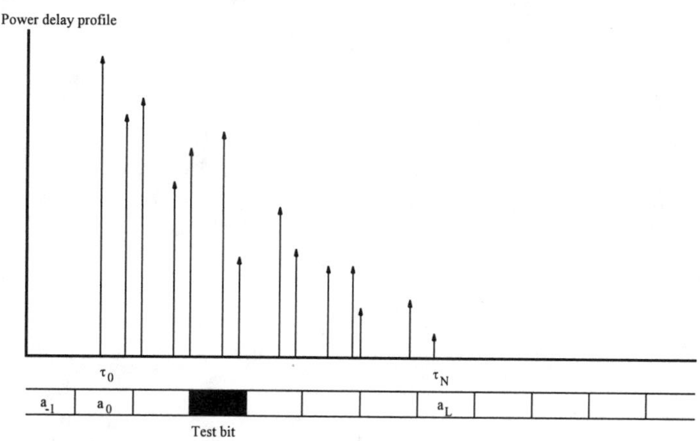

Figure 6.25 Paths contributing to the received signal of a test bit, and the other active bits in a multipath channel.

$$s_k = \sqrt{P}\beta_k T \cos\left(\theta_k - \phi_c\right)\left[\left(1 - \Delta_k\right)a_{i-1} + \Delta_k a_i\right] = a_{i-1}\kappa_{k,i-1} + a_i \kappa_{k,i} \tag{6.33}$$

with

$$\kappa_{k,i-1} = \sqrt{P}\beta_k \left(1 - \Delta_k\right) T \cos\left(\theta_k - \phi_c\right)$$

$$\kappa_{k,i} = \sqrt{P}\beta_k \Delta_k T \cos\left(\theta_k - \phi_c\right) \tag{6.34}$$

Δ_k indicates the fraction of bit a_i in the integration interval for path k

$$\Delta_k = \text{Frac}\left(\frac{\tau_k - \tau_s}{T}\right) \qquad \text{if}\tau_s < \tau_k$$

$$= \text{Frac}\left(\frac{\tau_s - \tau_k}{T}\right) \qquad \text{if}\tau_s > \tau_k \tag{6.35}$$

The total output signal of the matched filter for all $N + 1$ path contributions is given by

$$s_{\text{out}}(\boldsymbol{a}, \tau_s) = \boldsymbol{a}K(\tau_s) \tag{6.36}$$

Here $K(\tau_s)$ is the path matrix for sample time τ_s. Note that $K(\tau)$ is periodic in T.

$$K = \begin{pmatrix} \kappa_{0,-1} & 0 & 0 & .. & 0 & 0 \\ \kappa_{0,0} & \kappa_{1,0} & \kappa_{2,0} & .. & .. & .. \\ 0 & \kappa_{1,1} & \kappa_{2,1} & .. & .. & .. \\ 0 & 0 & 0 & .. & .. & .. \\ .. & .. & .. & .. & \kappa_{N-1,L-2} & .. \\ .. & .. & .. & .. & \kappa_{N-1.L-1} & \kappa_{N,L-1} \\ 0 & 0 & 0 & .. & 0 & \kappa_{N,L} \end{pmatrix} \tag{6.37}$$

The number of rows of K is equal to $L + 2$, which is equal to the number of active bits in the channel during one bit time. Row l, with $l \in \{-1, 0, \quad, L\}$, contains the contribution of bit a_l. The number of columns of K is $N + 1$, which is equal to the number of paths, where column vectors K_k contain the contribution of path k. K_k is defined as $K_k \stackrel{\Lambda}{=} (0,0,\ldots,0,\kappa_{k,i-1},\kappa_{k,i},0,\ldots,0)^T$, where superscript T indicates the transpose. Each column has only two successive nonzero elements, because only two bits are active during a bit time, and all preceding and following elements are zero. Now the received power from bit i, S_i, is written as

$$S_i(\tau_s) = \left|\sum_{k=0}^{N} \kappa_{k,i}(\tau_s)\right|^2 \tag{6.38}$$

Let the desired bit have the fixed value $a_d = +1$, then vector $\boldsymbol{a}$ can take on $V = 2^{L+1}$ different values, each resulting in a different output signal $s_{\text{out}}(\boldsymbol{a}_z|a_d = +1)$, with $\boldsymbol{a}_z \in \{\boldsymbol{a}_1, \ldots, \boldsymbol{a}_V\}$. The probability of an error, $P_{ez}(\varepsilon)$ for $\boldsymbol{a}_z$, is given by

$$P_{ez} = P(\varepsilon | a_d = +1, \mathbf{a}_z) = P(\text{sign})(s_{\text{out}}(\mathbf{a}_z | a_d = +1)) = -\text{sign}(s_d)$$

$$= \begin{cases} Q\left(\dfrac{|s_{\text{out}}(\mathbf{a}_z | a_d = +1)|}{\sigma_n}\right) & \text{if sign}(s_{\text{out}}) = \text{sign}(s_d) \\ 1 - Q\left(\dfrac{|s_{\text{out}}(\mathbf{a}_z | a_d = +1)|}{\sigma_n}\right) & \text{if sign}(s_{\text{out}}) = -\text{sign}(s_d) \end{cases} \tag{6.39}$$

with $Q(z) = \frac{1}{\sqrt{2\pi}} \int_z^\infty e^{-\lambda^2/2} d\lambda$ and σ_n^2 the received noise power. Sign (s_d) is determined with (6.32). BEP is calculated by averaging P_{ez} over all 2^{L+1} possible sequences $\mathbf{a}_z$

$$P_e(\varepsilon | a_d = +1) = \frac{1}{2^{L+1}} \sum_{z=1}^{2^{L+1}} P_{ez} \tag{6.40}$$

σ_n^2 is related to the average SNR γ as $\sigma_n^2 = S_{\text{av}} / \gamma$ with S_{av} is the average received signal power

$$S_{\text{av}} = P \sum_{k=0}^{N} \beta_k^2 \tag{6.41}$$

Now τ_s is optimized to one of the criterion mentioned earlier. In the following, we select optimum timing τ_s for minimum BEP. Therefore, BEP results given here are a lower bound. The optimum timing is calculated by evaluation of BEP when the sample time τ_s is varied with small steps over the bit duration T (note that the path matrix is periodic with T). At every τ_s, BEP is calculated for the desired bit, which has maximum energy as determined with (6.38).

BEP Results for Measured Indoor Channels

For a number of measured clusters as described in Section 6.3.2, BEP has been calculated as function of the bit rate at 2.4, 4.75, and 11.5 GHz. In the calculations, only paths less than 30 dB down compared with the dominant path are assumed to give a significant contribution to the received signal. Weaker paths are neglected, which makes computations less time-consuming. The SNR was taken as $\gamma = 10$ dB. Instead of S_{av}, the average power S_{avc} over six positions in the cluster is used as reference signal power to determine σ_n^2: $\sigma_n^2 = S_{\text{avc}} / \gamma$. S_{avc} is given by

$$S_{\text{avc}} = \frac{P}{6} \sum_{j=1}^{6} \sum_{k=1}^{N_j} \beta_{j,k}^2 \tag{6.42}$$

where j indicates the cluster position, $j = \{1, ..., 6\}$, and N_j is the number of relevant paths for position j. In Figures 6.26 to 6.28, BEP results as a function of the bit rate are shown for 2.4, 4.75, and 11.5 GHz, for three cluster at the following positions: OR1 → OR2 (OBS), CR (LOS), and CR (OBS). The vertical bold lines in the figure indicate the bit rate $R_{max} = 1/4\tau_{rms}$.

In general, τ_{rms} within a cluster does not show large variation. The calculated values of R_{max} for the different positions in a cluster are within 10% of the indicated average value of R_{max}. BEP results show a large variation for constant bit rate at different positions in a cluster, which do not differ more than a couple of wavelengths ($\lambda/2$ to 4λ) in distance. Over such a small area the PDP, and therefore also τ_{rms}, does not change significantly (Figure 6.15).

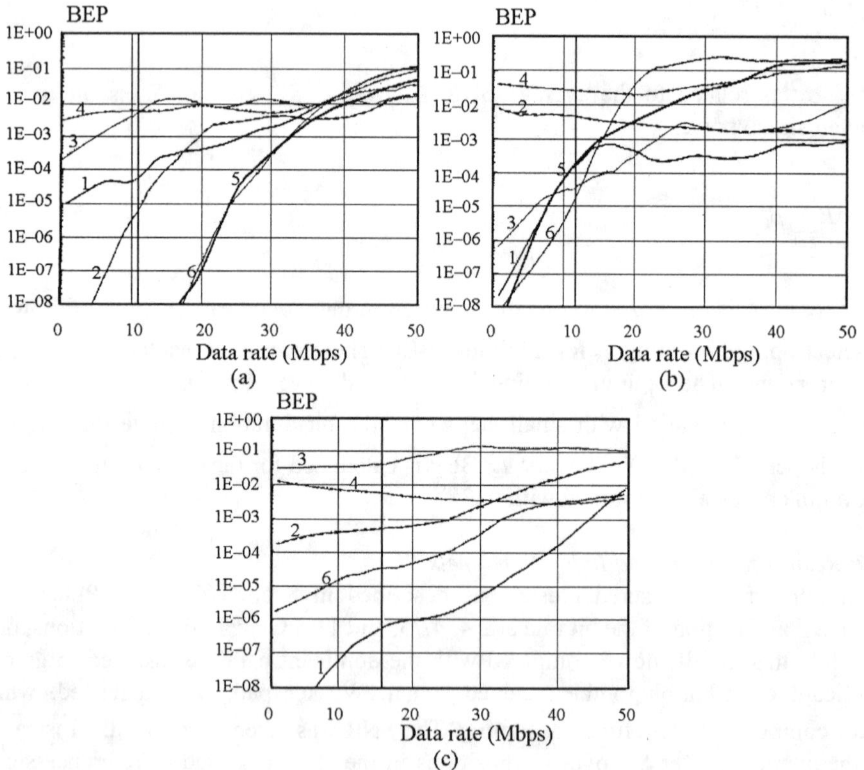

Figure 6.26 BEP as function of data rate in location OR1 → OR3, at (a) 2.4 GHz, (b) 4.75 GHz, and (c) 11.5 GHz. The different positions in the circular cluster are indicated as: 1, 2, 3, 4, 5, and 6. The bold vertical lines indicates $R_{max} = 1/4\,\bar{\tau}_{rms}$.

These BEP differences are therefore caused mainly by different combinations of path phases, which result in different frequency responses of the channel. In general, BEP increases with increasing bit rate as expected; however, for a significant number of positions BEP is nearly constant as a function of bit rate or even decreases with increasing bit rate. The reason for this unexpected behavior is that for those cases, paths with different delays and phases are more advantageous at the receiver for certain bit rates.

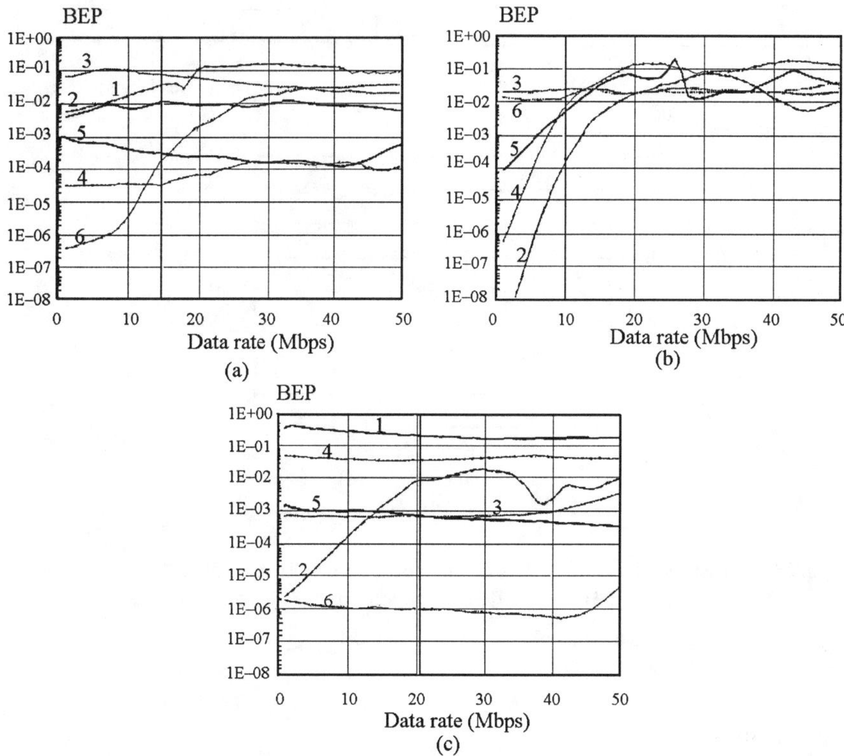

Figure 6.27 BEP as function of data rate in location CR (LOS), at (a) 2.4 GHz, (b) 4.75 GHz, and (c) 11.5 GHz. The different position in the circular cluster are indicated as: 1, 2, 3, 4, 5, and 6. The bold vertical lines indicates $R_{max} = 1/4\,\overline{\tau}_{rms}$.

From the BEP results, no clear difference is seen for different frequencies or even LOS and OBS situations. The large variation of channel performance, found when the antenna is moved over a distance in the order of the wavelength λ, is exploited by antenna selection diversity, where the antenna with the best performance is chosen.

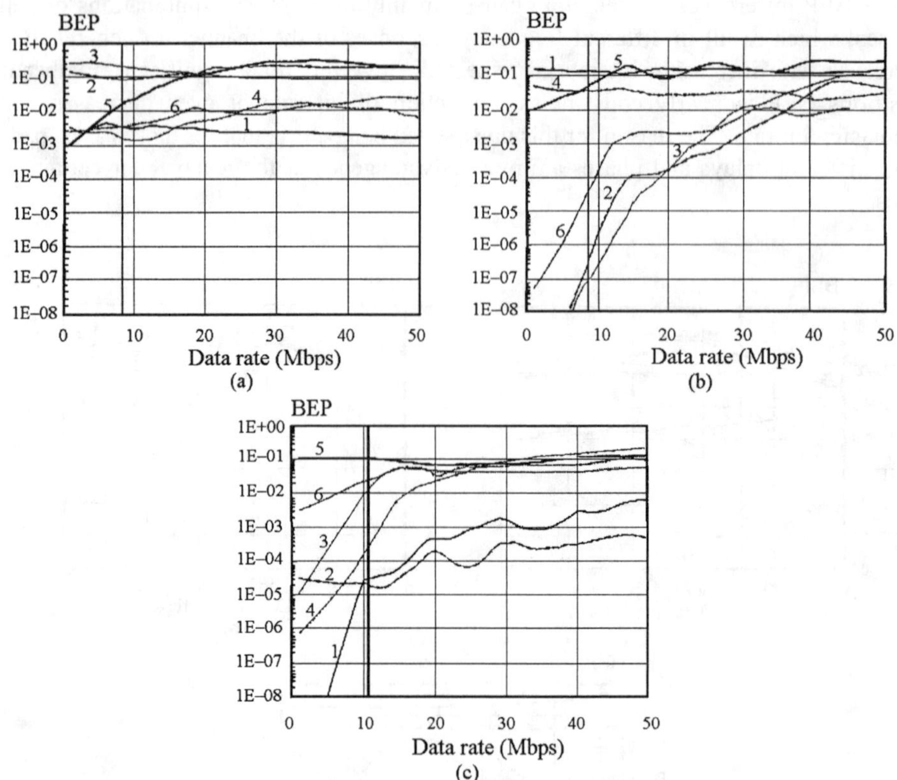

Figure 6.28 BEP as a function of data rate in location CR (OBS), at (a) 2.4 GHz, (b) 4.75 GHz, and (c) 11.5 GHz. The different positions in the circular cluster are indicated as 1, 2, 3, 4, 5, and 6. The bold vertical lines indicate $R_{max} = 1/4\,\bar{\tau}_{rms}$.

6.3.4 Conclusions

In this section, wideband channel measurement results for an indoor office environment three frequencies of 2.4, 4.75, and 11.5 GHz were presented, and an analytical BEP model for the stationary frequency selective multipath channel was derived.

Path loss. The parameters of a simple path-loss model were determined. We conclude that this model gives accurate results for LOS situations. The values for the path-loss law exponent found for LOS paths are within the range of 1.8 to 2.0 for all frequencies. In OBS situations, the path-loss law exponent increases with frequency and is 3.3, 3.8, and 4.5 at 2.4, 4.75, and 11.5 GHz, respectively. For OBS situations however, the attenuation predictions of the simple model are not very accurate.

Delay spread τ_{rms}. The delay spread τ_{rms} was determined for different types of rooms with different dimensions. τ_{rms} increases for larger rooms. In the HW, very strong fluctuations were found. At 11.5 GHz, τ_{rms} in most cases is significantly smaller, about 30%, than at 2.4 and 4.75 GHz, which are generally about the same value. OBS paths show increased τ_{rms} values when compared with LOS paths. In different situations, the median values of τ_{rms} at 2.4 and 4.75 GHz range from 10 to 20 ns. At 11.5 GHz, the median values τ_{rms} ranges from 5 to 15 ns.

Coherence bandwidth. The coherence bandwidth shows significant variation, also within a cluster of measurements. For the lower frequencies 40% to 45% of the clusters had at least one position with B_{coh} > 250 MHz. At 11.5 GHz, this percentage was even 65%. The relation $\alpha = 1/(\tau_{rms} . B_{coh})$ shows a large variation of α between 0.5 and 6. The median values of LOS and OBS are about 4 and 5, respectively.

Presence of people. The effects of people on path loss, when compared with the results for an empty room, is small. For LOS situations the effect is negligible. When people obstruct the direct path, 4 to 5.5 dB extra attenuation was found on the average. No clear difference in behavior was found for the three frequencies 2.4, 4.75, and 11.5 GHz. No influence of people on τ_{rms} was found.

BEP results. A model was developed to evaluate BEP in a stationary frequency-selective multipath channel for BPSK modulation. With this model, lower-bound BEP results were calculated for a conventional coherent BPSK receiver, as a function of bit rate up to 50 Mbps, at 2.4, 4.75, and 11.5 GHz. It was found that BEP for CBR varies significantly when the antenna is displaced over very small distances ($\lambda/2 - 4\ \lambda$). This difference is caused mainly by relative phase changes of the paths. In general, BEP increases with bit rate but this is not always the case. Many channels show a relatively constant BEP or even a decreasing BEP with increasing bit rate. The large variation in BEP within a small cluster can be exploited by using antenna diversity.

Combined channel impulse response measurements and BEP measurements are recommended to validate the BEP model presented in this chapter.

REFERENCES

[1] H. Zaghloul, G. Morrison, and M. Fattouche, "Frequency response and path-loss measurements of indoor channel," *Electron. Lett.*, Vol. 27, No. 12, pp. 1021–1022, June 1991.

[2] S.Y. Seidel and T.S. Rappaport, "Path-loss prediction in multifloored buildings at 914 MHz," *Electron. Lett.*, Vol. 27, No. 15, pp. 1384–1387, July 1991.

[3] D.M.J. Devasirvatham, C. Banerjee, R.R. Murray, and D.A. Rappaport, "Four-frequency radiowave propagation measurements of the indoor environment in a large metropolitan commercial building," *Proc. IEEE GLOBECOM '91*, Phoenix, AZ, pp. 1282–1286, December 1991.

[4] A.A.M. Saleh and R.A. Valenzuela, "A statistical model for indoor multipath propagation," *IEEE J. Selected. Areas in Comm.*, Vol. CSA-5, No. 2, pp. 128–137, February 1986.

[5] S.Y. Seidel and T.S. Rappaport, 'Path-loss prediction in multifloored buildings at 914 MHz," *Electron. Lett.*, Vol. 27, No. 15, pp. 1384–1387, July 1991.

[6] A.J. Motley and J.M.P. Keenan, "Personal communication radio coverage in buildings at 900 MHz and 1700 MHz," *Electron. Lett.*, Vol. 24, No. 12, pp. 763–764, June 1988.

[7] S.J. Howard and K. Pahlavan, "measurement and analysis of the indoor radio channel in the frequency domain," *IEEE Trans. Instrum. Meas.*, Vol. 39, No. 5, pp. 751–755, October 1990.

[8] S.J. Howard and K. Pahlavan, "Frequency domain measurements of indoor radio channels," *Electron. Lett.*, Vol. 25, No. 24, pp. 1645–1647, November 1989.

[9] G.J.M. Janssen and R. Prasad, "Propagation measurements in an indoor radio environment at 2.4 GHz, 4.75 GHz and 11.5 GHz," *Proc. IEEE VTS Conf.* '92, Denver, CO, pp. 617–620, May 10–13, 1992.

[10] M. Kavehrad and B. Ramamurthi, "Direct-sequence spread spectrum with DPSK modulation and diversity for indoor wireless communication," *IEEE Trans. Comm.*, Vol. COM-35, No. 2, pp. 224–236, February 1987.

[11] A. Zigic and R. Prasad, "Bit error rate of decision feedback equaliser for indoor wireless communication ," *Electron. Lett.*, Vol. 28, pp. 1949–1950, October 1992.

[12] R. Prasad, A. Kegel, and M.B. Loog, "Co-channel interference probability for picocellular system with multiple Rician faded interferers," *Electron. Lett.*, Vol. 28, pp. 2225–2226, November 1992.

[13] H.S. Misser, A. Kegel, and R. Prasad, "Monte Carlo simulation of direct sequence spread spectrum for indoor radio communication in a Rician fading channel," *IEE Proc.-I,* Vol. 139, No. 6, pp. 620–624, December 1992.

[14] J. C.-I. Chuang, "The effects of time delay spread on portable radio communications channels with digital modulation," *IEEE J. Selected Areas in Comm.*, Vol. SAC-5, No. 5, pp. 879–889, June 1987.

[15] F. van der wijk, A. Kegel, and R. Prasad, "Assessment of a pico-cellular system using propagation measurements at 11.9 GHz for indoor wireless

communications, *IEEE Trans. on Vehicular Technology*, Vol. 44, pp. 155–162, February 1995.

[16] G.J.M. Janssen, P.A. Stigter, and R. Prasad, "Wideband indoor channel measurements and BEP analysis of frequency selective multipath channels at 2.4, 4.75 and 11.5 GHz," *IEEE Trans. Comm.*, Vol. 44, pp. 1272–1288, October 1996.

[17] A. Verschoor, "Measurement of narrowband characteristics of the indoor radio channel," *PTT-Research DNL, Network Technology*, Leidschendam, pp. 27, June 1990 (in Dutch).

[18] W.J. Dixon and F. Massey, *Introduction to Statistical Analysis, 3rd ed.*, London, McGraw-Hill, p. 550, 1969.

[19] R. Prasad and A. Kegel, "Improved assessment of interference limits in cellular radio performance," *IEEE Trans. on Vehicular Technology*, Vol. 40, pp. 412–419, May 1991.

[20] R. Prasad and A. Kegel, "Effects of Rician faded and lognormal shadowed signals on spectrum efficiency in microcellular radio," *IEEE Trans. on Vehicular Technology*, Vol. 42, pp. 274–281, August 1993.

[21] H.L. van Trees, *Detection, Estimation and Linear Modulation Theory*, New York: John Wiley & Sons, 1986.

[22] E. Harreveld, "Samenvatting van resultaten verkregen uit simulatie van enige DECT-specificaties," *PTT-Research DNL, Network Technology*, Leidschendam, p.11, May 1991 (in Dutch).

[23] A. Kegel, F. van der Wijk, and R. Prasad, "Evaluation of a cellular structure in indoor systems," *Proc. IEEE Veh. Technol. Conf.*, Denver, Colorado, pp. 835–838, May 1992.

[24] W.C.Y. Lee, *Mobile Communication Design Fundamentals*, 1986.

[25] J. Wickman, "Analysis of the small scale fading distributions for indoor radio channels at 866 & 1800 MHz," *Swedish Telecom., COST 231 TD(91) 09 25*, Leidschendam, 1991.

[26] Abramowitz and Stegun, *Handbook of Mathematical Functions*, New York: Dover, 1970.

[27] L. Kuipers and R. Timman, *Handboek der Wiskunde I*, Amsterdam, Scheltema & Hollema n.v., p. 747, 1966.

[28] D.C. Cox and R.P. Leck, "Correlation bandwidth and delay spread multipath propagation statistics for 910 MHz urban mobile radio channels," *IEEE Trans. Comm.*, Vol. COM-23, pp. 1271–1280, 1975.

[29] B. Glance and L.J. Greenstein, "Frequency selective fading effects in digital mobile radio with diversity combining," *IEEE Trans. Comm.*, Vol. COM-31, No. 9, pp. 1085–1094, September 1983.

[30] P.F.M. Smulders and A.G. Wagemans, "Mm-wave biconical horn antennas for near uniform coverage in indoor pico-cells," *Electron. Lett.*, Vol. 28, pp. 679–681, March 1992.

[31] J.G. Proakis, *Digital Communications, 2nd edition*, New York: McGraw-Hill, 1989.

[32] W.C. Jakes, *Microwave Mobile Communications*, New York: John Wiley & Sons, 1974.

[33] G.J.M. Janssen, P.A. Stigter, and R. Prasad, "A model for bit error rate evaluation of indoor frequency selective channels using multipath measurements results at 2.4, 4.75 and 11.5 GHz," *Int. Zurich Seminar on Mobile Comm.*, Lect. Notes in Comp. Sc. 783, pp. 344–355, March 1994.

[34] G.J.M. Janssen, B.C.v. Lieshout, and R. Prasad, "BER performance of millimeter wave indoor communication systems using multiple antenna signals," *IEEE Proc. Commun. Theory Mini Conf. GLOBECOM'94*, San Francisco, CA, pp. 105–109, December 1994,.

[35] F.S. Gardner, *Phaselock Techniques, 2nd ed.*, New York: John Wiley & Sons, 1979.

APPENDIX 6A

Rician distribution is given by

$$f(A) = \frac{A}{\sigma^2} \exp\left(\frac{-A^2 + s^2}{2\sigma^2}\right) I_0\left(\frac{As}{\sigma^2}\right) \qquad 0 \le A < \infty \tag{6A.1}$$

The Rician distributed signal is the sum of a Rayleigh distributed amplitude, x, and a sinusoid, *s*:

$$A = s + x = \sqrt{(s_i + x_i)^2 + (s_q + x_q)^2} \tag{6A.2}$$

where s_i is the in-phase component of *s*, s_q is the quadrature component of *s*, x_i is the in phase component of *x*, and x_q is the quadrature component of *x*. This means that the average squared amplitude, $E[A^2]$, equals

$$E[A^2] = E[(s_i + x_i)^2 + (s_q + x_q)^2] \tag{6A.3}$$

For a Rayleigh process $E[x_i] = E[x_q] = 0$ and $E[x_i^2] = E[x_q^2] = 2\sigma^2$ [24, p. 173] rewriting (6A.3) results in:

$$E[A^2] = E[s_i^2 + s_q^2 + x_i^2 + x_q^2] + E[2s_i x_i + 2s_q x_q] = s^2 + 2\sigma^2 \tag{6A.4}$$

An expression is derived for the average amplitude $E[A]$ [25, p. 4]:

$$E[A]=\int_0^\infty A f(A)dA \tag{6A.5}$$

First, $E[x]$ is calculated, which is given by

$$x=A\frac{s}{\sigma^2}\Rightarrow\frac{dA}{dx}=\frac{\sigma^2}{s}\quad\Lambda\quad E[A]=\frac{\sigma^2}{s}E[x] \tag{6A.6}$$

Substituting (6A.6) in (6A.1), $f(x)$ is obtained as

$$f(x)=\frac{\sigma^2}{s^2}x\exp\left(-\frac{\sigma^2}{2s^2}x^2-\frac{s^2}{2\sigma^2}\right)I_0(x) \tag{6A.7}$$

With the help of $f(x)$, $E[x]$ is calculated:

$$E[x]=\int_0^\infty xf(x)dx=\frac{\sigma^2}{s^2}\exp\left(-\frac{s^2}{2\sigma^2}\right)\int_0^\infty x^2\exp\left(-\frac{\sigma^2}{2s^2}x^2\right)I_0(x)dx \tag{6A.8}$$

The following expression is obtained by partially integrating

$$E[x]=\frac{\sigma^2}{s^2}\exp\left(-\frac{s^2}{2\sigma^2}\right)\left[\left(\frac{1}{2b}+\frac{1}{8b^2}\right)C_1+\frac{1}{8b^2}C_2\right] \tag{6A.9}$$

with $b=\frac{\sigma^2}{2s^2}$ and C_1 and C_2 as [26, p.487]

$$C_1=\int_0^\infty \exp(-bx^2)I_0(x)dx=\sqrt{\frac{\pi}{4b}}\exp\left(\frac{1}{8b}\right)I_0\left(\frac{1}{8b}\right) \tag{6A.10}$$

$$C_2=\int_0^\infty \exp(-bx^2)I_2(x)dx=\sqrt{\frac{\pi}{4b}}\exp\left(\frac{1}{8b}\right)I_1\left(\frac{1}{8b}\right) \tag{6A.11}$$

$E[A]$ is found with the help of (6A.6), (6A.9) - (6A.11), as

$$E[A]=\frac{s^2}{\sqrt{2\sigma^2}}\sqrt{\pi}\left[\left(\frac{\sigma^2}{s^2}+\frac{1}{2}\right)I_0\left(\frac{s^2}{4\sigma^2}\right)+\frac{1}{2}I_1\left(\frac{s^2}{4\sigma^2}\right)\right]\exp\left(-\frac{s^2}{4\sigma^2}\right) \tag{6A.12}$$

This expression is verified by taking some limiting values for s and σ. When $\sigma = 0$, the Rayleigh component is canceled, leaving the dominant component with amplitude s. So $E[A]$ should become s: [26, p. 377]

$$\lim_{\sigma \to 0} E[A] = s \qquad \text{with} \lim_{x\to\infty} I_0(x) = \lim_{x\to\infty} I_1(x) \sim \frac{e^x}{\sqrt{2\pi x}} \tag{6A.13}$$

When $s = 0$, the Rayleigh distribution results with its well-known $E[A]$:

$$\lim_{s \to 0} E[A] = \sqrt{\frac{\pi}{2\sigma}} \text{ with } I_0(0) = 1 \;\wedge\; I_0(0) = 0 \tag{6A.14}$$

With (6A.4) and (6A.12) the relationship between $E[A]$, $E[A^2]$, and K follows as

$$\frac{E[A]}{\sqrt{E[A^2]}} = \sqrt{\frac{\pi}{4(1+K)}} e^{\frac{-K}{2}} \left[(1+K) I_0\left(\frac{K}{2}\right) + K I_1\left(\frac{K}{2}\right) \right] \tag{6A.15}$$

APPENDIX 6B

The conditional co-channel interference probability is defined as

$$F(\mathrm{CI}|n) = \Pr\left[\frac{p_d}{p_n} < \alpha\right] = \int_0^{\infty} \int_0^{\alpha p_n} f(p_d, p_n) dp_d dp_n \tag{6B.1}$$

where p_d is the desired instantaneous signal power, p_n is the interfering instantaneous signal power, α is the protection ratio, and $f(p_d, p_n)$ is the combined pdf of p_d and p_n. The desired and interfering signals are assumed to be independent, which means the outage probability is written as follows:

$$\Pr\left[\frac{p_d}{p_n} < \alpha\right] = \int_0^{\infty} \int_0^{\alpha p_n} f(p_d) dp_d f(p_n) dp_n \tag{6B.2}$$

First, the pdf of the interfering signal, $f(p_n)$, is calculated. This is an n times convolution of the individual exponential distributions (n equals the number of interferers). The exponential distribution is given by

$$f(p_i) = \frac{1}{p_{0i}} e^{-\frac{p_i}{p_{0i}}} \tag{6B.3}$$

where p_i is the received instantaneous signal power of the ith interferer and p_{0i} is the local mean power of the ith interferer.

In the Laplace domain, the convolution becomes a multiplication of the individual Laplace transforms. The following transform pair is obtained:

$$f(p_1) * f(p_2) * \ldots * f(p_n) \xleftarrow{\;L\;}\!\!\rightarrow \prod_{i=1}^{n} \frac{a_i}{(s + a_i)} \tag{6B.4}$$

where $a_i = p_{0i}^{-1}$. The Laplace transform of (6B.4) is written as

$$\prod_{i=1}^{n} \frac{a_i}{(s + a_i)} = \prod_{i=1}^{n} a_i \sum_{j=1}^{n} \frac{\mu_j}{(s + a_j)} \tag{6B.5}$$

where $\mu_j = \prod_{l=1 \wedge l \neq j}^{n} \frac{1}{(-a_j + a_l)}$

The inverse Laplace transformation of (6B.5) results in the following expression:

$$f(p_1) * f(p_2) * \ldots * f(p_n) = \prod_{i=1}^{n} a_1 \sum_{j=1}^{n} \mu_j e^{-a_j p_j}$$

$$= \prod_{i=1}^{n} a_i \left[\frac{\mu}{a_1} a_1 e^{-a_1 p_1} + \frac{\mu_2}{a_2} a_2 e^{-a_2 p_2} + \ldots + \frac{\mu_n}{a_n} a_n e^{-a_n p_n} \right] \tag{6B.6}$$

Equations (6B.3) and (6B.6) show that conditional CIP is calculated by summing conditional CIP caused by each individual interferer multiplied with a weight factor. Hence, conditional CIP is given by

$$F(\mathrm{CI}|n) = \sum_{i=1}^{n} \beta_i F(\mathrm{CI}|i)$$

with

$$\beta_1 = \prod_{j=1 \wedge j \neq i}^{n} \frac{1}{\left(1 - \frac{\mathrm{SI}_i}{\mathrm{SI}_j}\right)} \qquad \mathrm{SI}_i = \frac{P_{0d}}{P_{0i}} \tag{6B.7}$$

Now conditional CIP for the ith interfering signal $F(\mathrm{CI} \mid i)$, is calculated. When $f(p_i)$ substitutes $f(p_n)$ in (6B.2), we see that the inner integral is calculated first, which is given by

$$\int_0^{\alpha p_n} f(p_d)dp_d = \int_0^{\alpha p_n} \frac{K_d+1}{P_{0d}} \exp\left(-\frac{K_d+1}{P_{0d}} p_d - K_d\right) \left(\sqrt{\frac{4(K_d^2+K_d)p_d}{P_{0d}}}\right) dp_d \tag{6B.8}$$

This integral is written in a closed form with the help of Marcum's Q-function, which is given by [21, p. 394]

$$Q(\lambda,\mu) = \int_\mu^\infty x e^{-\frac{\lambda^2+x^2}{2}} I_0(\lambda x)dx \tag{6B.9}$$

Substituting the following in (6B.8),

$$\lambda = \sqrt{2K_d} \qquad \mu = \sqrt{2\alpha \;\; p_n \frac{K_d+1}{P_{0d}}} \tag{6B.10}$$

integral (6B.8) simplifies to

$$\int^{\alpha p_n} f(p_d)dp_d = 1 - Q\left(\sqrt{2K_d}, \sqrt{2\alpha \;\; p_n \frac{K_d+1}{P_{0d}}}\right) \tag{6B.11}$$

The last integral to be solved is the infinite integral over p_i. This integral simplifies to, with the help of [21, p. 395],

$$F(\mathrm{CI}|i) = \frac{1}{1+\frac{\mathrm{SI}_i}{(K_d+1)\alpha}} \exp\left[-\frac{K_d}{\frac{(K_d+1)\alpha}{\mathrm{SI}_i}+1}\right] \tag{6B.12}$$

The final expression for the total conditional interference probability is obtained with (6B.7) and (6B.12) as:

$$F(\mathrm{CI}|n) = \sum_{i=1}^{n} \frac{1}{1+\frac{\mathrm{SI}_i}{(K_d+1)\alpha}} \exp\left[-\frac{K_d}{\frac{(K_d+1)\alpha}{\mathrm{SI}_i}+1}\right] \prod_{j=1 \wedge j\neq i}^{n} \frac{1}{1-\frac{P_{0j}}{P_{0i}}} \tag{6B.13}$$

Chapter 7
Adaptive Equalization

7.1 INTRODUCTION

In Chapter 2 we discussed that in a frequency nonselective channel, the data bit duration is larger than the rms delay spread, whereas the data bit duration is smaller than the rms delay spread in a frequency selective channel. Therefore, the interference between several received symbols at the receiving antenna due to multiple propagation paths ISI, is negligible in a frequency nonselective channel. However, in a frequency selective channel, the data rate is increased because of the small data bit duration and the delay spread affects the system performance by causing ISI in the detection process. Wireless multimedia applications require very high data rates in the order of 155 Mbps, which are limited by multipath propagation. Several techniques are used to contact the multipath fading in order to enhance the system performance (i.e., antenna diversity, coding, spread spectrum, and adaptive equalization).

An overview of adaptive equalization techniques [1–33] is presented in Section 7.2. Section 7.3 discusses a decision feedback equalizer (DFE), and investigates how attractive it is to use the computationally complex fast Kalman algorithm instead of the conventional and computational simple least mean square (lms) algorithm as a training algorithm in DFE. In Section 7.4, simulations results are presented considering a 16-QAM system using square-root-raised cosine matched filtering with a decision feedback equalizer. Measured impulse responses of indoor radio channels, described in Chapter 6, are used to implement the multipath channel in the simulations. Two adaptive training algorithms are investigated: fast Kalman and lms. Conclusions are given in Section 7.5.

7.2 OVERVIEW OF ADAPTIVE EQUALIZATION TECHNIQUES

Adaptive equalization techniques were developed during the last three decades for high-speed digital data communications over telephone lines (modems) and radio channels such as microwave LOS or satellite links [2]. TDMA systems using indoor radio channels can use adaptive equalization to combat ISI resulting from time-variant multipath fading caused by the different propagation paths through the channel.

There is a large range of equalizers available. An overall categorization of adaptive equalization techniques is shown in Figure 7.1 [2]. As seen in Figure 7.1, the adaptive equalization techniques are characterized by:

1. Type (linear or nonlinear);
2. Structure (transversal or lattice);
3. Algorithm (lms, rls, etc.).

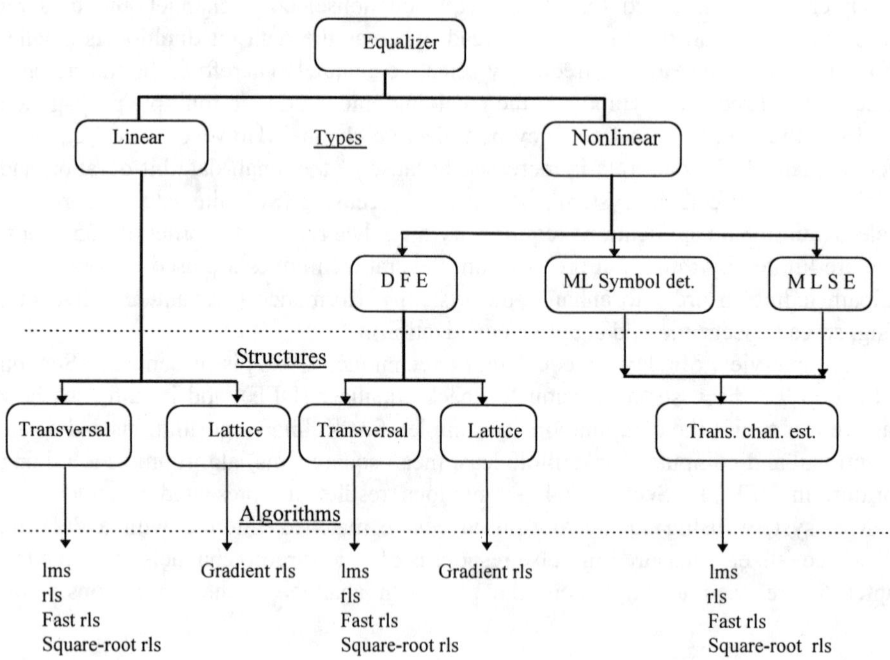

Figure 7.1 Overview of adaptive equalization.

There are two different types of adaptive equalizers: linear and nonlinear. In the linear case the adaptive equalizer consists of only one single linear (transversal or lattice) filter. This is not the case for the class of nonlinear adaptive equalizers. In this class the overall filters are nonlinear. There are two possibilities for nonlinear filter. First, the detected signal is fed back to the adaptive equalizer. This is the case for DFE. Second, knowledge about the characteristics of the channel and the statistical distribution of the noise corrupting the signal is needed. This is the case for maximum-likelihood (ML) symbol detection and maximum-likelihood sequence estimation (MLSE).

The second categorization is based on which type of structure is used for the digital filters within the adaptive equalizer. There are two different types of structures: transversal and lattice.

The transversal filter consists of array delay elements. All the outputs of the delay elements (also called taps) are multiplied by their individual filter coefficient. These filter coefficients determine the transfer function of the overall filter. The output of the filter is formed by the summation of all weighted taps. The structure of a transversal filter is shown in Figure 7.2

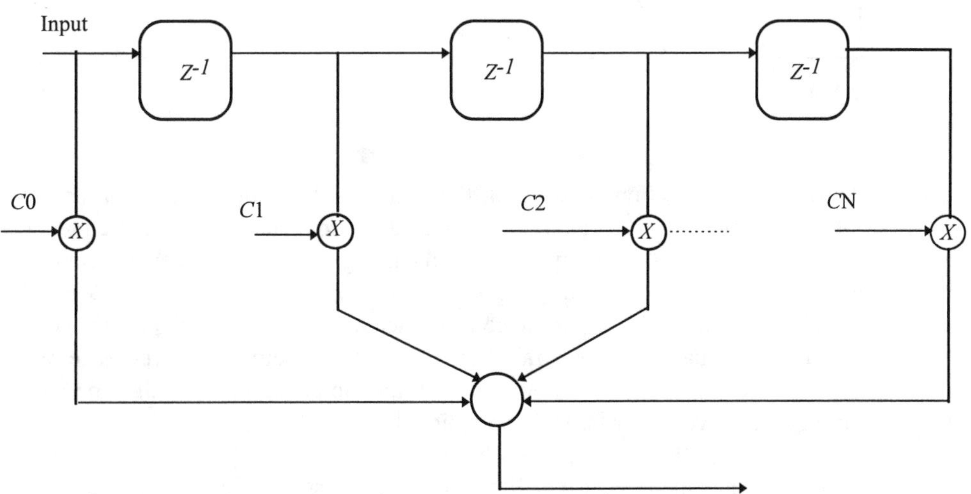

Figure 7.2 Transversal filter consisting of N sections.

Another approach is to create a filter out of several lattice sections. A lattice section consists of two inputs and two outputs. A filter is made by a chain of lattice sections where the outputs of a section are connected to the inputs of the following section. A lattice section is shown in Figure 7.3. In the lattice section the first input is delayed. The first output is obtained by the summation of the delayed input and the nondelayed weighted input. The second input is the summation of the weighted delayed input and the nondelayed input. The weighting factor in each section is equal for both the delayed and nondelayed input. The two outputs each have their own transfer function. One of these outputs is used as output of the overall filter.

It does not matter if an equalizer is constructed out of a transversal filter or a lattice filter. The filter performance will be exactly the same. However, with the use of lattice filter it is computationally simpler to adjust the tap gains of the filter.

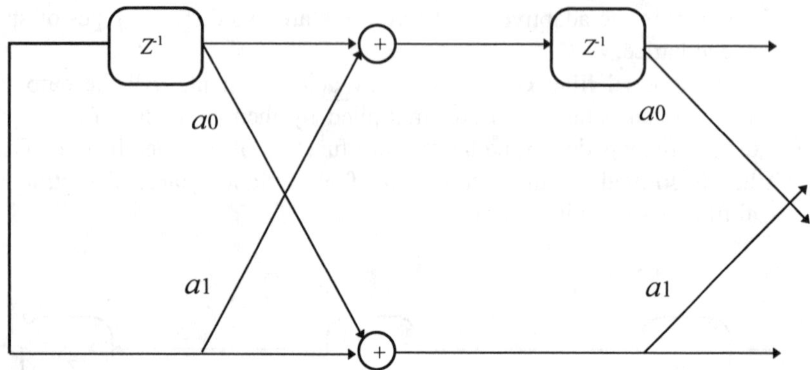

Figure 7.3 Lattice filter.

The third categorization is based on the training algorithm that adjusts the tap gains of the equalizer. The tap gains are adjusted to combat ISI or the frequency selective fading caused by the multipath channel. In Figure 7.1, a number of adaptive algorithms have been listed. The algorithms differ in convergence rate and computational complexity. The larger the convergence rate, the less symbols are needed to train the adaptive equalizer. In general, there is a trade-off between convergence rate and computational complexity for adaptive algorithms. Some algorithms might be very fast-converging, but have a very large computational complexity.

In this chapter, DFE is considered because nonlinear equalization is known to give better performance than linear equalization, and because no prior knowledge about the characteristics of the channel and the statistical distribution of the noise corrupting the signal is necessary in the receiver of the communication system. Since the choice of structure does not influence the performance of the filter and the simulations are not done in real time, the very simple transversal filter structure is used within the equalizer discussed in this chapter. Two adaptive algorithms are investigated:

1. lms;
2. Fast Kalman.

The lms algorithm is a conventionally-computationally simple but slow-converging algorithm. The fast Kalman algorithm is a much faster-converging algorithm, but is computationally more complex. The performance of both these algorithms in the adaptive equalizer is evaluated. For more information concerning adaptive equalization for TDMA mobile radio, the reader is referred to [2].

7.2.1 DFE

DFE consists of two transversal filters, the feed forward filter (FFF) and the feed backward filter (FBF). The input of the FFF is the sequence of received symbols and the input of the FBF is the detected symbol by the receiver.

The idea of using a backward filter is that future ISI caused by an already detected symbol can be subtracted from the incoming signal. This future ISI is estimated in the backward filter. The input of this FBF is the sequence of detected symbols. However, use of a backward filter has a drawback: error propagation. When an incorrect decision is made, it will propagate into the backward filter, affecting the estimation of the next few symbols. As shown later, the error propagation can be severe if the SNR is low. The general structure of DFE is shown in Figure 7.4 [9]. The adaptive algorithm in Figure 7.4 is the lms algorithm.

The c_ns are the adjustable tap coefficients. The summation of all the weightedtaps of both the forward filter and backward filter forms the predetection signal $\hat{I}_k$. This signal $\hat{I}_k$ is fed to a detector, which results in the detected signal $\tilde{I}_k$. The difference of the detected signal $\tilde{I}_k$ and the predetection signals $\hat{I}_k$ results in the error signal ε_k. The predetection signal $\hat{I}_k$ is expressed as

$$\hat{I}_k = \sum_{n=-K_1}^{0} c_n v_{k-n} + \sum_{n=1}^{K_2} c_n \tilde{I}_{k-n} \tag{7.1}$$

where $\{v_k\}$are the input values of the DFE and $\{\tilde{I}_k\}$ are the previously detected signals. The DFE has $K_1 + 1$ forward taps and K_2 backward taps.

When the symbol error rate (SER) is below 10^{-2}, the occasional error made by the detection has a negligible effect on the performance of the equalizer [2]. This is why a known training sequence is necessary in each data packet received by the equalizer, since in the beginning of the training the SER is larger than 10^{-2}.

The packet sequence consists of a known training sequence and a certain unknown data sequence. In the training mode, the error signal is equal to the difference of the detected signal and the known training sequence instead of the predetection signal $\hat{I}_k$, as is the case in the data mode.

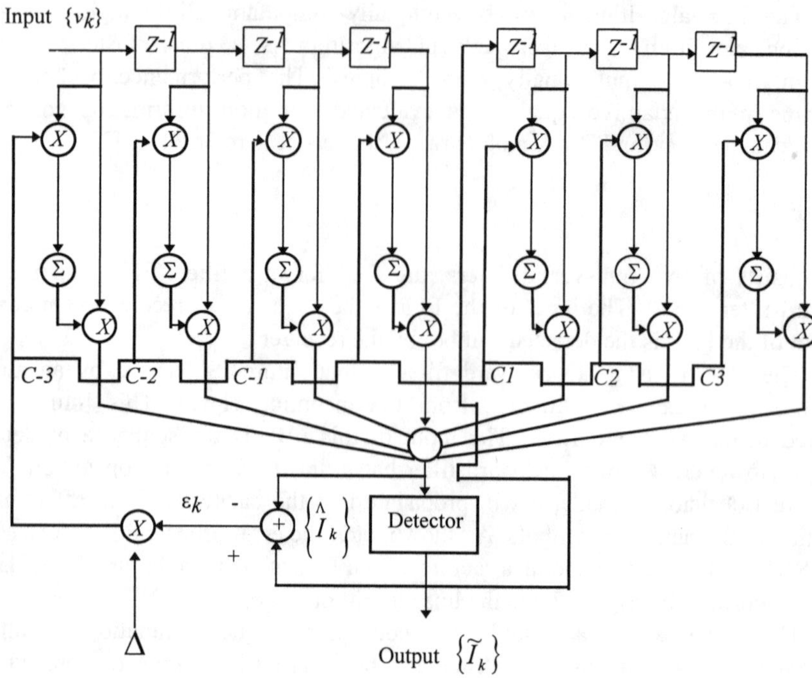

Figure 7.4 Example of a DFE.

The choice of the adaptive algorithm influences the minimum size of the training sequence required for convergence of the mean square error (MSE) in the adaptive equalizer which is needed for reliable data transmission. In this chapter the fast converging fast Kalman algorithm is used to adjust the tap gains of the DFE. Also the performance with the lms algorithm is evaluated. Both these algorithms are discussed in the next section.

7.2.2 Training Algorithms for Adaptive Equalization

For the training of the adaptive equalizer a training algorithm is required. The training algorithm uses the error signal ε_k as input to adjust the tap coefficients. This section describes the lms algorithm and the Fast Kalman algorithm.

Least Mean Square Algorithm

The lms algorithm [2] is based on MSE criterion. In MSE criterion, the tap weight coefficients $\{c_j\}$of the DFE are adjusted to minimize the mean square value of the error

$$\varepsilon_k = I_k - \hat{I}_k \tag{7.2}$$

where I_k is the information symbol in the kth signaling interval and $\hat{I}_k$ is the estimate of that signal at the output of the equalizer. The performance index for MSE criterion, denoted by J, is defined as

$$J = E\left[\left|\varepsilon_k\right|^2\right] = E\left[\left|I_k - \hat{I}_k\right|^2\right] \tag{7.3}$$

The minimization of the MSE gives the optimum Wiener filter solution for the tap coefficient vector:

$$C_{\text{opt}} = \Gamma^{-1}\xi \tag{7.4}$$

where Γ is the autocorrelation matrix of the vector of signal samples and ξ is the vector of cross-correlation between the desired data symbols in the equalizer taps.

An iterative procedure is used to determine C_{opt}. This algorithm, first proposed by Widrow and Hoff [33], is called the lms algorithm, which recursively adjusts the tap weight coefficients of the equalizer. This algorithm is based on the method of steepest decent. An initial choice is made for the tap coefficients vector corresponding to some point on the quadratic MSE surface in the $(K_1 + K_2 + 1)$ dimensional space coefficients. The coefficient update equation is the following:

$$C_{k+1} = C_k + \Delta \varepsilon_k X_k^* \tag{7.5}$$

where C_{k+1} is the vector of the updated equalizer coefficients at the kth iteration, X_k^* represents the complex conjugate of the signal vector for the signal samples stored in the taps of the DFE at the kth iteration, and Δ is the step-size parameter.

The step-size parameter Δ controls the rate of adaption and stability. The algorithm is stable when

$$0 < \Delta < \frac{2}{\lambda_{\max}} \tag{7.6}$$

where $\lambda_{\max}$ is the largest eigenvalue of the signal covariance matrix. The step-size parameter Δ also contributes to fluctuations in the output signal called self-noise. Overall, the choice of the step size parameter Δ is a trade-off between rapid convergence and the minimum MSE.

This concludes the description of the lms algorithm. The operation of the lms algorithm in the communication system is demonstrated in Section 7.3.

Fast Kalman Algorithm

The major advantage of the lms algorithm lies in its computational simplicity. The disadvantage, however, is the slow convergence rate. The lms algorithm has only a single adjustable parameter for controlling the convergence rate, namely the parameter Δ. Consequently, the slow convergence is due to this fundamental limitation.

A faster-converging adaptive algorithm, the fast Kalman algorithm [7,29,30] that involves additional parameters, is discussed. Instead of a statistical approach as is the case for the lms algorithm, a least-square approach is adapted. In the case of the lms algorithm, the expected value of the squared error is minimized. Here, we deal directly with the received data in minimizing the quadratic performance index. This means the performance index is expressed in terms of a time average instead of a statistical average.

In the following discussion a number matrix notations are used. The contents of the equalizer taps are defined as

$$x_N(n) = \begin{bmatrix} y(n-1) \\ y(n-2) \\ \cdot \\ \cdot \\ \cdot \\ y(n-N_1) \\ \cdots\cdots \\ d(n-1) \\ d(n-2) \\ \cdot \\ \cdot \\ \cdot \\ d(n-N_2) \end{bmatrix} \tag{7.7}$$

where $y(n)$ is the equalizer input at time n, $d(n)$ is the previously detected symbol or training symbol at time n (depending on whether the equalizer is in training mode or data mode), $N_1 = K_{1+1}$ is the number of forward taps, $N_2 = K_2$ is the number of backward taps, and $N = N_1 + N_2$ is the total number of taps. At every iteration the values in the x_N vector are shifted. This shifting property is crucial to the fast Kalman algorithm discussed later. For now, two vectors are defined, a vector $\xi_2(n)$ containing new elements that are inserted in the next iteration and a vector $\rho_2(n)$ containing the deleted elements of x_N:

$$\xi_2(n) = \begin{bmatrix} y(n) \\ d(n) \end{bmatrix} \tag{7.8}$$

$$\rho_2(n) = \begin{bmatrix} y(n-N_1) \\ d(n-N_2) \end{bmatrix} \tag{7.9}$$

Also defined is an extended vector $x_M(n)$ that contains the elements of $\xi_2(n)$ appended in the proper order to the elements of $x_N(n)$:

$$x_M(n) = \begin{bmatrix} y(n) \\ \cdots\cdots \\ y(n-1) \\ \cdot \\ \cdot \\ \cdot \\ y(n-N_1) \\ \cdots\cdots \\ d(n) \\ \cdots\cdots \\ d(n-1) \\ \cdot \\ \cdot \\ \cdot \\ d(n-N_2) \end{bmatrix} \tag{7.10}$$

In the algorithm, two permutation matrices are used that rearrange the elements of the extended vector x_M in the following way:

$$S_{MM} x_M(n) = \begin{bmatrix} \xi_2(n) \\ \cdots\cdots \\ x_N(n) \end{bmatrix} \tag{7.11}$$

$$Q_{MM} x_M(n) = \begin{bmatrix} x_N(n+1) \\ \cdots\cdots \\ \rho_2(n) \end{bmatrix} \tag{7.12}$$

The fast Kalman algorithm is based on the rls algorithm [9]. The performance criterion used for deriving the rls algorithm is the minimization of the exponentially weighted squared error at time n:

$$E_N = \sum_{k=1}^{n} \lambda^{n-k} \left[d(k) - C_N(n)^T x_N(k) \right]^2 \tag{7.13}$$

where λ is some positive number close to but less than 1. The minimizing vector is the solution of the Weiner-Hopf equation:

$$C_N(n) = R_{NN}(n)^{-1} \left[\lambda^{n-k} \sum_{k=1}^{n} d(k) x_N(k) \right] \tag{7.14}$$

where

$$R_{NN}(n) = \sum_{k=1}^{n} \lambda^{n-k} x_N(n) x_N(n)^T + \delta \lambda^n I_{NN} \tag{7.15}$$

and $\delta \geq 0$.

The addition of the matrix δI_{NN} is needed to make sure that the matrix $R_{NN}(n)$ is nonsingular. Finally, the time update equation of the vector containing the tap coefficients of the DFE for the rls algorithm is as follows:

$$C_N(n) = C_N(n-1) + k_N(n) e(n) \tag{7.16}$$

where $e(n)$ is the error at time n and $k_N(n)$ is the Kalman gain vector at time n:

$$k_N(n) = R_{NN}(n)^{-1} x_N(n) \tag{7.17}$$

This algorithm is called the conventional Kalman algorithm for equalizer adaption. A disadvantage of this algorithm is its high computational complexity. The conventional Kalman algorithm does not exploit the shifting property of the vector $X_N(n)$ mentioned above. The algorithm used for simulation that does exploit this shifting property is called the fast Kalman algorithm.

First, the initialization is be given. The following scalars, vectors, and matrices are stored and updated during each iteration:

1. N-by-2 matrices $A_{N2}(n)$ and $D_{N_2}(n)$ with initial values $A_{N_2}(0) = D_{N_2}(0) = 0_{N_2}$;
2. 2-by-2 matrix $E_{pp}(n)$ with initial value $E_{pp}(0) = \delta I_{22}$;
3. M-dimensional vector $k_M(n)$;
4. 2-dimensional vectors $\varepsilon_p(n)$, $\varepsilon_p(n)$, and $\mu_p(n)$;

5. N-dimensional vector $m_N(n)$;
6. N-dimensional vector $k_N(n)$ with initial value $k_N(1) = 0_N$.

Also, all $x(n) = 0$ for $n \leq 0$.
Then, starting with $n = 1$, $k_N(n+1)$ is updated with the following algorithm:

$$\varepsilon_2(n) = \xi_2(n) + A_{N2}(n-1)^H x_N(n) \tag{7.18}$$

$$A_{N2}(n) = A_{N2}(n-1) - k_N(n)\varepsilon_p(n)^H \tag{7.19}$$

$$\varepsilon_2(n)' = \xi_2(n) + A_{N2}(n)^H x_N(n) \tag{7.20}$$

$$E_{22}(n) = \lambda E_{22}(n-1) + \varepsilon_2(n)'\varepsilon_2(n)^H \tag{7.21}$$

$$k_M(n) = S_{MM}{}^T \begin{bmatrix} E_{22}(n)^{-1}\varepsilon_2(n)' \\ \cdots\cdots\cdots\cdots \\ k_N(n) + A_{N2}(n)E_{22}(n)^{-1}\varepsilon_2(n)' \end{bmatrix} \tag{7.22}$$

$$Q_{MM}k_M(n) = \begin{pmatrix} m_N(n) \\ \cdots\cdots \\ \mu_2(n) \end{pmatrix} \tag{7.23}$$

$$\eta_2(n) = \rho_2(n) + D_{N2}(n-1)^H x_N(n+1) \tag{7.24}$$

$$D_{N2}(n) = \left[D_{N2}(n-1) - m_N(n)\eta_2(n)^H\right]\left[I_{22} - \mu_2(n)\eta_2(n)^H\right]^{-1} \tag{7.25}$$

$$k_N(n+1) = m_N(n) - D_{N2}(n)\mu_2(n) \tag{7.26}$$

$$e(n+1) = d(n+1) - C_N(n)^T x_N(n+1) \tag{7.27}$$

$$C_N(n+1) = C_N(n) + k_N(n+1)e(n+1) \tag{7.28}$$

This concludes the description of the fast Kalman algorithm. The convergence rate of the fast Kalman algorithm is much larger than with the lms algorithm. The convergence rate of the conventional Kalman algorithm compared with the fast Kalman algorithm is the same [9]. The only difference is the computational complexity, which is higher in the case of the conventional Kalman algorithm. However, the fast Kalman algorithm is more sensitive to round-off noise. This means that more bits of number representation are needed to keep the algorithm stable than is the case for the conventional Kalman algorithm. Examples of the convergence rate of the fast Kalman algorithm compared with the convergence rate of the lms algorithm are shown in Section 7.3.

7.3 COMMUNICATION SYSTEM MODEL

The emphasis in this chapter is on the indoor channel model and the adaptive algorithm in the DFE. In this chapter, the complex measured indoor radio channels described in Chapter 6 are used for defining the channel model.

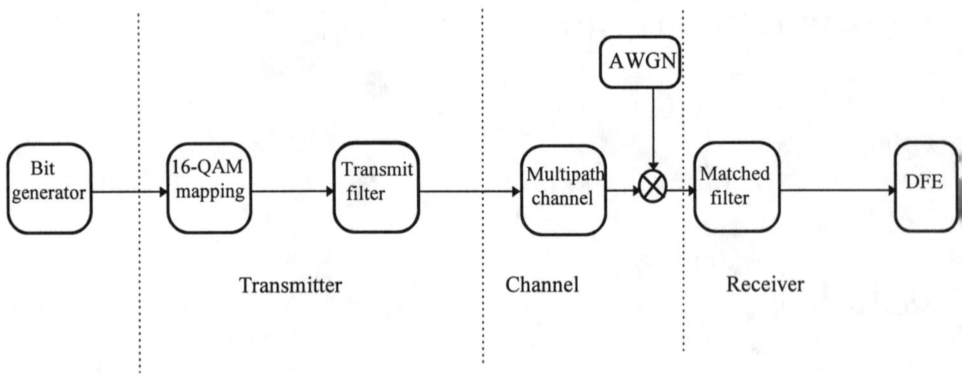

Figure 7.5 Block diagram of baseband communication system.

The equivalent baseband system that has been simulated is shown in Figure 7.5. The functional units (transmitter, channel, and receiver) are discussed in detail in the following sections. Section 7.3.1 describes the transmitter, Section 7.3.2 reviews the channel model, Section 7.3.3 describes the receivers used in the simulation.

7.3.1 Transmitter

The input of the transmitter is a sequence of random bits generated by a random generator. This sequence of bits is fed to the modulation mapping device.

The modulation type that is considered is 16-QAM. This device maps 4 bits at a time on the 16-QAM constellation. BEP can be minimized by the proper assignment of bits to the different symbols in the two-dimensional constellation [31]. For example, the assignment can be made such that its neighboring symbols differ in as few bits as possible. When for example Gray encoding is used, the neighboring symbols only differ in 1 bit. In Figure 7.6, an example of a 16-QAM constellation with Gray coding is given. The choice of a specific bit assignment does influence BEP, but not SER. Since in this book SER is evaluated, it does not matter in this simulation which 4 bits are assigned to which specific symbol.

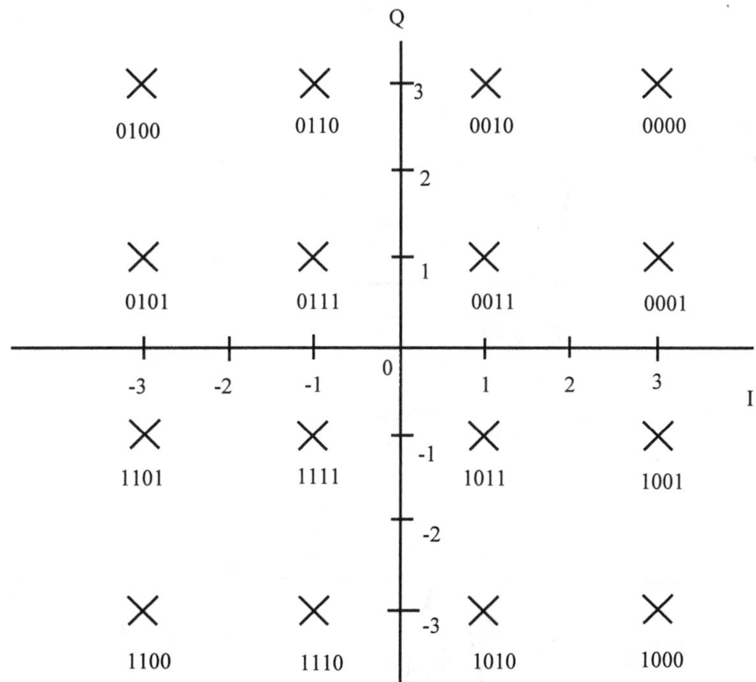

Figure 7.6 16-QAM signal constellation with Gray coding.

The output signal $u(t)$ of the 16-QAM mapping device is described as follows:

$$u\left(\frac{t}{T}\right) = \sum_n x_n \delta\left(\frac{t}{T} - n\right) \tag{7.29}$$

where x_n is the complex valued 16-QAM and $\delta(t/T)$ is a Dirac pulse.

It is possible in theory to design band-limited pulses that give no ISI on an ideal channel of bandwidth W, provided that the symbol rate does not exceed 2W [9]. This is called the Nyquist criterion for signal design with no ISI. There is a large variety of pulse shapes that have good spectral characteristics and are free of ISI on an ideal channel. A pulse shape that has found wide use in digital transmission on band-limited channels is one that has the raised cosine spectral characteristics. Figure 7.7 illustrates the raised cosine spectrum *P(f)* for the roll-off factor β=0.50. The raised cosine pulse shaped *P(t)* is shown in Figure 7.8 with β as a parameter. Here, *T* is the symbol duration, *t* is the time, and *f* is the frequency.

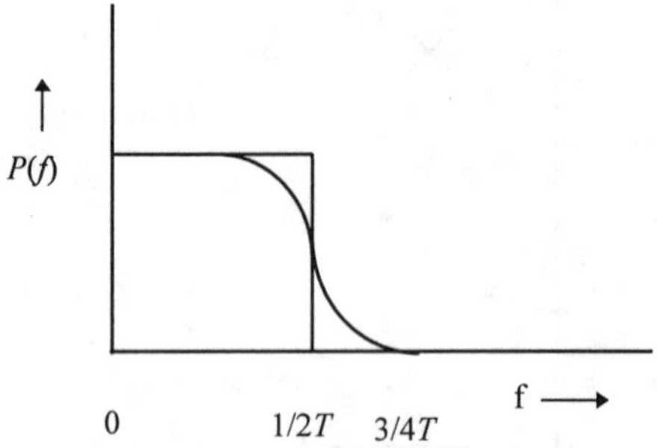

Figure 7.7 Raised cosine spectrum *P(f)* with β = 0.50.

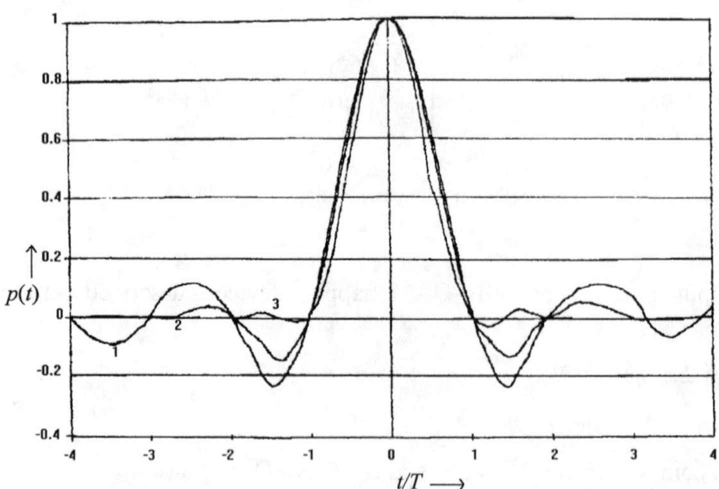

Figure 7.8 Raised cosine pulse shape for β as a parameter: (1) β=0.0, (2) β=0.5, and (3) β=1.0.

The spectrum of the filter at the transmitter and receiver is taken equal to the square root of the raised cosine spectrum to get the pulse shape after the matched filter in the receiver equal to the raised cosine pulse shape. The square-root-raised cosine pulse shape can be described by the following function *g*(*t*/*T*) [32]:

$$g\left(\frac{t}{T}\right) = C\frac{\cos\left((1+\beta)\frac{\pi t}{T}\right) + \sin\left((1-\beta)\frac{\pi t}{T}\right)\left(4\beta\frac{t}{T}\right)^{-1}}{1-\left(4\beta\frac{t}{T}\right)^{2}} \tag{7.30}$$

where C is a constant to normalize the filter, β is the roll-off factor, *T* is the symbol duration time, and *t* is the time. The shape of the pulse with roll-off factor 0.50 is shown in Figure 7.9.

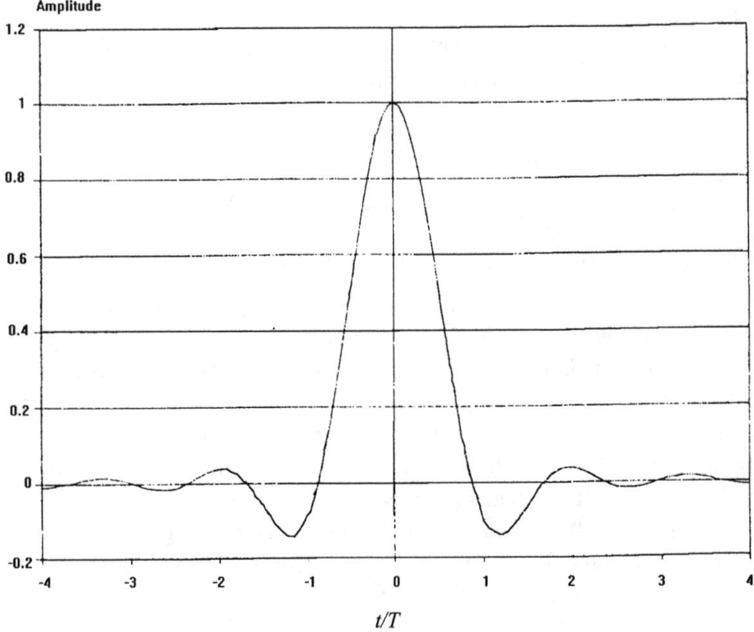

Figure 7.9 Square-root-raised cosine pulse shape with roll-off factor β = 0.50.

The signal after the first matched filter is represented by the following function $v(t/T)$:

$$v\left(\frac{t}{T}\right) = \sum_n x_n g\left(\frac{t}{T} - n\right) \tag{7.31}$$

where x_n is the complex valued 16-QAM signal and $g(t/T)$ is the square-root-raised cosine pulse shape.

7.3.2 Multipath Channel Model

The channel model used is based on the model proposed by Saleh and Valenzuela [34]. It is presumed that there are a number of propagation paths between the transmitter and the receiver, each with a different delay time, attenuation, and phase.

It is also presumed that the channel is time -invariant during the transmission of a packet. In practice, the channel characteristics are changing in time. However, these changes are slow and when the packets are not too long, the channel is considered static. For example, the transmission time of a packet consisting of 10,000 symbols at a data rate of 5 Msps is equal to 2 ms.

The time-invariant channel is characterized by the following baseband equivalent impulse response function [34]:

$$h\left(\frac{t}{T}\right) = \sum_k \alpha_k \delta\left(\frac{t-\tau_k}{T}\right) \exp(-j\Theta_k) \tag{7.32}$$

where α_k is the gain, Θ_k is the pulse, and τ_k is the delay time of the kth path.

The problem now is how to determine the α`$_n$s, $\Theta$`$_n$s and τ`$_n$s. It is assumed that there is a path whenever the derivative of the power delay profile changes from a positive sign to a negative sign. It is also assumed that the receiver is perfectly synchronized with the strongest path in the power delay profile. So, first the strongest path is determined, and from that instant of time the following paths are determined. Very small paths below a certain level were not taken into account. An example of a power delay profile with its paths is shown in Figure 7.10.

A parameter often used in the design of wireless communication systems to describe the radio channel is the rms time delay spread. This parameter is a measure of multipath severity and is defined by Saleh and Valenzuela [34] as

$$\tau_{\text{rms}} = \sqrt{\overline{\tau^2} - \left(\overline{\tau}\right)^2} \tag{7.33}$$

$$\overline{\tau^n} = \frac{\sum_{i=1}^{L} \tau_i^n |\alpha_i|^2}{\sum_{i=1}^{L} |\alpha_i|^2} \tag{7.34}$$

where L is the number of paths. The rms time delay spreads of the four measured channels are given in Table 7.1

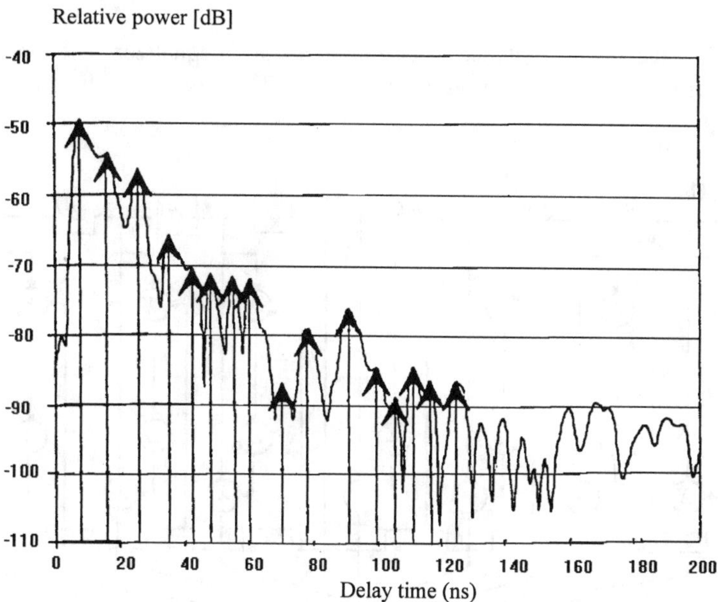

Figure 7.10 Example of power delay profile used in simulation, where paths are indicated by arrows.

Table 7.1
Rms Time Delay Spreads of Measured Impulse Responses Used in Simulation

Measured Indoor Radio Channel	Rms Time Delay Spread τ_{rms} (in ns)
LOS in office room	6.3
OBS in office room	14.6
LOS in conference room	13.0
OBS in conference room	20.2

After the channel, AWGN is added to the signal. The signal after noise addition is described with the following equation:

$$y\left(\frac{t}{T}\right) = \sum_k \sum_n x_k \alpha_n g\left(\frac{t-\tau_n}{T} - k\right) \exp(-j\Theta_n) + n\left(\frac{t}{T}\right) \tag{7.35}$$

where $n(t/T)$ is the additive noise signal.

7.3.3 Receiver

The receiver consists of a matched filter and an adaptive equalizer. The matched filter at the receiver is the same filter as at the transmitter. The type of equalizer used is DFE.

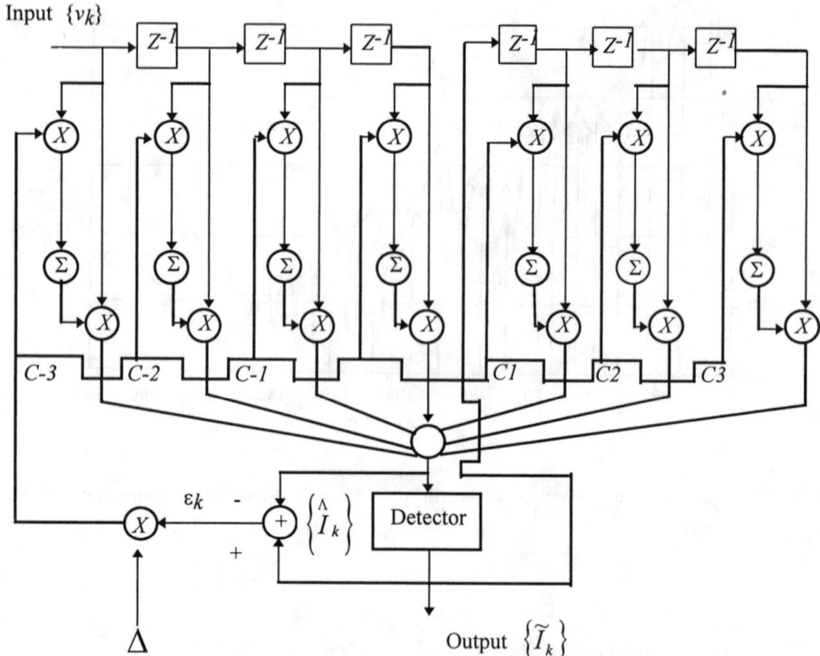

Figure 7.11 DFE with the lms algorithm.

The input signal of the equalizer is given as the following function:

$$z\left(\frac{t}{T}\right) = \sum_k \sum_n x_k \alpha_n p\left(\frac{t-\tau_n}{T} - k\right) \exp(-j\Theta_n) + n'\left(\frac{t}{T}\right) \tag{7.36}$$

where n' (t/T) is the additive noise after the matched filter at the receiver, and $p(t/T)$ is the raised-cosine pulse shape.

The signal is then sampled at the symbol rate at the input of the equalizer. The equalizer that is used is shown in Figure 7.11 [9].

The predetection signal $\hat{I}_k$ is described as follows:

$$\hat{I}_k = \sum_{j=-K_1}^{0} c_j v_{k-j} + \sum_{j=1}^{K_2} c_j \tilde{I}_{k-j} \tag{7.37}$$

where $\hat{I}_k$ is an estimate of the kth information symbol, $\{c_j\}$are the tap coefficients of the equalizer, $\{v_k\}$are the samples of signal $z(t/T)$, and $\tilde{I}_k$ are the previously detected symbols or the known training symbols during training. The equalizer has $K_1 + 1$ feed forward taps and K_2 feed backward taps.

After equalization, the detected 16-QAM signal is compared with the transmitted signal, and the number of symbol errors is divided by the total number of data symbols (total number of transmitted symbols minus the number of training symbols).

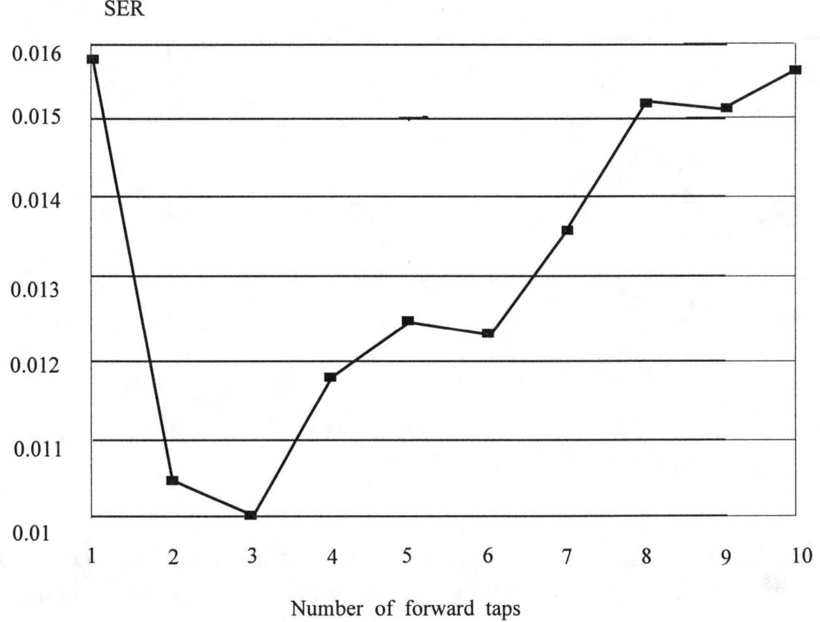

Figure 7.12 SER performance for a number of different forward taps with the number of backward taps fixed to three.

For the design of the DFE, the problem remains as to how many forward and backward taps to take. In Figures 7.12 and 7.13 the effect is shown for what happens if the number of backward and forward taps are changed. SER at first decreases with an increasing number of forward and backward taps, but after a certain minimum it increases. This is explained by the principle of implicit diversity. In the beginning, the equalizer takes advantage of the fact that it has more samples to choose from (time diversity), but after a certain point the equalizer cannot cope any more with the extra time samples, and the extra taps will only add noise to the detected signal. The minimum SER was searched for by changing the number of taps. The optimum situation that was found was three forward and three backward taps. Therefore, three forward and backward taps are used in the simulation.

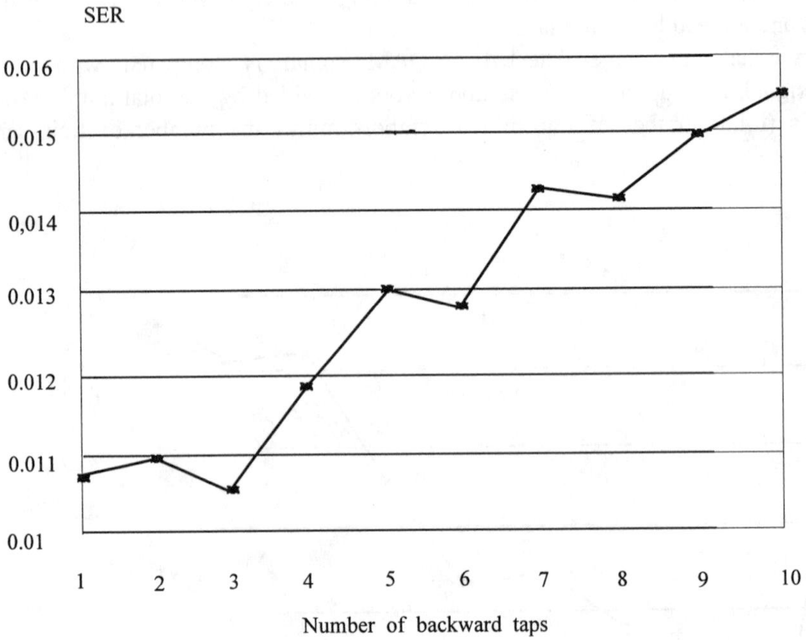

Figure 7.13 SER performance for a number of different backward taps with the number of forward taps fixed to three.

To train the DFE, a known training sequence is used in each packet. During training mode an error signal is formed by the difference between the known training sequence and the predetection signal $\hat{I}_k$. During data mode the error signal is equal to the detected signal $\tilde{I}_k$ and the predetection signal $\hat{I}_k$. This error signal is used by the lms and fast Kalman algorithms for adjusting both the forward tap gains and the backward tap gains of the DFE, whether the DFE is in training mode or data mode. The adaptive algorithms considered are lms and fast Kalman.

In the case of the lms algorithm, a certain step size Δ is chosen as described in Section 7.2. In this section, notice that the step size influences the convergence rate and the stability of the adaptive algorithm in the equalizer. The higher the step size, the faster the convergence, but the equalizer may become unstable. In Figure 7.14, MSE is plotted for different step sizes. To get this curve, 200 simulations of the initial training were performed for each step size and averaged. The number of forward and backward taps is taken as three. The SNR is 20 dB per symbol and the channel used for this simulation is the impulse response of the LOS office room measurement. It is seen in the plot that when the step size is small it will get a low MSE. This is because the self-noise is low. It is also clear that for higher step sizes the equalizer will converge faster, although when it is too high, it will get unstable.

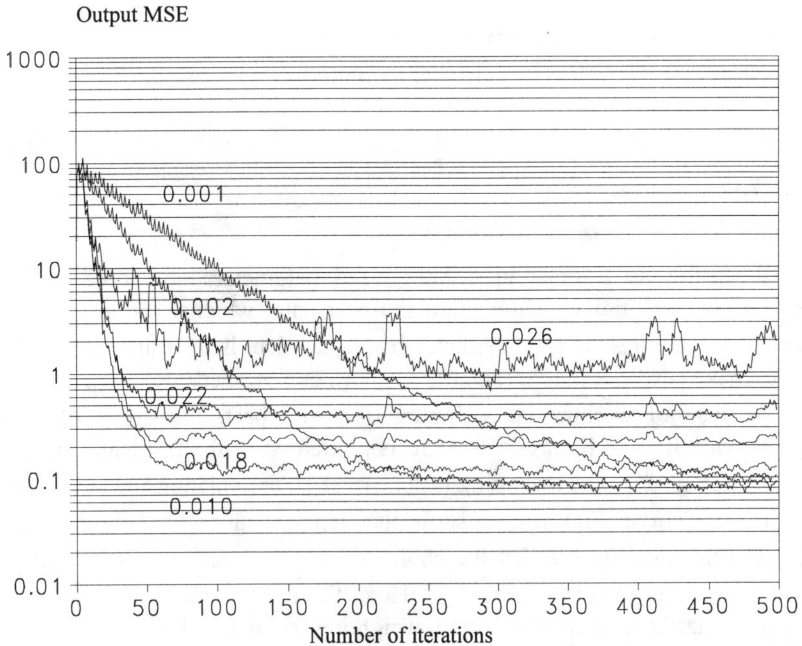

Figure 7.14 MSE with lms for different step sizes (δ) for after certain iterations.

In this simulation, a step size of 0.002 is used for both training and data modes. Figure 7.14 shows that after 200 iterations the MSE is almost at its minimum. There is a trade-off between minimizing the MSE (use more iterations and a smaller step size) and a smaller overhead (the part of the packet that is actual data). The packet format used in the evaluation using the LMS algorithm is shown in Figure 7.15.

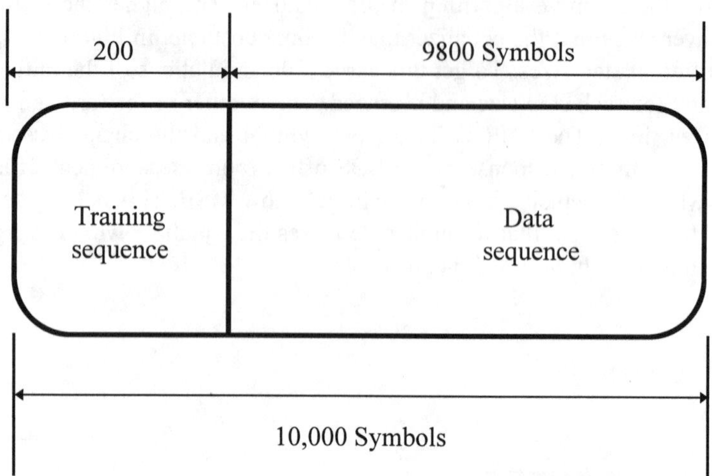

Figure 7.15 Packet format for lms evaluation.

When the fast Kalman algorithm is used, parameters have to be set: λ, a factor introducing exponential weighting into the past data, which is appropriate when the channel characteristics are time-variant, and δI_{22} , which is the initial value of the $R_n(t)$ matrix to make sure that the $R_n(t)$ matrix is not ill conditioned.

Because of exponential weighting into the past, the effect of adding δI_{22} dissipates with time. In Figure 7.16, MSE is plotted when the fast Kalman algorithm is used for different δI_{22}. The averaging is done over 100 simulations of the initial training for each MSE curve. Figure 7.16 show that it takes more time for a larger δI_{22} to dissipate in time. Also note that the choice of the δI_{22} matrix does not influence the minimum MSE, as seen in Figure 7.16. Here, the δI_{22} matrix is taken as equal to the 2×2 identity matrix, and the exponential weighting factor λ is 0.99.

In Figure 7.17, MSE is plotted for different λ values. For a smaller λ value, the convergence speed of the algorithm is slightly smaller and the minimum MSE is larger. This is explained by the fact that for smaller λ, values into the past have less influence on the training of tap coefficients. In other words, there is less averaging in time for smaller λ and thus the signal is more noisy, which results in a higher minimum MSE.

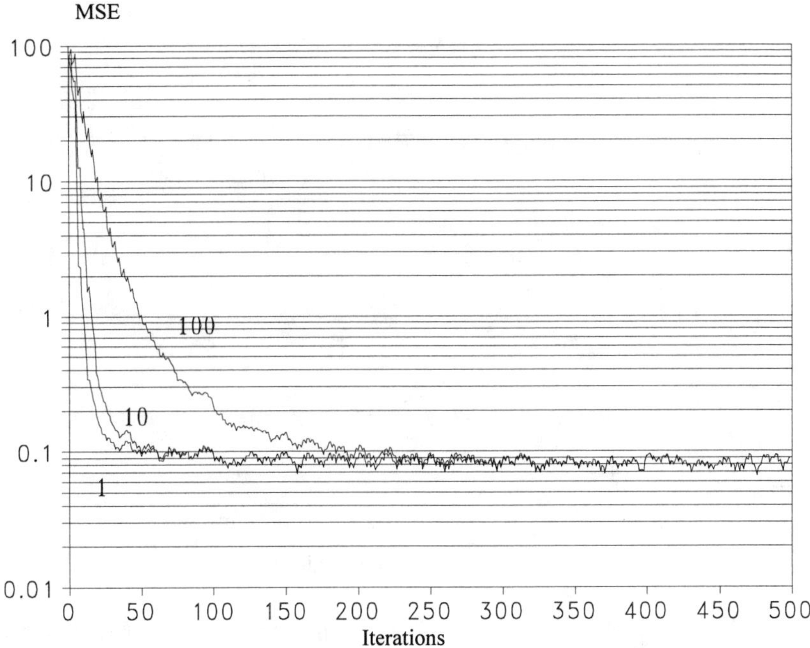

Figure 7.16 MSE with fast Kalman for different δI_{22} after a certain number of iterations.

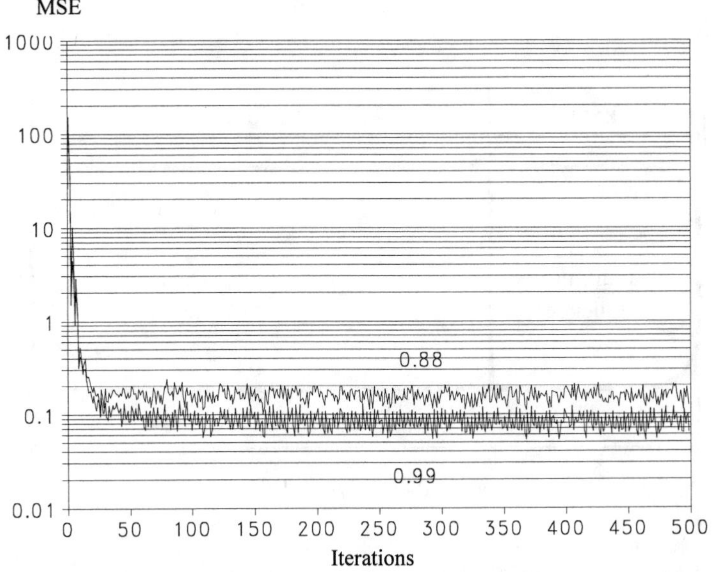

Figure 7.17 MSE with fast Kalman for different λ after a certain number of iterations.

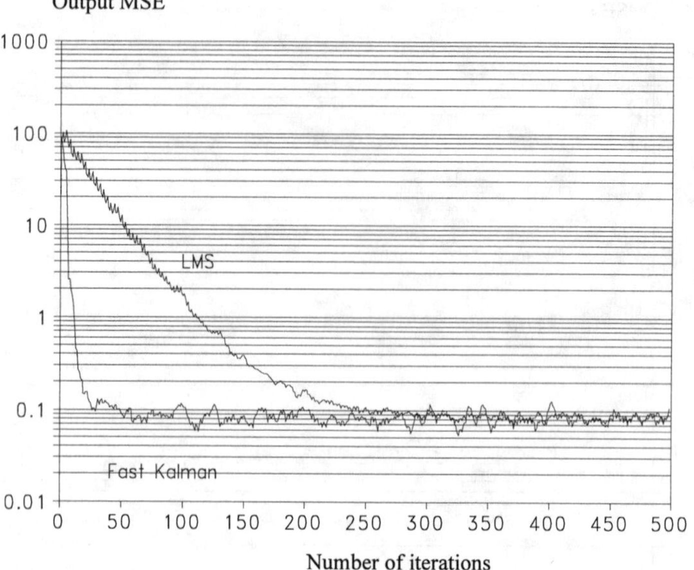

Figure 7.18 MSE with fast Kalman (λ=0.99 and δI_{22} =I) and lms (δ=0.002) algorithms after a certain number of iterations.

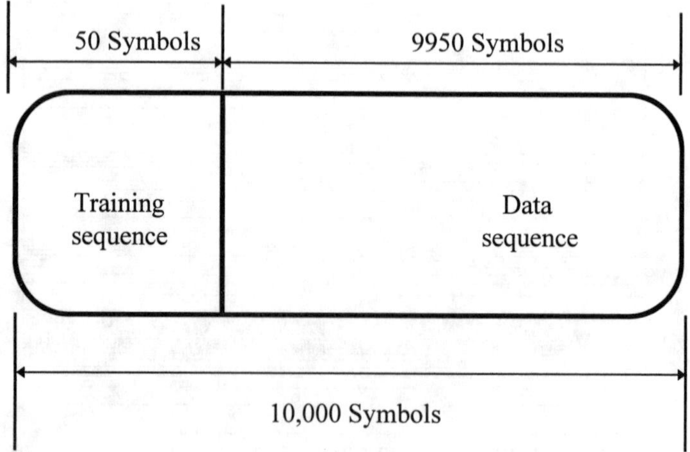

Figure 7.19 Packet format for fast Kalman evaluation.

In Figure 7.18, both the MSE of the lms algorithm and the fast Kalman algorithm are plotted during training. Clearly, the fast Kalman algorithm is much faster-converging than the lms algorithm. The number of training symbols in the evaluation using the fast Kalman algorithm is 50 symbols. The packet format used in the evaluation with the fast Kalman algorithm is shown in Figure 7.19.

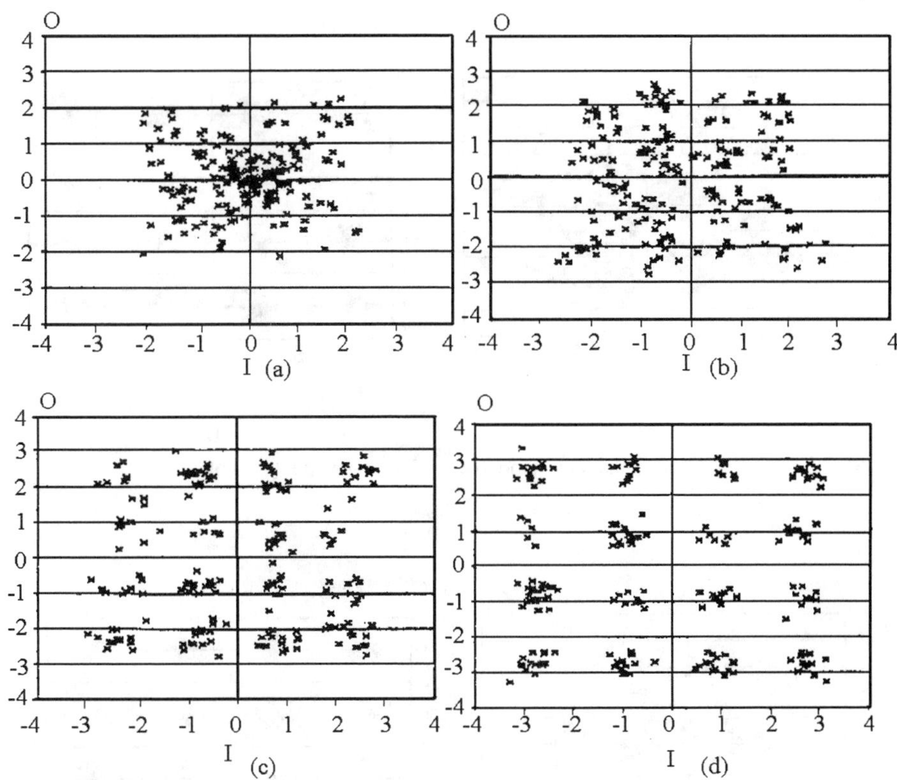

Figure 7.20 Snapshots of signal constellation during training (lms with δ=0.0005) after: (1) 200 iterations, (2) 400 iterations, (3) 600 iterations, and (4) 800 iterations.

An instructive way to investigate the training of the DFE is to look at the signal constellation during training. In Figure 7.20, the development of the signal constellation of the predetection signal during training is shown. Every snapshot shows the 200 symbols up to that time instant. It is seen that in the beginning the symbols start in the origin of the plot. This is explained by the fact that the initial tap gain coefficients are all set to 0, and from that value slowly change to their convergence value. In Figure 7.20, a very slow convergence rate was achieved by taking a small $\delta = 0.0005$ in the lms algorithm. An SNR of 20 dB and a symbol rate of 5 Msps were taken. The channel applied was the LOS office room impulse response.

Another way to look at the operation of training is to examine the signal constellation after training of the input signal of the equalizer and the predetection signal after equalization. In Figures 7.21(a), 7.21(b), 7.22(a), and 7.22(b), the signal constellations of the signals mentioned above for the LOS and OBS office room cases are given. The SNR is 20 dB and a symbol rate of 5 Msps is chosen.

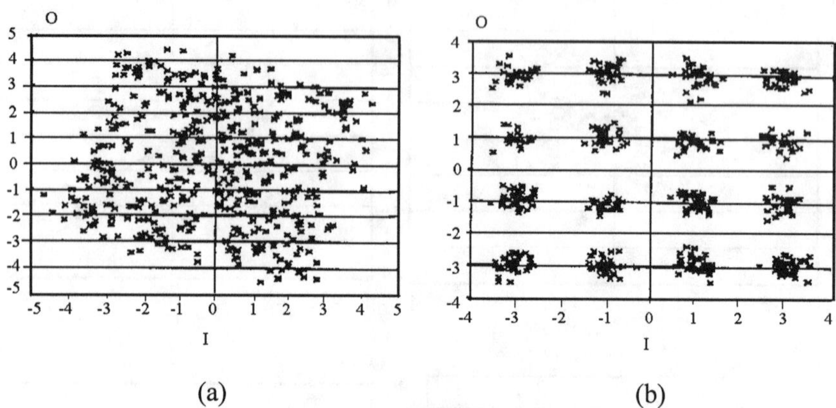

Figure 7.21 Signal constellations before (LOS office), and after (LOS office).

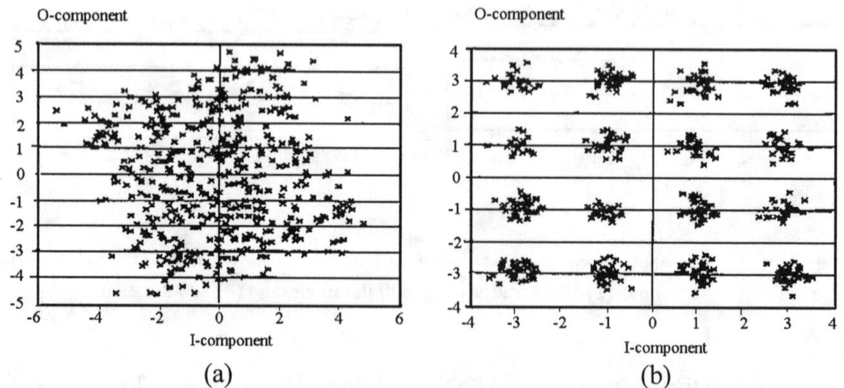

Figure 7.22 Signal constellations before (OBS office), and after (OBS office).

This concludes the description of the overall communication system model and the operation of DFE in this model. In the next section, the simulation procedure and the numerical results of the simulation are given.

7.4 SIMULATION RESULTS

In this section the simulations performed are explained and evaluated.

First, the procedure of simulation is discussed. Then, the results are presented with no equalization, with DFE for the lms algorithm and the fast Kalman algorithm with fixed data rate, and with DFE for both the adaptive algorithms with varying data rate.

7.4.1 Description of Simulation

The simulated communication system was discussed in the previous section. To get the SER performance, a large number of random generated packets were sent through the simulated system. SER is determined simply by comparing the received sequence of symbols with the transmitted sequence of symbols and dividing the number of symbol errors by the total number of transmitted data symbols. SER results of individual packets are averaged to get the final SER result. This is necessary because one packet only consisted of 10,000 symbols. To get an SER in the order of 10^{-5}, more than 100,000 symbols should be sent through the system.

The following SER performance simulations were performed:

1. AWGN channel without equalization;
2. Office room and conference room impulse responses;
3. LOS and OBS impulse responses;
4. Varying SNR;
5. Different data rates;
6. lms and fast Kalman algorithms.

7.4.2 Numerical Results

In this section, the numerical SER results of the simulations are presented. All results are compared with the theoretical SER performance without equalization and an ideal (no multipath) AWGN channel.

The theoretical SER performance for 16-QAM is expressed as follows [9]:

$$P_M = 2\left(1-\frac{1}{\sqrt{M}}\right)erfc\left(\sqrt{\frac{3}{2(M-1)}k\gamma_b}\right)\left[1-\frac{1}{2}\left(1-\frac{1}{\sqrt{M}}\right)erfc\left(\sqrt{\frac{3}{2(M-1)}k\gamma_b}\right)\right] \tag{7.38}$$

where P_m is the probability of a symbol error, $M = 16$ in the case of 16-QAM, k is the number of bits per symbol, and γ_b is the average SNR per bit. The complementary error function is *erfc* ( ), defined as

$$erfc(x) = \frac{2}{\sqrt{\pi}} \int_x^{\infty} e^{-t^2} dt \tag{7.39}$$

Without Equalization Technique

In Figure 7.23, SER performance is given for the no-equalization case with an AWGN channel. Figure 7.23 shows that the simulated results are very close to the theoretical results. It does not matter for which data rate the simulation is done, because there is no ISI with the AWGN channel.

In Figure 7.24, SER performance is shown for the four measured channel impulse responses with no equalization. The SNR in Figure 7.24 is defined as the SNR per symbol. The data rate is 5 Msps. If the no-equalization case is considered, SER performance is not sufficient for data communications. For reliable data communications the SER should be in the order of 10^{-5}.

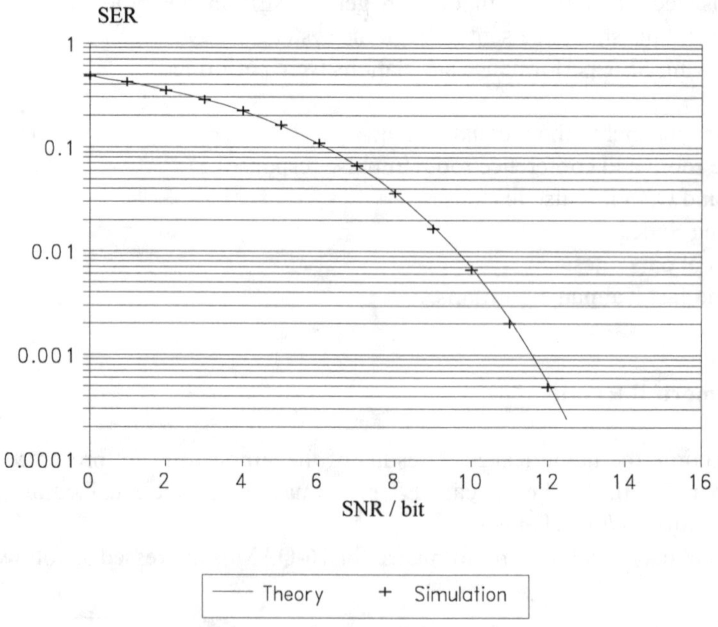

Figure 7.23 Both theoretical and simulated SER performance without equalization and an ideal AWGN channel.

LOS performance is also better than OBS performance for both the office room and the conference room. This is explained by the fact that the LOS impulse response is much faster-decaying than the OBS impulse response, and thus the interfering paths are more dominant in the OBS case.

The performance in the office room is better than that in the conference room because the maximum delay time in the conference room is much larger, and thus more interfering paths will occur.

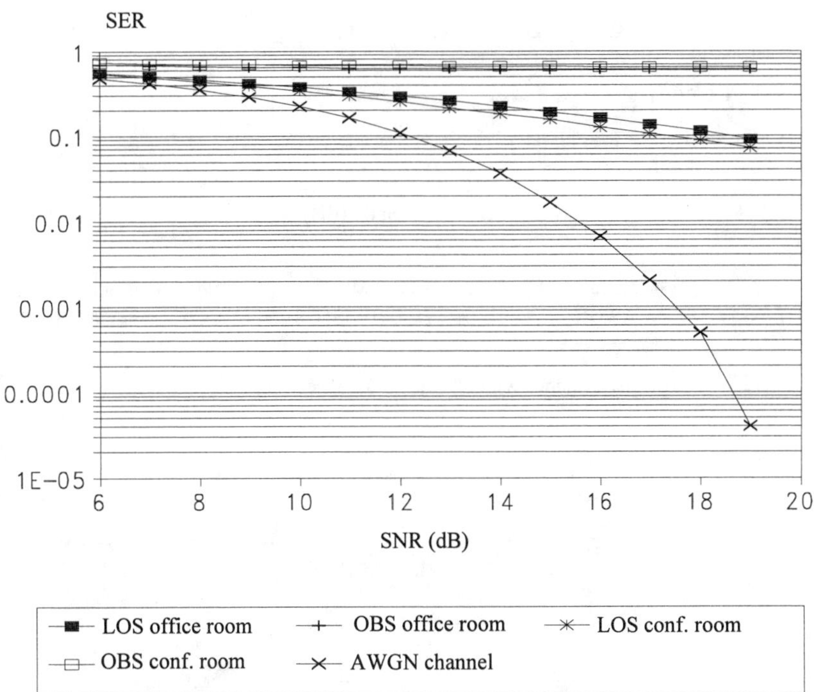

Figure 7.24 SER performance without equalization for the four measured channels.

If Table 7.1 is investigated together with the observations mentioned above, it is seen that the larger the rms time delay spread, the worse the SER performance.

Decision Feedback Equalization

In Figures 7.25, 7.26, 7.27 and 7.28, SER performance for the four measured channel impulse responses with DFE is shown. Both the results using the lms algorithm and the fast Kalman algorithm are plotted in these figures.

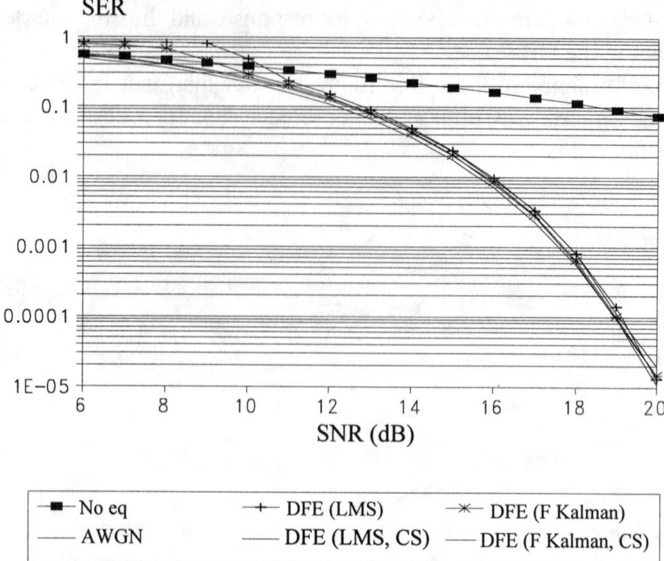

Figure 7.25 SER performance with DFE for LOS office room.

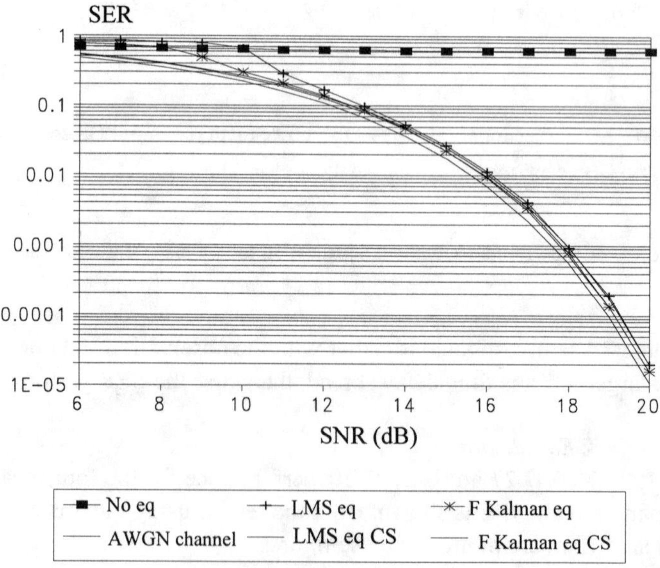

Figure 7.26 SER performance with DFE for OBS office room.

To investigate the error propagation in the backward filter of the DFE, two cases were separately investigated. In the first case, the detected symbols were fed back to the feed backward filter of the DFE, which happens in normal operation. In the second case, the correct symbols transmitted were fed back to the feed backward filter to eliminate error propagation caused by the backward filter. In reality this is of course not possible, but it gives an indication of the severity of error propagation caused by the backward filter. Both these cases are plotted in the figures.

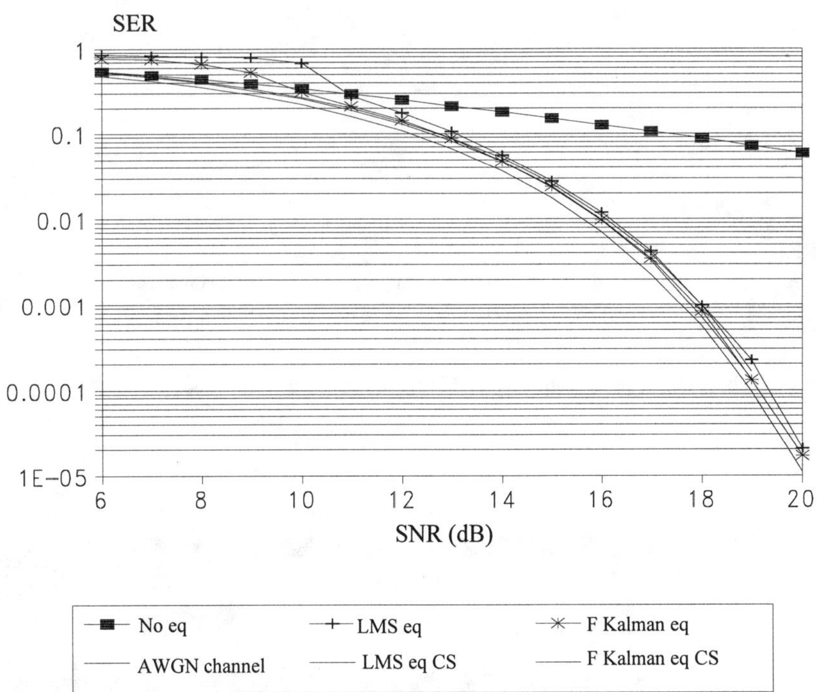

Figure 7.27 SER performance with DFE for LOS conference room.

From these figures we see that the use of DFE can improve SER performance dramatically. For example, at 20 dB SNR, performance is about 4 orders of magnitude better than in the no-equalization case.

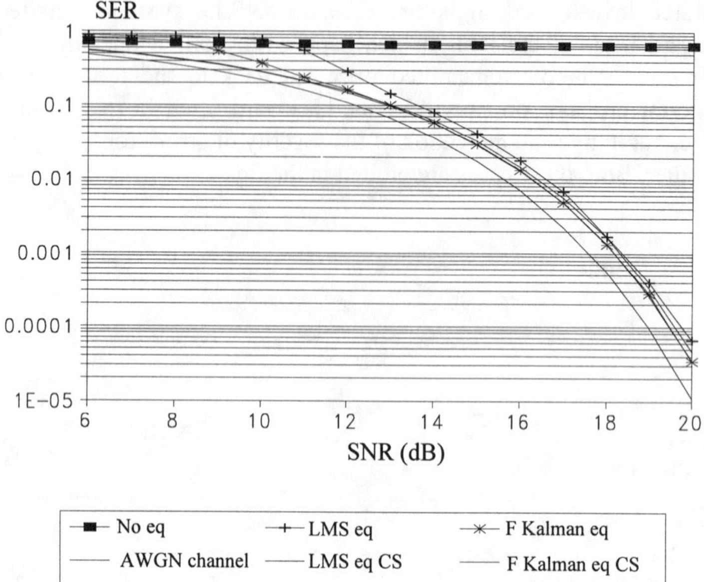

Figure 7.28 SER performance with DFE for the OBS conference room.

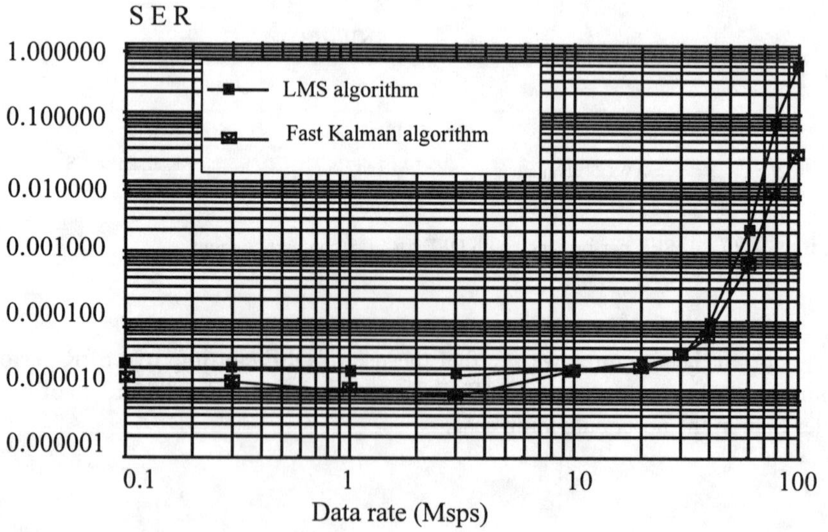

Figure 7.29 SER performance DFE for different data rates for the LOS office room.

At very low SNRs however, SER performance with the use of DFE will give worse results than in the no-equalization case. This is because of the error propagation in the backward filter of the DFE [9]. The error probability is too high and too many detected errors are fed back to the backward filter. When there is no error propagation, results are always better than in the no-equalization. With increasing SNR, the curves with feedback of the correct symbols and with feedback of the detected symbols slowly join each other. At high SNR, the influence of the errors made and fed back to the backward filter are neglected.

The best results were obtained in the LOS case and office room case. In these cases the interfering paths are the least dominant, and in the office room the propagation delays are smaller than in the conference room.

In Figures 7.29, 7.30, 7.31, and 7.32, the results are displayed for the four different measured channel impulse responses when the data rate is changed. The SNR is 20 dB. The simulation is done for both the lms and fast Kalman algorithms.

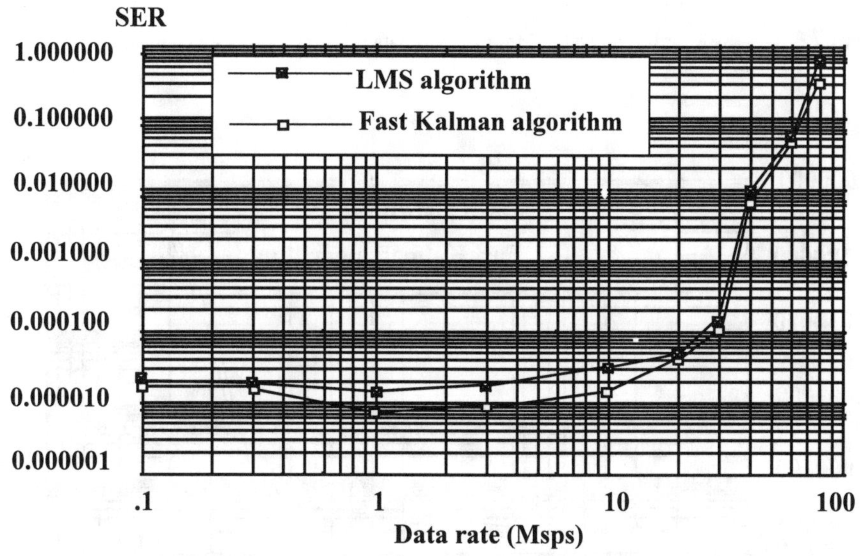

Figure 7.30 SER performance DFE for different data rates for the OBS office room.

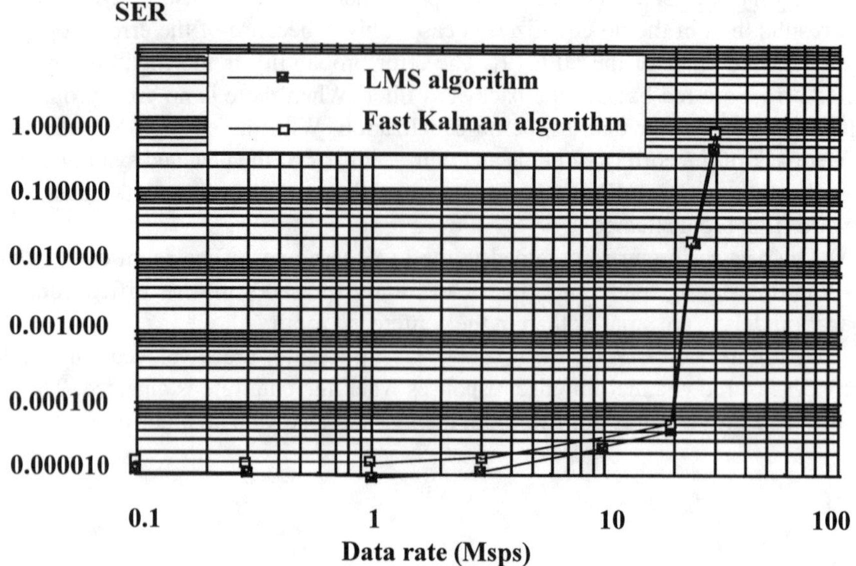

Figure 7.31 SER performance DFE for different data rates for the LOS conference room.

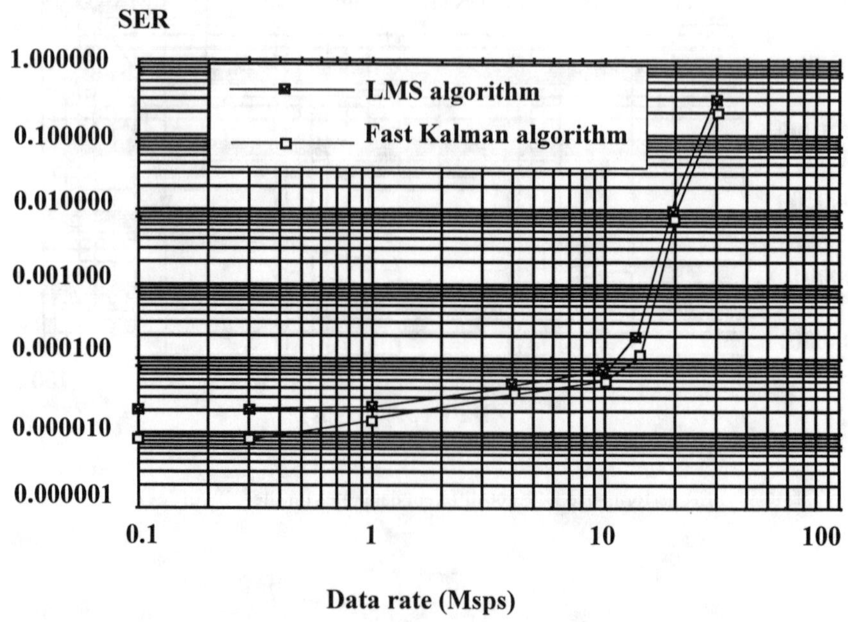

Figure 7.32 SER performance DFE for different data rates for the OBS conference room.

For all channels evaluated, the fast Kalman algorithm gives better performance than the lms algorithm, although in the lms case (200) more training symbols are needed to train the DFE than in the fast Kalman algorithm case (50). In theory this is expected, since the fast Kalman algorithm is superior as discussed in Section 7.2.

In Figures 7.33 and 7.34, the results are regrouped so that the different indoor radio channels can be compared. The results are shown for the lms algorithm and the fast Kalman algorithm.

As the data rate increases, after a certain threshold SER increases dramatically. This is because the equalizer takes advantage of the increasing ISI and uses that as a source of implicit diversity [23]. The office room with LOS allows the highest data rate and the conference room with OBS is the worst of the four channels. Again, this is very clear after inspection of the measured impulse responses and the rms time delay spreads of the channels.

For all the four channels, SER was in the order of 10^{-5} for data rates up to 10 Msps.

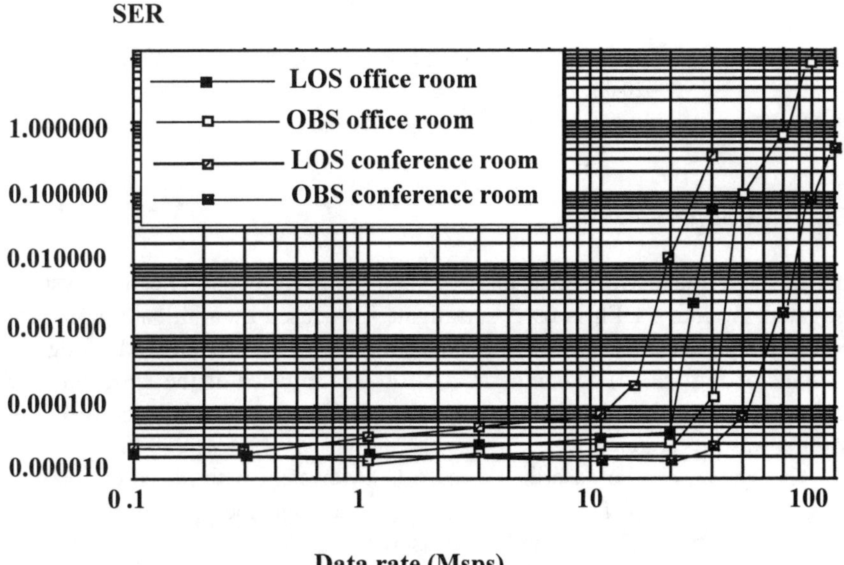

Figure 7.33 SER performance DFE with lms algorithm for different data rates, with SNR = 20 dB.

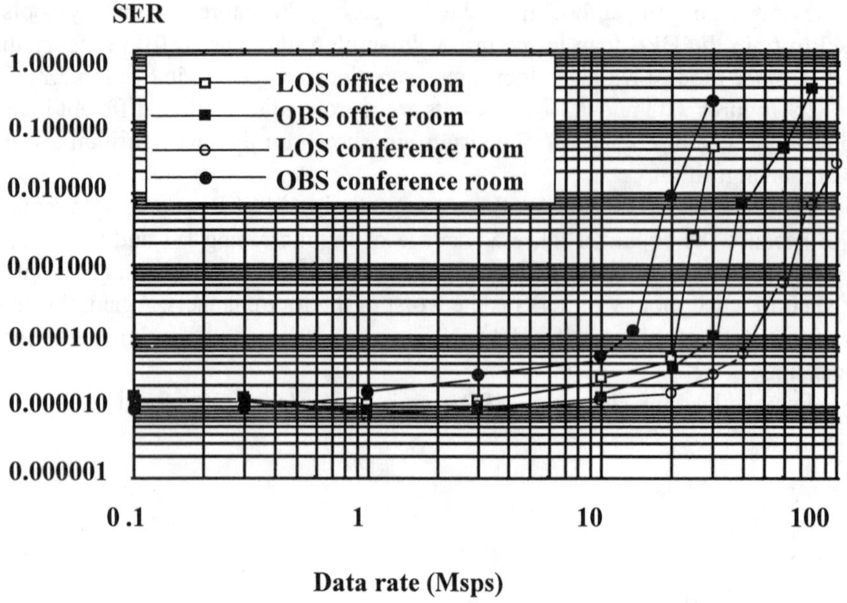

Figure 7.34 SER performance DFE with fast Kalman algorithm for different data rates, with SNR = 20 dB.

7.5 CONCLUSIONS

In this chapter, the performance of an indoor wireless communication system was investigated. Adaptive DFE is used to mitigate ISI caused by multipath propagation in the indoor radio channel. The fast Kalman algorithm is used for adjusting the tap coefficients of the DFE. For comparison, SER performance of the DFE with the lms algorithm as an adaptive algorithm was evaluated. Also, the convergence speed of both algorithms was investigated.

The system can be described as a 16-QAM system with raised-cosine matched filtering with a roll-off factor of 0.50. A realistic channel model is used in the simulations by interpreting complex measured impulse responses of indoor radio channels. The measurements used were done in different types of rooms and LOS and OBS situations at a frequency of 2.4 GHz. After the signals through the multipath channel, AWGN is added.

SER performance of radio connections in indoor wireless computer communication are improved substantially by using DFE with both the lms algorithm or the fast Kalman algorithm as a training algorithm. When the SNR was 20 dB and the data rate was 5 Msps (20 Mbps), SER performance improved with 4 orders of magnitude, compared with the situation where no adaptive equalization techniques were used. However, at very low SNRs (lower than 8 dB), SER performance is worse than without ISI compensation techniques. This is because of error propagation in the backward filter of the DFE. DFE only shows improvement at large SNRs (larger than 8 dB).

If the complex fast Kalman algorithm is used instead of the conventional and simple lms algorithm, SER performance improved by a factor of 2. The convergence speed, however, is much larger than with the lms algorithm. This means that the equalizer can be trained much faster, and thus less training symbols are required in the transmitted data packet (50 instead of 200 symbols). It is especially interesting to use the fast Kalman algorithm when it is necessary to transmit data as efficiently (less overhead) as possible. If the speed of training is not very important, it might be more interesting to use the lms algorithm, since this algorithm is mathematically much simpler and thus easier and cheaper to implement. There is a trade-off between efficiency and complexity of the receiver.

When SER performance was investigated for different data rates, the following was concluded. SER performance of the four investigated channels (LOS and OBS, office room, and conference room) all have a minimum at a certain data rate (around 1 Msps). At data rates larger than 20 Msps, SER increases dramatically and reliable data communication is not possible any more. The best SER performance was achieved in the office room with an LOS connection. We conclude that the lower the rms time delay spread, the better is SER performance. Overall, we conclude that SER is in the order of 10^{-5} for data rates up to 10 Msps (40 Mb/s).

REFERENCES

[1] A.P. Clark, *Equalizers for Digital Modems*, London: Pentech Press, 1985.

[2] J.G. Proakis, "Adaptive equalization for TDMA digital mobile radio," *IEEE Trans. on Vehicular Technology*, Vol. 40, No. 2, pp. 333–341, May 1991.

[3] A. Gersho, "Adaptive equalization of highly dispersive channels for data transmission ," *B.S.T.J.*, Vol. 48, No. 2, pp. 55–70, January 1969.

[4] J.G. Proakis and J.H. Miller, "An adaptive receiver for digital signalling through channels with inter symbol interference," *IEEE Trans. Inf. Theory*, Vol. 15, No. 4, pp. 484–497, July 1969.

[5] R.W. Lucky, J. Salz, and E.J. Weldon, *Principles of Data Communications*, New York: McGraw-Hill, 1968.

[6] D. Godard, "Channel equalization using a Kalman filter for fast data transmission," *IBM J. Res. Dev.*, Vol. 18, No. 3, pp. 267–273, May 1974.

[7] D.D. Falconer and L. Ljung, "Application of fast Kalman estimation to adaptive equalization," *IEEE Trans. on Commun.*, Vol. 26, No. 10, pp. 1439–1446, October 1978.

[8] E.H. Satorius and S.T. Alexander, "Channel equalization using adaptive lattice algorithms," *IEEE Trans. on Commun.*, Vol. 27, No. 6, pp. 899–905, June 1979.

[9] J.G. Proakis, *Digital Communications*, New York, McGraw-Hill, 1989.

[10] G.D. Forney, Jr. , "Maximum likelihood sequence estimation of digital sequences in the presence of inter symbol interference," *IEEE Trans. Inf. Theory*, Vol. 18, No. 3, pp. 363–378, May 1972.

[11] G.D. Forney, Jr., "The Viterbi algorithm," *Proc. IEEE*, Vol. 61, No.3, pp. 268–278, March 1973.

[12] S.U.H. Qureshi, "Adaptive equalization," *Proc. IEEE*, Vol. 73, No.9, pp. 1349–1387, September 1985.

[13] A.P. Clark, *Adaptive Detectors for Digital Modems*, London: Pentech Press, 1989.

[14] R.W. Lucky, "Techniques for adaptive equalization of digital communication systems," *Bell Syst. Tech. J.*, Vol. 45, No. 2, pp. 255–286, February 1966.

[15] D.D. Falconer, "Applications of passband decision feedback equalization in two-dimensional data communications systems," *IEEE Trans. Comm.*, Vol. 24, No. 10, pp. 1159–1166, October 1976.

[16] D.D. Falconer and F.R. Magee, "Evaluation of decision feedback equalization and Viterbi algorithm detection for voiceband data transmission," *IEEE Trans. Comm.*, Vol. 24, No. 10, pp. 1130–1138, October 1976, and Vol. 24, No. 11, pp. 1238–1245, November 1976.

[17] W.C.L. Wong and L.J. Greenstein, "Multipath fading modelling and adaptive equalizer in microwave digital radio," *IEEE Trans. Comm.*, Vol. 32, No. 8, pp. 928–934, August 1984.

[18] D.M. Brady, "An adaptive coherent diversity receiver for data transmission through dispersive media," *IEEE Int. Conf. on Commun. Proceedings*, pp. 21-35 to 21-39, June 1970.

[19] P. Monsen, "Adaptive equalization of the slow fading channel," *IEEE Trans. Comm.*, Vol. 22, No. 8, pp. 1064–1075, August 1974.

[20] N.W.K. Lo, D.D. Falconer, and A.U.H. Sheikh, "Adaptive equalization and diversity combining for mobile radio using interpolated channel estimates," *IEEE Trans. on Vehicular Technology*, Vol. 40, No. 3, pp. 636–645, August 1991.

[21] R.A. Valenzuela, "Performance of adaptive equalization for indoor radio communications," *IEEE Trans. Comm.*, Vol. 37, No.3, pp. 291–293, March 1989.

[22] T.A. Sexton and K. Pahlavan, "Channel modelling and adaptive equalization of indoor radio channels," *IEEE J. Selected Areas in Commun.*, Vol. 7, No. 1, pp. 114–120, January 1989.

[23] K. Pahlavan, S.J. Howard, and T.A. Sexton, "Decision feedback equalization of the indoor radio channel," *IEEE Trans. Comm.*, Vol. 41, No. 1, pp. 164–170, January 1993.

[24] A. Zigic, "Adaptive equalization for high-speed indoor wireless data communications," *Wireless Personal Communications—An International Journal*, Vol. 1, No. 3, pp. 229–251, 1995.

[25] A. Zigic, "Equalization of linear time dispersive intra and inter-building radio channel," Ph. D. Thesis, Delft University of Technology, The Netherlands, 1995.

[26] A. Zigic and R. Prasad, "Bit error rate of decision feedback equalizer for indoor wireless communication," *Electron. Lett.*, Vol. 28, No. 21, pp. 1949–1950, October 8, 1992.

[27] A. Zigic and R. Prasad, "Computer simulation of indoor data channels with a linear adaptive equalizer," *Electron. Lett.*, Vol. 26, No. 19, pp. 1596–1597, September 13, 1990.

[28] M.J. Krapels, R. Prasad, J. Wu, and A.H. Aghvami, "Performance evaluation of decision feedback equalization using measured indoor radio channels," *Proc. PIMRC'94*, Vol. 2, pp. 650–655, The Hague, The Netherlands, September 1994.

[29] G. D'Aria, R. Piermarini, and V. Zingarelli, "Fast adaptive equalizers for narrow-band TDMA mobile radio," *IEEE Trans. on Vehicular Technology*, Vol. 40, No. 2, May 1991.

[30] J.G. Proakis and F. Ling, "Recursive least squares algorithms for adaptive equalization of time-variant multipath channels," *Proc. PIMRC'94*, Vol. 2, pp. 650–655, The Hague, The Netherlands, September 1994.

[31] S.W. Cheung and A.H. Aghvami, "Performance of a 16-ARY DEQAM modem employing a baseband or RF predistorter over a regenerative satellite link," *IEE Proc.*, Vol. 135, Pt. F, No. 6, December 1988.

[32] I. Korn, *Digital Communications*, New York: van Nostrand Reinhold, 1985.

[33] B. Widrow and M.E. Holf, Jr., "Adaptive switching circuits," *IRE WESCON Comc. Rec.*, pt. 4, pp. 96–104, 1960.

[34] A.A.M. Saleh and R.A. Valenzuela, "A statistical model for indoor multipath propagation," *IEEE J. Comm.*, Vol. Sac-5, No. 2, pp. 128–137, February 1987.

Chapter 8
Basic CDMA Concepts

8.1 INTRODUCTION

CDMA protocols constitute a class of protocols in which the multiple access property is primarily achieved by means of coding. Each user is assigned a unique code sequence used to encode the information-bearing signal. The receiver, knowing the code sequences of the user, decodes a received signal after reception and recovers the original data. Since the bandwidth of the code signal is chosen to be much larger than the bandwidth of the information-bearing signal, the encoding process enlarges (spreads) the spectrum of the signal and is therefore also known as spread-spectrum modulation. The resulting encoded signal is also called a spread-spectrum signal, and the CDMA protocols are often denoted as spread-spectrum multiple access (SSMA) protocols.

It is the spectral spreading of the coded signal that gives the CDMA protocols their multiple access capability. It is therefore important to know the techniques to generate spread-spectrum signals and the properties of these signals.

The precise origin of spread-spectrum communications may be difficult to pin-point because modern spread-spectrum communication is the outcome of developments in many directions, such as high-resolution radars, direction finding, guidance, correlation detection, matched filtering, interference rejection, jamming avoidance, information theory, and secured communications [1–10].

The spread-spectrum modulation techniques were originally developed for use in military radar and communication systems because of their resistance against jamming signals and a low probability of detection. Only in recent years with new and cheap technologies emerging and a decreasing military market have the manufacturers of spread-spectrum equipment and researchers become interested in the civil applications.

To qualify as a spread-spectrum modulation technique, two criteria must be fulfilled [9]:

1. The transmission bandwidth must be much larger than the information bandwidth;
2. The resulting radio-frequency bandwidth is determined by a function other than the information being sent (so the bandwidth is independent of the information signal). This excludes modulation techniques like FM and pulse modulation (PM).

Therefore, spread-spectrum modulation transforms an information-bearing signal into a transmission signal with a much larger bandwidth. This transformation is achieved by encoding the information signal with a code signal that is independent of the data and has a much larger spectral width than the data signal. This spreads the original signal power over a much broader bandwidth, resulting in a low(er) power density. The ratio of transmitted bandwidth to information bandwidth is called the processing gain *PG* of the spread-spectrum system,

$$PG = \frac{B_t}{B_i} \tag{8.1}$$

where B_t is the transmission bandwidth and B_i is the bandwidth of the information-bearing signal.

The receiver correlates the received signal with a synchronously generated replica of the code signal to recover the original information-bearing signal. This implies that the receiver must know the code signal used to modulate the data.

Because of the coding and the resulting enlarged bandwidth, spread-spectrum signals have a number of properties that differ from the properties of narrowband signals. The most interesting from a communication systems point of view are discussed below. Each property is briefly explained with the help of illustrations, if necessary, by applying direct-sequence spread-spectrum techniques.

1. *Multiple access capability*. If multiple users transmit a spread-spectrum signal at the same time, the receiver can still distinguish between the users, provided each user has a unique code that has a sufficiently low cross-correlation with the other codes. Correlating the received signal with a code signal from a certain user will then only despread the signal of this user, while the other spread-spectrum signals will remain spread over a large bandwidth. Thus, within the information bandwidth, the power of the desired user is much larger than the interfering power, provided there are not too many interferers, and the desired signal can be extracted. The multiple access capability is illustrated in Figure 8.1. In Figure 8.1(a), two users generate a spread-spectrum signal from their narrowband data signals. In Figure 8.1(b) both users

transmit their spread-spectrum signals at the same time. At the receiver only the signal of user 1 is "despread" and the data recovered.

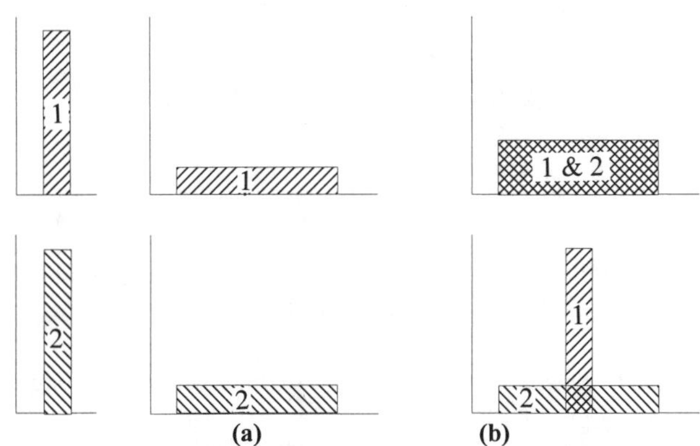

Figure 8.1 Principle of spread-spectrum multiple access.

2. *Protection against multipath interference.* In a radio channel there is not just one path between a transmitter and receiver. Due to reflections (and refractions), a signal is received from a number of different paths. The signals of the different paths are all copies of the transmitted signal but with different amplitudes and phases. Adding these signals at the receiver is constructive at some of the frequencies and destructive at others. In the time domain, this results in a dispersed signal. Spread-spectrum modulation can combat this multipath interference; however, the way in which this is done depends very much on the type of modulation used. In the next section where CDMA protocols based on different modulation methods are discussed, we show for each protocol how multipath interference rejection is obtained.
3. *Privacy.* The transmitted signal can only be despread and the data recovered if the code is known to the receiver.
4. *Interference rejection.* Cross-correlating the code signal with a narrowband signal spreads the power of the narrowband signal, thereby reducing the interfering power in the information bandwidth. This is illustrated in Figure 8.2. The spread-spectrum signal (s) receives a narrowband interference (i). At the receiver the spread-spectrum signal is despread while the interference signal spreads, making it appear as background noise compared with the despread signal.
5. *Anti-jamming capability, especially narrowband jamming.* This is more or less the same as interference rejection except the interference is now willfully inflicted on the system. It is this property together with the next one that makes spread-spectrum modulation attractive for military applications.
6. *Low probability of interception (LPI) or covert operation.* Because of its low power density, the spread-spectrum signal is difficult to detect.

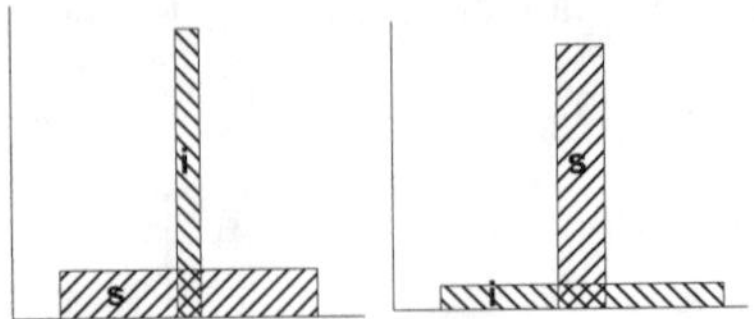

Figure 8.2 Interference rejection.

There are a number of modulation techniques that generate spread-spectrum signals. We briefly discuss the most important ones:

Direct-sequence (DS) spread-spectrum. The information-bearing signal is multiplied directly by a fast code signal.
Frequency hopping (FH) spread-spectrum. The carrier frequency at which the information-bearing signal is transmitted is rapidly changed according to the code signal.
Time-hopping (TH) spread-spectrum. The information-bearing signal is not transmitted continuously. Instead, the signal is transmitted in short bursts where the times of the bursts are decided by the code signal.
Chirp modulation. This kind of spread-spectrum modulation is almost exclusively used in military radars. The radar continuously transmits a low-power signal whose frequency is (linearly) varied (swept) over a wide range.
Hybrid modulation. Two or more of the above-mentioned spread-spectrum modulation techniques can be used together to combine the advantages and, it is hoped, to combat their disadvantages.

In Section 8.2, the above-mentioned modulation techniques are used to obtain the multiple access capability that we want for CDMA (SSMA) protocols. Section 8.3 presents the code sequences and the properties of these sequences in detail.

8.2 SPREAD-SPECTRUM MULTIPLE ACCESS

We can classify the SSMA or CDMA protocols in two different ways: by concept or by modulation method. The first classification gives us two protocol groups, averaging systems and avoidance systems. The averaging systems reduce the interference by averaging the interference over a wide time interval. The avoidance systems reduce the interference by avoiding it for a large part of the time.

Classifying by modulation gives us five protocols, direct-sequence (or pseudo-noise), frequency hopping, time-hopping, protocols based on chirp modulation, and hybrid methods. Of these, the first (DS) is an averaging spread-spectrum protocol, while the hybrid protocols can be averaging protocols depending on whether DS is used as part of the hybrid method. All the other protocols are of the avoidance type. Table 8.1 summarizes both ways of classification.

Table 8.1
Classifying SSMA Protocols

	DS	TH	FH	Chirp	Hybrid
Averaging	x				x
Avoidance		x	x	x	x

In the following sections, CDMA protocols are discussed where a division has been made that is based on the modulation technique.

8.2.1 DS

In the DS-CDMA protocols, the modulated information-bearing signal (the data signal) is directly modulated by a digital code signal. The data signal can be either an analog or digital signal. In most cases it will be a digital signal. What we often see in the case of a digital signal is that the data modulation is omitted and the data signal is directly multiplied by the code signal and the resulting signal modulates the wideband carrier. It is from this direct multiplication that the DS-CDMA protocol gets its name.

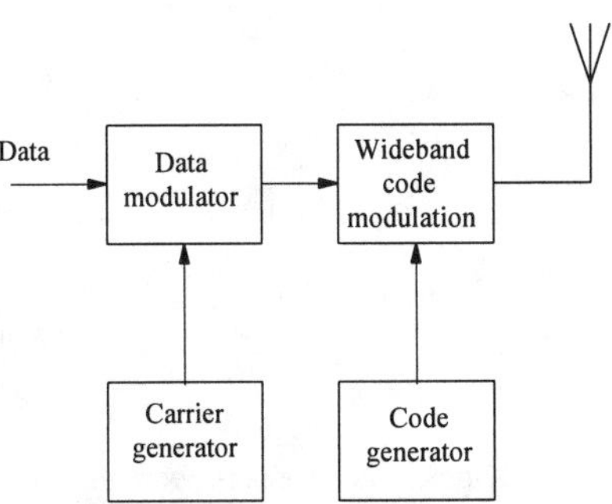

Figure 8.3 Block diagram of a DS-SSMA transmitter.

In Figure 8.3, a block diagram of a DS-CDMA transmitter is given. The binary data signal modulates a radio frequency (RF) carrier. The modulated carrier is then modulated by the code signal. This code signal consists of a number of code bits or "chips" that can be either +1 or −1. To obtain the desired spreading of the signal, the chip rate of the code signal must be much higher than the chip rate of the information signal. For the code modulation various modulation techniques can be used but usually some form of PSK such as BPSK, differential BPSK (D-BPSK), quadrature PSK (QPSK), or MSK is employed.

If we omit the data modulation and use BPSK for the code modulation, we get the block diagram given in Figure 8.4.

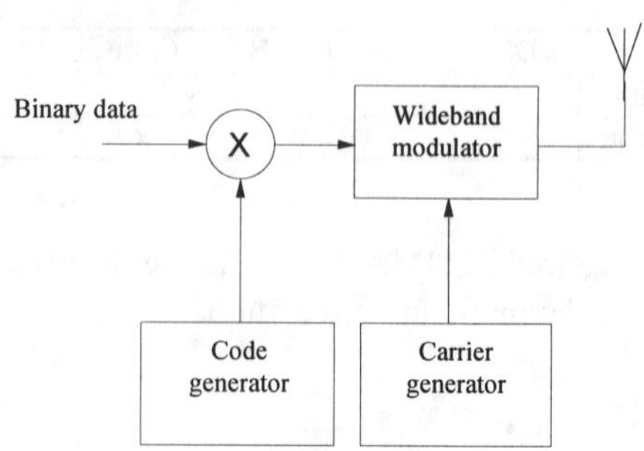

Figure 8.4 Modified block diagram of a DS-SS transmitter.

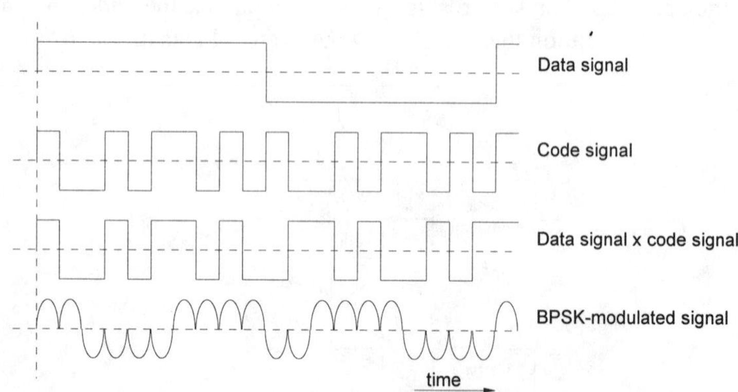

Figure 8.5 Generation of a BPSK-modulated spread-spectrum signal.

The DS-SS signal resulting from this transmitter is shown in Figure 8.5. In this figure, ten code signals per information signal are transmitted (the code chip rate is 10 times the information chip rate), so the processing gain is equal to 10. In practice, the processing gain is much larger (in the order of 10^2 to 10^3).

After transmission of the signal, the receiver (seen in Figure 8.6) uses coherent demodulation to despread the spread-spectrum signal using a locally generated code sequence. To perform the despreading operation, the receiver must not only know the code sequence used to spread the signal, but the codes of the received signal and the locally generated code must also be synchronized.

This synchronization must be accomplished at the beginning of the reception and maintained until the whole signal is received. The synchronization/tracking block performs this operation. After despreading, a data-modulated signal results and after demodulation the original data are recovered.

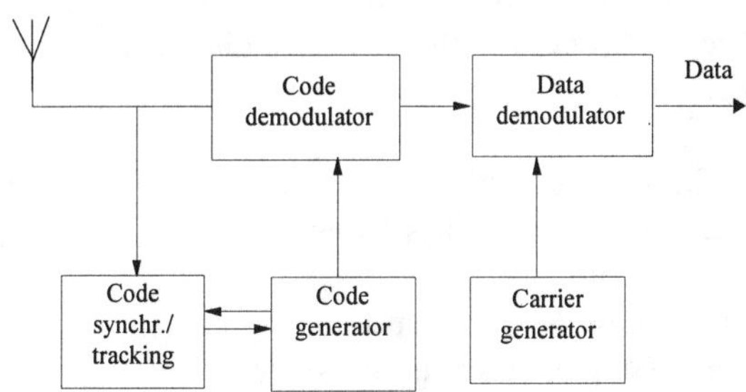

Figure 8.6 Receiver of a DS-SS signal.

In the previous section, a number of advantageous properties of spread-spectrum signals were mentioned. The most important of those properties from the viewpoint of CDMA protocols is multiple access capability, multipath interference rejection, narrowband interference rejection, and, with respect to secure/private communication, LPI. We explain these four properties in relation to DS-CDMA.

- *Multiple access*. If multiple users use the channel at the same time, multiple DS signals will overlap in time and frequency. At the receiver, coherent demodulation is used to remove the code modulation. This operation concentrates the power of the desired user in the information bandwidth. If the cross-correlation between the code of the desired user and the code of the interfering user is small, coherent detection will only put a small part of the power of the interfering signals into the information bandwidth.
- *Multipath interference*. If the code sequence has an ideal autocorrelation function, the correlation function is zero outside the interval $[-T_c, T_c]$ where T_c is the chip duration. This means that if the desired signal and a version that is delayed for more than $2T_c$ are received, coherent demodulation treats the delayed version as an interfering signal, putting only a small part of the power in the information bandwidth.
- *Narrowband interference*. The coherent detection at the receiver involves a multiplication of the received signal by a locally generated code sequence.

However, as we saw at the transmitter, multiplying a narrowband signal with a wideband code sequence spreads the spectrum of the narrowband signal so that its power in the information bandwidth decreases by a factor equal to the processing gain.

- *LPI.* Because the direct sequence signal uses the whole signal spectrum all the time, it has a very low transmitted power per hertz. This makes it very difficult to detect a DS signal.

Apart from the above-mentioned properties, the DS-CDMA protocols have a number of other specific properties that we can divide into advantageous (+) and disadvantageous (–) behavior:

+ The generation of the coded signal is easy. It is done by simple multiplication.
+ Since only one carrier frequency has to be generated, the frequency synthesizer (carrier generator) is simple.
+ Coherent demodulation of the spread-spectrum signal is possible.
+ No synchronization among the users is necessary.
– It is difficult to acquire and maintain the synchronization of the locally generated code signal and the received signal. Synchronization has to take place within a fraction of the chip time.
– For correct reception, the locally generated code sequence and the received code sequence must be synchronized within a fraction of the chip time. This combined with the nonavailability of large contiguous frequency bands practically limits the spread bandwidth to 10 to 20 MHz.
– The power received from users close to the base station is much higher than that received from users further away. Since a user continuously transmits over the whole bandwidth, a user close to the base will constantly create a lot of interference for users far from the base station, making their reception impossible. This near-far effect is solved by applying a power control algorithm so that all users are received by the base station with the same average power. However, this control proves to be quite difficult.

8.2.2 FH

In FH-CDMA protocols, the carrier frequency of the modulated information signal is not constant but changes periodically. During time intervals T, the carrier frequency remains the same but after each time interval the carrier hops to another (or possibly the same) frequency. The hopping pattern is decided by the code signal. The set of available frequencies the carrier can attain is called the hop-set.

The frequency occupation of an FH-spread-spectrum system differs considerably from a DS-spread-spectrum system. A DS system occupies the whole frequency band when it transmits, whereas an FH system uses only a small part of the bandwidth when it transmits but the location of this part differs in time.

Suppose an FH system is transmitting in frequency band 2 during the first time period (see Figure 8.7). A DS system transmitting in the same time period spreads its signal power over the whole frequency band so the power transmitted in frequency band 1 will be much less than that of the FH system. However, the DS system transmits in frequency band 1 during all time periods, while the FH system only uses this band part of the time. On average both systems will transmit the same power in the frequency band.

In a DS system, the narrowband interference is reduced by a factor of *PG*. In an FH system however, it is the (raw) chip error rate that is reduced by *PG*. For strong interference this can be a significant difference, where a single interferer may "overload" a DS receiver but only cause a single chip error now and then in an FH system (which the forward error control (FEC) scheme may easily handle).

The difference between FH-spread-spectrum and DH-spread-spectrum frequency usage is illustrated in Figure 8.7.

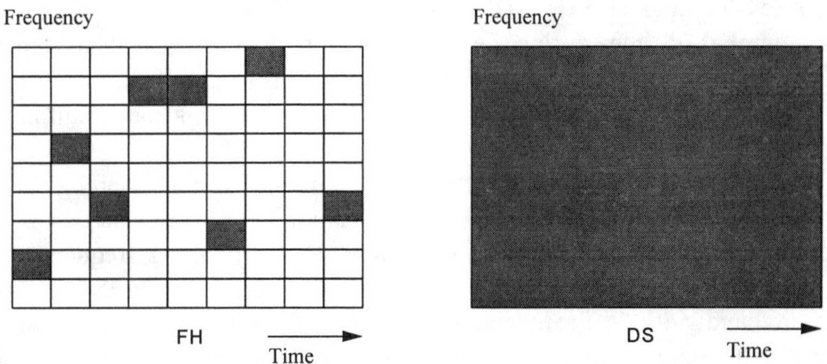

Figure 8.7 Time/frequency occupancy of FH and DS signals.

The block diagram for an FH-CDMA system is given in Figure 8.8.

Figure 8.8 shows the block diagram of an FH-CDMA transmitter and receiver. The data signal is baseband-modulated on a carrier. Several modulation techniques can be used for this but it does not really matter which one is used for the application of frequency hopping. Usually FM modulation is used for analog signals and frequency shift keying (FSK) modulation for digital signals. Using a fast-frequency synthesizer controlled by the code signal, the carrier frequency is converted up to the transmission frequency.

The inverse process takes place at the receiver. Using a locally generated code sequence, the received signal is converted down to the baseband-modulated carrier. The data are recovered after (baseband) demodulation. The synchronization/tracking circuit ensures that the hopping of the locally generated carrier synchronizes to the hopping pattern of the received carrier so that correct despreading of the signal is possible.

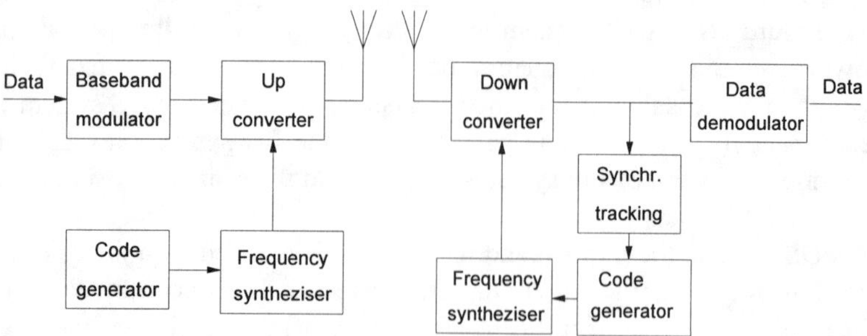

Figure 8.8 Block diagram of an FH-CDMA transmitter and receiver.

Within the FH-CDMA protocols, a distinction is made based on the hopping rate of the carrier. If the number of hops is (much) greater than the data rate, it is called fast-frequency-hopping (F-FH) CDMA protocol. In this case, the carrier frequency changes a number of times during the transmission of 1 bit, so that 1 bit is transmitted in different frequencies. If the number of hops is (much) smaller than the data rate, it is called slow-frequency-hopping (S-FH) CDMA protocol. In this case, multiple bits are transmitted at the same frequency.

The occupied bandwidth of the signal on one of the hopping frequencies depends not only on the bandwidth of the information signal, but also on the shape of the hopping signal and the hopping frequency. If the hopping frequency is much smaller than the information bandwidth (which is the case in S-FH), then the information bandwidth is the main factor that decides the occupied bandwidth. If however, the hopping frequency is much greater than the information bandwidth, the pulse shape of the hopping signal will decide the occupied bandwidth at one hopping frequency. If this pulse shape is very abrupt (resulting in very abrupt frequency changes), the frequency band will be very broad, limiting the number of hop frequencies. If we make sure that the frequency changes are smooth, the frequency band at each hopping frequency will be about $1/T_h$ times the frequency bandwidth, where T_h is equal to the hopping frequency. We can make the frequency changes smooth by decreasing the transmitted power before a frequency hop and increasing it again when the hopping frequency has changed.

As was done with the DS-CDMA protocols, we discuss the properties of FH-CDMA with respect to multiple access capability, multipath interference rejection, narrowband interference rejection, and probability of interception.

- *Multiple access*. It is quite easy to visualize how the F-FH and S-FH CDMA protocols obtain their multiple access capability. In the F-FH protocol, 1 bit is transmitted in different frequency bands. If the desired user is the only one to transmit in most of the frequency bands, the received power of the desired signal is much higher than the interfering power and the signal is received correctly.

In the S-FH protocol, multiple bits are transmitted at one frequency. If the probability of other users transmitting in the same frequency band is low enough, the desired user is received correctly most of the time. For those times that interfering users transmit in the same frequency band, error-correcting codes are used to recover the data transmitted during that period.

- *Multipath interference.* In the F-FH CDMA protocol the carrier frequency changes a number of times during the transmission of 1 bit. Thus, a particular signal frequency is modulated and transmitted on a number of carrier frequencies. The multipath effect is different at the different carrier frequencies. As a result, signal frequencies that are amplified at one carrier frequency will be attenuated at another carrier frequency and vice versa. At the receiver, the responses at the different hopping frequencies are averaged, thus reducing the multipath interference. This is not as effective as the multipath interference rejection in a DS-CDMA system but it still gives quite an improvement.
- *Narrowband interference.* Suppose a narrowband signal is interfering on one of the hopping frequencies. If there are PG hopping frequencies (where PG is the processing gain), the desired user will (on the average) use the hopping frequency where the interferer is located $1/PG$ percent of the time. The interference is therefore reduced by a factor PG.
- *LPI.* The difficulty in intercepting an FH signal lies not in its low transmission power. During a transmission, it uses as much power per hertz as does a continuous transmission. But the frequency at which the signal is going to be transmitted is unknown and the duration of the transmission at a particular frequency is quite small. Therefore, although the signal is more readily intercepted than a DS signal, it is still a difficult task to perform.

Apart from the above-mentioned properties, FH-CDMA protocols have a number of other specific properties that we can divide into advantageous (+) and disadvantageous (–) behavior:

+ Synchronization is much easier with FH-CDMA than with DS-CDMA. With FH-CDMA, synchronization has to be within a fraction of the hop time. Since spectral spreading is not obtained by using a very high hopping frequency but by using a large hop-set, the hop time will be much longer than the chip time of a DS-CDMA system. Thus, an FH-CDMA system allows a larger synchronization error.
+ The different frequency bands that an FH signal can occupy do not have to be contiguous, because we can make the frequency synthesizer easily skip over certain parts of the spectrum. Combined with the easier synchronization, this allows much higher spread-spectrum bandwidths.

+ Because FH-CDMA is an avoidance spread-spectrum system, the probability of multiple users transmitting in the same frequency band at the same time is small. If a user far from the base station transmits, it is received by the base station even if users close to the base station are transmitting, since those users are probably transmitting at other frequencies. Thus, the near-far performance is much better than that of DS.
+ Because of the larger possible bandwidth a FH system can employ, it offers a higher possible reduction of narrowband interference than a DS system.
- A highly sophisticated frequency synthesizer is necessary.
- An abrupt change of the signal when changing frequency bands lead to an increase in the frequency band occupied. To avoid this, the signal turned off and on when changing frequency.
- Coherent demodulation is difficult because of the problems in maintaining phase relationship during hopping.

8.2.3 TH

In the TH-CDMA protocols, the data signal is transmitted in rapid bursts at time intervals determined by the code assigned to the user.

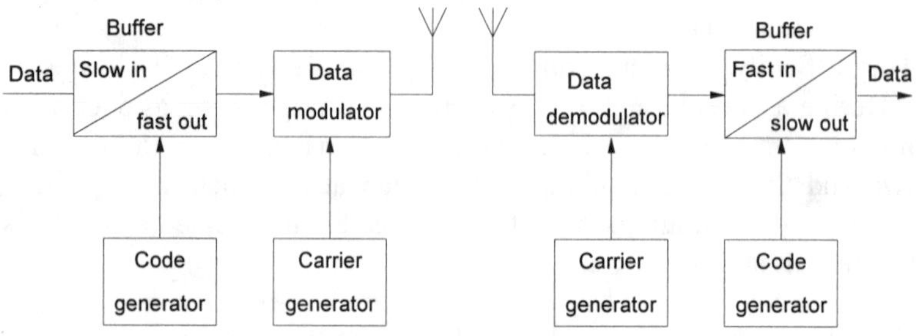

Figure 8.9 Block diagram of a TH-CDMA transmitter and receiver.

The time axis is divided into frames and each frame is divided into M time slots. During each frame the user will transmit in one of the M time slots. Which of the M time slots is transmitted depends on the code signal assigned to the user. Since a user transmits all of its data in 1, instead of M time slots, the frequency it needs for its transmission has increased by a factor M. A block diagram of a TH-CDMA system is given in Figure 8.9.

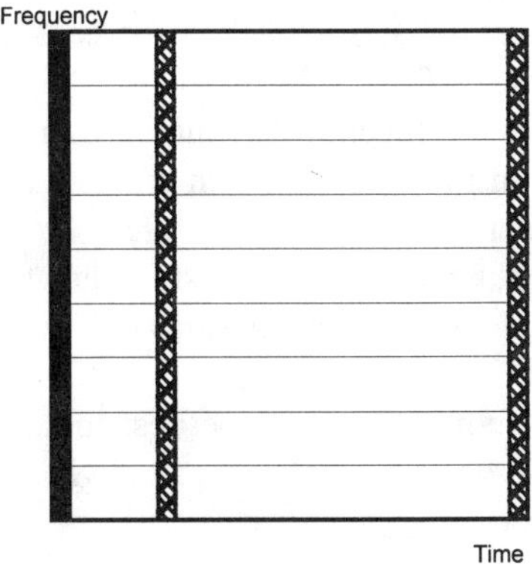

Figure 8.10 Time-frequency plot of the TH-CDMA protocol.

Figure 8.10 shows the time-frequency plot of the TH-CDMA systems. Comparing Figure 8.10 with Figure 8.7, we see that the TH-CDMA protocol uses the whole wideband spectrum for short periods instead of parts of the spectrum all of the time.

Following the same procedure as for the previous CDMA protocols, we discuss the properties of TH-CDMA with respect to multiple access capability, multipath interference rejection, narrowband interference rejection, and probability of interception.

- *Multiple access*. The multiple access capability of TH-SS signals is acquired in the same manner as that of the FH-SS signals, namely by making the probability of users' transmissions in the same frequency band at the same time small. In the case of time hopping, all transmissions are in the same frequency band so the probability of more than one transmission at the same time is small. This is again achieved by assigning different codes to different users. If multiple transmissions do occur, error-correcting codes ensure that the desired signal can still be recovered.
  If there is synchronization among the users, and the assigned codes are such that no more than one user transmits at a particular slot, then the TH-CDMA protocol reduces to a TDMA protocol where the slot in which a user transmits is not fixed but changes from frame to frame.
- *Multipath interference*. In the TH-CDMA protocol, a signal is transmitted in reduced time. The signaling rate therefore increases and dispersion of the signal will lead to overlap of adjacent bits much sooner. Therefore, no advantage is gained with respect to multipath interference rejection.
- *Narrowband interference*. A TH-CDMA signal is transmitted in reduced time. This reduction is equal to $1/PG$ where PG is the processing gain. At the receiver we only

receive an interfering signal during the reception of the desired signal. Thus, we only receive the interfering signal $1/PG$ percent of the time, reducing the interfering power by a factor PG.

- *LPI.* With TH-CDMA, the frequency at which a user transmits is constant but the times at which a user transmits are unknown and the durations of the transmissions are very short. Particularly when multiple users are transmitting, this makes it difficult for an intercepting receiver to distinguish the beginning and end of a transmission and to decide which transmissions belong to which user.

Apart from the above-mentioned properties, the TH-CDMA protocols have a number of other specific properties that we can divide into advantageous (+) and disadvantageous (–) behavior:

+ Implementation is simpler than that of FH-CDMA protocols.
+ It is a very useful method when the transmitter is average-power limited but not peak-power limited since the data are transmitted is short bursts at high power.
+ As with the FH-CDMA protocols, the near-far problem is much less of a problem since TH-CDMA is an avoidance system, so most of the time a terminal far from the base station transmits alone, and is not hindered by transmissions from stations close by.
– It takes a long time before the code is synchronized, and the time is short in which the receiver has to perform the synchronization.
– If multiple transmissions occur, a lot of data bits are lost so a good error-correcting code and data interleaving are necessary.

8.2.4 Chirp Spread Spectrum

Although chirp spread spectrum is not yet adapted as a CDMA protocol, for the sake of completeness a short description is given here.

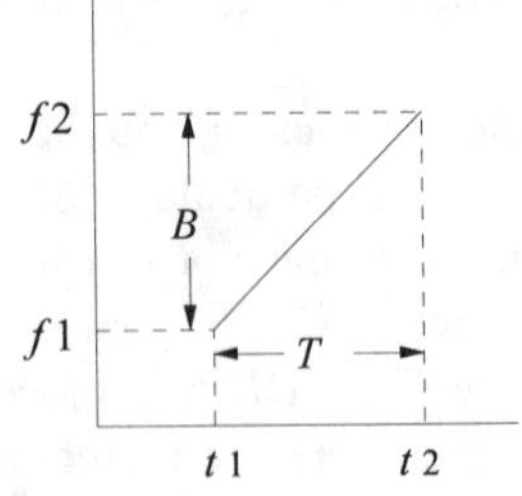

Figure 8.11 Chirp modulation.

A chirp spread-spectrum system spreads the bandwidth by linear frequency modulation of the carrier. This is shown in Figure 8.11.

The processing gain PG is the product of the bandwidth B over which the frequency is varied and the duration T of a given signal waveform:

$$PG = BT \tag{8.2}$$

8.2.5 Hybrid Systems

The hybrid CDMA systems include all CDMA systems that employ a combination of two or more of the above-mentioned spread-spectrum modulation techniques. If we limit ourselves to DS, FH, and TH modulations, we have four possible hybrid systems: DS/FH, DS/TH, FH/TH, and DS/FH/TH.

The idea of the hybrid system is to combine the specific advantages of each of the modulation techniques. If we take, for example, the combined DS/FH system, we have the advantage of the antimultipath property of the DS system combined with the favorable near-far operation of the FH system. Of course, the disadvantage lies in the increased complexity of the transmitter and receiver. For illustration purposes, we give a block diagram of a combined DS/FH CDMA transmitter in Figure 8.12.

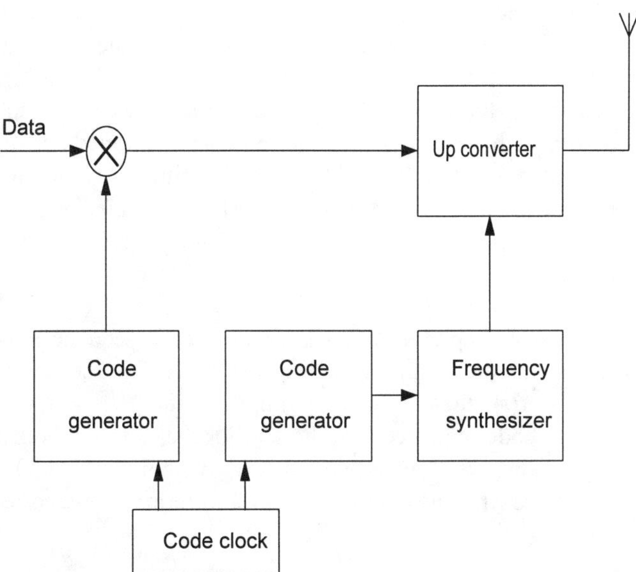

Figure 8.12 Hybrid DS-FH transmitter.

The data signal is first spread using a DS code signal. The spread signal is then modulated on a carrier whose frequency hops according to another code sequence. A code clock ensures a fixed relation between the two codes.

8.3 DESIGN OF PSEUDONOISE SEQUENCES

The objective of this section is to present an overview of the design and properties of code sequences for CDMA systems. The choice of the type of code sequence for CDMA systems is important with respect to the resistance against both multipath and multiuser interference.

8.3.1 Basics

To combat these types of interference, two properties are very important: (1) Each code sequence in the set must be easy to distinguish from a time-shifted version of itself, and (2) each code sequence in the set must be easy to distinguish from (a possibly tune-shifted version of) every other signal in the set.

The first property is important with respect to multipath propagation effects that occur in mobile outdoor and indoor radio environments The second property plays an important role with respect to the multiple access capability of the communications system. In this chapter, we describe three types of code sequences: maximum length, Gold, and kasami.

Before describing the construction of these code sequences and discussing the correlation properties with respect to the requirements mentioned above, we present in the next section some basic definitions of correlation functions and some information about shift register sequences. Finally, some numerical examples of the aforementioned code sequences are given.

Correlation functions

Primarily because system implementation is simpler, code sequences used for CDMA communications systems are required to be periodic. If T is the period of the code sequence denoted as $X(t)$, then periodicity implies that $X(t) = X(t+T)$ for all time instants t and for each code sequence X in the set. For the code symbols of X we use the notation X_m. Furthermore, we assume that the code symbols are $\{-1,1\}$ The distinction between code sequences $X(t)$ and $Y(t)$ is measured in terms of the correlation function defined as

$$r_{X,Y}(\tau) = \int_0^T X(t)Y(t+\tau)dt \tag{8.3}$$

This expression is shown as the cross-correlation function if $X \neq Y$ and as the autocorrelation function if $X = Y$. Code sequence of interest are those consisting of a number of time-limited pulses called chips. Then, the signal $X(t)$ is written as

$$X(t) = \sum_{i=-\infty}^{\infty} a_x^i \varphi(t - iT_c) \tag{8.4}$$

where a_x^i is the ith code symbol user of x, $\varphi(t)$ is the basic pulse wave form, and T_c is the time duration of the pulse, termed chip duration and

$$\int_0^{T_c} \varphi(t - iT_c)\varphi(t - jT_c)dt = 0 \qquad \text{if } i \neq j \tag{8.5}$$

If the delay τ is taken as a multiple of the chip duration ($\tau = lT_c$), then (8.3) is written as

$$r_{X,Y}(l) = \lambda \sum_{i=0}^{N-1} a_x^i a_x^{i+1} \tag{8.6}$$

where

$$\lambda = \int_0^{T_c} \varphi^2(t)dt \tag{8.7}$$

which equals T_c if $\varphi(t)$ is a rectangle pulse duration T_c with amplitude.

In a DS-CDMA system, each data bit is multiplied with a user-specific code sequence. For the current data bit and the previous data bit of a user with code sequence Y, we use the notation b_y^0 and b_y^{-1}, respectively.

Now consider the situation in a multipath radio environment where X is the code sequence of the reference user and Y is the code sequence of a interfering user (or a delayed part of code sequence X). The length of a code sequence is denoted by N_c. In the receiver of the reference user a partial correlation operation is performed. This situation is shown in Figure 8.13

Now we define the following two partial correlation functions

$$R_{XY}(\tau) = \int_0^{\tau} X(t)Y(t - \tau)dt \tag{8.8}$$

$$\hat{R}_{XY}(\tau) = \int_{\tau}^{N_c T_c} X(t)Y(t - \tau)dt \tag{8.9}$$

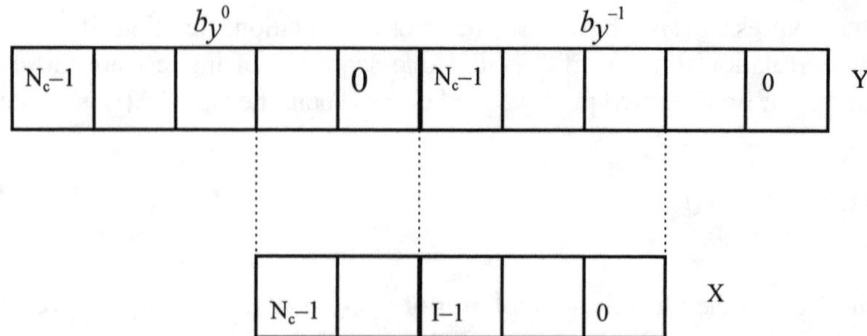

Figure 8.13 Cross-correlation of code sequence X of the reference user, and code sequence Y being the code sequence of an interfering user or a delayed version of the reference code sequence.

Assuming that τ is a multiple of the chip duration T_C implying $\tau = lT_C$, these partial correlation functions are written as

$$R_{xy}(l) = \sum_{i=0}^{l-1} a_x^i a_y^{i-1} \tag{8.10}$$

$$\hat{R}_{xy}(l) = \sum_{l=1}^{N_c-1} a_x^l a_y^{l-1} \tag{8.11}$$

If $X = Y$, then (8.10) and (8.11) are the partial autocorrelation functions and in the case $X \neq Y$, (8.10) and (8.11) are the partial cross-correlation functions.

With respect to the sign of b_y^0 and b_y^{-1}, there are two possibilities: either b_y^0 and b_y^{-1} have the same sign or they have the opposite sign. For the first case we have the *periodic* correlation function:

$$\theta_{xy}(l) = R_{xy}(l) + \hat{R}_{xy}(l) = \sum_{m=0}^{N_c-1} a_x^l a_y^{l-1} \tag{8.12}$$

Also, $\theta_{xy}(l)$ satisfies

$$\theta_{x,y}(N_c - l) = \theta_{y,x}(l) \tag{8.13}$$

which implies that $\theta_{xy}(l)$ is an even function with respect to N.

If b_y^0 and b_y^{-1} have the opposite sign, then we have an *a-periodic* correlation function

$$\hat{\theta}_{xy}(l) = R_{xy}(l) - \hat{R}_{xy}(l) = \sum_{i=0}^{l-1} a_x^i a_y^{i-1} - \sum_{n=l}^{N_c-1} a_x^i a_y^{i-1} \tag{8.14}$$

which is odd with respect to N_C, that is,

$$\hat{\theta}_{x,y}(N_c - l) = -\hat{\theta}_{y,x}(l) \tag{8.15}$$

For analysis and comparison of code sequences it is convenient to define peak correlation functions. For a set χ of periodic sequences, the peak periodic cross-correlation magnitude θ_C is defined as

$$\theta_c = \max\{|\theta_{xy}(l)| : 0 \le l \le N_c - 1, x \in \chi, y \in \chi, x \ne y\} \tag{8.16}$$

The peak out-of-phase-periodic autocorrelation magnitude is defined as

$$\theta_a = \max\{|\theta_x(l)| : 0 \le l \le N_c - 1, x \in \chi\} \tag{8.17}$$

Now the larger of θ_C and θ_a is denoted by $\theta_{\max}$.

Analogously, we define the peak a-periodic cross-correlation magnitude $\hat{\theta}_c$ and the peak out-of-phase a-periodic autocorrelation magnitude $\hat{\theta}_a$, respectively

$$\hat{\theta}_c = \max\left\{\left|\hat{\theta}_{xy}(l)\right| : 0 \le l \le N_c - 1, x \in \chi, y \in \chi, x \ne y\right\} \tag{8.18}$$

and

$$\hat{\theta}_a = \max\left\{\left|\hat{\theta}_x(l)\right| : 0 \le l \le N_c - 1, x \in \chi\right\} \tag{8.19}$$

The larger of $\hat{\theta}_c$ and $\hat{\theta}_a$ is denoted as $\hat{\theta}_{\max}$.

The two requirements mentioned in the introduction are now equivalently reformulated: (1) For each sequence in the set, both the periodic out-of-phase autocorrelation function and the a-periodic out-of-phase autocorrelation function must be small for $1 \le l \le N_C - 1$. (2) For each pair of sequences x and y, both the periodic autocorrelation and the a-periodic cross-correlation function must be small for all l.

Linear Shift Registers

Code sequences must have noiselike properties to meet the requirements mentioned in the preceding section. However, because of implementation problems, the code sequences are generated according to a periodic deterministic scheme. Periodic code sequences having noiselike properties are called PN sequences. A proper way to generate these code sequences is by using linear binary shift registers. An example for such a register is presented in Figure 8.14. In this configuration, an XOR operation is performed on the contents of register 2 and register 0 and the result is fed back to the input of register 4. The shift direction of the shift register is from left to right (4-3-2-1-0).

The register in Figure 8.14 with five sections generates a code sequence with length N_C=2^5–1=31. We show in the next section that the feedback connections are chosen very carefully in order to generate a code sequence with satisfactory correlation properties. In general, the configuration of a linear binary shift register of *n* sections is described by a generator polynomial, which is a binary polynomial of degree *n*. The number *n* is the number of sections of the shift register:

$$h(x) = h_0 x^n + h_1 x^{n-1} + \ldots + h_{n-1} x + h_n \qquad \left(h_i \in \{0,1\}\right) \tag{8.20}$$

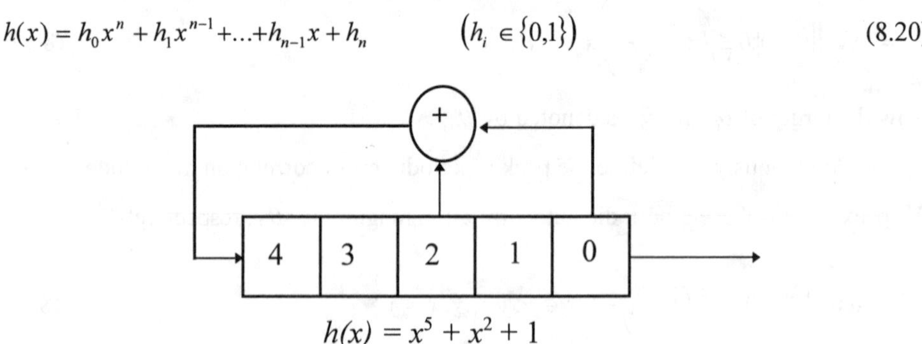

Figure 8.14 Two-tap linear binary shift register.

We use the notation used by Sarwate and Pursley [11] where $h_0 = h_n = 1$ and $h_i = 1$ for i ≠ 0 and i ≠ n if there is a feedback connection from the *i*th cell. It is convenient and conventional to represent the polynomial *h*(*x*) by a binary vector $h = (h_0, h_1, \ldots, h_n)$ and to express this vector in octal notation. For example, the shift register in Figure 8.14 is represented by generator polynomial $h(x) = x^5 + x^2 + 1$. The binary representation of this polynomial is 100101 and the octal representation is 45. If the sequence generated by *h*(*x*) is denoted as *u*, then a circular shifted version of *u* is denoted as *Tu* where i is the number of shifts. For instance, if *u* = 10011101 then Tu = 1001110. Note that $T^N u = T^0 u = u$.

8.3.2 PN Codes

Three basic classes of code sequences suitable for CDMA applications are discussed in this section:

- maximum length sequences;
- Gold sequences;
- Kasami sequences.

Maximum length sequences
Maximum length code sequences are generated by a single linear shift register. As the name suggests, maximum length code sequences are precisely the sequences of maximum possible period ($N_c = 2^n$-1) from an n-stage binary shift register with linear feedback. To generate a maximum length code sequence, the generator polynomial must be a *primitive polynomial* of degree *n*. Before explaining the term primitive polynomial, first the following terms are defined:

Irreducible polynomial
This is a polynomial that can not be decomposed into other polynomials. Example: The polynomial $x^2 + 2x + 1$ is not irreducible since $x^2 + 2x + 1 = (x + 1)(x + 1)$. The polynomial $x^2 + x + 1$ is irreducible.

Exponent
The smallest positive integer *p* for which the polynomial $h(x)$ is a divisor of $(1 - x^p)$ is called the exponent of $h(x)$.

Primitive polynomial
A primitive polynomial of degree *n* is an irreducible polynomial with exponent $2^n - 1$.

It has been proven that the periodic autocorrelation for a maximum length sequence *u* is given by

$$\theta_u(l) = \begin{cases} N_c & l = 0 \mod N_c \\ -1 & l \neq 0 \mod N_c \end{cases} \tag{8.21}$$

Thus, binary maximum length-sequence has a two-valued periodic autocorrelation function.

Golomb [12] observed that if $n \neq 0 \mod 4$, there exists pairs of maximum length sequences with three-valued cross-correlation functions, where the three values are $\{-1, -t(n), t(n) - 2\}$with

$$t(n) = \begin{cases} 1+2^{(n+1)/2} & n \text{ odd} \\ 1+2^{(n+2)/2} & n \text{ even} \end{cases} \tag{8.22}$$

A cross-correlation function taking on these values is called a preferred three-valued cross-correlation function and the corresponding pair of maximum length sequences (polynomials) is called a preferred pair of maximum length sequences. For this preferred pair of maximum length sequences we find

$$\theta_a = \theta_c = t(n) \tag{8.23}$$

Figures 8.15 to 8.17 show preferred pairs of maximum length sequences for periods 31, 63, and 127, respectively.

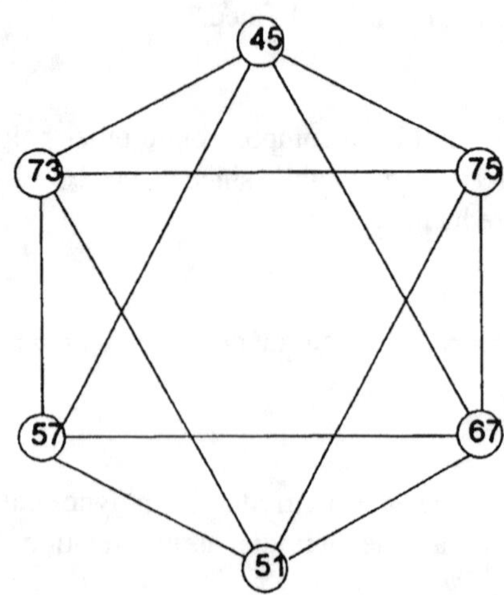

Figure 8.15 Preferred pairs of maximum length sequences of period 31. The vertices of every triangle form a maximal connected set.

It is clear from these figures that only very small sets of maximum length sequences can have good periodic cross-correlation properties. For multiple access systems, it is desirable to obtain larger sets of sequences of period $N_C = 2^n - 1$ which have the same bound $\theta \leq t(n)$ on the peak periodic cross-correlation as do the maximal connected sets. Although there are no analytical results that give the values of the maximal a-periodic cross-correlation $\hat{\theta}_c$ for maximum length sequences generated by preferred pairs, Massey and Uhran [13] have obtained bounds on $\hat{\theta}_c$. They state that if the maximum of the periodic cross-correlation and the periodic autocorrelation θ_{max} equals $t(n)$, then the bound for the a-periodic cross-correlation is

$$\hat{\theta}_c = \begin{cases} 2^{n-1} + 2^{n/2} + 1 & n \text{ even} \\ 2^{n-1} + 2^{(n-1)/2} + 1 & n \text{ odd} \end{cases} \tag{8.24}$$

We did not find a bound in the literature for out-of-phase a-periodic autocorrelation.

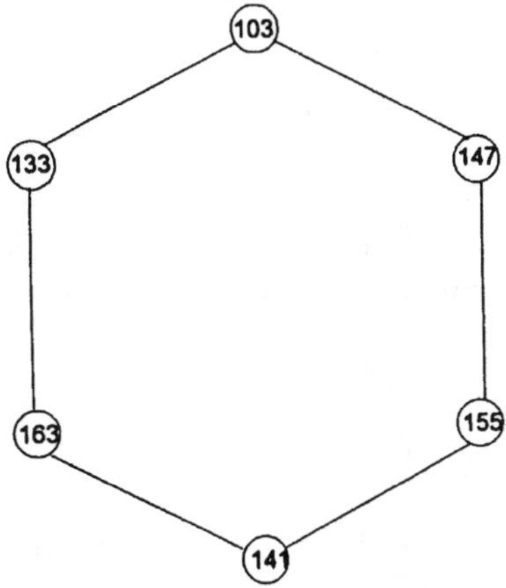

Figure 8.16 Preferred pairs of maximum length sequences of period 63. Every pair of adjacent vertices is a maximal connected set.

Gold Sequences

One important class of periodic sequences that provides larger sets of sequences with good periodic cross-correlation is the class of Gold sequences. A set of Gold sequences of period $N = 2^{n-1}$ consists of $(N_C + 2)$ sequences for which the maximum periodic cross-correlation $\theta_C = t(n)$ with $t(n)$ given by (8.22). A set of Gold sequences is constructed from appropriately selected maximum length sequences. Suppose a shift register polynomial $f(x)$ factors into $h(x)\,\hat{h}(x)$ where $h(x)$ and $\hat{h}(x)$ have no factors in common. Then the set of all sequences generated by $f(x)$ is just the set of all sequences of the form $a \oplus b$, where a is some sequence generated by $h(x)$, b is some sequence generated by $\hat{h}(x)$, and a and b are not necessarily nonzero sequences. Now suppose that $h(x)$ and $\hat{h}(x)$ are two different primitive binary polynomials of degree n that generate the maximum length sequences u and v, respectively, of period $N_c = 2^n - 1$. If y denotes a nonzero sequence generated by $f(x) = h(x)\,\hat{h}(x)$, then either

$$y = T^i u \tag{8.25}$$

or

$$y = T^j v \tag{8.26}$$

or

$$y = T^i u \oplus T^j v \tag{8.27}$$

where $0 \le i, j \le N_c - 1$ and where, as before, $T^i u \oplus T^j v$ denotes the sequence whose kth element is $u_{i+k} \oplus v_{j+k}$. From this it follows that y is some phase of some sequence in the set $G(u,v)$ defined by

$$G(u,v) \overset{\Delta}{=} \{u, v, u \oplus v, u \oplus Tv, u \oplus T^2 v, \ldots, u \oplus T^{N_c - 1} v\} \tag{8.28}$$

Note that $G(u, v)$ contains $N_c + 2 = 2^n + 1$ sequences of period N_c.

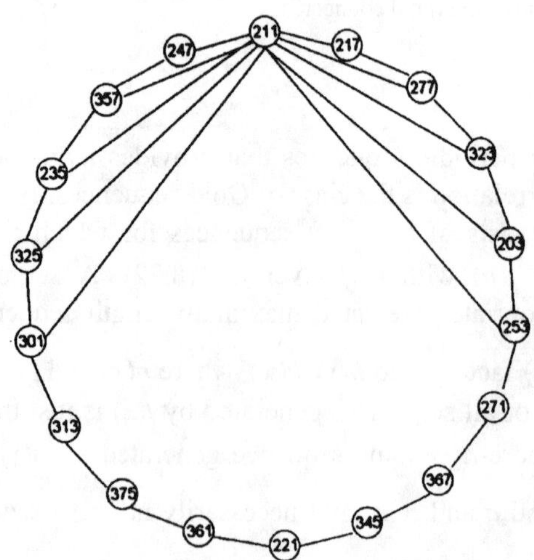

Figure 8.17 Preferred pairs of maximum length sequences of period 127. Every set of six consecutive vertices is a maximal connected set.

Figure 8.18 shows a possible configuration to generate a Gold sequence of length $N_C = 63$. The octal representation of the shift registers producing u and v are 141 and 163, respectively ($h(x) = 1 + x^5 + x^6$ and $\hat{h}(x) = 1 + x + x^4 + x^5 + x^6$).

In 1969, Gold [14] published the construction $G(u,v)$. In particular, he showed that if (u,v) is any preferred pair of maximum length sequences, then $G(u,v)$ has peak periodic correlation parameter $\theta_C = t(n)$. For all the Gold codes that have $\theta_C = t(n)$, the bound for the a-periodic cross-correlation is

$$\hat{\theta}_c = \begin{cases} 2^{n-1} + 2^{n/2} + 1 & n \text{ even} \\ 2^{n-1} + 2^{(n-1)/2} + 1 & n \text{ odd} \end{cases} \tag{8.29}$$

We did not find a bound in the literature for out-of-phase a-periodic autocorrelation.

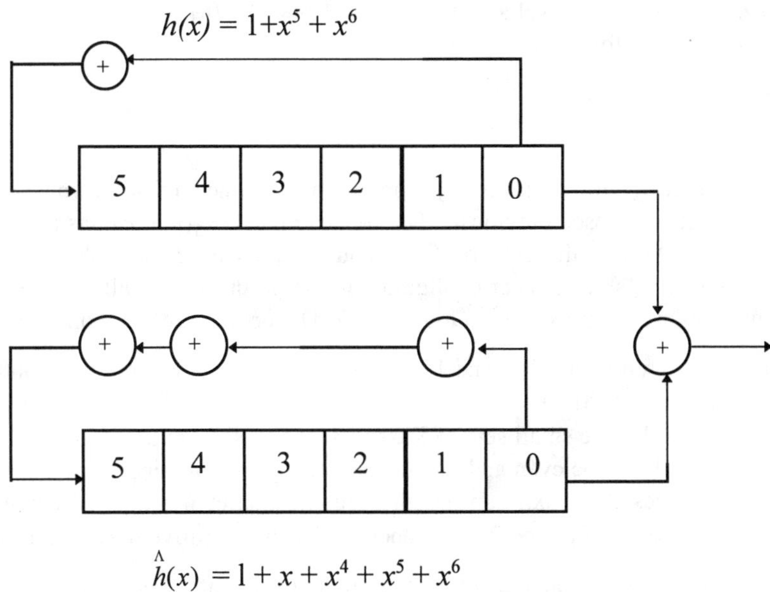

Figure 8.18 Shift register configuration for set of Gold codes with length $N_c = 63$.

Kasami Sequences

Let n be even and u denote an maximum length sequence of period $N_C = 2^n - 1$ generated by $h(x)$. Consider the sequence w of period $2^{n/2} - 1$, which is generated by the polynomial $\hat{h}(x)$ whose roots are the $s(n)$th powers of the roots of $h(x)$, where

$$s(n) = 1 + 2^{n/2} \tag{8.30}$$

Furthermore, since $\hat{h}(x)$ is shown to be a polynomial of degree $n/2$, w is a maximum length sequence of period $2^{n/2} - 1$. Now consider the sequences generated by the polynomial $h(x)\,\hat{h}(x)$ of degree $3n/2$. Clearly, any such sequence must be of one of the forms $T^i u$, $T^j w$ or $T^i u \oplus T^j w$, where $0 \le i \le 2^n - 1$ and $0 \le j \le 2^{n/2} - 1$. Thus, any sequence y of period $2^n - 1$ generated by $h(x)\,\hat{h}(x)$ is one of the sequences in the set $G(u, w)$ defined by

$$G(u,w) \stackrel{\wedge}{=} \left\{u, w, u \oplus w, u \oplus Tw, u \oplus T^2 w, \ldots, u \oplus T^{2^{n/2}} w\right\} \tag{8.31}$$

This set of sequences is called the *small set of Kasami sequences* denoted by $K_S(u)$ in honor of Kasami, who discovered that the correlation functions for sequences belonging to $K_S(u)$ take on the values in the set $\{-1, -s(n), s(n) - 2\}$ with $s(n)$ given by (8.30). Consequently, for the set $K_S(u)$,

$$\theta_{\max} = s(n) = 1 + 2^{n/2} \tag{8.32}$$

Notice that $\theta_{\max}$ for the set $K_S(u)$ is approximately one half of the value of $\theta_{\max}$ achieved by the sets discussed previously. On the other hand, $K_S(u)$ contains only $2^{n/2} = (N_C+1)^{1/2}$ sequences, while the sets of Gold code sequences contain $N_C + 2$ sequences. An example for a shift register configuration that produces a small set of Kasami codes of length $N_C = 63$ is presented in Figure 8.19. The octal representation of the shift registers producing u and v are 103 and 15, respectively, $h(x) = 1 + x + x^6 = 1000011$ and $\hat{h}(x) = 1 + x^2 + x^3 = 1101$.

Besides the small sets of Kasami sequences, there are also large sets of Kasami sequences. Let n be even and let $h(x)$ denote a primitive binary polynomial of degree n that generates the maximum length sequence u. Let $w = u[s(n)]$ denote a maximum length sequence of period $2^{n/2} - 1$ generated by the primitive polynomial $h'(x)$ of degree $n/2$ and let $\hat{h}(x)$ denote the polynomial of degree n that generates $u[t(n)]$.

Then, the set of sequences of period N_C generated by $h(x)\,h'(x)\,\hat{h}(x)$, called the large set of Kasami sequences and denoted by $K_L(u)$, is as follows:

1. If $n = 2 \bmod 4$, then

$$K_L(u) = G(u,v) \bigcup \left[\bigcup_{i=0}^{2^{n/2}-2} \left\{T^i w \oplus G(u,v)\right\}\right] \tag{8.33}$$

where $v = u[t(n)]$ and $G(u,v)$ is defined in (8.28).

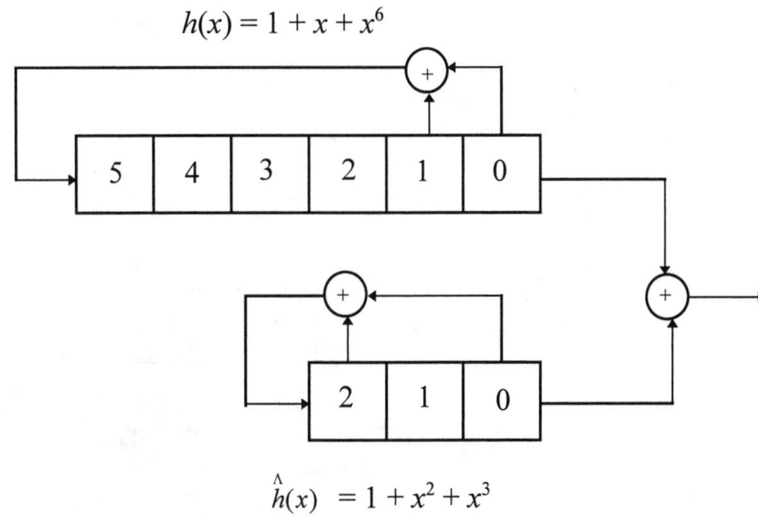

Figure 8.19 Shift register configuration for the small set of Kasami codes with length $N_C = 63$.

2. If $n = 0 \bmod 4$ then

$$K_L(u) = H_{t(n)}(u) \bigcup \left[\bigcup_{i=0}^{2^{n/2}-2} \left\{ T^i w \oplus H_{t(n)}(u) \right\} \right]$$

$$\bigcup \left\{ v^{(j)} \oplus T^k w : 0 \le j \le 2, 0 \le k < \tfrac{1}{3}\left(2^{n/2} - 1\right) \right\} \quad (8.34)$$

where $v(t)$ is the result of decimating $T^i(u)$ by $t(n)$ and $H_{t(n)}(u)$ is defined as

$$H_{t(n)}(u) = \begin{Bmatrix} u, u \oplus v^{(0)}, & u \oplus Tv^{(0)}, & \ldots, & u \oplus T^{N_c/3-1} v^{(0)} \\ u, u \oplus v^{(1)}, & u \oplus Tv^{(1)}, & \ldots, & u \oplus T^{N_c/3-1} v^{(1)} \\ u, u \oplus v^{(2)}, & u \oplus Tv^{(2)}, & \ldots, & u \oplus T^{N_c/3-1} v^{(2)} \end{Bmatrix} \quad (8.35)$$

In either case, the correlation functions for $K_L(u)$ take on the values in the set {–1,–t(n), t(n) –2, –s(n), s(n) –2} and $\theta_{\max}$ = t(n). If $n = 2 \bmod 4$ then $K_L(u)$ contains $2^{n/2}(2^n + 1)$ sequences while if $n = 0 \bmod 4$, $K_L(u)$ contains $2^{n/2}(2^n + 1)$ sequences.

An example of a shift register configuration that produces a large set of Kasami codes of length $N_C = 63$ is presented in Figure 8.20. The octal representation of the shift registers producing the sequences u, v, and w are 103, 15, and 147, respectively, ($h(x) = 1 + x + x^6$, $\hat{h}(x) = 1 + x^2 + x^3$, and $h'(x) = 1 + x + x^2 + x^5 + x^6$).

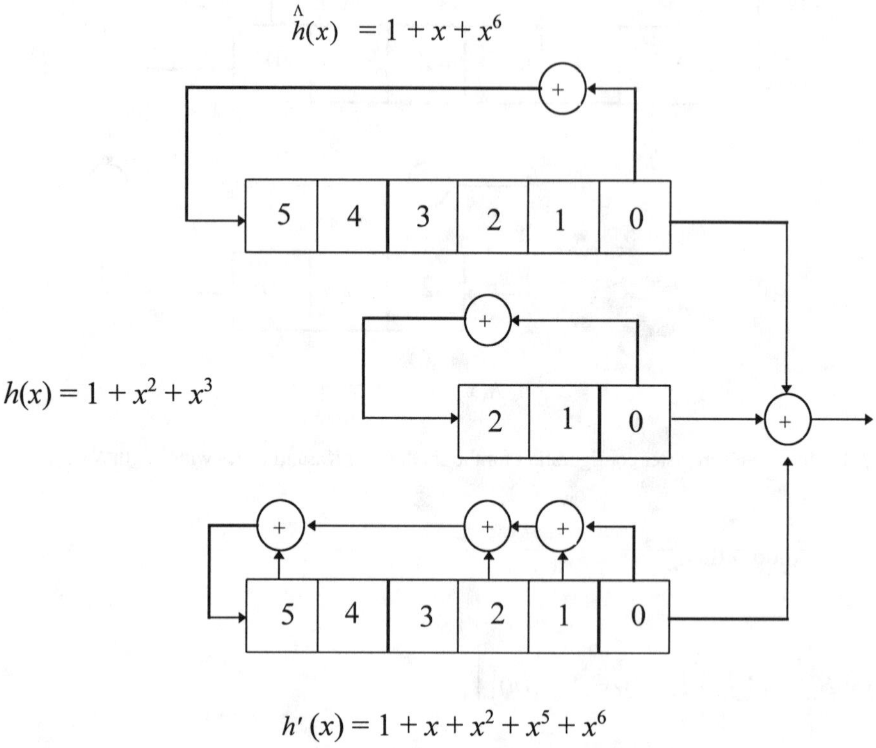

Figure 8.20 Shift register configuration for the large set of Kasami codes with length $N_C = 63$.

Numerical Examples

Table 8.2 presents a number of examples of PN code sequences. Three types of sequences are considered. Gold sequences (G), small-set Kasami sequences (K_S), and large-set Kasami sequences (K_L). The first column denotes the code length N_C, the third column shows the maximum number of code sequences per set, and in the fourth column the values that periodic cross-correlation can take are given. The last three columns show the octal representation of the polynomials describing the linear binary shift registers used to generate the desired code sequence. Table 8.2 only shows some examples for pseudo-random code sequences.

Table 8.2

Examples of Gold Codes, Small Sets of Kasami Codes, and Large Sets of Kasami Codes and Their Basic Properties

N_C	Code	No.	Periodic Cross-Correlation	Pol. 1	Pol. 2	Pol. 3
31	G	33	{−9, −1, −7}	45	67	-
63	G	65	{−17, −1,15}	141	163	-
	K_S	8	{−9, −1,7}	103	15	-
	K_L	520	{−17, −9, −1, 7,15}	103	15	147
127	G	129	{−17, −1,15}	211	277	-
255	K_S	16	{−17, −1,15}	435	023	-
	K_L	4111	{−33,−17, −1, 15,31}	435	023	675
511	G	513	{−33, −1,31}	1021	1333	-

G, Gold codes; K_S, small sets of Kasami codes; K_L, large set of Kasami codes.

From Table 8.2 it is clear that for a code sequence with length N_C = 63, the small set of Kasami codes has better correlation properties than the set of Gold codes. However, the number of code sequences in the set is less (8 instead of 33 in the case of Gold codes). The large set of Kasami sequences has the same cross-correlation properties as the set of Gold codes, but the number of code sequences in the set is higher (520).

We plotted the correlation functions of some Gold and Kasami code sequences. For the autocorrelation we used the code generated by the unshifted registers. The cross-correlation function was produced by considering one code generated by unshifted registers and one code generated by shifting one of the registers 1 chip period.

Figures 8.21(a) and 8.21(b) show the periodic autocorrelation and cross-correlation of a Gold code with length N_C=127. We considered the codes $x = u \oplus w$ and $y = u \oplus T^1 w$. It is clear that the periodic out-of-phase autocorrelation and cross-correlation have the values {−17, −1, 15}. Figures 8.22(a) and 8.22(b) show the a-periodic out-of-phase autocorrelation and cross-correlation for the same code sequence. The results of these figures are consistent with Table 8.2.

It is not possible to find preferred pairs with sequence-length N_C = 255, and hence Gold codes with length N_C = 255 having a three-valued periodic cross-correlation function do not exist. Figures 8.23(a) and 8.23(b) show the periodic autocorrelation and cross-correlation for a code sequence generated by the XOR-sum of two not preferred maximum length sequence with length $N_c = 255$.

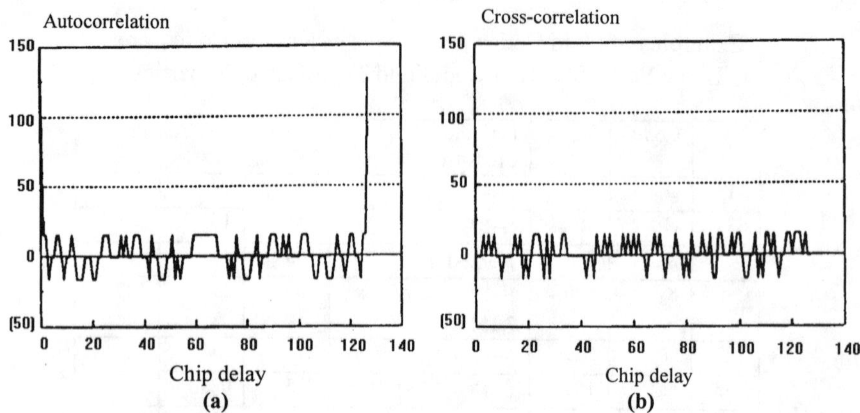

Figure 8.21 (a) Periodic autocorrelation and (b) cross-correlation for Gold code with code length $N_c = 127$. Generator polynomials 211 and 277 (octal).

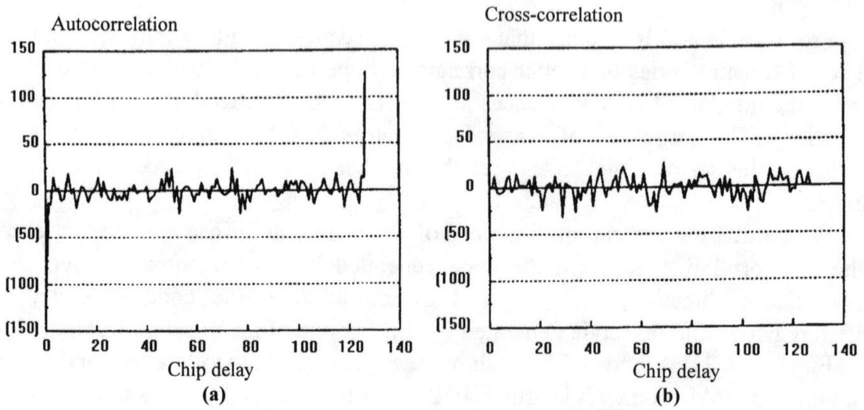

Figure 8.22 (a) a-Periodic autocorrelation and (b) cross-correlation for Gold code with code length $N_c = 127$. Generator polynomials 211 and 277 (octal).

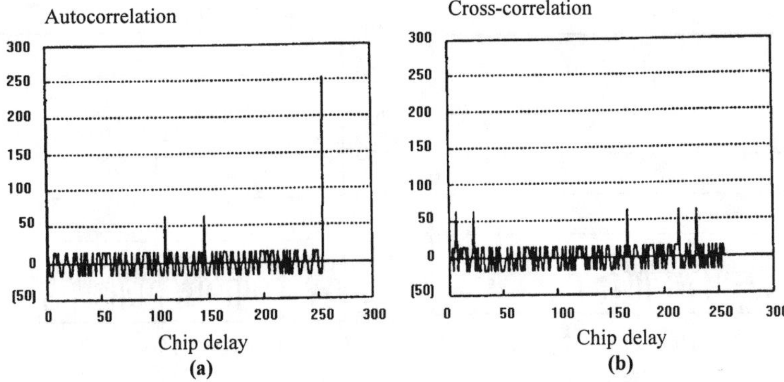

Figure 8.23 (a) Periodic autocorrelation and (b) cross-correlation for Gold code with $N_c = 255$. Generator polynomials 717 and 765 (not preferred sequences).

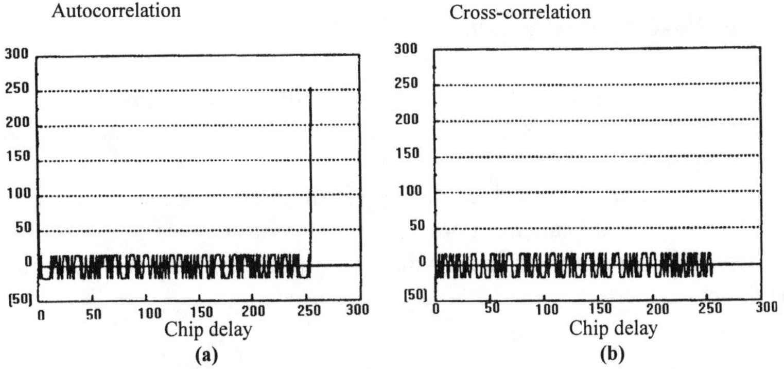

Figure 8.24 (a) Periodic autocorrelation and (b) cross-correlation functions for small-set Kasami code with N_c=255. Generator polynomials 435 and 023.

Code sequences with length N_c=255 that have better correlation properties are the Kasami sequences. Figures 8.24(a) and 8.24(b) show the autocorrelation and cross-correlation of a small set of Kasami codes. Figures 8.25(a) and 8.25(b) show the autocorrelation and cross-correlation of a large set of Kasami codes.

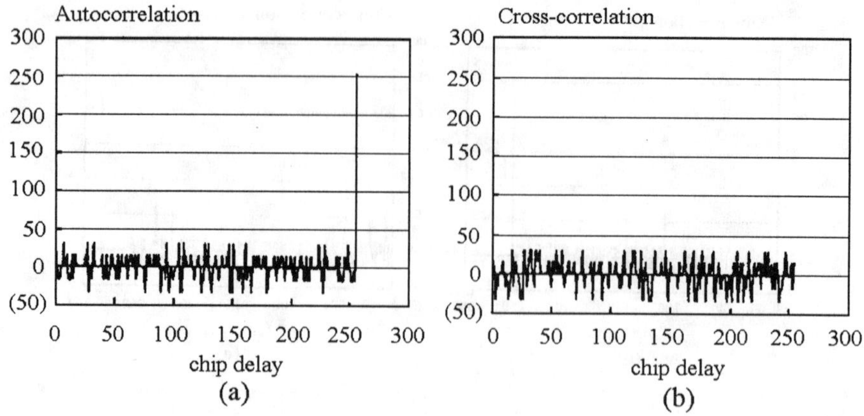

Figure 8.25 (a) Periodic autocorrelation and (b) cross-correlation functions for large Kasami code with $N_c = 255$. Generator polynomials 435, 023, and 675.

8.3.3 Random Wave Approximation

In a multipath propagation environment such as the indoor radio environment, there is overlap between shifted data bits of the reference user and of other users. This situation is depicted in Figure 8.26.

This implies that the receiver suffers from interference caused by both the desired signal and by the signals from other users. The interference power is determined by the characteristics of the multipath channel (time delay, phase, and amplitude) and by the correlation properties of the code sequences used.

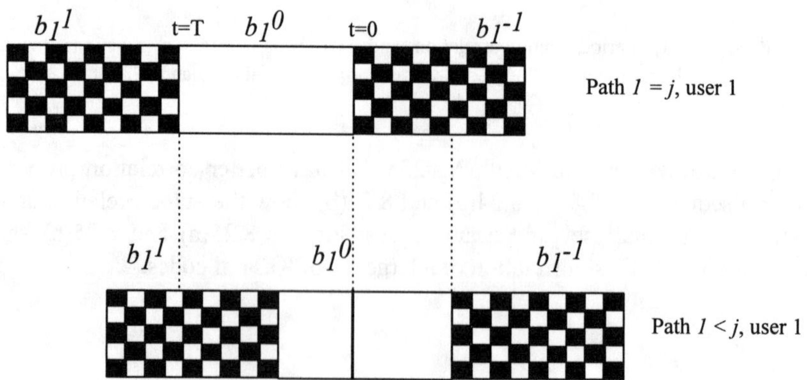

Figure 8.26 Overlap of bits (and code sequences) due to multipath interference.

The ACF determines the amount of self-interference, while the cross-correlation between the desired code sequence and code sequences of other users determines the amount of multiuser interference. To simplify these performance calculations, an approximation of the code sequence characteristics is applied. The starting point for this approximation is the expression for the correlation variance defined as

$$\psi_a = E\left[b_1^{-1} R(\tau) + b_1^0 \hat{R}(\tau)\right]^2 \tag{8.36}$$

It is worth mentioning here that the interference power is related to this variance. In (8.36), b_1^{-1} and b_1^0 are, respectively, the previous and the next bit with values $\{-1,1\}$ having equal probability of occurrence. The functions $R(\tau)$ and $\hat{R}(\tau)$ are the partial correlation functions. If the delay is a multiple of the chip duration ($\tau = pT_C$), then the expressions for the discrete correlation functions are

$$R(p) = \frac{1}{N_c} \sum_{i=0}^{p} a_1^i a_k^{i-p} \tag{8.37}$$

$$\hat{R}(p) = \frac{1}{N_c} \sum_{i=p+1}^{N_c-1} a_1^i a_k^{i-p} \tag{8.38}$$

where a_k^i is the ith code symbol of the sequence of user k and N_C is the length of the code sequence. To compute the variance in (8.36), we average over all possible time delays and over all possibilities for b_1^{-1} and b_1^0. The time delays in a multipath environment are obviously not multiples of the code symbol duration. However, if we know the discrete correlation function, then the correlation for a non discrete time delay is very easy to calculate by considering that the correlation function is linear between two discrete time delays, as shown in Figure 8.27.

To find an expression for the variance in (8.36), we approximate the Gold and Kasami sequences by a random code sequence consisting of N_C code symbols. For a random code sequence, the code symbols $a_k^i = 1$ and $a_k^i = -1$ occur with equal probability. Assuming that the data bits are -1 or 1 with equal probability, the variance in (8.36) is written as

$$\psi_a = \frac{1}{2} E\left[R(\tau) + \hat{R}(\tau)\right]^2 + \frac{1}{2} E\left[R(\tau) - \hat{R}(\tau)\right]^2 \tag{8.39}$$

To simplify this expression further, it is instructive to observe one chip of normalized duration overlapped by another chip with delay τ. This situation is depicted in Figure 8.28.

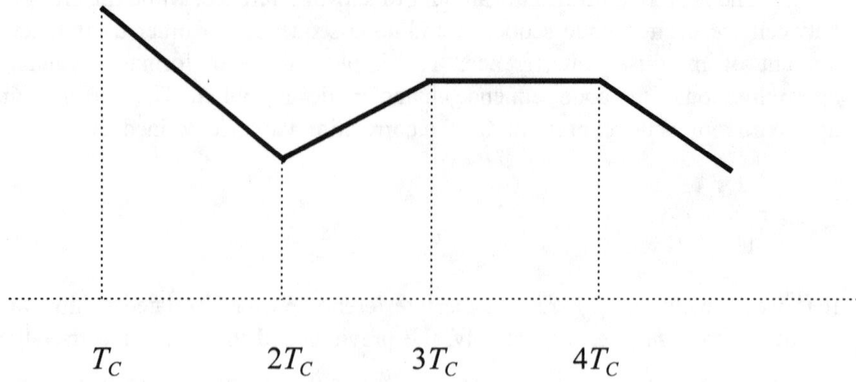

Figure 8.27 Part of a correlation function.

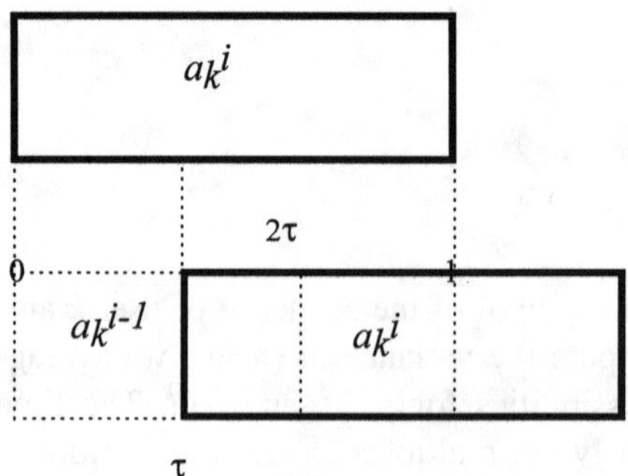

Figure 8.28 Overlap of two chips, a_k^i and a_k^{i-1}.

For the two chips a_k^i and a_k^{i-1} there are two possibilities: either they have the same sign or they have opposite signs. Both situations occur with the same probability in case of a random sequence. The overlap area denoted here by A_o determines the contribution of chip a_k^i to the correlation function (normalized on the code length N_C). In the case of equal bits, the overlap area, normalized on the code length N is $1/N$, while in the case of unequal bits, the part between 0 and τ cancels the part between τ and 2τ, implying that the overlap area is $(1-2\tau)/N_C$. In terms of the overlap area:

$$A_0 = \frac{1}{N_c} \qquad if \;\; a_k^{i-1} = a_k^i$$

$$A_0 = \frac{1-2\tau}{N_c} \qquad if \;\; a_k^{i-1} = a_k^i \tag{8.40}$$

Now, the expectation of A_0^2 is calculated by averaging over all possible delays

$$E\left[A_0^2\right] = \frac{1}{2}\frac{1}{N_c^2} + \frac{1}{2}\int_0^1 \frac{(1-2\tau)^2}{N_c^2} d\tau = \frac{2}{3N_c^2} \tag{8.41}$$

Now, the final simplified expression for the variance in (8.35) is

$$\psi_a = N_c E\left[A^2\right] = \frac{2}{3N_c} \tag{8.42}$$

Before applying this approximation, we verify the approximation for a number of practical Gold sequences and Kasami sequences. Therefore, a large number of these code sequences has been generated, and for each combination of code sequences the expectation of the correlation variance has been calculated by averaging over all possible delays. The results are shown in Figures 8.29(a) to 8.29(e). To explain these figures, we use the Gold code of length 31 as an example. The plot shows a number of bars and a line. The line represents the approximated value for the correlation variance given by (8.42), and each bar represents the simulated correlation variance between two code sequences of the same length. Earlier, we found for example that it is possible to generate 65 code sequences of length 63. For all figures, we only show 50 bars for clarity of the plot. In each plot we first observe a large peak, which is actually the average correlation variance of the autocorrelation function. All the other bars present the average correlation variance of cross-correlations with other codes. By observing Figures 8.29(a) to 8.29(e), it is reasonable to conclude that the approximation of (8.36) is rather good. To strengthen this conclusion, quantitative results are shown in Table 8.3.

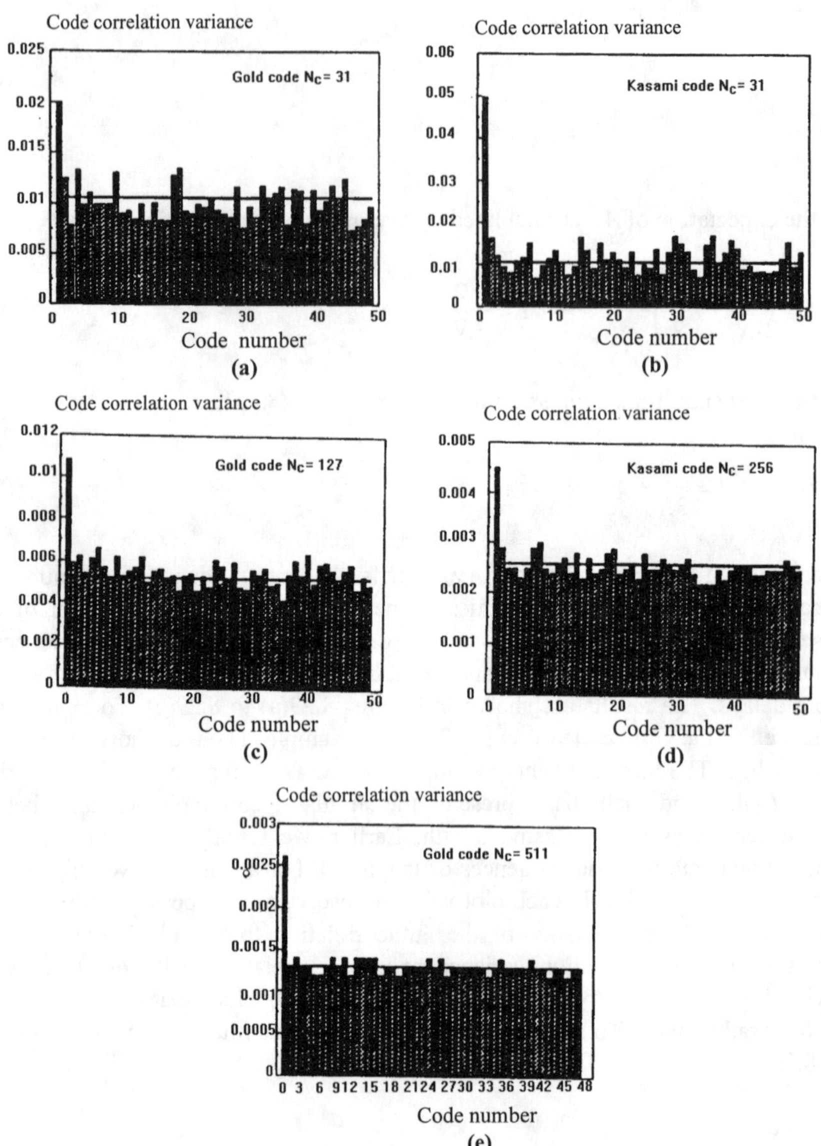

Figure 8.29 (a) Gold code (N_c = 31), (b) Kasami code (N_c = 31), (c) Gold code (N_c = 127), (d) Kasami code (N_c = 255), and (e) Gold code (N_c = 511).

Table 8.3

Quantitative Comparison of Simulation Results and Approximation for Correlation Variance for Several Practical Code Sequences

	Approximation $\psi_a = \dfrac{2}{3N_c}$	Sample Mean m_s	Sample Standard Deviation σ_S
Gold code $N_C = 63$	0.010582	0.010406	0.002182
Kasami code $N_C = 63$	0.010582	0.0119	0.005654
Gold code $N_C = 127$	0.005249	0.005553	0.000722
Kasami code $N_C = 255$	0.002614	0.0025691	0.000289

8.3.4 Conclusions

In this section, a description was given of three types of pseudo-noise sequences suitable for DS-CDMA systems. Gold codes and Kasami code sequences are especially suitable, since it is possible to generate a relatively large number of code sequences with bounded periodic correlation functions. The basic properties of the code sequences were discussed and shift register configurations to generate the sequences were shown. The periodic correlation functions are important when the desired data bit overlaps with a previous or next data bit having the same sign. In the situation where the previous or next overlapping bit has the opposite sign, the so-called a-periodic correlation function is very important. For this a-periodic correlation function, well-defined bounds are found in literature. However, by simulation, we saw that the peaks in the a-periodic correlation function can be much larger than the bounded periodic correlation function.

Furthermore, it was shown that Gold code sequences and Kasami code sequences cannot be generated with arbitrary length. In the first place, the length of a code sequence can always be written as $2^n - 1$ where n is an integer. This implies that we can generate sequences with lengths 31, 63, 127, 255, and so forth. Gold sequences cannot be generated for integer values n being a multiple of 4. This implies that we cannot generate Gold sequences with length $N_C = 255$ since this would imply $n = 8$, which is a multiple of 4. On the other hand, Kasami sequences cannot be generated for odd values of n. For example, Kasami code sequences with length $N_C = 127$ and $N_C = 511$ are not possible. Code sequences with these lengths must be generated by a Gold shift register generation.

PASCAL software was developed to generate Gold sequences and Kasami sequences. For simplification of performance calculations, an approximation of the correlation variance is desirable. A very simple expression for this correlation variance results if we approximate real code sequences of length N_C by random sequences of length N_C. This approximation has been verified by simulation and the results show that the approximation is quite good.

REFERENCES

[1] R. Prasad, *CDMA for Wireless Personal Communications*, Norwood, MA: Artech House, 1996.

[2] Special issue on spread-spectrum communication, *IEEE Trans. Commun.*, Vol. COM-30, May 1982.

[3] M.K. Simon, J.K. Omura, R.A. Scholtz, and B.K. Levitt, "Spread-spectrum communications, Vol. I, II, III, *Comp. Sci*., 1985.

[4] R.A. Scholtz, "The spread-spectrum concept," *IEEE Trans. Commun.*, Vol. COM-25, pp. 748–755, August 1977.

[5] D.J. Torrieri, *Principles of Secure Communication Systems*, Norwood, MA: Artech House, 1985.

[6] G.R. Cooper and C.D. McGillem, *Modern Communication and Spread-Spectrum*, New York: McGraw-Hill Company, 1986.

[7] S.G. Glisic and P.A. Leppanen (Eds.), *Code Division Multiple Access Communications*, Boston: Kluwer Academic Publishers, 1995.

[8] A.J. Viterbi, *CDMA Principles of Spread-Spectrum Communications*, Reading, MA: Addison-Wesley Publishing Company, 1995.

[9] R.C. Dixon, *Spread-Spectrum Systems*, New York: John Wiley & Sons, 1984.

[10] A.M. Viterbi and A.J. Viterbi, "Erlang capacity of a power controlled CDMA system," *IEEE J. Selected Areas in Comm*., Vol. 11, pp. 892–899, August 1993.

[11] D.P. Sarwate and M.B. Pursley, "Crosscorrelation properties of pseudo-random and related sequences," *Proc. IEEE*, Vol. 68, No. 5, May 1980.

[12] S.W. Golomb, *Shift Register Sequences*, San Francisco, CA: Holden-Day, 1967.

[13] J.L. Massey and J.J. Uhran, Jr., "Final report for multipath study," *Contract NASS-10786*, Univ. Notre Dame, IN, 1969.

[14] R. Gold, "Optimal binary sequences for spread spectrum multiplexing," *IEEE Trans. Inf. Theory*, Vol. IT-13, pp. 619–621, October 1967.

Chapter 9
Multiple Access Protocols

9.1 INTRODUCTION

Agreement among users on the means of communication is known as protocol. When users use a common medium for communications, it is called multiple access. Thus, the multiple access protocol is defined as the agreement and set of rules among users for successful transmission of information using a common medium. Whenever some resource is used and thus accessed by more than one independent user, the need for a multiple access protocol arises. In the absence of such a protocol, conflicts occur if more than one user tries to access the resource at the same time. Therefore, the multiple access protocol should avoid or at least resolve these conflicts. Thus, the multiple access technique is defined as the function sharing (limited) common transmission resource among (distributed) terminals in a network.

The multiple access protocols addressed in this chapter are those used in communication systems in which the resource to be shared is the communication channel [1–27]. In this case, the reason for sharing the resource is mainly the connectivity environment. In wireless communication systems, an added reason is the scarceness of the resource; there is only one ether [1]. When instead of a common medium with an access for every user, there is a network that consists of point-to-point links (different medium), but one of these links is simultaneously shared by many users, the multiplexing technique is needed. Multiplexing is the transmission of different informations in the same physical link.

Table 9.1 shows the difference between multiple access and multiplexing.

Table 9.1

Difference Between Multiple Access and Multiplexing

	Multiple Access	*Multiplexing*
Resource	Network	Link
Terminal connectivity	Matrix	Point-to-(multi-) point
Switching	Yes (≤ 3)	No (OSI ≤ 2)
Possible	(OSI: open systems interconnection)	
Topologies	- Bus - Star - Ring - Tree	Path(s)
Control	- Central - Distributed	Terminal

The design of a protocol is usually accomplished with a specific goal (environment) in mind, and the properties of the protocol are mainly determined by the design goal. But if we rule out the environment-specific properties, we can still address a number of properties that any good multiple access protocol should possess:

- The first and foremost task of the multiple access protocols addressed here is to share the common transmission channel among the users in the system. To do this, the protocol must control the way in which the users access (transmit onto) the channel by requiring that the users conform to certain rules. The protocol controls the allocation of channel capacity to the users.
- The protocol should perform the allocation such that the transmission medium is used efficiently. Efficiency is usually measured in terms of channel throughput and the delay of the transmissions.
- The allocation should be fair toward individual users; that is, not taking into account any priorities that might be assigned to the users, each user should (on the average) receive the same allocated capacity.
- The protocol should be flexible in allowing different types of traffic (e.g., voice and data).
- The protocol should be stable. This means that if the system is in equilibrium, an increase in load should move the system to a new equilibrium point. With an unstable protocol an increase in load will force the system to continue to drift to even higher load and lower throughput.

- The protocol should be robust with respect to equipment failure and changing conditions. If one user does not operate correctly, this should affect the performance of the rest of the system as little as possible.

In this book, the wireless mobile environment is what we are most interested in. In such an environment, we can be more specific about some of the protocol properties, especially on the robustness with respect to changing conditions. In the wireless mobile environment, the protocol should be able to deal with:

- The hidden terminal problem [two terminals are out of range (hidden from) of each other by a hill, a building, or some physical obstacle opaque to UHF signals but both are within the range of the central or base station].
- The near-far effect (transmissions from distant users are more attenuated than transmissions from users close by).
- The effects of multipath fading and shadowing experienced in radio channels.
- The effects of co-channel interference in cellular wireless systems caused by use of the same frequency band in different cells.

Many of the protocol properties mentioned above are conflicting and a trade-off has to be made during the protocol design. The trade-off depends on the environment and the specific use for the protocol one has in mind.

This chapter is organized as follows. Section 9.2 presents the classification of multiple access protocols. Slotted ALOHA, unslotted nonpersistent ISMA, slotted nonpersistent ISMA, slotted 1-persistent ISMA, and slotted ISMA with collision detection are investigated in Section 9.3. In Section 9.4, capture effect is introduced. Throughput analysis of these protocols is carried out in a single-cell radio environment in Section 9.5. Section 9.6 extends the investigation of BPSK modulation. An analytical model is developed in Section 9.7 to evaluate the performance of a cellular DS-CDMA system.

9.2 CLASSIFICATION OF MULTIPLE ACCESS PROTOCOLS

Starting in 1970 with the ALOHA protocol (discussed later in this section), a number of multiple access protocols have been developed. Numerous ways have been suggested to divide these protocols into groups [14,16]. In this book, multiple access protocols are classified into three main groups (Figure 9.1): the contentionless protocols, the contention protocols, and the class of CDMA protocols [1].

The contentionless (or scheduling) protocols avoid the situation in which two or more users access the channel at the same time by scheduling the transmissions of the users. This is either done in a fixed fashion where each user is allocated part of the transmission capacity, or in a demand-assigned fashion where the scheduling only takes place between the users that have something to transmit.

With the contention (or random access) protocols, a user cannot be sure that a transmission will not collide because other users may be transmitting (accessing the channel) at the same time. Therefore, these protocols need to resolve conflicts if they occur.

The CDMA protocols do not belong to either the contentionless or the contention protocols. As explained in Chapter 8, CDMA falls between the two groups. In principle, it is a contentionless protocol where a number of users are allowed to transmit simultaneously without conflict. However, if the number of simultaneously transmitting users rises above a threshold, contention occurs.

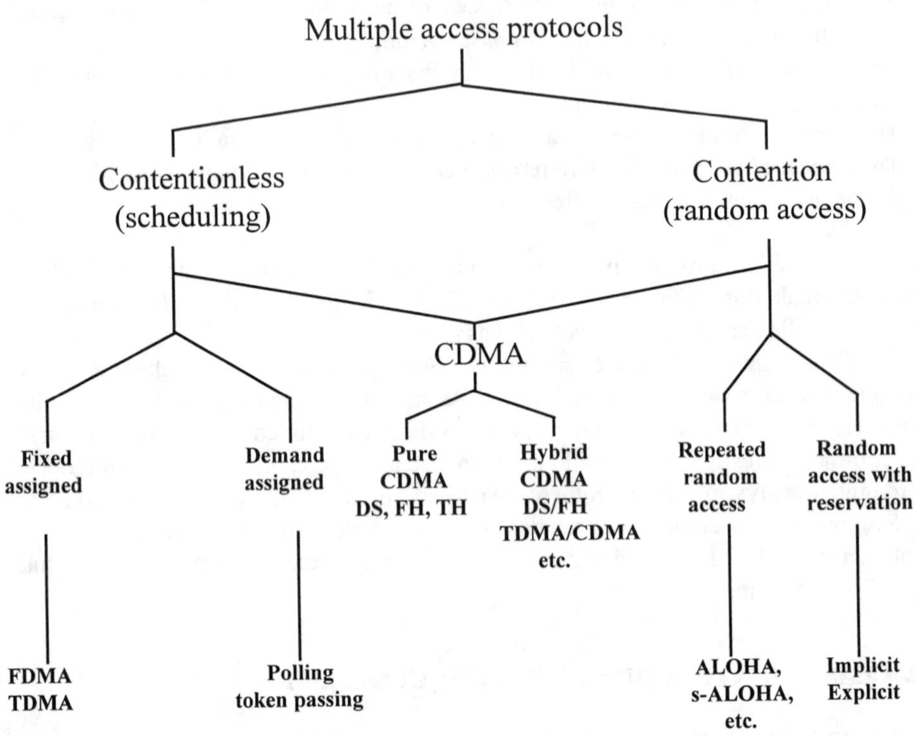

Figure 9.1 Classification of the multiple access protocols.

The contention protocols are further subdivided into repeated random protocols and random protocols with reservation. In the latter, the initial transmission of a user uses a random access method to get access to the channel. However, once the user accesses the channel, further transmissions of that user are scheduled until the user has nothing more to transmit. Two major types of protocols are known as implicit and explicit reservations. Explicit reservation protocols use a short reservation packet to request transmission at scheduled times. Implicit reservation protocols are designed without the use of any reservation packet.

CDMA protocols are subdivided into pure CDMA and hybrid CDMA. With CDMA forming a separate group and taking into account the subgroups of the contentionless and contention protocols, we end up with six categories.

9.2.1 Contentionless (Scheduling) Multiple Access Protocols

The contentionless multiple access protocols avoid the situation in which multiple users try to access the same channel at the same time by scheduling the transmissions of all users. The users transmit in an orderly scheduled manner so every transmission will be a successful one. The scheduling can take two forms:

1. *Fixed assignment scheduling*. With these types of protocols, the available channel capacity is divided among the users such that each user is allocated a fixed part of the capacity, independent of its activity. The division is done in time or frequency. The time division results in the TDMA protocol, where transmission time is divided into frames and each user is assigned a fixed part of each frame, not overlapping with parts assigned to other users. TDMA is illustrated in Figure 9.2(a). The frequency division results in the FDMA protocol, where the channel bandwidth is divided into nonoverlapping frequency bands and each user is assigned a fixed band. Figure 9.2(b) illustrates FDMA.
2. *Demand assignment scheduling*. A user is only allowed to transmit if it is active (if it has something to transmit). Thus, the *active* (or ready) users transmit in an orderly scheduled manner. Within the demand assignment scheduling, we distinguish between centralized control, and distributed control. With centralized control, a single entity schedules the transmissions. An example of such a protocol is the roll-call polling protocol. With distributed control, all users are involved in the scheduling process and such a protocol is the token-passing protocol.

9.2.2 Contention (Random) Multiple Access Protocols

With contention multiple access protocols there is no scheduling of transmissions. This means that a user getting ready to transmit does not have exact knowledge of when it can transmit without interfering with the transmissions of other users. The user may or may not know of any ongoing transmissions (by sensing the channel), but it has no exact knowledge about other ready users. Thus, if several ready users start their transmissions more or less at the same time, all of the transmissions will fail. This possible transmission failure makes the occurrence of a successful transmission a more or less random process. The random access protocol should resolve the contention that occurs when several users transmit simultaneously .

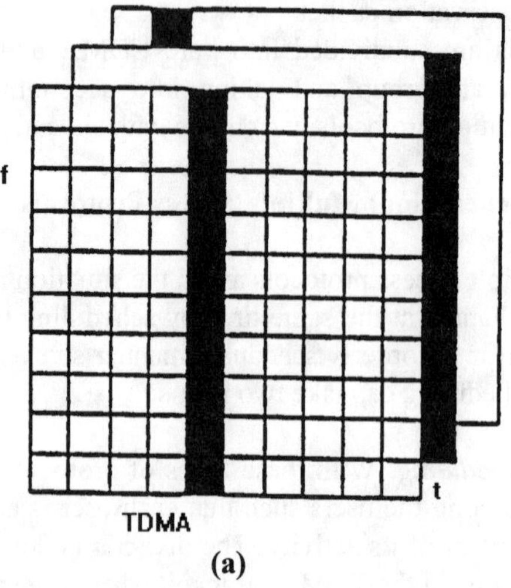

(a)

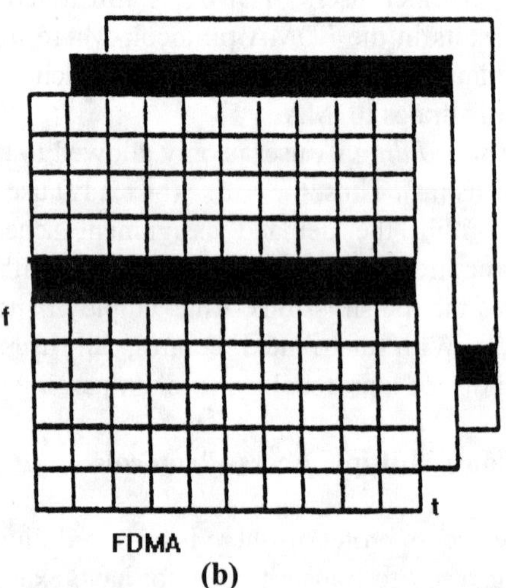

(b)

Figure 9.2 (a) TDMA, and (b) FDMA.

We subdivide contention multiple access protocols into two groups, the repeated random access protocols (e.g., pure (p)-ALOHA, slotted (s)-ALOHA, carrier sense multiple access (CSMA), ISMA, etc.), and random access protocols with reservation (e.g., reservation ALOHA (r-ALOHA), packet reservation multiple access (PRMA), etc.). With the former protocols, every transmission a user makes is as described above. With every transmission there is a possibility of contention. With the latter protocols, only in its first transmission does a user not know how to avoid collisions with other users. However, once a user has successfully completed its first transmission (once the user has access to the channel), future transmissions of that user will be scheduled in an orderly fashion so that no contention can occur. Thus, after a successful transmission, part of the channel capacity is allocated to the user and other users will refrain from using that capacity. The user loses its allocated capacity if, for some time, it had nothing to transmit.

9.2.3 CDMA Protocols

This section deals with the class of CDMA protocols, which rely on coding to achieve their multiple access property. In this section the basic principles of CDMA protocols are discussed and placement of the protocols between the contentionless (scheduling) and contention (random access) protocols is explained.

CDMA protocols do not achieve their multiple access property by a division of the transmissions of different users in either time or frequency, but instead make a division by assigning each user a different code. This code is used to transform a user's signal into a wideband signal (spread-spectrum signal). If a receiver receives multiple wideband signals, it uses the code assigned to a particular user to transform the wideband signal received from that user back to the original signal. During this process, the desired signal power is compressed into the original signal bandwidth, while the wideband signals of the other users remain wideband signals and appear as noise when compared with the desired signal.

As long as the number of interfering users is not too large, the SNR is large enough to extract the desired signal without error. Thus, the protocol behaves as a contentionless protocol. However, if the number of users rises above a certain limit, the interference becomes too large for the desired signal to be extracted and contention occurs, making the protocol interference limited. Therefore, the protocol is basically contentionless unless too many users access the channel at the same time. This is why we place CDMA protocols between contentionless and contention protocols.

There are several ways to divide CDMA protocols into a number of groups. The most common is the division based on the modulation method used to obtain the wideband signal. This division leads to four protocol types: DS-CDMA, FH-CDMA, TH-CDMA, and hybrid CDMA, where the last group of protocols uses a combination of the modulation methods of the other protocols. A general classification of CDMA protocol is shown in Figure 9.3. Future air interface multiple access scheme for FPLMTS/IMT-2000/UMTS is expected to be one of the CDMA protocols. A description of CDMA is given in Chapter 8. Multicarrier (MC)-CDMA and multitone (MT-CDMA) are introduced in Chapter 11.

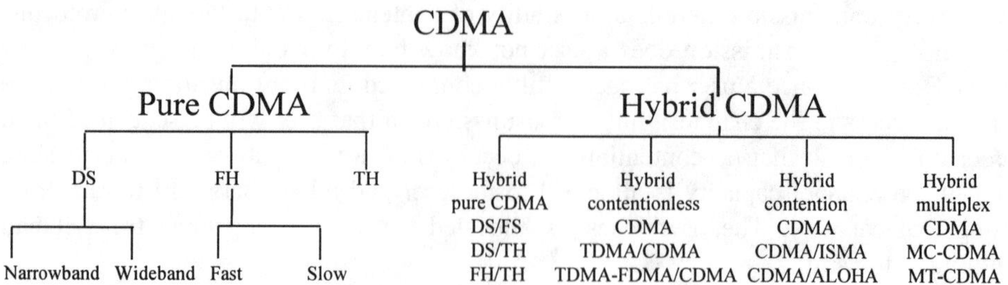

Figure 9.3 Classification of CDMA protocols.

9.3 SOME RANDOM ACCESS PROTOCOLS

We consider mobile terminals communicating with a centrally located base station using packet radio. The best known method for random access packet communication, ALOHA [2], suffers heavily due to collisions among packets. The CSMA protocol [3,4] offers higher capacity, but its performance is affected by the problem of "hidden terminals." A third category (ISMA) [5–11], in which the central base station controls the flow of packets from the mobile terminals, reduces these two problems of collisions among data packets and hidden terminals.

The inhibit delay fraction [5–11] is a decisive parameter in designing a packet communication systems using ISMA, and is defined as the fraction of the packet duration necessary to inhibit other mobile packet transmissions by a broadcast outbound signaling channel (base -to- terminal) in the event that the inbound multiple access channel (terminal -to- base) is already busy.

ISMA is a class of protocols which we can divide into two subclasses: the nonpersistent ISMA protocols and the p-persistent ISMA protocols.

In the nonpersistent ISMA protocols, if the inbound channel is idle, a user will transmit a packet, otherwise the user will wait a random time and then try again. At first sight it may seem that this protocol totally avoids collisions, however, due to the propagation delay between two users, it is possible that a user considers the inbound channel idle and starts transmission while another transmission is already in progress.

A user is informed of a collision by the absence of an acknowledgment packet from the receiving station. Upon detecting the collision, the packet is rescheduled for transmission a random time later.

A special case of the p-persistent ISMA protocols is the 1-persistent ISMA protocol. The protocol is the same as the nonpersistent ISMA protocol except when a user knows the inbound channel is busy. In this case the transmission is not rescheduled a random time later. Instead, the user keeps monitoring the channel until it becomes idle and then immediately transmits its packet.

As a result, all users that become ready during a busy channel will transmit as soon as the channel becomes idle, which leads to a high probability of collision at the end of a successful transmission.

To avoid the collision of packets accumulated while the channel was busy, the start of the transmission times of the accumulated packets can be randomized. This can be done by letting all users that generate a packet during a busy channel transmit as soon as the channel becomes idle with a probability p. With a probability 1 minus p, they will defer their transmission for τ seconds (with τ being the maximum propagation delay between any two users in the system). After the τ seconds, the deferred terminal will sense the channel again and apply the same algorithm as before.

With this p-persistent ISMA protocol the probability of more than one ready user transmitting when the channel becomes idle can be made quite small by choosing a small value for p.

With the nonpersistent and p-persistent ISMA protocols, a user will not learn about a collision until after its whole packet has been transmitted. The reason for this is, of course, that an acknowledgment packet will only be sent after the complete packet has been received by the receiving user. Since a collision can only occur within the propagation delay after the start of the transmission, it is a waste of time to transmit more of the packet if a collision has occured within this period. For this reason the ISMA with collision detect (ISMA-CD) protocols have been developed. With these protocols a user keeps monitoring the channel while it is transmitting. If it detects a collision, it aborts its transmission as soon as possible thus saving time.

9.3.1 Slotted ALOHA

In 1970, the University of Hawaii proposed the pure ALOHA system [2,12]. The pure ALOHA random access protocol is very simple: let the number of packets generated and retransmitted in the network be Poisson distributed, with a mean generation rate of λ packets per second. Every packet will have the same duration τ. Then, the mean offered channel traffic G in packets per time slot is

$$G = \tau\lambda \tag{9.1}$$

The probability that n packets are generated is given by

$$\Pr[n] = \frac{G^n}{n!}\exp(-G) \tag{9.2}$$

For zero packet, the probability is just $\exp(-G)$. In an interval two packets-time long, the mean number of packets generated is $2G$. So the probability of no other traffic being initiated during the entire vulnerable period is thus given by $P_0 = \exp(-2G)$. Using $S = GP_0$ we get the throughput

$$S = G\exp(-2G) \tag{9.3}$$

With a simple modification of this access algorithm, namely that the messages are required to be sent in the slot time between two synchronization pulses, and can be started only at the beginning of a time slot, the rate of collisions can be reduced by one half [13]. This version is called the slotted ALOHA. The throughput is now

$$S = G\exp(-G) \tag{9.4}$$

In Figure 9.4, a comparison of the two throughputs are shown.

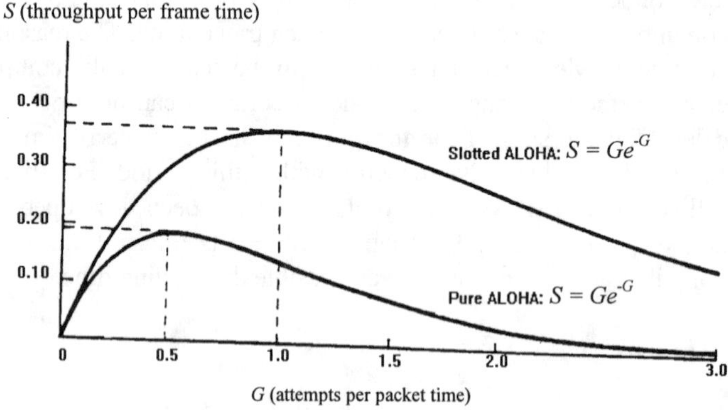

Figure 9.4 Throughput of pure ALOHA and slotted ALOHA.

9.3.2 Unslotted Nonpersistent ISMA

CSMA [3,4] reduces the probability of collision by allowing terminals to sense the carrier due to other users' transmissions. But CSMA cannot avoid collisions if two terminals are out of range of each other or if they are separated by some physical obstacle opaque to UHF signals. Two such terminals are known as hidden from each other. Therefore, we instead consider another category known as ISMA [5–11].

ISMA reduces the two problems of collision among data packets and hidden terminals. To prevent collisions among data packets transmitted from mobile terminals to a common base station, the inbound multiple access channel is supplemented by a broadcast inhibit-signaling channel (base-to-terminals). In the latter, inhibit bits indicate the state of the inbound channel: busy or idle. As soon as the base station receives an inbound packet, the outbound signaling channel broadcasts the "busy" condition to all terminals. A fraction of the constant packet length is necessary to inhibit the mobile packet transmissions, defined as the inhibit delay fraction *d*, a dimensionless quantity. The time fraction required to reverse this condition is assumed to have the same value too. Thus, mobile terminals are inhibited from transmission until the inbound channel is free, preventing most packet collisions. The inhibited packets are rescheduled according to a retransmission distribution. An illustration of ISMA timing is given in Figure 9.5.

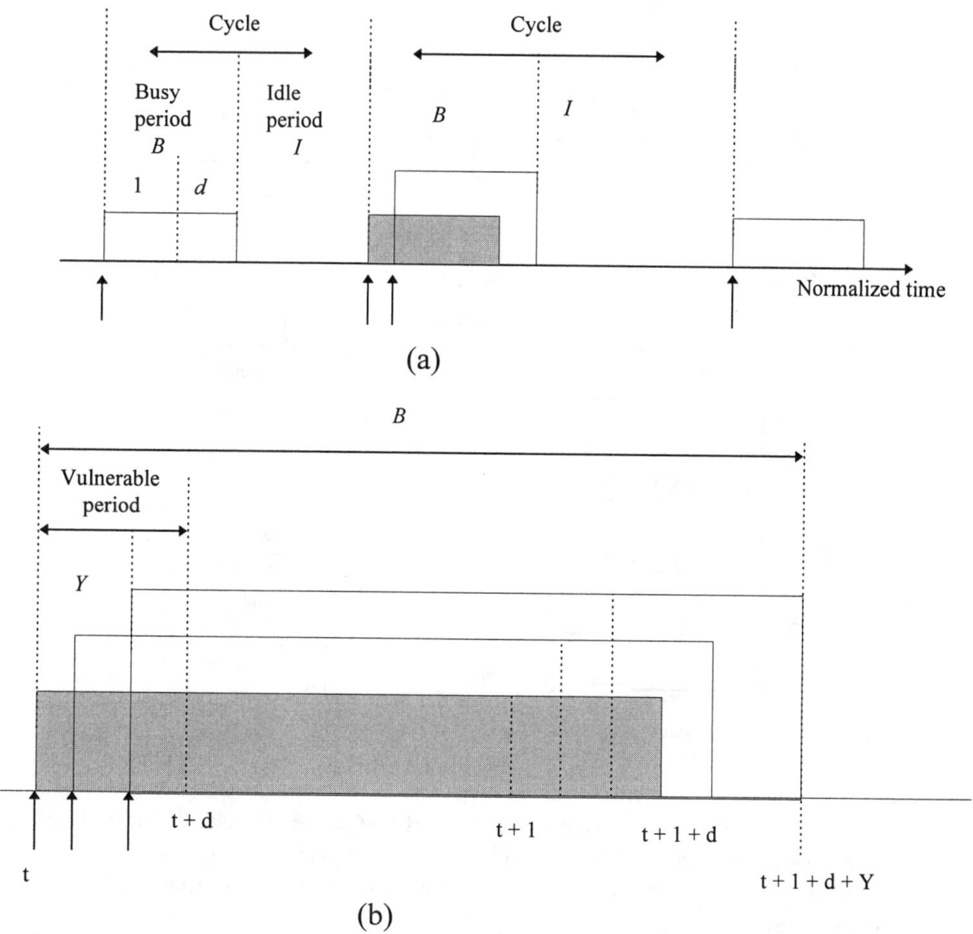

Figure 9.5 Unslotted nonpersistent ISMA timing: (a) cycle structure, and (b) unsuccessful transmission period.

Figure 9.6 shows the terminal transmission of unslotted nonpersistent ISMA. As with ALOHA, we assume that the number of packets generated in the network be Poisson distributed, with a mean generation rate λ. The mean offered traffic is G. The packets are of identical length, normalized to have length 1.

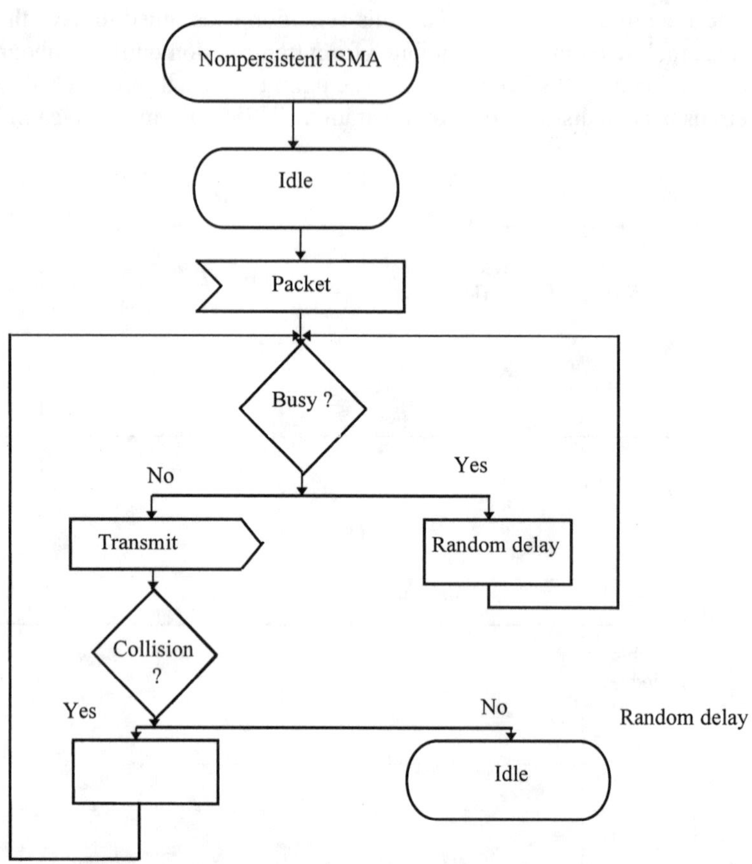

Figure 9.6 Terminal transmission of unslotted nonpersistent ISMA.

Let t be the time of arrival of a packet that senses the channel idle. Because of the inhibit delay fraction d, there should be only one packet arriving between t and $t + d$ (vulnerable period). If there is more than one packet, the transmission will be unsuccessful.

In case of a collision, the channel will therefore be busy for some (random) duration between $1 + d$ and $1 + 2d$. This period in which a transmission takes place is referred to as the transmission period. Suppose that $t + Y$ is the time of occurrence of the last packet arriving between t and $t + d$. The transmission of all packets arriving in $(t, t + Y)$ is completed at $t + Y + 1$. And because of the inhibit delay d, the channel is sensed idle after $t + Y + 1 + d$.

The expression for throughput is

$$S = \frac{U}{B + I} \tag{9.5}$$

where U denotes the time during a cycle that the channel is used without conflicts, B is the expected duration of the busy period, and I is the expected duration of the idle period.

The probability that the channel is used without conflicts is the probability that no other terminal senses the channel during the vulnerable period d. Therefore,

$$U = \exp(-dG) \tag{9.6}$$

The average duration of an idle period is $1/G$. The duration of a busy interval is $1 + Y + d$, where Y is the expected value of Y.

The distribution of Y is

$$F_Y(y) \underline{\Delta} \Pr\left\{\text{no arrival occurs in an interval of length } d - y\right\}$$

$$= \exp\{-G(d - y)\} \qquad (y \le d) \tag{9.7}$$

The average of Y is therefore given by

$$\overline{Y} = E(Y) = d - \frac{1}{G}(1 - \exp(-dG)) \tag{9.8}$$

So, using the above-derived expressions, the throughput is

$$S = \frac{\exp(-dG)}{(1 + 2d) + \frac{1}{G}\exp(-dG)} \tag{9.9}$$

where the term in the denominator is called the expected length of a cycle ($\overline{L}_c$).

9.3.3 Slotted Nonpersistent ISMA

We assume that all terminals have the same characteristics except their relative positions. Data packets arrive at the terminals generated by the users. The transmission occurs only when the terminal gets permission to transmit. When there is a busy signal, transmission attempts are unsuccessful (see Figure 9.7). Such packets are rescheduled for later by putting them into retransmitting buffer, which transmits again at the start of the next idle time slot, and the duration of a time slot is assumed to be exactly the same as the transmission time of a single packet. Thus, there is either no collision or complete collision of the packets. Since there is no collision detection, a higher-level process is needed to determine which packets were lost due to this complete collision and need to be retransmitted.

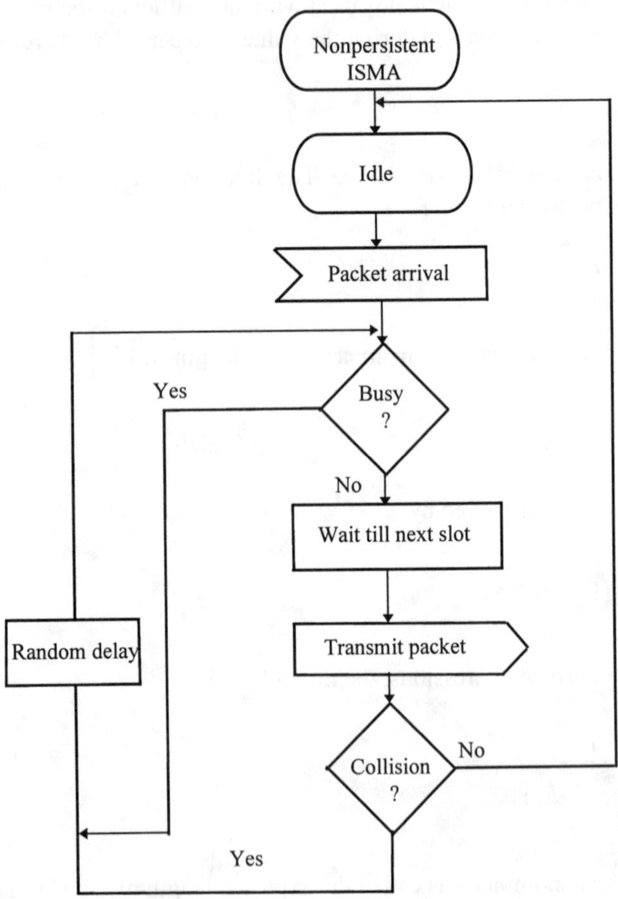

Figure 9.7 Terminal transmission of slotted nonpersistent ISMA protocol.

A terminal in the blocked state will listen to the busy signal transmitted by the base station to determine when the channel becomes idle.

Each terminal of slotted nonpersistent ISMA is restricted to start transmission only at the beginning of a time slot, and the duration of a time slot is assumed to be exactly the same as the transmission time of a single packet. Thus, there is either no collision or complete collision of the packets, in which case an unsuccessful packet is subsequently retransmitted after a random number of slots. To prevent these collisions among data packets transmitted from mobile terminals to common base station, the inbound multiple access channel can be supplemented by a broadcast inhibit-signaling channel. The inhibit bits indicate the state of inbound channel: busy or idle. The moment when the base station receives an inbound packet, the outbound signaling channel broadcasts the busy condition to all terminals. A fraction of constant packet length is needed to inhibit the mobile packet transmissions, the inhibit delay fraction *d*, a dimensionless quantity. Thus, mobile terminals are inhibited from transmission until the inbound channel is free, preventing collisions. These packets are rescheduled according to retransmission distribution [8–10]. The successful and unsuccessful transmissions in ISMA are illustrated in Figure 9.8. Inhibit delay fraction is shown in Figure 9.8 by notation *d*, ratio of propagation delay (T_d), and packet timing (T_p).

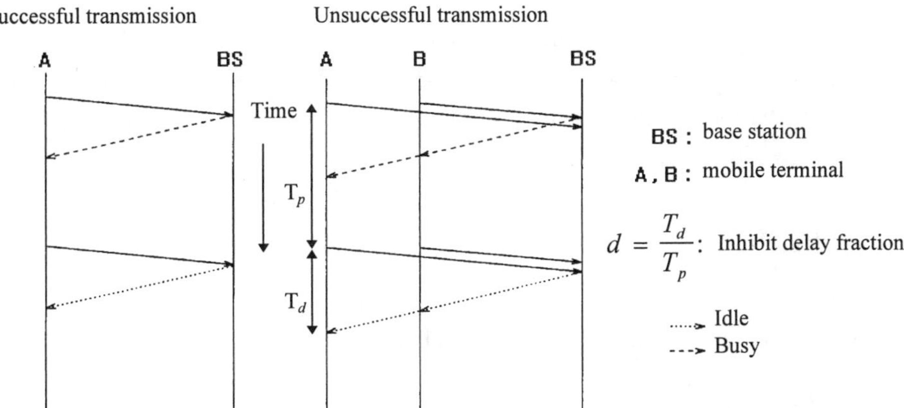

Figure 9.8 Slotted Non-persistent ISMA packet timing.

From Figure 9.8, the length of an idle period is at least one time slot. When the idle period is only one slot long, it means there is at least one arrival in the first slot of the idle period. For the period to be two slots long means that there is no arrival at the first slot, but there is at least one arrival in its second slot. Continuing the reasoning and considering the Poisson scheduling process, we have

$$I = \frac{d}{1-\exp(-dG)} \tag{9.10}$$

A collision might occur if two or more packets arrive within the same slot and are scheduled for transmission in the next slot. A busy period will contain k transmission periods if there is at least one arrival in the slot of each of the first $k-1$ transmission periods, and no arrival in the last slot of the kth transmission period. Thus, the busy period is

$$B = \frac{1+d}{\exp(-dG)} \tag{9.11}$$

The expected useful time is found as

$$U = \frac{B}{1+d} P_{\text{suc}} \tag{9.12}$$

where P_{suc} is the probability of a successful transmission period. We have

$$P_{suc} = \frac{\text{Prob}\left[\text{single arrival within a slot}\right]}{\text{Prob}\left[\text{more arrivals within a slot}\right]} \tag{9.13}$$

$$P_{suc} = \frac{dG\exp(-dG)}{1-\exp(-dG)} \tag{9.14}$$

Putting all these together, we get the throughput

$$S = \frac{dG\exp(-dG)}{1+d-\exp(-dG)} \tag{9.15}$$

9.3.4 Slotted 1-Persistent ISMA

In nonpersistent ISMA there are situations in which the channel is idle, although one or more users have packets to transmit. The 1-persistent ISMA is an alternative to nonpersistent ISMA that avoids such situations. When the outbound channel sends the busy signal to the terminal, it persists to wait and transmits as soon as the channel becomes idle (see Figure 9.9). Thus, the channel is always used if there is a user with a packet.

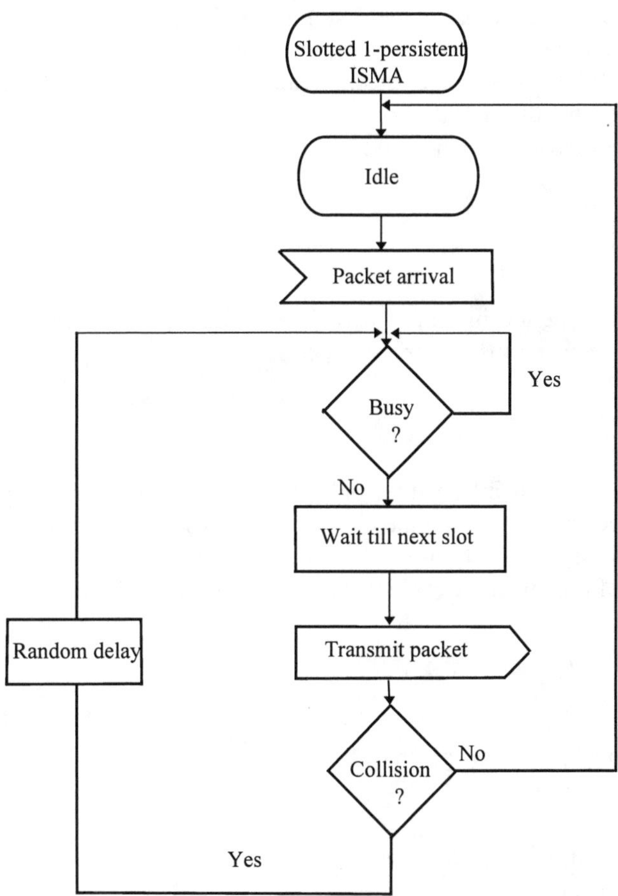

Figure 9.9 Terminal transmission of slotted 1-persistent ISMA protocol.

The analysis of slotted 1-persistent ISMA is similar to the slotted nonpersistent ISMA. The mean of idle period is same as in slotted nonpersistent ISMA (Figure 9.8). The busy period will contain k transmission periods, if at least one packet arrives in each of the first $k-1$ transmission periods and no packets arrive in the kth transmission period. So, the busy period B is

$$B = \frac{1+d}{\exp\left(-G(1+d)\right)} \tag{9.16}$$

The probability of success in the first transmission period in a busy period, P_{suc1}, is different from the success probability in any other transmission period within the busy period, P_{suc2}. For the first transmission period in a busy period to be successful, we need the last slot of the idle period to contain exactly one arrival, taking in account that there is at least one arrival there, since it is the last slot. Hence,

$$P_{suc_1} = \frac{dG\exp(-dG)}{1-\exp(-dG)} \tag{9.17}$$

For any transmission period in a busy period to be successful, we must have exactly one arrival during the previous transmission period,

$$P_{suc_2} = \frac{G(1+d)\exp[-G(1+d)]}{1-\exp[-G(1+d)]} \tag{9.18}$$

The channel carries useful information only during successful transmission periods. P_{suc1} is the expected amount of time the channel carries useful information during these periods and P_{suc2} is the probability of success in each of these transmission periods. The expected amount of time within a cycle that the channel carries useful information is

$$U = P_{suc_1} + \frac{B-(1+d)}{1+d} P_{suc_2} \tag{9.19}$$

Therefore, the throughput can be given by

$$S = \frac{G\exp[-(1+d)G][1+d-\exp(-dG)]}{(1+d)[1-\exp(-dG)]+d\exp[-(1+d)G]} \tag{9.20}$$

9.3.5 Slotted ISMA With Collision Detection

In this group of protocols the throughput is the ratio between the expected useful time spent in a cycle to the cycle duration itself. To improve the throughput, the cycle length must therefore be reduced. A cycle is composed of a transmission period followed by an idle period. Shortening the idle period is possible by means of 1-persistent ISMA protocols, which do not perform very well under most loads. The duration of the successful transmission periods should not be changed, for this is the time the channel is used best. Hence, performance can be improved by shortening the duration of unsuccessful transmission periods.

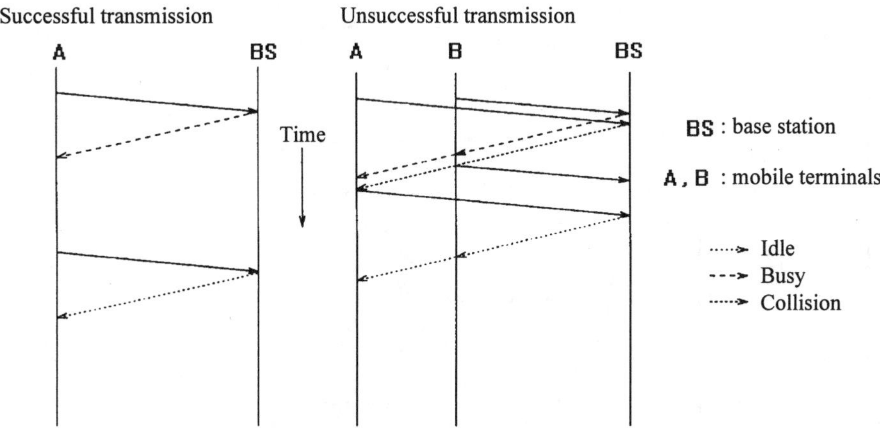

Figure 9.10 Collision detection timing.

In all ISMA protocols, a transmission that is initiated when the inbound channel is idle, and the outbound channel sends an idle signal that reaches all the users after at most one inhibit delay fraction, *d*. Beyond this time, the inbound channel is sensed busy. Figure 9.10 describes the behavior of the slotted ISMA/CD protocols when a collision occurs.

Slotted Nonpersistent ISMA\CD

The operation of slotted ISMA\CD protocols is identical to the operation of the corresponding ISMA protocols, expect that if a collision is detected during transmission by the base station, the transmission is aborted and the packet is scheduled for later transmission (see Figure 9.11).

The nonpersistent ISMA\CD time alternates between busy periods (which includes both successful and unsuccessful transmission periods) and idle periods. The length of the busy period is B (see Figure 9.12):

$$B = \frac{P_{\text{suc}}(1+d) + (1-P_{\text{suc}})(\gamma' + d)}{\exp(-dG)} \tag{9.21}$$

where P_{suc} is the probability of a successful transmission period

$$P_{\text{suc}} = \frac{Gd\exp(-dG)}{1-\exp(-dG)} \tag{9.22}$$

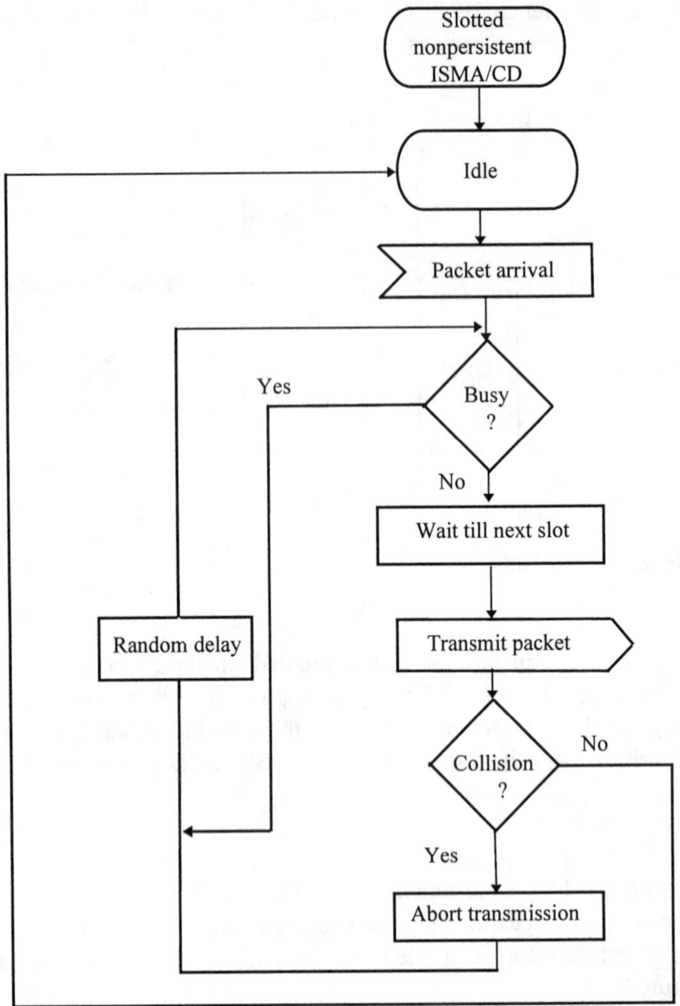

Figure 9.11 Slotted nonpersistent ISMA/CD.

and γ' is the time taken to complete a successful transmission of a user.

$$\gamma' = 2d + d_{CD} + d_{cr} \tag{9.23}$$

d_{CD} is the time taken by the base station to detect a collision, and d_{cr} is the time taken to let the other user know that a collision took place.

The distribution of the idle period is identical to that computed for slotted nonpersistent ISMA, thus, the expected length of the idle period is

$$I = \frac{d}{1-\exp(-dG)} \tag{9.24}$$

Amount of time within a cycle during which the channel carries useful information is

$$U = \frac{P_{suc}}{\exp(-dG)} \tag{9.25}$$

Combining the equations, throughput can be analyzed

$$S = \frac{U}{B+I} = \frac{dG\exp-(dG)}{dG\exp(-dG)+[1-\exp(-dG)-dG\exp(-dG)]\gamma'+d} \tag{9.26}$$

$\gamma' = 1$, throughput is identical to slotted nonpersistent ISMA.

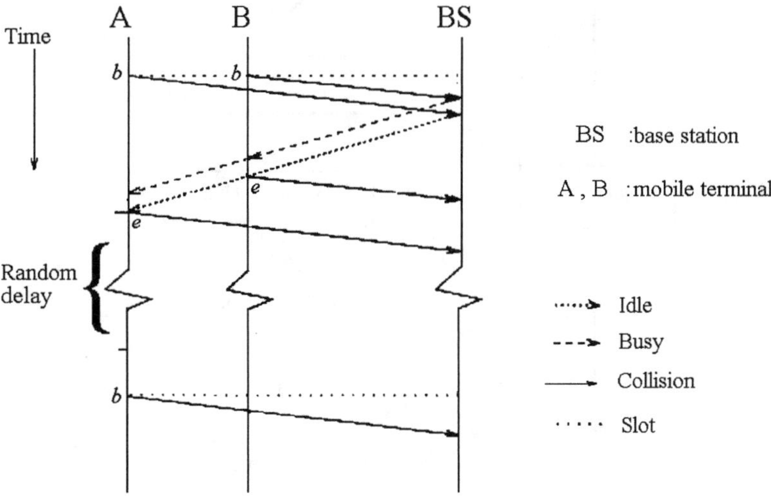

Figure 9.12 Slotted nonpersistent ISMA/CD Packet Timing.

Slotted 1- persistent ISMA/CD

The operation of this protocol is similar to that of 1-persistent ISMA, except that if a collision is detected during transmission by the base station, the transmission is aborted and the packet is scheduled for later transmission (see Figure 9.13).

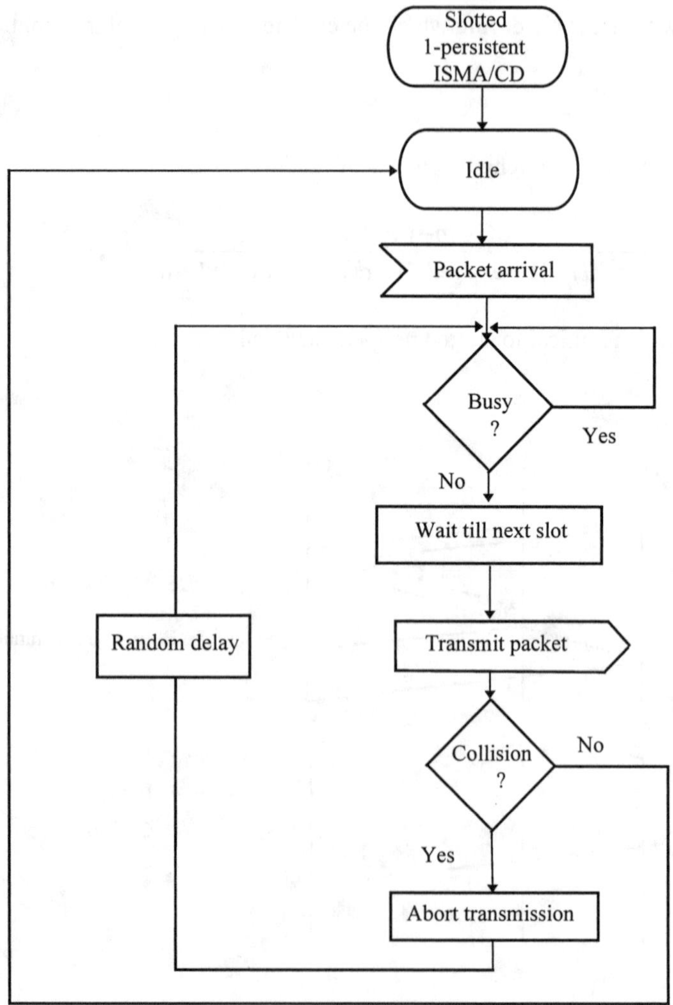

Figure 9.13 Terminal transmission of slotted 1 persistent ISMA/CD.

With the 1-persistent ISMA/CD, the time also alternates between busy periods (containing successful and unsuccessful transmission periods) and the idle periods, and a cycle in a busy period followed by an idle period (see Figure 9.14). A success or failure of a transmission period is the busy period depends (only) on the length of the preceding transmission period, except for the first transmission period that depends on the arrivals during the preceding slot.

The average length of an idle period is *d*/1− exp(− *dG*), the throughput is given by [14]

$$S = \frac{U}{B+I} = \frac{U(d)}{B(d) + \dfrac{d}{1-\exp(-dG)}} \tag{9.27}$$

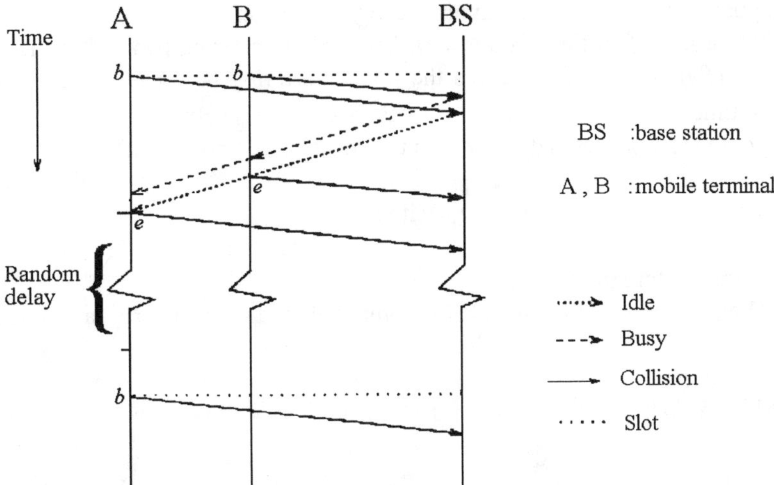

Figure 9.14 Slotted 1-persistent ISMA/CD Packet Timing.

9.4 CAPTURE - MODEL

The development of packet communication protocols has been based mainly on the assumption that when two or more packets collide (overlap in time) all packets are lost. However, a realistic radio receiver is able to be captured by the strongest of the overlapping packets and thus receive this packet correctly. This is called the capture effect. The probability that one packet captures the common receiver in the presence of several other (weather) contenders depends on the power differences with which the packets are received. In mobile radio channels these power differences with capture probabilities, are increased in a natural way as a result of near-far effects and fading. The influence of capture effects on the performance of the slotted ALOHA and the unslotted ISMA protocols have been reported in numerous papers, e.g. [9, 18].

In this section, the behavior of the capture model is given for slotted ALOHA, unslotted nonpersistent ISMA, slotted nonpersistent ISMA, slotted 1-persistent ISMA, slotted nonpersistent ISMA/CD, and slotted 1-persistent ISMA/CD.

9.4.1 Slotted ALOHA With Capture

As we see in Section 9.3.1, if there is more than one transmission, all packets are destroyed because of overlapping. In this section, a more realistic model is given: the capture model.

When data packets competing for access to a common radio receiver arrive from different distances and with independent fading levels, it is no longer certain that all colliding packets will always be annihilated by each other.

We assume that there is a test packet and n interfering packets. If the power of the former (P_s) sufficiently exceeds the joint interfering power (P_n) during a certain section of time slot of duration t_w ($0 < t_w \leq \tau$), capture is assured.

So, the test packet is destroyed in the collision if (and only if)

$$P_s / P_n < z_0 \qquad \text{during } t_w \text{ with } n > 0 \tag{9.28}$$

with z_0 as the capture ratio [18].

The probability of being able to capture the receiver in an arbitrary time slot is

$$P_{\text{capt}}(Z_0) = 1 - \sum_{n=1}^{\infty} R_n \operatorname{Prob}\{P_s / P_n < z_0\} \tag{9.29}$$

Using the capture probability (8.29), the channel throughput is stated as

$$S = G\left[1 - \sum_{n=1}^{\infty} R_n F_{Z_n}(z_0)\right] \tag{9.30}$$

with $F_{zn}(z_0) = \text{Prob}\{P_s/P_n < z_0\}$.

Defining the signal to interference ratio for the test packet and n contenders

$$Z_n \triangleq \frac{P_s}{P_n} \qquad 0 \leq Z_n < \infty \tag{9.31}$$

and the random variable

$$W \underline{\Delta} P_n \qquad 0 \le W < \infty \tag{9.32}$$

we may write the two-dimensional pdf

$$f_{Zn,W}(zw) \underline{\Delta} f_{P_s,P_n}(p_s, p_n) \left| \frac{\delta(p_s, p_n)}{\delta(z, w)} \right| \tag{9.33}$$

By virtue of the stochastic independent P_s and P_n, this becomes

$$f_{Z_n,W}(z, w) = f_{P_s}(zw) f_{P_n}(w) w \tag{9.34}$$

so the pdf for Z_n

$$f_{Z_n}(z) = \int_0^\infty f_{P_s}(zw) f_{P_n}(w) w dw \tag{9.35}$$

and the corresponding distributed function

$$F_{Z_n}(z_0) = \int_0^{z_0} dz \int_0^\infty f_{P_s}(zw) f_{P_n}(w) w dw \tag{9.36}$$

9.4.2 Unslotted Nonpersistent ISMA With Capture

In Section 9.3.2, nonpersistent ISMA without capture is discussed. Now we take the capture effect into consideration, which means that the colliding packets are not always destroyed, but have a certain probability of being received successfully by the receiver.

The throughput of nonpersistent ISMA with capture is defined as the ratio of the probability of success of the test packet and expected length of the cycle, assuming that there is an infinite population of users with the total arrival rate of (new plus inhibited, rescheduled) packets, G, is Poisson distributed. Thus, the throughput is:

$$S \underline{\Delta} \frac{P_{\text{succes}}(z_0)}{\overline{L}_c} \tag{9.37}$$

where the expected length of a cycle is defined as the sum of a busy period and the following idle period, as we see in Section 9.3.2

$$\overline{L}_c \underline{\Delta} 1 + 2d + \frac{1}{G} \exp(-dG) \tag{9.38}$$

The probability of success is

$$P_{\text{succes}}(z_0) = \sum_{n=0}^{\infty} R_n (n+1)\left[1 - \text{Prob}\{P_s / P_n < z_0\}\right] \tag{9.39}$$

and R_n, the probability of n interfering packets overlapping a test packet, is given by

$$R_n = \frac{(dG)^n}{n!} \exp(-dG) \tag{9.40}$$

So finally the throughput is

$$S = \frac{\sum_{n=0}^{\infty} (n+1) R_n \left[1 - \text{Prob}\left\{\frac{P_s}{P_n} < z_0\right\}\right]}{1 + 2d + \frac{1}{G}\exp(-dG)} \tag{9.41}$$

9.4.3 Slotted Nonpersistent ISMA With Capture

Using the capture effect on the results derived in Section 9.3.3, now not all the packets are annihilated by collision. Expression for the expected useful time is given by (9.12):

$$U = \frac{B}{1+d} P_{\text{suc}} \tag{9.42}$$

Because we are now taking the capture effect in consideration, the probability of a successful transmission period is

$$P_{\text{suc}} = \frac{dG}{1 - \exp(-dG)} \sum_{n=0}^{\infty} R_n (n+1)\left[1 - F_{Z_n}(z_0)\right] \tag{9.43}$$

Using (9.5), (9.10), (9.42), and (9.43) we finally get the throughput for the slotted nonpersistent ISMA with capture

$$S = \frac{dG}{1 + d - \exp(-dG)} \sum_{n=0}^{\infty} R_n (n+1)\left[1 - F_{Z_n}(z_0)\right] \tag{9.44}$$

9.4.4 Slotted 1-Persistent ISMA With Capture

Using the capture effect on the results derived in Section 9.3.4, the probability of a successful transmission period in a busy period is

$$P_{\text{suc}_1} = \frac{dG\exp(-dG)}{1-\exp(-dG)}\sum_{n=1}^{N} R_n\left[1-F_{zn}(z_0)\right] \tag{9.45}$$

and the success probability in any other transmission period with the busy period is

$$P_{\text{suc}_2} = \frac{G(1+d)\exp[-G(1+d)]}{\left[1-\exp\left(-G(1+d)\right)\right]}\sum_{n=1}^{N} R_n\left[1-F_{zn}(z_0)\right] \tag{9.46}$$

Thus, giving the throughput for the slotted 1-persistent ISMA with capture

$$S = \frac{G\exp\left[-(1+d)G\right]\left[1+d-\exp(-dG)\right]}{(1+d)\left[1-\exp(-dG)\right]+d\exp\left[-(1+d)G\right]}\sum_{n=1}^{N} R_n\left[1-F_{zn}(z_0)\right] \tag{9.47}$$

9.4.5 Slotted Nonpersistent ISMA/CD With Capture

Taking capture effect into consideration, means that the colliding packets are not always be destroyed, but have a certain probability of being received successfully by the receiver.

The throughput of slotted nonpersistent ISMA/CD with capture is defined as ratio of the probability of success of the test packet, and expected length of the cycle assuming that there is N population of users with the total arrival rate of packets, G. Thus, the throughput is obtained using the capture effects on the results derived in Section 9.3.5

$$S = \frac{U}{B+I} = \frac{dG\exp-(dG)}{dG\exp(-dG)+\left[1-\exp(-dG)-dG\exp(-dG)\right]\gamma'+d}\sum_{n=1}^{N} R_n\left[1-F_{zn}(z_0)\right] \tag{9.48}$$

9.4.6 Slotted 1-persistent ISMA/CD With Capture

The throughput of slotted 1-persistent ISMA\CD with capture is derived using the results of Section 9.3.5

$$S = \frac{U(d)}{B(d) + \dfrac{d}{1 - \exp(-dG)}} \sum_{n=1}^{N} R_n \left[1 - F_{zn}(z_0)\right] \tag{9.49}$$

9.5 THROUGHPUT ANALYSIS CONSIDERING RAYLEIGH FADING, LOGNORMAL SHADOWING, AND NEAR-FAR EFFECT

This section presents the throughput study of unslotted nonpersistent ISMA protocol in a wireless environment having Rayleigh fading, lognormal shadowing, and near-far effect using Chapter 4 and Section 9.4.2. The near-far model is developed in [9].

9.5.1 Pure Rayleigh Fading

With pure Rayleigh fading, one obtains the packet success probability [9]

$$P_{\text{success}}(z_0) = \sum_{n=0}^{\infty} \left[\frac{(dG)^n}{n!} \frac{(n+1)}{(nz_0 + 1)} \exp(-dG) \right] \tag{9.50}$$

Using (9.37), (9.38), and (9.50), the throughput is written as

$$S = \frac{G \sum_{n=0}^{\infty} \left[\dfrac{(dG)^n}{n!} \dfrac{(n+1)}{(nz_0 + 1)} \right]}{1 + G(1 + 2d)\exp(dG)} \tag{9.51}$$

Zdunek et al. [6] derived the expression for the throughput for the incoherence interference cumulation, which is given as

$$S = \frac{G\left(1 + \dfrac{dG}{z_0 + 1}\right) \exp\left(\dfrac{dG}{z_0 + 1}\right)}{1 + G(1 + 2d)\exp(dG)} \tag{9.52}$$

Figure 9.15 compares the throughput(s) against packet arrival rate G characteristics for coherent (9.51) and incoherent (9.52) interference cumulation for Rayleigh fading only with $d = 0.05$ and $z_0 = 4$.

Figure 9.15 shows that coherent interference produces slightly higher throughput than the incoherent case. The throughput characteristics for the ideal ISMA channel ($z_0 \rightarrow \infty$) is also included in Figure 9.15 for the sake of comparison, which is expressed as $G \exp(-Gd) / [G(1 + 2d) + \exp(-Gd)]$.

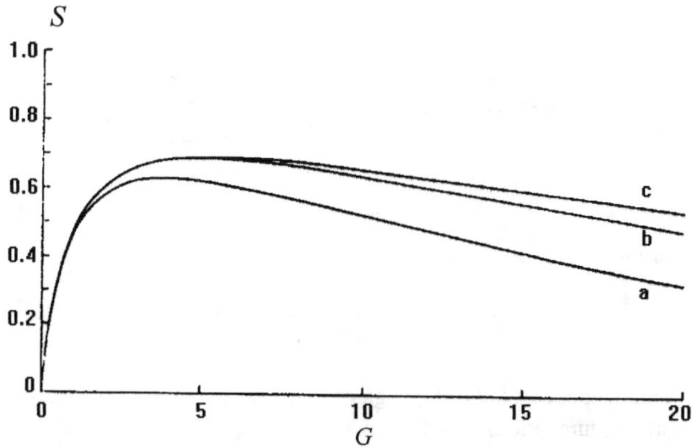

Figure 9.15 Throughput curves. (a) Ideal ISMA (capture ratio $z_0 = \infty$), (b) Rayleigh incoherent fading, and (c) Rayleigh coherent fading with inhibit delay fraction $d = 0.05$ and capture ratio $z_0 = 4.0$.

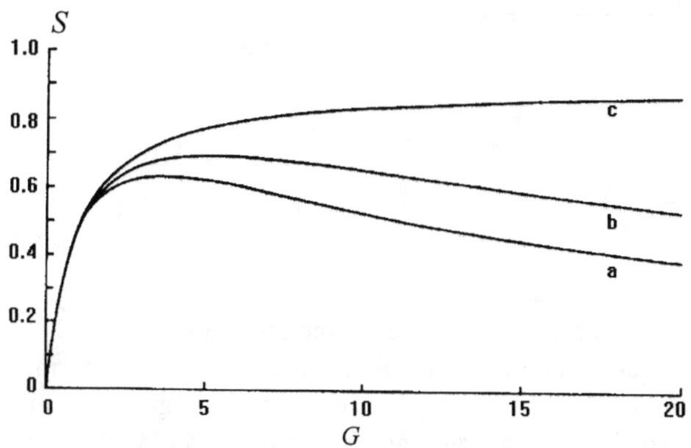

Figure 9.16 Throughput curves for Rayleigh coherent fading with capture ratio $z_0 = 4.0$. (a) Inhibit delay fraction $d = 0.1$, (b) $d = 0.05$, and (c) $d = 0.0$.

Figure 9.16 illustrates the effect of inhibit delay fraction on the throughput characteristics. The higher the delay fraction, the lower the throughput.

The influence of capture ratio (z_0) is shown in Figure 9.17. The perfect capture effect ($z_0 = 1$) shows the highest throughput and the higher capture effect corresponds to the lower throughput.

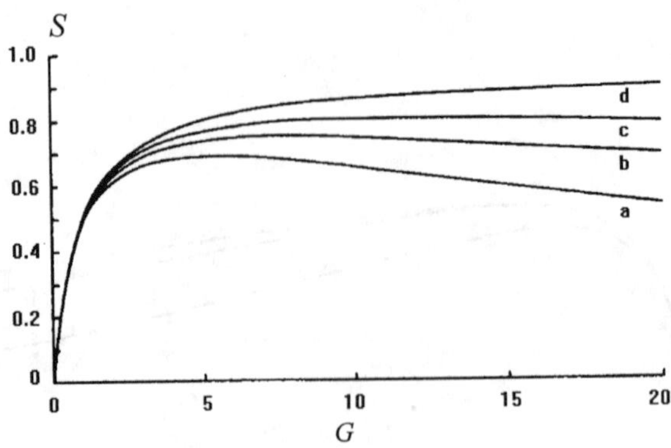

Figure 9.17 Throughput curves for Rayleigh coherent fading with inhibit delay fraction $d = 0.05$. (a) Capture ratio $z_0 = 4.0$, (b) $z_0 = 2.0$, (c) $z_0 = 1.41$, and (d) $z_0 = 1.0$.

9.5.2 Pure Lognormal Shadowing

In the case of pure lognormal shadowing [9,27]

$$P_{\text{success}}(z_0) = \sum_{n=0}^{\infty}\left[\frac{(n+1)}{\sqrt{2\pi}}\frac{(dG)^n}{n!}\exp(-dG)\int_{-\infty}^{L_1}\exp\left(-\frac{u^2}{2}\right)du\right] \tag{9.53}$$

where

$$L_1 = \frac{[m_d - m_u - \ln(z_0)]}{(\sigma^2 + \sigma_u^2)} \tag{9.54}$$

Here, $\xi_d = \exp(m_d)$ is the area mean power of the test packet (desired signal), σ_u^2 and $\xi_u = \exp(m_u)$ are the logarithmic variance and the area mean of a lognormal variable that is approximately equivalent to the sum of n independent lognormal interferers. The m_u and σ_u^2 are determined using Schwartz and Yeh's method [28,29]. Hence, using (9.37), (9.38), (9.53), and (9.54)

$$S = \frac{G\sum_{n=0}^{\infty}\left[\frac{(n+1)}{\sqrt{2\pi}}\frac{(dG)^n}{n!}\int_{-\infty}^{L_1}\exp\left(-\frac{u^2}{2}\right)du\right]}{1 + G(1+2d)\exp(dG)} \tag{9.55}$$

The effect of lognormal shadowing is shown in Figure 9.18 for $\sigma = 6$ dB, 8 dB and 12 dB, $d = 0.05$, and $z = 4.0$. Figure 9.18 shows that the throughput increases with increase in variance.

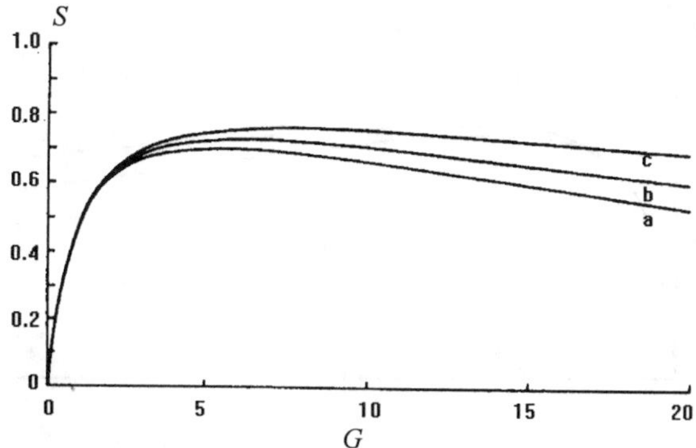

Figure 9.18 Throughput curves for lognormal shadowing with inhibit delay fraction $d = 0.05$ and capture ratio $z_0 = 4.0$. (a) Shadowing spread $\sigma = 6$ dB, (b) $\sigma = 8$ dB, and (c) $\sigma = 12$ dB.

9.5.3 Combined Rayleigh Fading and Shadowing

The packet success probability for the combined Rayleigh fading and shadowing is obtained [9]

$$p_{\text{success}}(z_0) = \sum_{n=0}^{\infty}\left[\frac{(n+1)}{\sqrt{\pi}}\frac{(dG)^n}{n!}\exp(-dG)\int_{-\infty}^{\infty}\exp(-w^2)f(w)dw\right] \tag{9.56}$$

where

$$f(w) = 1 - \left[1 + \exp\left\{m_d - \ln(z_0) - m_u - \left[2(\sigma^2 + \sigma_u^2)\right]^{\frac{1}{2}} w\right\}\right]^{-1} \tag{9.57}$$

Equations (9.37), (9.38), (9.56), and (9.57) yield the throughput

$$S = \frac{G\sum_{n=0}^{\infty}\left[\frac{(n+1)}{\sqrt{\pi}}\frac{(dG)^n}{n!}\int_{-\infty}^{\infty}\exp(-w^2)f(w)dw\right]}{1 + G(1+2d)\exp(dG)} \tag{9.58}$$

Figure 9.19 depicts the throughput characteristics for combined Rayleigh fading and lognormal shadowing for $d = 0.05$, $z_0 = 4.0$, and $\sigma = 6$ dB. The plot for lognormal shadowing only for $\sigma = 6$ dB is also included for comparison. The combined effect presents the higher throughput than the individual effect alone.

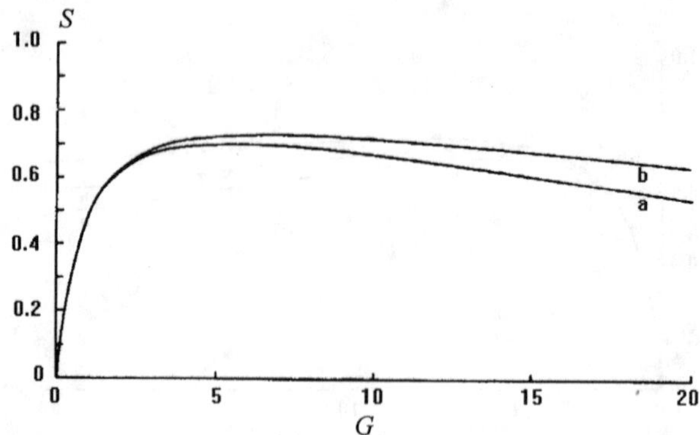

Figure 9.19 Throughput curves. (a) Lognormal shadowing only with shadowing spread $\sigma = 6$ dB, and (b) Lognormal shadowing ($\sigma = 6$ dB) plus Rayleigh with inhibit delay fraction $d = 0.05$ and capture ratio $z_0 = 4.0$.

9.5.4 Combined Shadowing and Near-Far Effect

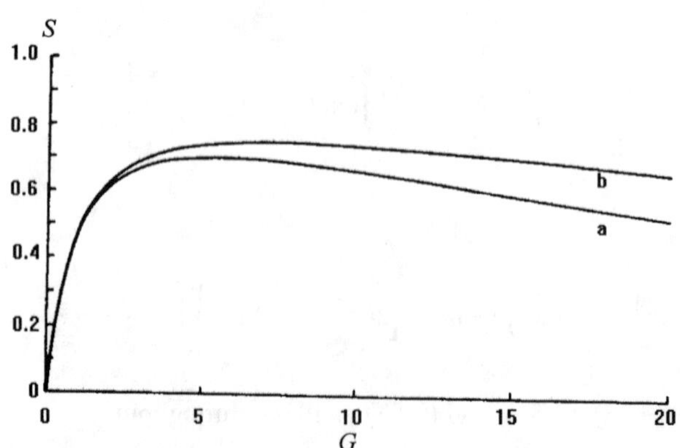

Figure 9.20 Throughput curves for lognormal shadowing plus near-far effect with inhibit delay fraction $d = 0.05$, capture ratio $z_0 = 4.0$, and shadowing spread $\sigma = 6$ dB. (a) Spatial spread $\sigma_d = 0$ dB, and (b) $\sigma_d = 8$ dB with $\gamma = 4$.

In the event of combined shadowing and near-far effect, the expression of the packet success probability and throughput are similar to (9.53) and (9.55), respectively, with changes

$$\sigma_u^2 \to \sigma_{u1}^2$$
$$m_u \to m_{u1} \tag{9.59}$$

σ_{u1}^2 and m_{u1} are calculated using Schwartz and Yeh's method [28,29] with the variance σ_1^2 of individual interferer.

The combined lognormal shadowing plus near-far effect is compared with pure lognormal shadowing in Figure 9.20 for σ = 6 dB and σ_d = 8 dB with γ = 4. As expected, the combined effect enhances the throughput.

9.5.5 Combined Rayleigh Fading, Shadowing, and Near-Far Effect

In the generalized case of combined multipath fading, shadowing and near-far effect $f_n(p_{ni})$ is obtained

$$f_n(p_{ni}) = \int_0^{\infty} \frac{1}{\sqrt{(2\pi)}\sigma_{u1} p_{0ni}^2} \exp\left[-\frac{p_{ni}}{p_{0ni}} - \frac{(\ln p_{0ni} - m_{u1})^2}{2\sigma_{u1}^2} \right] dp_{0ni} \tag{9.60}$$

where $p_{0ni} = \sum_{i=1}^{n} p_{0i}$

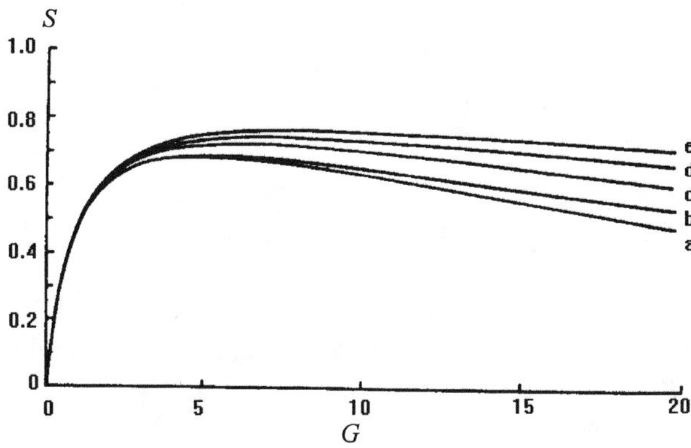

Figure 9.21 Throughput curves. (a) Ideal ISMA, (b) Rayleigh coherent fading only, (c) lognormal shadowing only (shadowing spread σ = 8 dB), (d) Rayleigh fading plus lognormal shadowing (σ = 8 dB), and (e) combined Rayleigh fading, lognormal shadowing (σ = 8 dB), and near-far effect (spatial spread σ_d = 8 dB with γ = 4) with inhibit delay fraction d = 0.05 and capture ratio z_0 = 4.0.

We obtain $F_p(p_d > wz_0)$

$$F_p(p_d > wz_0) = \int_{wz_0}^{\infty} dp_d \int_0^{\infty} \frac{1}{\sqrt{(2\pi)}\sigma_t p_{0d}^2} \exp\left[-\frac{p_d}{p_{0d}} - \frac{(\ln p_{0d})^2}{2\sigma_t^2}\right] dp_{0d} \tag{9.61}$$

The packet success probability and throughput are in the form of (9.56) and (9.58), respectively, with the simple substitution of (9.59).

Figure 9.21 presents the overall throughput characteristics for $\sigma = 8$ dB, $\sigma_d = 8$ dB with $\gamma = 4$, $d = 0.05$, and $z_0 = 4$, and also includes a plot for particular cases with the same parameters, for comparison. The combined effect of Rayleigh fading, lognormal shadowing, and near-far effect offers the higher channel capacity.

9.5.6 Comparison Between Slotted ALOHA, CSMA, and ISMA

In combined Rayleigh fading, shadowing, and near-far effect, the throughput of a slotted ALOHA network, considering coherent cumulation of interference, is obtained using [9,30]

$$S_{ac} = G\left[1 - \sum_{n=1}^{\infty} P_n F_{Z_n}(z_0)\right] \tag{9.62}$$

where

$$F_{z_n}(z_0) = \frac{1}{\sqrt{(\pi)}} \int_{-\infty}^{\infty} \exp(-V^2) f(V) dV \tag{9.63}$$

$$f(V) = \left[1 + \exp\left\{m_d - \ln(z_0) - m_{u1} - \left[2(\sigma_t^2 + \sigma_{u1}^2)\right]^{1/2} V\right\}\right]^{-1} \tag{9.64}$$

Note that for $\sigma_d \to 0$, (9.62) reduces to the throughput for Rayleigh fading plus shadowing with $f(V)$ given in [30]

$$f(V) = \left[1 + \exp\left\{m_d - \ln(z_0) - m_u - \left[2(\sigma^2 + \sigma_u^2)\right]^{1/2} V\right\}\right]^{-1} \tag{9.65}$$

and for $\sigma_d \to 0$, $\sigma \to 0$ and all packet powers equal, one obtains from (9.62) the throughput for Rayleigh fading, only with $F_{zn}(z_0)$ expressed as in [9,30]

$$F_{z_n}(z_0) = \frac{nz_0}{1 + nz_0} \tag{9.66}$$

In the event of lognormal shadowing only [22], (9.62) simplifies to

$$S_{ac} = G\left[1 - \sum_{n=1}^{\infty} P_n Q(l_n)\right] \tag{9.67}$$

where

$$Q(l_n) \Delta \frac{1}{\sqrt{2\pi}} \int_{l_n}^{\infty} \exp\left(-\frac{t^2}{2}\right) dt \tag{9.68}$$

$$l_n \Delta \{m_d - \ln(z_0) - m_u\} / (\sigma^2 + \sigma_u^2)^{1/2} \tag{9.69}$$

In the limit $z_0 \to \infty$, $F_{zn}(z_0) \equiv 1$, (9.62) reduces to the classical expression for slotted ALOHA in an ideal channel given by $G \exp(-G)$.

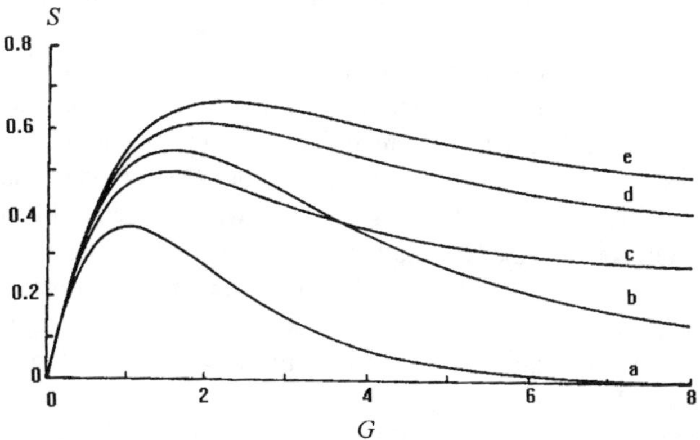

Figure 9.22 Throughput curves for slotted ALOHA. (a) Ideal channel, (b) Rayleigh coherent fading only, (c) Lognormal shadowing (shadowing spread σ = 8 dB), (d) Rayleigh fading plus lognormal shadowing (σ = 8 dB), and (e) combined Rayleigh fading, lognormal shadowing, (σ = 8 dB) and near-far effect (spatial spread σ_d = 8 dB with γ = 4).

Figure 9.22 shows the throughput characterizations for slotted ALOHA for σ = 8 dB, σ_d = 8 dB with γ = 4, and z_0 = 4.0. For completeness, the influence of various propagation conditions on the throughput of slotted ALOHA is also included in Figure 9.22. Like ISMA network, the combined presence of Rayleigh fading, shadowing, and near-far effect in the slotted ALOHA channel results in higher channel capacity. Comparing Figures 9.21 and 9.22, it is found that the channel capacity of ISMA network is always higher than that of slotted ALOHA. With regard to CSMA, it is well known that there is a serious degradation in its throughput performance due to hidden terminal problems. However, the results presented here for ISMA are also applicable for those CSMA systems where the hidden terminal problem is negligible. Otherwise, a separate investigation is required for modeling a CSMA system in mobile radio environments, including the effect of hidden terminals.

9.6 CAPTURE PROBABILITY AND THROUGHPUT ANALYSIS IN A RICIAN / RAYLEIGH ENVIRONMENT USING BPSK MODULATION

The performance of mobile radio communication systems is analyzed in terms of the capture probability and throughput, for the slotted ALOHA and the unslotted nonpersistent ISMA protocol with receiver capture. The influence of Rayleigh faded interference and AWGN is studied on the performance of the two protocols considering BPSK modulated desired test signal, which suffers from Rician fading and the near-far effect. BCH (Bose-Chaudhuri-Hocquenghem) error correction codes consisting of packets with equal numbers of bits are introduced to improve the performance of the system, considering both fast and slow multipath fading.

In this section, the test terminal is assumed to be relatively close to the base station, so LOS component is present, and thus, the test terminal also suffers from Rician fading. The interfering terminals are assumed to be either (1) near to the border of the cell and therefore have larger propagation distances without LOS component and suffer from Rayleigh fading, or (2) close to the base station with Rayleigh fading due to the nonexistence of LOS between the base station and interfering terminal antennas. Thus, the signal envelope distribution of the interfering terminals is always Rayleigh-faded irrespective of their positions with respect to the base station. The above assumptions are made for the simplicity of analysis to obtain initial results for the random multiple access protocols in a microcellular system.

In this section, the performance of mobile radio communication systems using the slotted ALOHA and the unslotted nonpersistent ISMA protocol is investigated in terms of the capture probability and the throughput. The capture probability is defined as the probability that a packet, consisting of a number of bits, is successfully transmitted, while the throughput is defined as the number of packets that are received correctly within a certain time. The effect of using BCH error correction codes on the performance of mobile communication systems with random access techniques, investigated earlier (e.g., in [11, 32, 33]) is also investigated.

9.6.1 Channel Model

In the model presented in this section, the test terminal is assumed to be very near to the base station, so the test terminal is in LOS of the base station. In this way, the amplitude of the test signal can be modeled by a Rician pdf [34],

$$f_{\rho_0}(\rho_0|\overline{p}_0) = \frac{\rho_0}{\sigma^2}\exp\left(-\frac{\rho_0^2/2+s^2/2}{\sigma^2}\right)I_0\left(\frac{s\rho_0}{\sigma^2}\right) \tag{9.70}$$

where r_0 is the amplitude of the received instantaneous test signal, σ^2 is the scattered power, s is the peak value of the specular component, $\overline{p}_0$ is the local-mean power of the test signal, with $\overline{p}_0 = s^2/2+\sigma^2$ and $I_0()$ is the modified Bessel function of the first kind and zero-th order [35]. An important parameter is the Rician parameter K, where K is defined as the ratio of the power of the specular (line-of-sight) component and the power of the scattered (reflected) component, thus $K = s^2/2\sigma^2$. In this way it is possible to express (9.70) in terms of the instantaneous power p_0 of the test signal (with $p_0 = ½r_0^2$) and the Rician parameter K

$$f_{p_0}(p_0|\overline{p}_0) = \frac{(1+K)}{\overline{p}_0}\exp\left(-K-\frac{p_0(1+K)}{\overline{p}_0}\right)I_0\left(\sqrt{\frac{4p_0K(1+K)}{\overline{p}_0}}\right) \tag{9.71}$$

Shadowing, consisting of very slow fluctuations of the local mean power, is not considered in this section. The interfering terminals are assumed to be situated near the border of the cell, so LOS between any of the mobile terminals and base station is absent (i.e., $K = 0$). In that case, the amplitude r_j of the jth signal ($j = 1,\ldots, i$, where i is the number of interfering signals) is Rayleigh distributed, while the instantaneous power p_j is then negative-exponentially distributed about the local mean power $\overline{p}_j$

$$f_{p_j}(p_j|\overline{p}_j) = \frac{1}{\overline{p}_j}\exp\left(-\frac{p_j}{\overline{p}_j}\right) \tag{9.72}$$

The local mean powers of both test signal and interfering signals are determined by path-loss effects during propagation, which means that the power of the transmitted signal decreases with increasing distance to a certain propagation loss exponent b. This propagation loss exponent b depends on the environment where the mobile communication system is situated. Here, b is assumed to be fixed; that is, $b = 4$. In this way, the local mean power $\overline{p}_j$ of the jth signal (both test signal and interfering signals) at a normalized distance r_j (normalized to the cell radius, so $0 < r_j \leq 1$), with r_j defined as the distance from terminal j to the base station, is given by

$$\overline{p}_j = \frac{1}{r_j^4} \tag{9.73}$$

A radio channel with a total traffic load G, expressed in packets per unit time, or packet per time slot is considered. The terminals are assumed to move randomly in the area around the base station. If the position of the mobile terminal is unknown, the local mean power $\overline{p}_j$ has the following pdf:

$$f_{\overline{p}_j}(\overline{p}_j) = \frac{2\pi}{G} G(r_j) r_j \left| \frac{dr_j}{d\overline{p}_j} \right| \tag{9.74}$$

The average number of packet transmissions per unit time and per unit area at normalized distance r_j from the base station is denoted by $G(r_j)$. The offered traffic $G(r)$ is given by a quasi-uniform spatial distribution function [18]

$$G(r) = \frac{G}{\pi} \exp\left(-\frac{\pi}{4} r^4 \right) \tag{9.75}$$

which is an approximation of the exactly uniform distribution [36], resulting in the fact that almost all traffic arrives from normalized distances $0 < r < 1$. The distribution function of the received local mean power $\overline{p}_j$ can then, after substituting (9.73) and (9.75) in (9.74), be given by

$$f_{\overline{p}_j}(\overline{p}_j) = \frac{\overline{p}_j^{-\frac{3}{2}}}{2} \exp\left(-\frac{\pi}{4\overline{p}_j} \right) \tag{9.76}$$

The pdf of the local mean total interference power p_t, caused by i interfering signals, with $\overline{p}_t = \sum \overline{p}_j$, is determined after i-fold convolution of the pdf of the local mean power, given by (9.76), [18]

$$f_{\bar{p}_t}(\bar{p}_t) = \left[f_{\bar{p}_j}(\bar{p}_j)\right]^{\otimes i} = \frac{i}{2}\bar{p}_t^{-\frac{3}{2}} \exp\left(-\frac{\pi i^2}{4\bar{p}_t}\right) \tag{9.77}$$

9.6.2 Capture Probability

In this section, only BPSK modulation is considered. The receiver input signal $y(t)$ is on the form [36,37]

$$y(t) = \rho_0 k_0 \cos\,(\omega_c t + \phi_0) + \sum_{j=1}^{n} \rho_j k_j \cos\,(\omega_c t + \phi_j) + n(t) \tag{9.78}$$

where k_j represents phase reversals due to BPSK modulation of the jth signal and $n(t)$ represents AWGN. The received energy per bit is $E_b = p_0 T_b = ½ r_0^2 T_b$. For each signal, the amplitude r_j and phase f_j are assumed to remain constant during 1 bit duration T_b. Further, perfect bit synchronization is assumed; for each of the interfering signals, exactly overlapping bit periods are assumed. The receiver is also assumed to be locked to the test signal. In the receiver, $y(t)$ is multiplied by a locally generated cosine ($2\cos\omega_C\, t$) and integrated over the entire bit duration T_b. In a Rayleigh-fading channel, the in-phase carrier components of the i interfering signals are all independent Gaussian variables, with variance $\bar{p}_j$ [37]. BEP for a receiver locked to the BPSK-modulated test signal in the presence of multiple interferers with local mean power and AWGN is on the form [36, 37]

$$P_{be}(be|p_0, \bar{p}_t) = \frac{1}{2}\operatorname{erfc}\left(\sqrt{\frac{p_0 T_b}{N_0 + \bar{p}_t T_b}}\right) \tag{9.79}$$

where erfc represents the complementary error function [35] and N_0 is the (one-sided) spectral power density of the AWGN.

The radio signals are transmitted in packets, each having a number of L bits. The capture probability is defined as the probability that a packet of L bits is received correctly, so all L bits are received without error. A capture model is discussed in Section 8.4, where a test packet captures the receiver if and only if the ratio of the power of the test packet and the total power of i interfering packets exceeds a certain value called the capture ratio. Here, this capture ratio is assumed to be chosen in such a way that the receiver locks to the test packet despite the interfering packets. Capture probability is investigated in this section, where two different types of multipath fading [32] are distinguished: (1) fast multipath fading, where the signal amplitude is assumed independent from bit to bit. The bit errors in successive bits are assumed independent and have equal BEP, and (2) slow multipath fading, where the amplitude and phase of each signal are assumed constant for the entire duration of the packet LT_b.

In the following analysis it is assumed that the location of the test terminal as well as the interfering terminals follow the same spatial distribution given by (9.75). The performance analysis is simplified by assuming that the test and interfering signals are always Rician and Rayleigh faded, respectively, irrespective of their locations with respect to the base station. Rayleigh distribution is a particular case for Rician distribution with no LOS (i.e., Rician parameter $K = 0$).

A. Fast Multipath Fading

Because the instantaneous power p_0 of the test signal is averaged over the pdf (9.71) to take the (Rician) fading statistics into account, and each of the L bits of the packet has equal BEP, the capture probability for fast multipath fading is given by

$$P_{\text{cap}}\left(fast|\bar{p}_0,\bar{p}_t\right)=\left[1-\int_0^\infty P_{\text{be}}\left(be|p_0,\bar{p}_t\right)f_{p_0}\left(p_0|\bar{p}_0\right)dp_0\right]^L \tag{9.80}$$

If the quasi-uniform spatial distribution of the position of both test terminal and i interfering terminals, given by (9.76) and (9.77), respectively, is also taken into account, the following expression for the capture probability for the fast multipath fading channel is obtained

$$P_{\text{cap}}\left(\text{fast}|\bar{p}_0,\bar{p}_t\right)=\int_0^\infty\int_0^\infty\left[1-\int_0^\infty P_{\text{be}}\left(be|p_0,\bar{p}_t\right)f_{p_0}\left(p_0|\bar{p}_0\right)dp_0\right]^L f_{\bar{p}_0}\left(\bar{p}_0\right)f_{\bar{p}_t}(\bar{p}_t)d\bar{p}_0 d\bar{p}_t \tag{9.81}$$

Introducing the variables x, y, and z, with

$$x=\frac{p_0(1+K)}{\bar{p}_0}\qquad y=\sqrt{\frac{\pi i^2}{4\bar{p}_t}}\qquad\text{and}\qquad z=\sqrt{\frac{\pi}{4\bar{p}_0}}$$

$$dx=\frac{(1+K)}{\bar{p}_0}dp_0\qquad dy=-\frac{2y^3}{\pi i^2}d\bar{p}_t\qquad\text{and}\qquad dz=-\frac{2z^3}{\pi}d\bar{p}_0 \tag{9.82}$$

finally, the capture probability for the fast multipath fading channel becomes, after some manipulations,

$$P_{\text{cap}}(\text{fast}|\bar{p}_0,\bar{p}_t) = \int_0^\infty\int_0^\infty \left[I - \frac{\exp(-K)}{2}\int_0^\infty \exp(-x) I_0(\sqrt{4Kx})\,\text{erfc}\left(\sqrt{\frac{\pi x}{z^2(1+K)\left(\frac{4N_0}{T_b}+\frac{\pi i^2}{y^2}\right)}}\right)dx\right]^L$$

$$.\frac{4}{\pi}\exp\left(-y^2-z^2\right)dydz \qquad (9.83)$$

B. Slow Multipath Fading

After averaging p_0 over the pdf (9.71), the capture probability for the slow multipath fading channel is given by

$$P_{\text{cap}}(\text{slow}|\bar{p}_0,\bar{p}_t) = \int_0^\infty \left[1 - P_{\text{be}}(be|p_0,\bar{p}_t)\right]^L f_{p_0}(p_0|\bar{p}_0)\,dp_0 \qquad (9.84)$$

In the same way as for the fast multipath fading channel, the capture probability for the slow multipath fading channel is found by taking the quasi-uniform spatial distribution of the position of both test terminal and i interfering terminals, (9.76) and (9.77), respectively, into account

$$P_{\text{cap}}(\text{slow}|\bar{p}_0,\bar{p}_t) = \exp(-K)\int_0^\infty\int_0^\infty\int_0^\infty \left[1 - \frac{1}{2}\text{erfc}\left(\sqrt{\frac{\pi x}{z^2(1+K)\left(\frac{4N_0}{T_b}+\frac{\pi i^2}{y^2}\right)}}\right)\right]^L$$

$$.\frac{4}{\pi}\exp\left(-x-y^2-z^2\right)I_0\left(\sqrt{4Kx}\right)dxdydz \qquad (9.85)$$

Computational results for the capture probability in both fast and slow multipath fading channels are given in Figures 9.23 to 9.25. In Figure 9.23, the capture probability for the fast multipath fading channel as a function of the normalized distance is given for several values of the number of interferers. The capture probability for fast multipath fading is given in (9.83); however in Figure 9.23, it is assumed that only the positions of the i interfering terminals have a quasi-uniform spatial distribution (9.77), while the test terminal is at distance r from the base station. This means that the capture probability (9.80) only has to be averaged over the quasi-uniform spatial pdf of $\bar{p}_t$ (9.77). In Figure

9.23, $K = 7$ dB, SNR $\overline{p}_0/(N_o/T_b)$ is equal to 10 dB and $L = 63$ bits. From Figure 9.23, we see that the number of interfering terminals has a large influence on the capture probability: the capture probability decreases as the number of interferers increases.

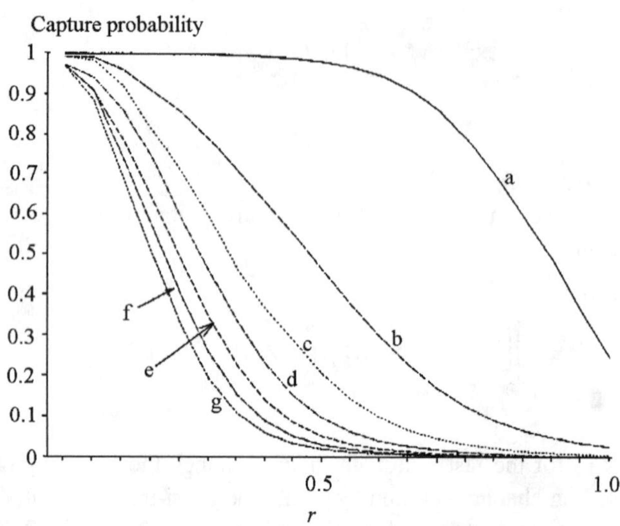

Figure 9.23 Capture probability as a function of normalized distance for fast multipath fading, $K = 7$ dB, SNR = 10 dB, and $L = 63$ bits for $i = 0,1,..., 6$ interferers (a-g, respectively).

Figure 9.24 shows the influence of SNR and packet length on the capture probability for fast multipath fading (9.83) as a function of the number of interferers i. $K = 7$ dB is also assumed in Figure 9.24. An increasing number of interfering terminals causes the capture probability to decrease dramatically. From Figure 9.24, we see that SNR has a larger influence on the capture probability than L, especially for large SNR and small i. It is worth mentioning that SNR represents the localmean SNR $\overline{p}_0/(N_o/T_b)$. Therefore, a large SNR yields enhanced throughput, because of the reduced BER. For $i > 6$, the number of interfering signals has a negligible influence on the capture probability, because the capture probability remains almost constant.

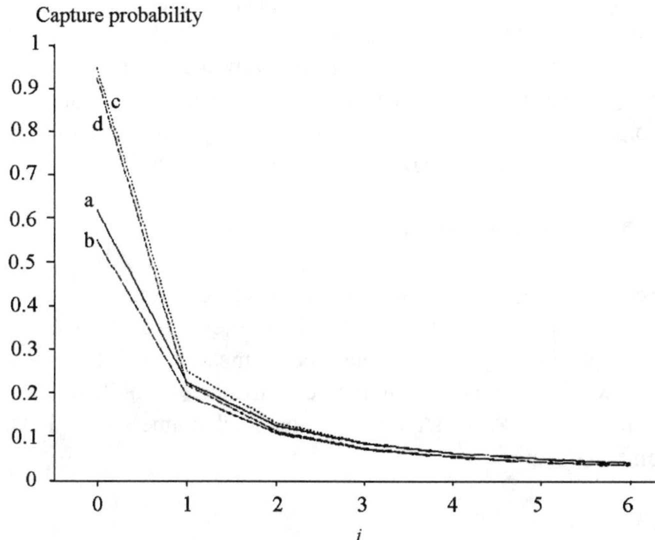

Figure 9.24 Capture probability as a function of the number of interferers for fast multipath fading and $K = 7$ dB for $L = 63$ bits ((a) SNR = 10 dB, (c) SNR = 40 dB), and $L = 127$ bits ((b) SNR = 10 dB, (d) SNR = 40 dB).

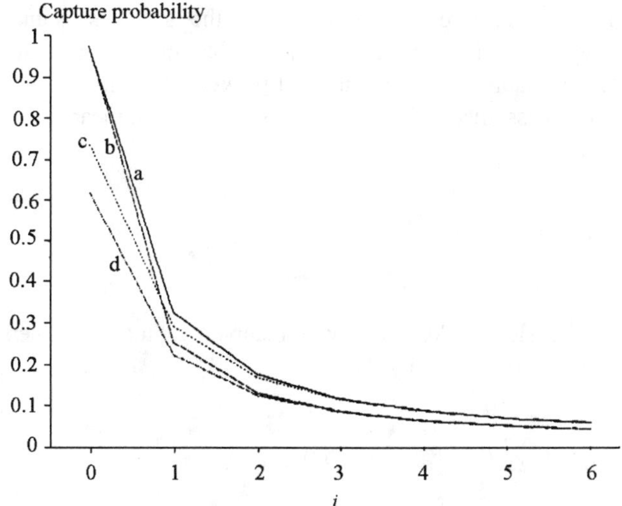

Figure 9.25 Capture probability as a function of the number of interferers, with $K = 7$ dB and $L = 63$ bits, for slow ((a) SNR = 40 dB, (c) SNR = 10 dB), and fast multipath fading ((b) SNR = 40 dB, (d) SNR = 10 dB).

A comparison between the capture probabilities for fast and slow multipath fading channels, given by (9.83) and (9.85), respectively, is given in Figure 9.25. In Figure 9.25, the capture probability is plotted as a function of the number of interfering terminals for different SNR. The difference between fast and slow multipath fading (slow multipath fading gives better performance) is relatively large for small i and small SNR, as seen in Figure 9.25. For $i > 3$, the performance for fast and slow multipath fading is almost equal, and only SNR has a small influence on the capture probability.

9.6.3 Random Access Systems With Receiver Capture

Random access techniques are introduced in mobile communications to facilitate the users. In random access systems, users transmit whenever they want to, with possibly a few constraints depending on the particular access method used. Several random access protocols are known, but in this section, the performance of mobile radio communication systems using the slotted ALOHA and the unslotted nonpersistent ISMA protocol is studied and compared.

Slotted ALOHA

Using the slotted ALOHA protocol, terminals transmit their packets regardless of other terminals that might already be transmitting a packet in the same time slot, where a time slot has the same length as the length of the packet. If two (or more) packets are transmitted at the same time, the packets get lost. An unsuccessful transmission of a packet is discovered by the terminal, if after a certain time, no acknowledgment from the transmitter is received. After a random time, the terminal retransmits the same packet. On channels with receiver capture, the packet with the strongest signal has still the probability of being captured by the receiver, also when more packets are transmitted at the same time. The probability $P_i(i)$ that the test packet suffers from i interfering packets in the same time slot is assumed to be Poisson distributed, with mean G,

$$P_i(i) = \frac{G^i}{i!}\exp(-G) \tag{9.86}$$

The throughput S of slotted ALOHA with receiver capture can then be found by multiplying the capture probability by $P_i(i)$

$$S(\text{slottedALOHA}) = G\sum_{i=0}^{\infty} P_i(i) P_{\text{cap}}(\text{fast / slow}|\overline{p}_0, \overline{p}_t)$$

$$= G\sum_{i=0}^{\infty} \frac{G^i}{i!}\exp(-G). P_{\text{cap}}(\text{fast / slow}|\overline{p}_0, \overline{p}_t) \tag{9.87}$$

The throughput for the fast and the slow multipath fading channel is found by substituting the appropriate capture probability in (9.87), given by (9.83) and (9.85), respectively.

Computational results for the throughput of slotted ALOHA are given in Figures 9.26 and 9.27. In Figure 9.26, the throughput is given as a function of the total traffic load G, for fast and slow multipath fading and different values of SNR. In Figure 9.26, $K = 7$ dB and $L = 63$ bits are assumed. From Figure 9.26, we see that slow multipath fading gives a much higher performance than fast multipath fading. The SNR only influences the throughput for $G < 10$.

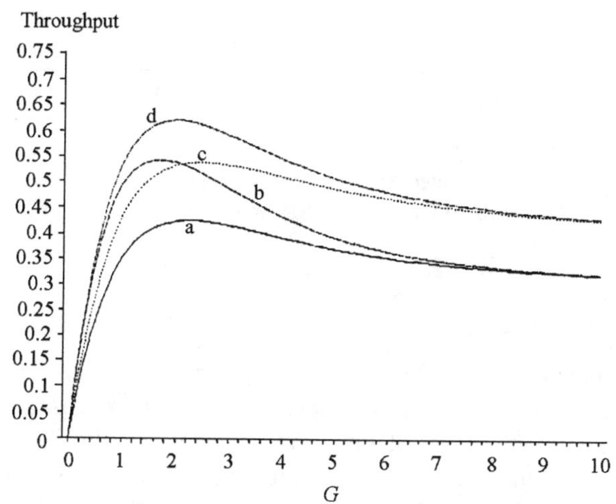

Figure 9.26 Throughput of slotted ALOHA as a function of G, for $L = 63$ bits and $K = 7$ dB, for fast multipath fading ((a) SNR = 10 dB, (b) SNR = 40 dB), and slow multipath fading ((c) SNR = 10 dB, (d) SNR = 40 dB).

Figure 9.27 shows the effect of K on the throughput. Propagation measurements in several environments, for instance in [38], results in several values of K ($K = 0$ corresponds to Rayleigh fading, i.e., the LOS component is absent). In Figure 9.27, the throughput for fast multipath fading is given as a function of G, with $L = 63$ bits and SNR = 10 dB. The Rician parameter K has a very large influence on the throughput. For large K, the performance is much better than for $K = 0$, thus, if the test terminal is in LOS of the base station, the performance is much better than when there is no LOS. This means that if the test terminal is very near to the base station (a strong LOS component), the probability that a packet is successfully transmitted (or captured by the receiver) is higher than for the case the distance between test terminal and base station is larger (a smaller LOS component).

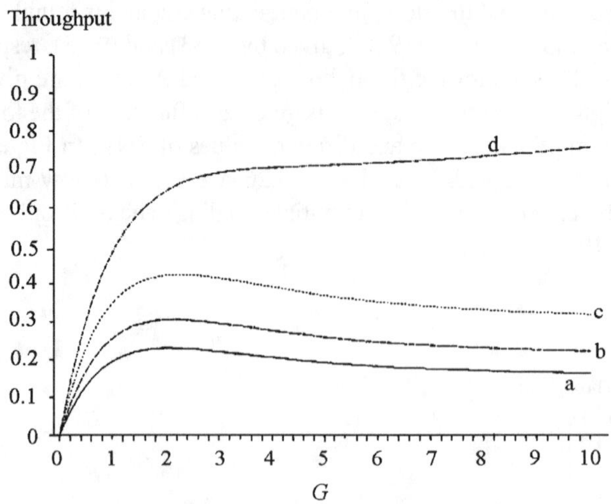

Figure 9.27 Throughput of slotted ALOHA as a function of G, for fast multipath fading, $L = 63$ bits and SNR = 10 dB for (a) $K = 0$ (Rayleigh fading), (b) $K = 3$ dB, (c) $K = 7$ dB, and (d) $K = 10$ dB.

Unslotted nonpersistent ISMA

According to the nonpersistent ISMA protocol, each terminal listens to the channel before transmitting. Because the protocol is nonpersistent, the terminals do not continuously sense, but wait a random period of time. An advantage of the ISMA protocol is that ISMA eliminates the hidden-terminal problem occurring in the CSMA protocol, where terminals might not be able to sense a transmission by another terminal. The base station transmits a busy signal on the outbound channel as soon as an inbound packet is received. After complete reception of the packet, the busy signal is removed so that the channel is idle. In ISMA, collisions may occur due to new packet transmissions during the transmission delay of the inhibit signal d, necessary to switch from busy to idle (and vice versa). This delay is normalized to the packet length, resulting in $0 \le d < 1$.

The probability U that the channel, during a cycle (one cycle consists of an idle period plus a transmission period), is used without conflicts is defined as the probability that a test packet is overlapped by i interfering packets $P_i(i)$, multiplied by the probability that the test packet is received correctly in the presence of i interfering packets

$$U = \sum_{i=0}^{\infty} P_i(i)(i+1) P_{\text{cap}}(\text{fast / slow} | \overline{p}_0, \overline{p}_t)$$

$$\text{with } P_i(i) = \frac{(dG)^i}{i!} \exp(-dG) \tag{9.88}$$

The term $(i + 1)$ is included in (9.88), because when inhibit bits are in the outbound channel and a test packet is subjected to i interfering packets, $i + 1$ packets are presented during the busy period. One of these $i + 1$ packets is capable of capturing the receiver with a certain probability. One must take care of the fading type of the test and interfering packets while evaluating the success probability U using (9.88). In the case of Rayleigh-faded test and interfering packets, P_{cap} (fast/slow/ $\overline{p}_0, \overline{p}_t$) is calculated using (9.83) and (9.85) by substituting $K = 0$. If the test packet is Rician-faded and the interfering packets are Rayleigh-faded, then a proper value of K is used in (9.83) and (9.85) to calculate the success probability. Thus, it is clear that the expression for the success probability changes with the change in fading type.

The throughput for unslotted nonpersistent ISMA with capture is then found by dividing U by the sum of the expected duration of the transmission period T, given by

$$T = 1 + 2d - \frac{1}{G}\left[1 - \exp(-dG)\right] \tag{9.89}$$

and the expected duration of the idle period I, with $I = 1/G$, thus

$$S(\text{unslotted nonpersistent ISMA}) = \frac{U}{T+I}$$

$$= \frac{\sum_{i=0}^{\infty} \frac{(dG)^i}{i!} \exp(-dG)(i+1)}{(1+2\text{d}) + \frac{1}{G}\exp(-dG)} P_{\text{cap}}(\text{fast/ slow}|\overline{p}_0, \overline{p}_t) \tag{9.90}$$

where again the throughput for fast and slow multipath fading channels is found by substituting the capture probability (9.83) and (9.85), respectively, in (9.90).

Computational results for the throughput of unslotted nonpersistent ISMA are given in Figure 9.28. In Figure 9.28, the throughput for both fast and slow multipath fading is given as a function of G for different values of SNR, with $K = 7$ dB, $L = 63$ bits, and $d = 0.05$. From Figure 9.28, we see that also for ISMA, slow multipath fading gives a higher throughput than fast multipath fading. A higher value of SNR gives also a higher throughput for all values of G.

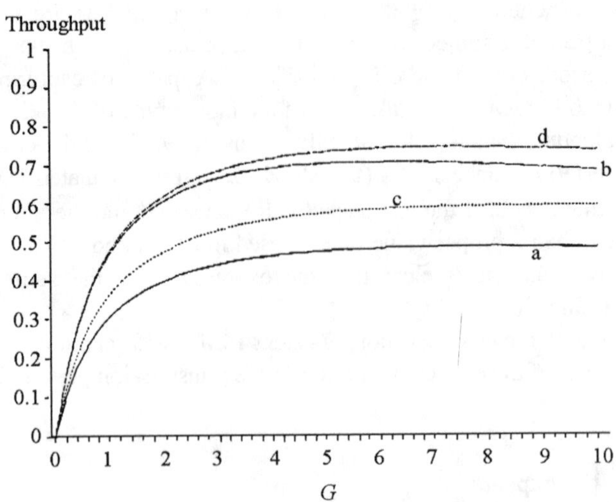

Figure 9.28 Throughput of unslotted nonpersistent-ISMA as a function of G, for $L = 63$ bits, $K = 7$ dB and $d = 0.05$ for fast multipath fading ((a)) SNR = 10 dB, (b) SNR = 40 dB), and slow multipath fading ((c) SNR = 10 dB, (d) SNR = 40 dB).

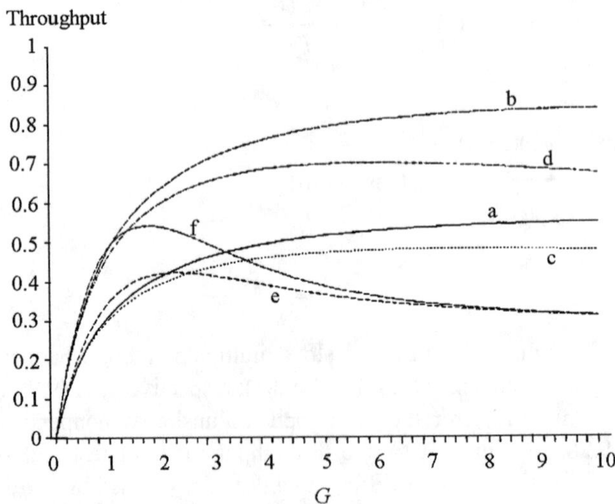

Figure 9.29 Throughput as a function of G, for fast multipath fading, $L = 63$ bits and $K = 7$ dB for unslotted nonpersistent ISMA with $d = 0.01$ ((a) SNR = 10 dB, (b) SNR = 40 dB), and $d = 0.05$ ((c) SNR = 10 dB, (d) SNR = 40 dB) and slotted ALOHA ((e) SNR = 10 dB, (f) SNR = 40 dB).

Figure 9.29, compares the throughput for the slotted ALOHA protocol and the unslotted nonpersistent ISMA protocol. The throughput for fast multipath fading is given as a function of G. In Figure 9.29, again $L = 63$ bits and $K = 7$ dB are assumed. From Figure 9.29, we see that, at least for high values of G, the unslotted nonpersistent ISMA protocol gives a much higher throughput than the slotted ALOHA protocol. We see that a larger value of d gives a lower performance.

9.6.4 Error Correction Coding

In Section 9.6.2, it was discussed that all L bits of the packet have to be received correctly to have a successful reception of the packet. At this point error correction coding is added to the system to improve the performance. Using error correction coding some of the L bits may be received incorrectly, still resulting in a successful reception of the packet. In this chapter, only linear BCH error correction codes with parameters (L,k,t) [39] are used, where L is the packet length, k is the number of actual information (user) bits in the packet (or codeword), and t is the error correction capability. This error correction capability offers the possibility of correcting up to t erroneous bits. This means that to receive the packet successfully, only up to t bits may be received incorrectly (instead of $t = 0$, when no error correction coding is used). For a higher error correction capability, the number of parity-check bits increases, resulting in a lower code efficiency because k diminishes. Thus, a tradeoff between high error correction capability and high code efficiency is made. Usually, error correction codes also offer a certain error detection capability, denoting the number of erroneous bits the receiver can detect but not correct. The effect of the error detection capability on the performance, studied for example in [33], is not considered here. The probability of successful reception of a packet is given by [33]

$$P(\text{success}|\, p_0, \overline{p}_t) = \sum_{m=0}^{t} \binom{L}{m} \left[P_{\text{be}}\left(\text{be}|\, p_0, \overline{p}_t\right)\right]^m \left[1 - P_{\text{be}}\left(\text{be}|\, p_0, \overline{p}_t\right)\right]^{L-m} \tag{9.91}$$

The expressions for the throughput of the two protocols now change because of the introduction of error correction codes.

Slotted ALOHA

The probability of successful reception of a packet (9.91) is equal to the capture probability if the appropriate BEP is substituted into (9.91). In this way the following expression for the throughput of slotted ALOHA is given by

$$S(\text{slottedALOHA}) = G \sum_{i=0}^{\infty} \sum_{m=0}^{t} \binom{L}{m} \frac{G^i}{i!} \exp(-G) P_{\text{cap}}(\text{fast / slow}|\, \overline{p}_0, \overline{p}_t) \tag{9.92}$$

where the capture probability for fast and slow multipath fading with coding is given by

$$P_{\text{cap}}(\text{fast}|\bar{p}_0,\bar{p}_t) = \int_0^{\infty}\int_0^{\infty}\left(1-\int_0^{\infty} P_{\text{be}}(\text{be}|p_0,\bar{p}_t)f_x(x)dx\right)^{L-m}$$

$$\cdot\left(\int_0^{\infty} P_{\text{be}}(\text{be}|p_0,\bar{p}_t)f_x(x)\right)^{m} f_y(y)f_z(z)dydz \tag{9.93}$$

and

$$P_{\text{cap}}(\text{slow}|\bar{p}_0,\bar{p}_t) = \int_0^{\infty}\int_0^{\infty}\int_0^{\infty}\left(1-P_{\text{be}}(\text{be}|p_0,\bar{p}_t)\right)^{L-m}\left(P_{\text{be}}(\text{be}|p_0,\bar{p}_t)\right)^{m}$$

$$\cdot f_x(x)f_y(y)f_z(z)dxdydz \tag{9.94}$$

respectively.

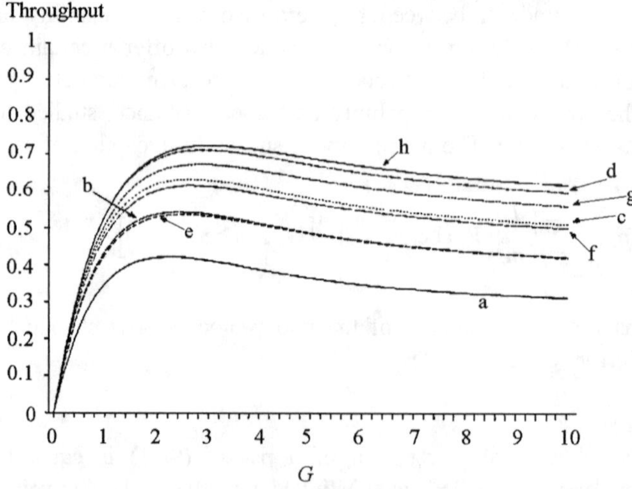

Figure 9.30 Throughput of slotted ALOHA with coding as a function of G, for SNR = 10 dB, K = 7 dB and L = 63 bits, for fast multipath fading with t = 0, 1, 2, and 3 ((a-d), respectively) and slow multipath fading with t = 0, 1, 2, and 3 ((e-h), respectively).

Computational results for the throughput of slotted ALOHA with error correction coding are shown in Figure 9.30. Figure 9.30 shows the effect of the error correction capability t on the throughput of the fast multipath fading channel. The throughput is plotted as a function of G, with $K = 7$ dB, $L = 63$ bits and SNR = 10 dB. From Figure 9.30, we see that a larger value of t gives also a higher throughput. For larger values of t, the difference between the performance of fast and slow multipath fading decreases; however, slow multipath fading still gives a better performance than fast multipath fading.

Unslotted Nonpersistent ISMA

As for the slotted ALOHA protocol, the expression for the throughput of unslotted nonpersistent ISMA changes due to the introduction of error correction codes

$$S(\text{unslotted nonpersistent } ISMA) = \frac{\displaystyle\sum_{i=0}^{\infty}\sum_{m=0}^{t}\binom{L}{m}\frac{dG^i}{i!}\exp(-dG)(i+1)}{(1+2d)+\dfrac{1}{G}\exp(-dG)} P_{\text{cap}}(\text{fast / slow}|\overline{p}_0, \overline{p}_t) \quad (9.95)$$

where the capture probability for fast and slow multipath fading channels with coding is given by (9.93) and (9.94), respectively.

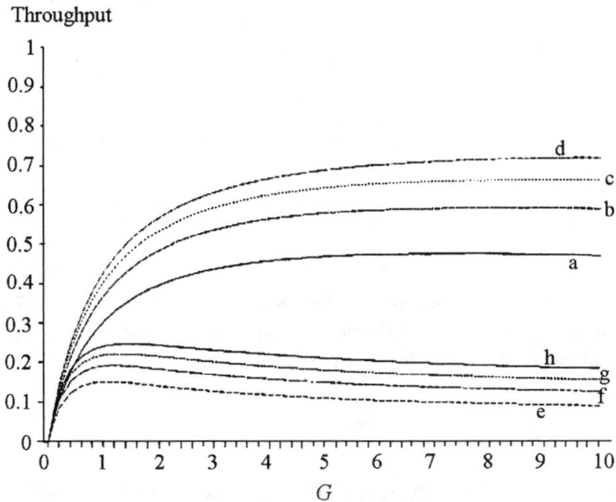

Figure 9.31 Throughput of unslotted nonpersistent-ISMA with coding as a function of G, for fast multipath fading, SNR = 10 dB, $L = 63$ bits and $K = 7$ dB, for $d = 0.05$ with t = 0, 1, 2, and 3 ((a-d), respectively) and $d = 1$ with t = 0, 1, 2, and 3 ((e-h), respectively).

Computational results for the throughput of unslotted nonpersistent ISMA with error correction coding are given in Figures 9.31 and 9.32. The throughput for fast multipath fading as a function of G is plotted in Figure 9.31, for several values of d and t. In Figure 9.31, again $K = 7$ dB, $L = 63$ bits and SNR = 10 dB are assumed. As for the slotted ALOHA protocol, a higher value of t gives a higher throughput, while again we see that a higher value of d gives a small throughput.

Finally, Figure 9.32 shows a comparison between the performance of slotted ALOHA and unslotted nonpersistent ISMA. In Figure 9.32, the throughput for both protocols is plotted as a function of G, for both fast and slow multipath fading, with $K = 7$ dB, $L = 63$ bits, SNR = 10 dB, $t = 2$ and $d = 0.05$. From Figure 9.32, we see that unslotted nonpersistent ISMA gives a higher throughput than slotted ALOHA, provided G is relatively large. For the used parameters, we also see that the difference in performance between fast and slow multipath fading is larger for slotted ALOHA than for unslotted nonpersistent ISMA.

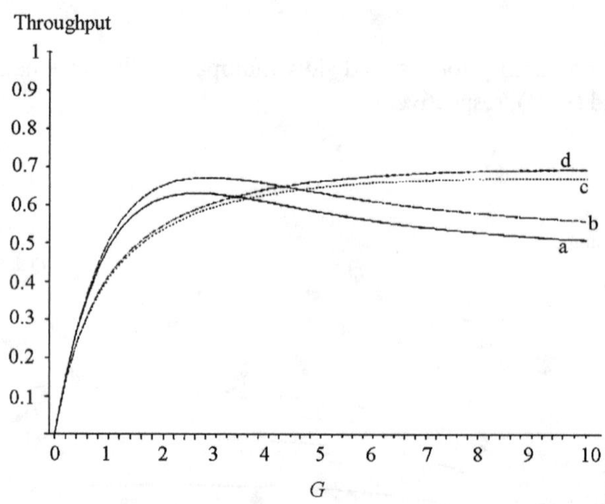

Figure 9.32 Throughput as a function of G, for SNR =10 dB, $k = 51$ bits, $K = 7$ dB, and t = 2, for slotted ALOHA ((a) fast, (b) slow multipath fading), and unslotted nonpersistent ISMA with $d = 0.05$ ((c) fast, (d) slow multipath fading).

9.6.5 User Data Throughput

The user throughput S_u (i.e., the throughput of user data) is defined as the product of the throughput S, the user bits in a packet or a codeword k, and the inverse of the packet length L.

$$S_u \underline{\Delta} S \frac{k}{L} \tag{9.96}$$

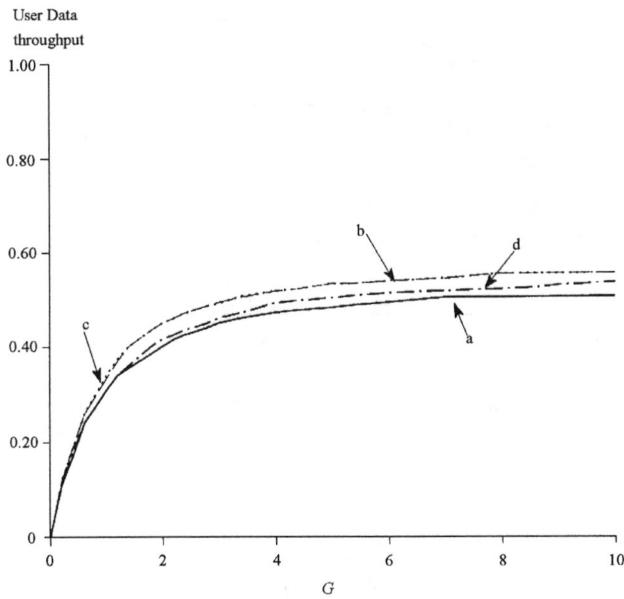

Figure 9.33 User data throughput of unslotted nonpersistent-ISMA with coding as a function of G for fast multipath fading, SNR = 10 dB, K = 7 dB, for d = 0.05 with k = 63, 57, 51, and 45 ((a–d) respectively).

The user data throughput S_u of nonpersistent ISMA protocol is calculated using the BCH (L,k,t) error correction codes and depicted in Figures 9.33 and 9.34. Figure 9.33 shows that with the increase in error correction capability t, the user data throughput S_u first increases (curves (a) and (b)), then remains almost constant (curves (b) and (c)), and finally decreases (curves (c) and (d)). In contrast to the user data throughput S_u, the throughput S always increases with the increase in t (Figure 9.31). Figure 9.34 compares the throughput S with the user data throughput S_u. We see from Figure 9.34 that the throughput S is higher than the user data throughput S_u, and this is because k is always smaller than L for any value of t except for t = 0, S_u and S are equal. Table 9.2 shows some values for L, k, t for BCH codes.

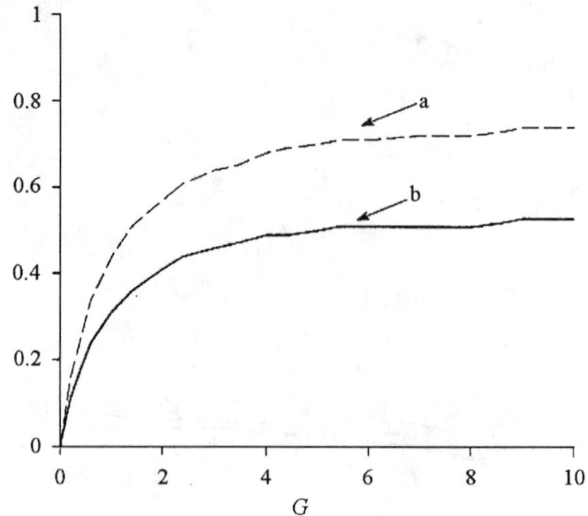

Figure 9.34 (a) Throughput S and (b) user data throughput S_u as a function of G, for SNR = 10 dB, $k = 45$ bits, $K = 7$ dB, and t = 3 for fast multipath fading and nonpersistent-ISMA.

Table 9.2

Selected BCH (L, k, t) Codes

L	k	t
63	63	0
63	57	1
63	51	2
63	45	3
31	21	2
15	7	2

9.6.6 Conclusions

In this section, a model was developed for the performance evaluation of mobile radio communication systems using the slotted ALOHA and the unslotted nonpersistent ISMA protocol with receiver capture. Compared with earlier studies, the system here is assumed to be in a Rician/Rayleigh environment with AWGN, where the test terminal is very near to the base station and the interfering terminals are close to the border of the cell. The amplitude of the BPSK-modulated test signal is modeled by a Rician pdf, because the test terminal is assumed to be in LOS of the base station, while the signals of the interfering terminals are Rayleigh-distributed without LOS propagation. The performance of mobile radio communication systems is measured in terms of the capture probability and the throughput. Both fast and slow multipath-fading channels were considered. The effect of BCH error correction codes on the performance was also investigated.

We conclude that the capture probability decreases dramatically if the number of interfering signals increases; however, for $i > 6$, the capture probability is very small and remains almost constant. The packet length L also influences the capture probability: the capture probability decreases as L increases. Slow multipath-fading channels offer a better performance than fast multipath-fading channels. We also conclude that for a higher value of the Rician parameter K when the LOS component is stronger or when the test terminal is very close to the base station, the performance is very much increased. Unslotted nonpersistent ISMA gives a higher throughput than slotted ALOHA, provided the transmission delay d is small and the traffic load G is large. For larger values of d, the performance of unslotted nonpersistent ISMA decreases very rapidly. Error correction codes enhance the performance of the system because a larger error correction capability t gives a higher throughput.

We recommend extending this present study by developing a model that includes Rician-faded interfering signals and the influence of multiple cells. The results obtained in this book may be useful for both outdoor and indoor data communications.

9.7 CELLULAR DS-CDMA SYSTEM WITH IMPERFECT POWER CONTROL AND IMPERFECT SECTORIZATION

In this section, a general analytical model is presented that is used to evaluate the influence of both imperfect power control and imperfect sectorization on the performance of two types of CDMA systems, namely, voice oriented and data oriented systems. For the first type, the maximum user capacity is an important performance measure, while for the second type, throughput and delay characteristics are appropriate performance parameters. An analytical model is developed assuming that the data oriented system is slotted and simultaneous transmission of two or more packets is permitted. If the actual number of transmitting packets exceeds the value of the maximum capacity, then all packets are destroyed. In both systems, only the reverse link (i.e., from mobile to base station) is considered.

9.7.1 Cellular DS-CDMA

In DS-CDMA the data signal at the transmitter is multiplied by a pseudorandom, user-specific code sequence waveform. At the receiver, the original data is recovered by correlating the received spread-spectrum signal with the correct code sequence. If code sequences with good correlation properties are used, after the correlation operation the unwanted signals appear as noiselike signals with a very low power density spectrum. To establish the spreading effect, the duration T_C of a code symbol (chip) is chosen much smaller than the duration T_b of a data symbol. Therefore, the bandwidth B_{SS} of the transmitted spread-spectrum signal is much larger than the bandwidth B_S of the original data signal. The ratio between transmitted and original bandwidth is defined as the processing gain:

$$PG \stackrel{\wedge}{=} \frac{B_{ss}}{B_s} \tag{9.97}$$

In practice, this number depends on the specific modulation method. However, we will use the following approximation:

$$PG \stackrel{\wedge}{=} \frac{B_{ss}}{B_s} \approx \frac{T_b}{T_c} \tag{9.98}$$

where T_b is the bit duration and T_C is the chip duration.

Since two code sequences with a relative delay of more than two chip times usually have a low correlation value compared with the fully synchronized situation, DS-CDMA offers the possibility of distinguishing between paths with a relative delay of more than two chip times. This is called the inherent diversity of DS-CDMA, implying that it is possible to resolve a number of paths separately using only one receiver. This property makes DS-CDMA suitable for applications in mobile radio environments usually corrupted with severe multipath effects.

It is well known that one of the most serious problems faced by a DS-CDMA system is the multiuser interference. Because all users are transmitting in the same frequency band and the cross-correlations of the codes are rarely zero, the signal-to-interference ratio and hence the performance decreases as the number of users increases, which shows that DS-CDMA is an interference-limited rather than a noise-limited system. The near-far effect plays an especially important role when considering multiuser interference. The near-far effect is explained by considering the reverse link. Due to the path-loss law (which implies that the received power decreases as the transmitter-receiver distance increases), a close user dominates over a user located at the cell boundary. The path-loss law is usually described by assuming that the received power P_r is proportional to the distance d to the power γ:

$$P_r \sim d^{-\gamma} \tag{9.99}$$

where γ is the path-loss law exponent. In the case of free space propagation, the value of γ is 2. In a practical mobile radio environment the value of γ is in the range of 3 to 5, which is caused by the fact that the radio waves are reflected and partially absorbed by objects between receiver and transmitter and by the surface of the earth. Power control is used to combat the near-far effect. There are two other important techniques to reduce the multiuser interference in a cellular DS-CDMA system. The first technique is sectorization, which is established by using sectorized antennas at the base station, implying that only interference signals are received from within a limited angle. The second technique is voice activity monitoring. A system using voice activity monitoring prevents a mobile from radiating power during speech pauses. A complication in a cellular system is the fact that a base station not only receives interference from users within its own cell (intracell interference) but also from users in other cells (intercell interference).

A. Power Control

Power control is established by letting the base station continuously transmit a (wideband) pilot signal monitored by all mobile terminals. According to the power level detected by the mobile, the mobile adjusts its transmission power. Hence, mobiles near the cell boundary transmit at a lower power level than mobiles located close to the base station. This is open-loop power control. It is also possible to use closed-loop power control as suggested in [40,41]. In this case, the base station measures the energy received from a mobile and controls the transmit power of the mobile by sending a command over a (low data rate) command channel. In a practical power-control system, power-control errors occur implying that the average received power at the base station may not be the same for each user signal.

The performance of a power-control system depends on the power-control algorithm, speed of the adaptive power-control system, dynamical range of the transmitter, and spatial distribution of users and propagation statistics (such as fading and shadowing). All these factors influence the pdf of the received power. Since the objective here is to investigate the influence of power-control imperfections on the system performance, the explicit influence of the factors mentioned above on the pdf of the average received power is not investigated.

Investigations on DS-CDMA have shown that the pdf of the received power P due to the combined influence is assumed to be lognormal [42,43].

$$f(P) = \frac{1}{\sqrt{2\pi}\sigma P}\exp\left[-\frac{\ln^2(P)}{2\sigma^2}\right] \tag{9.100}$$

where the imperfection in the power-control system is determined by the logarithmic standard deviation σ of the lognormal power distribution of the received signal. In the case of perfect power control, the logarithmic standard deviation is 0 dB.

There are a number of motivations for choosing a lognormal distribution for the individual received power. First, the received power at the base station depends on a lot of independent factors. We assume that each factor gives a contribution in decibels to the received power. Using the central limit theorem, it is clear that the logarithm of the received power is Gaussian-distributed, implying that the received power is lognormally distributed. The second motivation is that this model provides a means to investigate the influence of imperfect power control analytically. A third motivation is data provided in [40,41], where it is shown that the received SNR is lognormally distributed with standard deviation between 1 and 2 dB. Assuming constant noise power, this implies that the received power can be modeled as a lognormal variable.

If a number of k mobiles is transmitting and if the power of each mobile is controlled individually, then the total received interference power P_I is the summation of k-independent, identically lognormally distributed random variables denoted by P_i:

$$P_I = \sum_{i=1}^{k} P_i \tag{9.101}$$

Fenton [44] showed that the pdf of P_I for k users is approximately lognormal with the following logarithmic mean $m_I(k)$ and logarithmic variance $\sigma_I^2(k)$:

$$\sigma_I^2(k) = \ln\left(\frac{1}{k}e^{\sigma^2} + \frac{k-1}{k}\right) \tag{9.102}$$

$$m_I(k) = \ln(k) + m + \frac{\sigma^2}{2} - \frac{1}{2}\ln\left(\frac{k-1}{k} + \frac{1}{k}e^{\sigma^2}\right) \tag{9.103}$$

This method is valid for a logarithmic standard deviation σ less than 4 dB.

B. Sectorization

Multiuser interference received at the base station is reduced by dividing a cell into a number of sectors by means of directional antennas. In the case of perfect directional antennas there is a sharp separation between the sectors. Due to overlap and sidelobe anomalies of practical antennas, the base station still receives some interference from users in other sectors. This effect is studied by modeling the real antenna gain pattern for instance by a parabolic function.

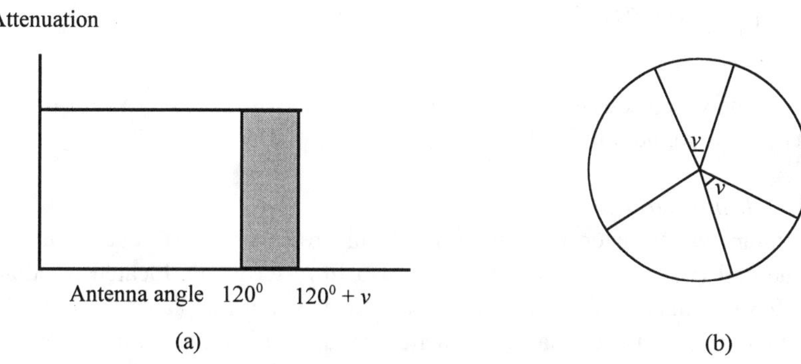

Figure 9.35 (a) Radiation pattern model for imperfect directional antenna with opening angle 120 deg, and (b) sector coverage with an imperfect directional antenna with opening angle 120 deg.

However, for simplicity we assume the simplified antenna radiation pattern shown in Figure 9.35(a) and (b) where the antenna imperfections are modeled by the overlap angle v. It is suggested to transform the real antenna gain patterns to this uniform model, analogously as the method for describing noise by the noise equivalent bandwidth. If D is the number of sectors and v is the overlap angle, then a relation is derived between the ratio of the total interference power received in a sectorized system and the total interference power received in a nonsectorized system denoted by F_S.

$$F_s \stackrel{\wedge}{=} \frac{P_{\text{sectorized}}}{P_{\text{nonsectorized}}} = \left(\frac{1}{D} + \frac{2v}{360}\right) \tag{9.104}$$

where D is an integer. From (9.104) it is clear that $v = 0$ corresponds to perfect sectorization, and the combination $D = 1$ and $v = 0$ corresponds to the situation without sectorization.

C. Voice Activity Monitoring

Voice activity monitoring implies that the transmitter is not active during silent periods in human speech. It is possible to detect a silent period in the speech signal and let the transmitter stop transmitting during this period. Voice activity factors between 35% and 40% are reported in [45,46]. Studies done in Europe suggest that the total activity due to voice and background noise is higher in a mobile environment than in a wireline system and can have values between 50% and 60% [47]. When using voice activity monitoring, the probability that k out of n interferers are active is described by a binomial distribution:

$$P(n,k)=\binom{n}{k}a^k(1-a)^{n-k} \tag{9.105}$$

where a is the voice activity factor. The effect of the voice activity factor on the CDMA capacity is investigated in Section 9.7.4.

D. Intercell Interference

In a cellular system, each base station not only receives interference from mobiles in the home cell (intracell interference) but also from terminals located in adjacent cells (Figure 9.36). This is called intercell interference. The principle of the CDMA protocol allows each cell to use the same frequency band, eliminating the need for a mobile to change its frequency when moving into another cell (soft handover). To model the interference received from terminals in other cells, we investigate here the interference power received by the base station in the home cell (B_H) from mobile terminals in another cell (B_0) with base station separation distance d (Figure 9.37).

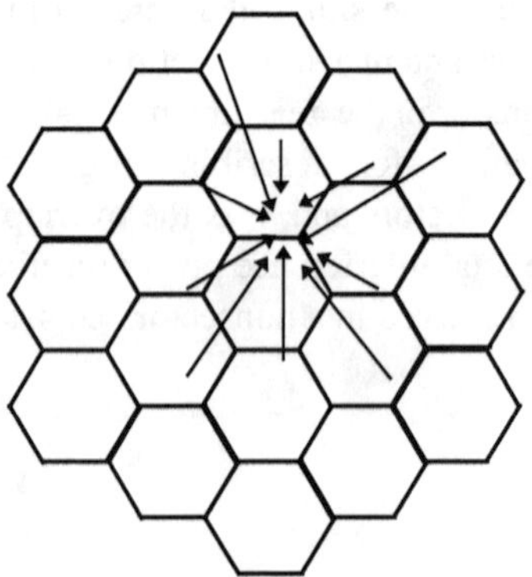

Figure 9.36 Intercell interference in a cellular system.

We adopt a generalized version of the approach as described in [48]. For analytical convenience the hexagonal cell structure is approximated by circles with radius R. The power of terminals in cell B_H and cell B_0 is controlled by the base stations in B_H and B_0, respectively.

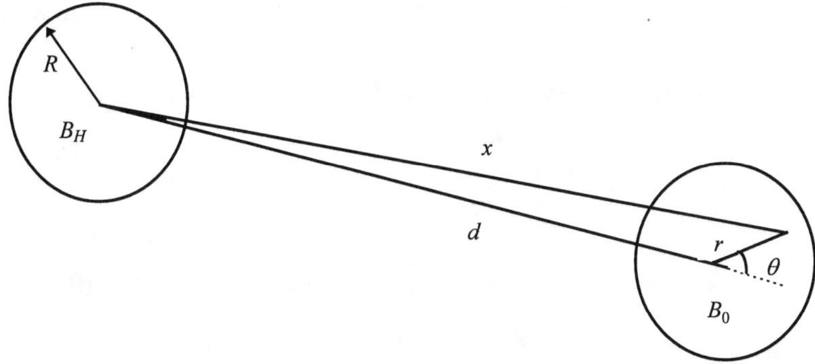

Figure 9.37 Interference from terminals in a distant cell.

In the case of perfect power control, the power received by a base station from each mobile terminal in its service area is constant. This power is denoted by S_p. We assume that users are uniformly distributed in each cell with user density

$$\eta = \frac{N}{\pi R^2} \tag{9.106}$$

where N is the number of users per cell and R is the cell radius. The power control strategy is such that a terminal at a distance r from the base station transmits with power

$$P_T(r) = S_p r^\gamma \tag{9.107}$$

implying that the power received by the base station is

$$P_R(r) = S_p r^\gamma r^{-\gamma} = S_p \tag{9.108}$$

where γ is the path-loss law exponent, which is 2 in case of free space loss and lies between 3 and 5 for mobile systems. In [48], the special case of γ = 4 is considered while we derive an expression valid for all values of γ. From Figure 9.37, we see that the power received by the base station denoted by $P_{RB}(d)$ from mobile users in a cell at distance d is obtained by integrating over the cell area (A):

$$P_{RB}(d) = \int S_p \left(\frac{r}{x}\right)^\gamma \eta dA \tag{9.109}$$

where

$$x = \sqrt{d^2 + r^2 + 2dr\cos\theta} \tag{9.110}$$

Using Figure 9.37 and substituting (9.106) and (9.110), (9.109) is written as

$$P_{RB}(d) = \frac{2NS_p}{\pi R^2}\int_0^R dr\, r^{\gamma+1}\int_0^\pi \frac{d\theta}{(d^2 + r^2 + 2dr\cos\theta)^{\gamma/2}} \tag{9.111}$$

In [41] a closed expression is found for the case γ = 4, which we obtain from (9.111)

$$P_{RB}(d) = 2NS_p\left[2\left(\frac{d}{R}\right)^2 \ln\left(\frac{(\frac{d}{R})^2}{(\frac{d}{R})^2 - 1}\right) - \frac{4(\frac{d}{R})^4 - 6(\frac{d}{R})^2 + 1}{2\left((\frac{d}{R})^2 - 1\right)^2}\right] \tag{9.112}$$

If we now consider a hexagonal cellular structure (Figure 9.36), it is possible to compute the total received interference from all the cells in the system. The interference correction factor due to multiple cell interference F_m is defined as the ratio of the interference power received from the outer cells (I_m) and the interference power generated by users in the home cell (I_h). Using (9.108), F_m is written as

$$F_m \stackrel{\wedge}{=} \frac{I_m}{I_h} = \frac{I_m}{(N-1)S_p} \tag{9.113}$$

where N is the number of users per cell and S_p is the power received from a user in the case of perfect power control.

The value of I_m depends on the value of γ and on the number of cell tiers considered. From (9.111), it follows that for a large number of users (N >> 1), the ratio F_m is a constant because I_m is proportional to N. Figure 9.38 and Table 9.3 show the interference correction factor F_m as function of the path-loss law exponent γ considering several number of cell tiers around the home cell. It is clear that for larger values of γ, the influence of the outer cell tiers is decreasing with increasing tier number. For γ = 4, no significant difference is found when considering 10 or 15 tiers. In both cases, we found F_m = 0.326. This value is less than the value of 0.66 reported in [46] that was obtained by simulation assuming lognormal shadowing. The value 0.326 is found analytically for the situation without shadowing and is confirmed by results presented in [48]. The value 0.326 is used in all the calculations in Section 9.7.4.

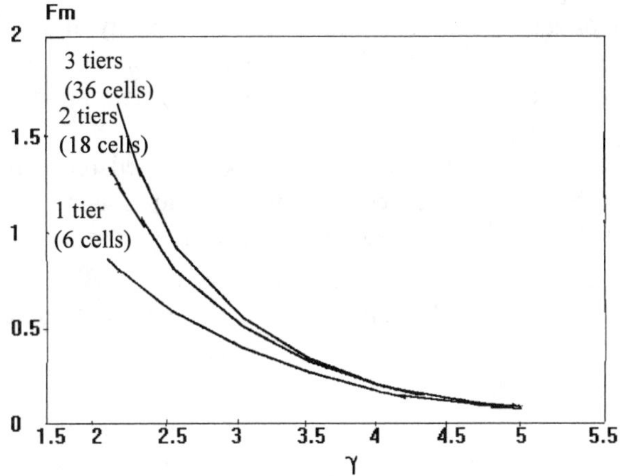

Figure 9.38 Multiple cell interference reduction factor F_m as function of the path-loss exponent γ considering different numbers of interfering cells.

Table 9.3

F_m as a Function of γ

	Interference Correction Factor F_m			
Number of Tiers	$\gamma = 2$	$\gamma = 3$	$\gamma = 4$	$\gamma = 5$
1	0.904	0.417	0.284	0.191
2	1.365	0.579	0.312	0.199
3	1.669	0.625	0.319	0.200
5	2.079	0.667	0.323	0.201
10	2.665	0.701	0.326	0.201
15	3.018	0.714	0.326	0.201

9.7.2 Capacity Analysis

In a speech-oriented DS-CDMA system, after setting up a call each user is allowed to transmit data continuously. A very important design issue for such a system is the user capacity. A service provider is interested in the maximum number of users that can be served simultaneously by the system with a predefined performance in a certain geographic area. In a digital communication system such as DS-CDMA, BEP is an appropriate performance measure. BEP calculations offer the possibility of detailed system considerations, but often require complex models that can sometimes disturb the basic understanding of the system. To obtain a general idea about the influence of power control, sectorization, and voice activity on the capacity of a cellular DS-CDMA system, a simplified model is proposed in this book. It is better to rely on BEP models for a more detailed system investigation since the simplified model does not consider the effects of multipath in general.

The performance model used in this book is based on a minimum required signal-to-interference ratio at the receiver in order to obtain a prespecified BEP performance. In [43] it is mentioned that irrespective of modulation method, the minimal signal to interference ratio after correlation for a BEP better than 10^{-3} is $E_b/N_0 = 5$ (7 dB). The threshold value V for the signal-to-interference ratio before correlation is

$$V = \frac{1}{PG}\left(\frac{E_b}{N_0}\right)_{\min} \tag{9.114}$$

where PG is the processing gain equivalent to the number of chips per bit in this case. In a system with multiple cells and (imperfect) sectorization, the signal-to-interference ratio at the receiver is

$$\frac{C}{I} = \frac{P_i}{P_I\left(1+F_m\right)F_s} \tag{9.115}$$

with

P_i = Received power from desired terminal in home cell;
P_I = Intra cell interference power from users in home cell (single isolated cell without sectorization);
F_S = Interference correction factor due to imperfect sectorization (9.104);
F_m = Interference correction factor due to intercell interference (9.113);
The failure probability P_f is given by

$$P_f = \Pr\left(\frac{P_i}{P_I} < V\left(1+F_m\right)F_s\right) \tag{9.116}$$

In the case of imperfect power control, the received power P_i from each mobile is modeled as a lognormal variable with zero logarithmic mean and standard deviation σ. The pdf of the total interference signal produced by (k - 1) other active terminals is approximated by a lognormal variable with logarithmic mean $m_I(k-1)$ and standard deviation $\sigma_I(k-1)$ ((9.102) and (9.103), respectively) computed according to the method of Fenton [44]. After some mathematical manipulations, the failure probability for k active users is written as

$$P_f(k) = \frac{1}{2} + \frac{1}{2}\text{erf}\left(\frac{\ln\left[V(1+F_m)F_s\right] - m_I(k-1)}{\sqrt{2(\sigma^2 + \sigma_I^2(k-1))}}\right) \tag{9.117}$$

where erf (.) is the error function, V is the threshold ratio for the signal-to-interference ratio, F_m is the interference correction factor due to multiple cell interference, F_S is the interference correction factor due to sectorization, and $m_I(k-1)$ and $\sigma_I(k-1)$ are the logarithmic mean and the logarithmic standard deviation, respectively, of the total power received from $(k - 1)$ users in the case of imperfect power control with individual power-control error σ. Assuming that there are n users per cell ($n \geq k$), each with voice activity a, then the failure probability is

$$P_f^a(n) = \sum_{k=1}^{n} P_f(k)\binom{n}{k} a^k (1-a)^{n-k} \tag{9.118}$$

The user capacity is now defined as the maximum number of users per cell with voice activity a such that the failure probability is less than 0.01.

9.7.3 Throughput and Delay Analysis

For transmission of computer data, a packet communications schedule is more efficient than using a circuit-switched protocol. In this section, we consider a slotted system, implying that each user is allowed to transmit packets only at fixed time instants. In a packet network, throughput and delay are appropriate parameters rather than maximum user capacity. The throughput determines the average number of successfully received packets per time slot given a certain amount of traffic. For a certain amount of throughput, it is important to know the average delay of a packet. The throughput is defined as the average number of successfully received packets per time slot [49–52]:

$$S = \sum_{k=1}^{N} kP_t(k)P_s(k) \tag{9.119}$$

where N is the number of users per cell, $P_S(k)$ is the packet success probability and $P_t(k)$ is the probability of a packet transmitted with $k-1$ other packets.

We assume that a total of n_f independent failures occur within one time slot. In general, this number depends on the length of a time slot T_{slot} and the average time between two failures T_{av} and is defined as:

$$n_f \stackrel{\wedge}{=} \frac{T_{\text{slot}}}{T_{\text{av}}} \tag{9.120}$$

Now, the packet success probability $P_S(k)$ defined as the probability of a test packet not being destroyed by $(k-1)$ interfering packets is

$$P_s(k) = \left(1 - P_f(k)\right)^{n_f} \tag{9.121}$$

with $P_f(k)$ being the failure probability given by (9.117). Although it is difficult to implement power control to packet transmission due to short packet duration, the present analysis is based on power control for simplicity.

Because the number of simultaneous terminals is limited, the offered traffic is assumed to have a binomial distribution. The probability of a packet transmitted with $k-1$ other packets P_k is then given by

$$P_t(k) = \binom{N}{k}\left[\frac{G}{N}\right]^k\left[1 - \frac{G}{N}\right]^{N-k} \tag{9.122}$$

where G is the average number of offered packets per time slot. Here it is assumed that the offered traffic consists of new packets and retransmission packets.

Another important performance parameter for a slotted packet network is the delay, which is defined as the average time between the generation and successful reception of a packet. Assuming a positive acknowledgment scheme and no transmission errors in the acknowledgment packets, the expression for the delay in a slotted DS-CDMA system is [50–52]:

$$D = 1.5 + \left[\frac{G}{S} - 1\right]\left(\lfloor \delta + 1 \rfloor + 1\right) \tag{9.123}$$

where δ is the mean of the retransmission delay, which is uniformly distributed over the range from which the retransmission delay is selected, S is the throughput, and G is the offered traffic. The delay in (9.123) is normalized on the slot time.

9.7.4 Results

All figures presented is this section are valid for the reverse link. First, the results are given for the capacity (i.e., the maximum number of users per cell). For all results, arbitrary PN sequences with 255 chips per bit are assumed, implying a processing gain PG = 255. Figure 9.39 shows the influence of power-control imperfections on the capacity of a DS-CDMA system with only one cell and for a system with multiple cells. In the multiple cell case, the intercell interference is modeled as a fraction of the intracell interference using (9.113), where we assume a path-loss exponent $\gamma = 4$. In the case of perfect power control, the maximum capacity shows a decrease of 25% when considering multiple cells instead of a single cell. For lower values of γ, the decrease in capacity due to multiple cell interference is higher.

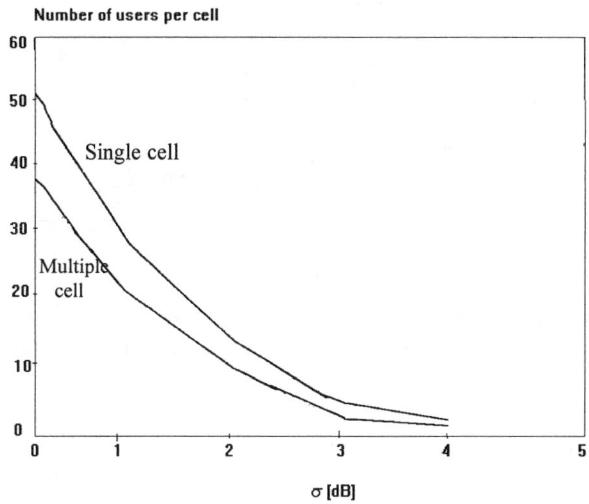

Figure 9.39 Capacity as function of powercontrol imperfection for a single cell and multiple cell system. Processing gain PG = 255, no voice activity monitoring (a = 1), no sectorization (D = 1).

Figure 9.39 also shows that the system is very sensitive to power-control errors; a power-control error of only 1 dB gives a capacity reduction of about 60%. A power-control error of 1 dB is interpreted as the standard deviation of 1 dB around the desired power level. Reference [40] concludes that a standard deviation of 2.5 dB causes a reduction in capacity of only 20%. However, [40] assumes that an outage occurs if the interference power density exceeds the background noise level by 10 dB, whereas here an outage occurs if the SNR after correlation exceeds a predefined value ($E_b/N_0 = 5$ (7 dB)). This explains the difference in capacity reduction.

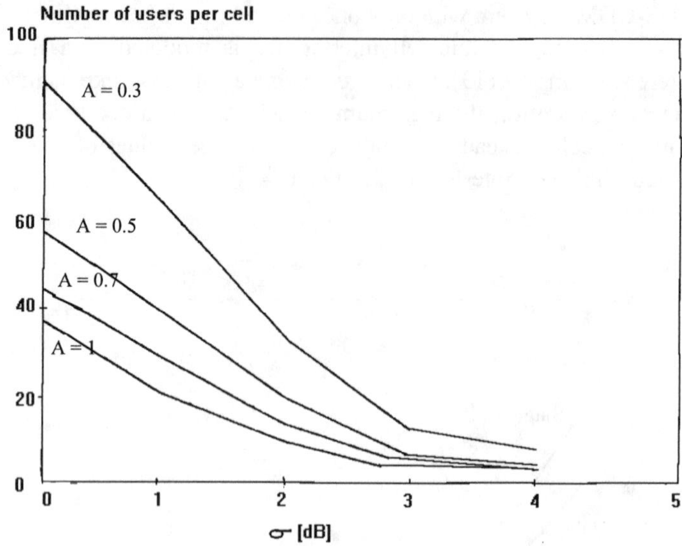

Figure 9.40 Capacity as a function of power-control imperfection for several values of the voice activity factor (a) Processing gain $PG = 255$, no sectorization ($D = 1$) multiple cells.

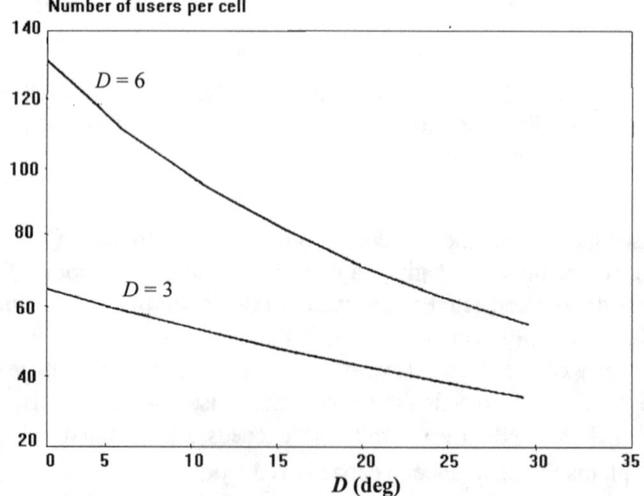

Figure 9.41 Capacity as a function of overlap angle for sectorization with sectorization degree $D = 3$ and $D = 6$. Processing gain $PG = 255$, power control imperfection $\sigma = 1$ dB, no voice activity monitoring ($a = 1$), multiple cells.

Figure 9.40 shows that capacity decreases as the voice activity factor increases. If the processing gain is increased (i.e., if the number of chips per bit is increased), then the maximum number of users increases at the cost of more required bandwidth (since the chip duration must decrease to keep the data rate fixed). As expected, the capacity increases with the increase in the degree of sectorization, which is shown in Figure 9.41. Furthermore, we see from Figure 9.41 that the capacity decreases as the overlap angle due to imperfect sectorized antennas increases. The model with the uniform antenna gain pattern and the overlap angle is a drastic simplification of real antenna gain patterns, but it simplifies the calculations tremendously. It suggests transforming real antenna gain patterns into uniform patterns analogously to modeling thermal filtered noise by the noise-equivalent bandwidth.

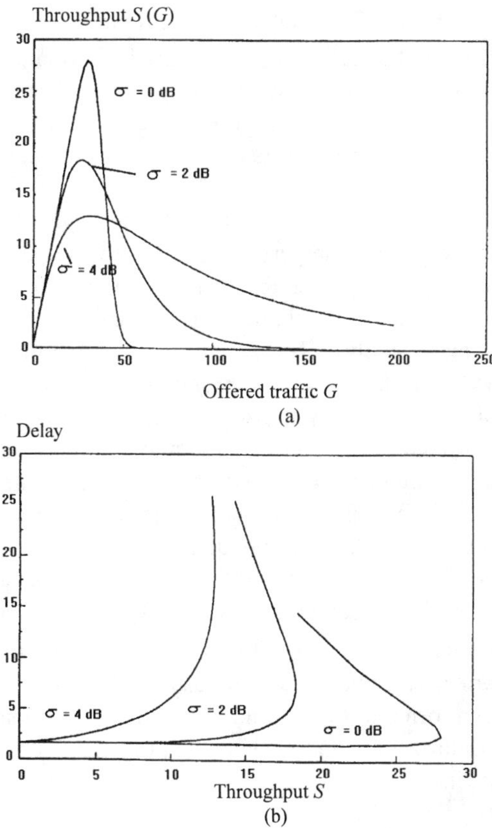

Figure 9.42 (a) Influence of power-control error on throughput for processing gain $PG = 255$, 200 terminals per cell, and no sectorization, (b) influence of power-control error on the delay for processing gain $PG = 255$, 200 terminals per cell and no sectorization.

In a packet-oriented DS-CDMA system, the throughput and delay characteristics are more appropriate performance parameters than the maximum user capacity. For all the following figures, we assume that on the average 1 fade per packet occurs ($n_f = 1$) and that the processing gain PG again is 255. The throughput is given per cell and the delay is measured in time slot units. For the delay calculations we assume a mean retransmission delay δ = 10 slot times. Figures 9.42(a,b) show the influence of power-control imperfections on the throughput and delay, respectively. It is clear that again the performance decreases as the power-control error increases. A decrease in performance implies that the maximum throughput is lower and the delay corresponding to a throughput value is higher.

From Figure 9.42(a) we see that for high traffic loads, the throughput increases for higher power-control errors. This is explained by considering that in the case of perfect power-control, the failure probability always equals 1 if the number of users exceeds the maximum capacity. This is not the case if we have imperfect power control ($\sigma \geq 0$ dB). Because of the tail of the lognormal distribution, there is always a probability that a few users are successful although the total number of users exceeds the maximum capacity. Then, the failure probability decreases as the variance in the received power due to power-control errors increases. This effect occurs in the overload region.

The influence of sectorization (perfect) on the throughput and delay is depicted in Figures 9.43(a,b). As expected, the performance increases with increasing number of sectors. From Figure 9.43(a), we see that the maximum throughput is better than proportional (i.e., the maximum throughput for $D = 3$ and $D = 6$ is higher than 3S at $D = 1$ and 6S at $D = 1$, respectively). To explain this, we first observe that the average number of offered packets at the maximum throughput is below the maximum capacity (maximum number of simultaneously transmitting users). This caused by the fact that for values of average offered traffic less than the maximum capacity, there is already a probability that the actual number of offered packets exceeds the maximum capacity. The maximum capacity of the cell is of course exactly proportional to the number of sectors in the cell. Since the total user population for each cell is the same for the sectorization schemes we considered ($N = 200$), the expression for the probability of having k simultaneous packets in one entire cell ($P_t(k)$) is the same. The probability of increasing a maximum user capacity of 50 users is then higher than the probability of increasing a maximum user capacity of 150 users. This implies that the difference between the maximum offered traffic and the maximum capacity decreases with increasing degree of sectorization, and explains that the maximum throughput increases more than proportional in our model.

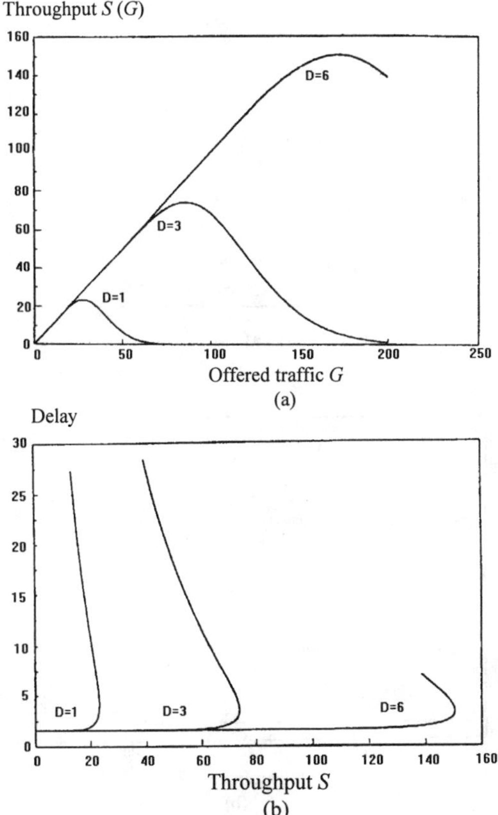

Figure 9.43 (a) Influence of sectorization on throughput for power-control error $\sigma = 1$ dB, processing gain PG = 255, 200 terminals per cell, and perfect sectorization, and (b) influence of sectorization on delay for power-control error $\sigma = 1$ dB, processing gain $PG = 255$, 200 terminals per cell and perfect sectorization.

Finally, Figures 9.44(a,b) show the influence of antenna imperfections on the throughput and delay of a slotted DS-CDMA system. The performance decreases as the overlap angle due to imperfect directional antennas increases.

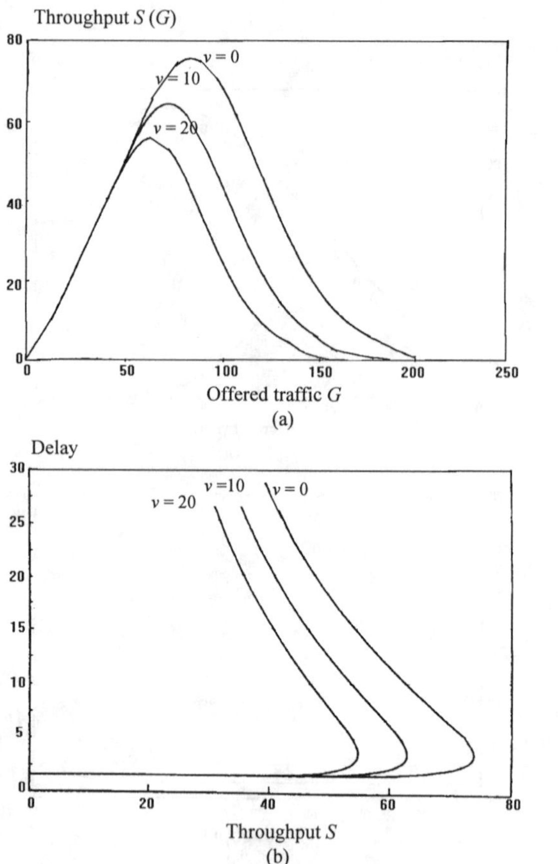

Figure 9.44 (a) Influence of overlap angle (in degrees) on throughput for power-control error σ =1 dB, processing gain PG = 255, 200 terminals per cell, and D = 3 sectors, and (b) influence of overlap angle (in degrees) on delay for power-control error σ =1 dB, processing gain PG = 255, 200 terminals per cell and D = 3 sectors.

9.7.5 Conclusions and Recommendations

A general model was presented in this section that is suitable to evaluate the reverse link performance of a cellular DS-CDMA system in terms of maximum user capacity, throughput, and delay. The multiple-cell interference is modeled analytically for an environment with path-loss exponent γ. To keep the model comprehensible and to investigate global effects of power control, sectorization and voice activity monitoring, effects of shadowing, and multipath propagation were not considered. Results are presented in this book for a path-loss exponent $\gamma = 4$. Voice-oriented as well as data-oriented systems were evaluated and the effects of imperfect power control and imperfect sectorization were studied. It was shown that a power control error of 1 dB leads to a capacity loss of 50% to 60%. Furthermore, an overlap degree of 15deg due to imperfect sectorization in a system with three sectors per cell causes a capacity reduction of 20%. This decrease in performance is due to the lower maximum user capacity and lower maximum throughput and higher delay.

The model described in this section is appropriate to investigate general system properties in a very clear way, and can be used to obtain a general idea of the effects of voice activity monitoring, power control and sectorization on performance. However, to obtain more detailed information, we recommend including shadowing effects, multipath statistics, and consideration of BEP performance [1].

REFERENCES

[1] R. Prasad, *CDMA for Wireless Personal Communication* , Norwood, MA: Artech House, 1996.

[2] N. Abramson, "The ALOHA system — another alternative for computer communications," *Proceedings Fall, Joint Computer Conference (AFIPS)* 37, pp. 281–285, 1970

[3] L. Kleinrock and F.A. Tobagi, "Packet switching in radio channels, part I-carrier sense multiple access nodes and their throughput — delay characteristics," *IEEE Trans. Comm.*, Vol. 23, pp. 1400–1416, December 1975.

[4] F.A. Tobagi and L. Kleinrock, "Packet switching in radio channels: part II — the hidden terminal problem in carrier sense multiple-access and the busy tone solution," *IEEE Trans. Comm.* Vol. 23, pp. 1417–1433, December 1975.

[5] J. Krebs and T. Freeburg, "Method and apparatus for communicating variable length messages between a primary station and remote stations at a data communication system," U.S. Patent No. 4519068, 1985.

[6] K.J. Zdunek, D.R. Ucci, and J.L. Locicero, "Throughput of non-persistent inhibit sense multiple access with capture," *Electron. Lett.*, pp. 30–32, January 1989.

[7] R. Prasad and J.C. Arnbak, "Capacity analysis of non-persistent inhibit sense multiple access in channels with multipath fading and shadowing," *Proc. 1989 Workshop on Mobile and Cordless Telephone Communications, IEEE*, King's College, University of London, pp. 129-134, September 1989.

[8] R. Prasad, "Throughput analysis of non-persistent inhibit sense multiple access in multipath fading and shadowing channels," *European Trans. on Telecommunications and Related Technologies*, Vol. 2, pp. 313–317, May–June 1991.

[9] R. Prasad, "Performance analysis of mobile packet radio networks in real channels with inhibit sense multiple access ," *IEE Proc.-I*, Vol. 138, No. 5 pp. 458–464, October 1991.

[10] R. Prasad and C.Y. Liu, "Throughput analysis of some mobile packet radio protocols in Rician fading channels," *IEE Proc.-I,* Vol. 139, No. 3, pp. 297–302, June 1992.

[11] I. Widipangestu, A.J. 't Jong, and R. Prasad, "Capture probability and throughput analysis of slotted ALOHA and unslotted np-ISMA in a Rician/Rayleigh environment," *IEEE Trans. on Vehicular Technology.,* Vol. 43, pp. 457–465, August 1994.

[12] N. Abramson, "The throughput of packet broadcasting channels," *IEEE Trans. Comm.*, Vol. 25, pp. 117–128, January 1977.

[13] A.S. Tanenbaum, *Computer Networks*, Englewood Cliffs, N.J.: Prentice-Hall, 1989.

[14] R. Rom and M. Sidi, *Multiple Access Protocols Performance and Analysis,* New York: Springer-Verlag, 1990.

[15] J.D. Spragins, J.L. Hammond, and K. Pawlikowski, *Telecommunications Protocols and Design*, Reading, MA:, Addison-Wesley Publishing Co., 1991.

[16] C.A. Sunshine, *Computer Network Architecture and Protocols*, New York: Plenum Press, 1989.

[17] M. Schwartz, *Computer-Communications Network Design and Analysis,* Englewood Cliffs, N.J.:Prentice-Hall, 1977.

[18] J.C. Arnbak and W. van Blitterswijk, "Capacity of slotted ALOHA in Rayleigh - fading channels," *IEEE J. Selected Areas in Comm.*, Vol. SAC-5, No.2, pp. 261–269, February 1987.

[19] B.S. Tsybakov, "Survey of USSR contributions to random multiple access communications," *IEEE Trans. Inf. Theory,* Vol. IT-31, No. 2, pp. 143–165, March 1985.

[20] J.A.M. Nijhof, R.D. Vossenaar, and R. Prasad, "Stack algorithm in mobile radio channels," *Proc. IEEE 44th Vehicular Technology Conference,* Stockholm, Sweden, pp. 1193–1197, June 1994.

[21] L. G. Roberts, "Dynamic allocation of satellite capacity through packet reservation," *Proc. National Computer Conference (AFIPS),* 42, pp. 711–716, June 1973.

[22] W. Crowther et. al., "A system for broadcast communications: reservation ALOHA," *Proc. Sixth Hawaii International Conference on System Sciences*, pp. 371–374, January 1973.

[23] R. Binder, R. Rettberg, D. Walden, S. Ornstein, and F. Heart, "A dynamic packet-switching system for satellite broadcast channels," *Proc. IEEE International Conference on Communications*, San Francisco, CA, pp. 41.1–41.5, June 1975.

[24] D.J. Goodman and S.X. Wei, "Factors affecting the bandwidth efficiency of packet reservation multiple access," *Proc. 39th IEEE Vehicular Technology Conference*, San Francisco, CA, pp. 292–299, May 1989.

[25] C. van den Broek, D. Sparreboom, and R. Prasad, "Performance evaluation of packet reservation multiple access protocol using steady-state analysis of a Markov chain model," *Proc. COST 227/231 Workshop*, Limerick, Ireland, pp. 454–462, September 6–10, 1993.

[26] C. van den Broek and R. Prasad, "Effect of capture on the performance of the PRMA protocol in an indoor radio environment with BPSK modulation," *Proc. IEEE 44th Vehicular Technology Conference,* Stockholm, Sweden, pp. 1223–1227, June 1994.

[27] R. Prasad and J.C. Arnbak, "Enhanced throughput in packet radio channels with shadowing," *Electron. Lett.*, 24, pp. 986–988, August 1988.

[28] S.C. Schwartz and Y.S. Yeh, "On the distribution function and moments of power sums with log-normal components," *Bell Syst. Tech. J.*, 24, pp. 1441–1462, September 1982.

[29] R. Prasad and J.C. Arnbak, "Comments and analysis for spectrum efficiency in single cell trunked and cellular mobile radio," *IEEE Trans.*, VT-37, pp. 220–222, November 1998.

[30] R. Prasad and J.C. Arnbak, "Enhanced packet throughput in radio channels with fading and shadowing," *Proc. Canadian Conf. on Electrical and Computer Eng.,* Vancouver, pp. 78–80, November 1988.

[31] N.R. Prasad and J.A.M. Nijhof, "Indoor wireless communication using slotted non-persistent ISMA, 1-persistent ISMA and non-persistent ISMA/CD," *Proc. IEEE 47th VTC-97*, Phoenix, AZ, pp. 1513–1517, May 1997.

[32] I.M.I. Habeb, M. Kavehrad, and C.E.W. Sundberg, "ALOHA with capture over slow and fast fading radio channels with coding and diversity," *IEEE J. Selected Areas in Comm.,* Vol. 7, pp. 79–88, January 1989.

[33] J.P.M.G. Linnartz and J.J.P. Werry, "Error correction and error detection coding in a fast fading narrowband slotted ALOHA network with BPSK modulation," in *Proc. 1st Symp. Commun. Theory Appl.*, Crieff Hydro Hotel, Scotland, Paper 37, September 9–13, 1991.

[34] J.G. Proakis, *Digital Communications*, New York: McGraw-Hill, 1983.

[35] M. Abramowitz and I.A. Stegun, *Handbook of Mathematical Functions*, New York: Dover 1965.

[36] J.P.M.G. Linnartz, R. Hekmat, and R.J. Venema, "Near-far effects in land mobile random access networks with narrowband Rayleigh fading channels," *IEEE Trans. on Vehicular Technology,* Vol. 41, February 1992.

[37] J.P.M.G. Linnartz, "Effect of fading and interference in narrowband land-mobile networks," Ph.D. dissertation, Delft Univ. Technol., The Netherlands, December 1991.

[38] R.J.C. Bultitude and G.K. Bedal, "Propagation characteristics on microcellular urban mobile radio channels at 910 MHz," *IEEE J. Selected Areas in Comm.,* Vol. 7, pp. 31–39, January 1989.

[39] S. Lin and D.J. Costello, Jr., *Error Control Coding, Fundamentals and Applications,* Englewood Cliffs, NJ: Prentice-Hall, 1983.

[40] A.M. Viterbi and A.J. Viterbi, "Erlang capacity of a power controlled CDMA system," *IEEE J. Select. Areas in Comm.*, Vol. 11, pp. 892–899, August 1993.

[41] A.J. Viterbi, A.M. Viterbi, and E. Zehavi, "Performance of power controlled wideband terrestrial digital communication," *IEEE Trans. Commun.*, Vol. 41, pp. 559–569, April 1993.

[42] R. Prasad, M.G. Jansen, and A. Kegel, "Capacity analysis of a cellular direct sequence code division multiple access system with imperfect power control," *IEICE Trans. Commun.*, Vol. E76-B, No. 8, pp. 894–905, August 1993.

[43] R. Prasad, M.G. Jansen and A. Kegel, "Effect of imperfect power control on a cellular code division multiple access system," *Electron. Lett.*, Vol. 28, No. 9, pp. 848–849, April 23, 1992.

[44] L.F. Fenton, "The sum of a log-normal probability distribution in scattered transmission systems," *IRE Trans.*, Vol. CS-8, pp. 57–67, March 1960.

[45] W.C.Y. Lee, "Overview of cellular CDMA," *IEEE Trans. Comm.*, Vol. 40, pp. 291–302, May 1991.

[46] K.S. Gilhousen, J.M. Jacobs, R. Padovani, A.J. Viterbi, L.A. Weaver, and C.E. Wheatly, III, "On the capacity of a cellular CDMA system," *IEEE Trans. Comm.*, Vol. 40, pp. 302–312, May 1991.

[47] H.J. Braun, G. Cosier, D. Freeman, A. Gillorie, D. Sereno, C.B. Southcott, and A. Van der Krogt, "Voice control of the Pan-European digital mobile radio system," *CSELT Tech. Rep.*, Vol. XVIII, No. 3, pp. 183–187, June 1990.

[48] K.I. Kim, "CDMA cellular engineering issues," *IEEE Trans. on Vehicular Technology*, Vol. 42, pp. 345–349, August 1993.

[49] M.G. Jansen and R. Prasad, "Throughput analysis of a slotted CDMA system with imperfect power control," *Proc. Inst. Elect. Eng. Colloquium on Spread Spectrum Techniques for Radio Commun. Syst.*, pp. 8.1–8.4, London, April 27, 1993.

[50] L. Kleinrock and F.A. Tobagi, "Packet switching in radio channels: part I–carrier sense multiple access modes and their throughput-delay characteristics," *IEEE Trans. Comm.*, Vol. COM-23, pp. 1400–1416, December 1975.

[51] R. Prasad, C.A.F.J. Wijffels, and K.L.A. Sastry, "Performance analysis of slotted CDMA with DPSK modulation diversity and BCH-coding in indoor radio channels," *AEU*, Vol. 46, No. 6, pp. 375–382, November 1992.

[52] C.A.F.J. Wijffels, H.S. Misser, and R. Prasad, "A micro-cellular CDMA system over slow and fast Rician fading channels with forward error correcting coding and diversity," *IEEE Trans. on Vehicular Technology*, Vol. 42, pp. 570–580, November 1993.

Chapter 10
OFDM

10.1 INTRODUCTION

Over the past few years, there has been increasing emphasis on extending the services available on wired public telecommunication networks to mobile/movable nonwired telecommunication users. At present, in addition to voice services, only low-bit-rate data services are available to mobile users. However, demands of WBMCS are anticipated within both the public and private sectors. Wired networks are not sufficient to support the extension to wireless networks with mobility, because mobile radio channels are more contaminated than wired data transmission channels, we also cannot preserve the high quality of service (QoS) required in wired networks [1].

The mobile radio channel is characterized by multipath reception: the signal offered to the receiver contains not only a direct LOS radio wave, but also a large number of reflected radio waves that arrive at the receiver at different times. Delayed signals are the result of reflections from terrain features such as trees, hills, mountains, vehicles or buildings. These reflected delayed waves interfere with the direct wave and cause ISI, which in turn causes significant degradation of network performance. A wireless network should be designed in such a way that adverse effects are minimized.

To create broadband multimedia mobile communication systems, it is necessary to use high bit-rate transmission with at least several megabits per second. However, if digital data are transmitted at the rates of several megabits per second, the delay time of the delayed waves is greater than 1 symbol time. To use adaptive equalization techniques at the receiver is one of the candidates to equalize these signals. There are practical difficulties in operating this equalization at several megabits per second with compact, low-cost hardware.

To overcome such a multipath-fading environment with low complexity and achieve WBMCS, this chapter presents an overview of the OFDM transmission scheme. OFDM is one of the applications of a parallel data transmission scheme, which reduces the influence of multipath-fading and makes complex equalizers unnecessary. First, the concept of the parallel data transmission scheme is described in Section 10.2. Then, Section 10.3 presents the history of OFDM transmission. After that, we show the configuration of the OFDM transmitter and receiver in Section 10.4 by using recent publications. In Section 10.5, simple computer simulation results about comparison between single carrier and OFDM broadband transmission methods are indicated. Finally, conclusions are discussed in Section 10.6.

10.2 CONCEPT OF PARALLEL TRANSMISSION SCHEME

The main problem with reception of radio signals is fading caused by multipath propagation environment, in which there are not only direct transmission signals but also many reflected signals that arrive at the receiver at different times in time domain. Delayed signals are the result of reflections from terrain features such as trees, hills, or mountains, or objects such as people, vehicles, or buildings. These direct and delayed waves received through multipath propagation at the receiver interfere with each other and create a fading environment, which is called multipath-fading. This multipath-fading is characterized by the channel impulse response, which includes the information of relative time when the delayed signals arrive at the receiver signal power, and signal phase as compared with direct wave. In Figure 10.1, one typical example of the impulse responses of multipath-fading is indicated in time domain. When we observe a multipath-fading environment from the perspective of frequency domain, a characteristic of multipath-fading is that some frequencies are enhanced whereas others are attenuated. Figure 10.1 also illustrates one of the impulse responses of multipath-fading in frequency domain. If there is a mobile reception, then the relative lengths and attentions of the various reception paths will change with time. A narrowband signal will vary in quality as the peaks and the troughs of the frequency response move around in frequency domain. There is also a noticeable variation in phase response that affects all systems using phase as a means of signaling.

Consider single-carrier, serial high-speed wireless data transmission. As mentioned in Section 10.1, if digital data are transmitted at the rate of several megabits per second, the delay time of the delayed waves is greater than 1 symbol time. Figure 10.1 also illustrates the waveforms of a single-carrier, serial high-speed wireless data transmission scheme in time and frequency domain when delay time of the delayed wave is larger than 1 symbol period of data transmission.

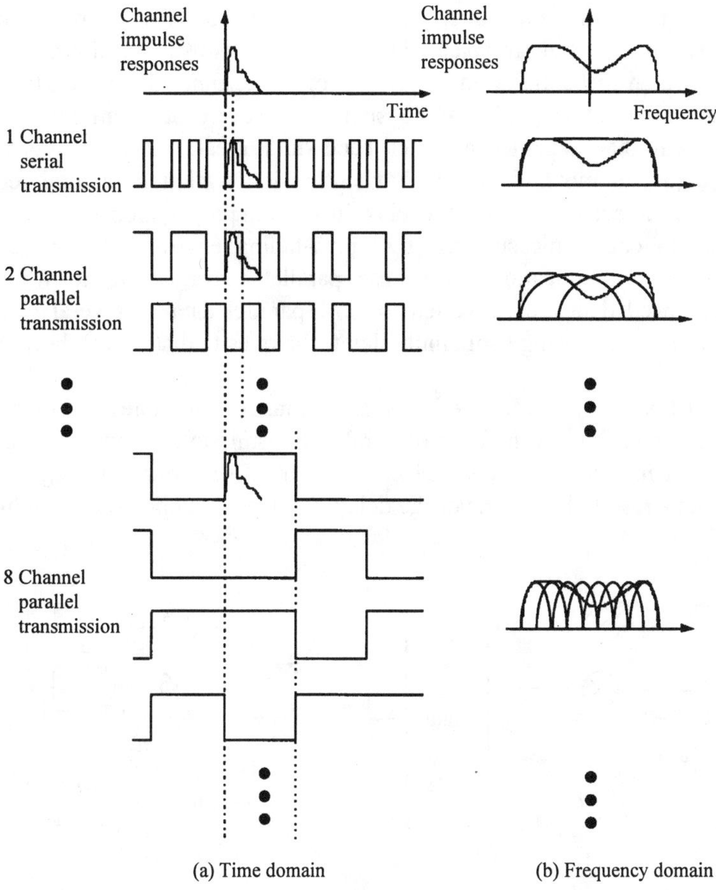

Figure 10.1 Concept of parallel transmission.

To use adaptive equalization techniques, which estimate the channel impulse response and multiply the complex conjugate of the estimated impulse response by the received data signal, the receiver is one candidate to equalize this signal. However, there are practical difficulties in operating such equalization at several megabits per second with high-speed, compact, and low-cost hardware, because, if we recover the transmitted data from the received data as shown in Figure 10.1 we must store several successive symbols and equalize the received data sequentially. In addition, with frequency domain, when the transmitted signal suffers from multipath-fading, some parts of the signal suffer from constructive interference and are enhanced in levels, whereas others suffer from destructive interference and are attenuated, sometimes to the point of extinction. In general, frequency components that are closed together will suffer well-correlated variations in signal strength. The correlation (or coherence) is used as a measure of this phenomenon. There is no standard definition of the correlation bandwidth. For a narrowband signal, distortion is usually minimized if the bandwidth is less than the correlation bandwidth of the channel. However, there is a significant chance that the signal will be subject to severe attenuation on some occasions. A signal that occupies a wider bandwidth greater than the correlation bandwidth is subject to more distortion, but suffers less variation in total received power, even if it is subject to significant levels of multipath propagation.

One way to countermeasure the multipath-fading environment and achieve broadband mobile communications is to use parallel transmission, in which the transmitted high-speed data are converted to slow parallel data in several channels. These data are multiplexed using some multiplex techniques to distinguish between the subchannels.

Figure 10.1 shows the effect of a parallel transmission scheme. For a given overall data rate, increasing the number of parallel transmission channels reduces the data rate that each individual subchannel conveys. In other words, it lengthens the symbol period. As a result, delay time of the delayed waves is suppressed to within one symbol time.

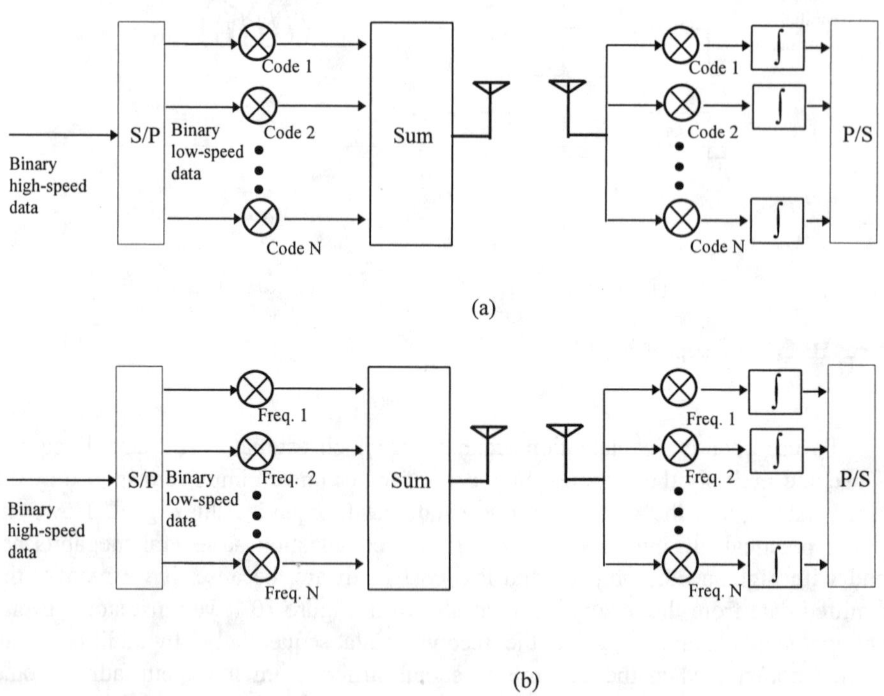

Figure 10.2 Configurations of (a) multicode and (b) multicarrier transmission schemes.

Frequency division multiplex and code division multiplex are often used to distinguish between subchannels. Frequency division multiplex is called multicarrier transmission and code division multiplex is called multicode transmission. Figure 10.2 illustrates the configurations of both parallel transmission schemes.

OFDM is one of the multicarrier transmission techniques, which is the most efficient multicarrier technique. There are two schools of thought about OFDM. The first is that OFDM is introduced as a countermeasure to multipath. The second says that OFDM transforms frequency-selective fading into flat fading: we do not get rid of the multipath problem, but change it slightly (from an error floor to a BEP decreasing inversely proportional to the SNR). It is in fact the (intercarrier) channel-coding techniques applied to OFDM that really do the trick.

10.3 HISTORY OF OFDM TRANSMISSION

The OFDM transmission scheme is the optimum version of the multicarrier transmission scheme. The concept of using parallel data transmission and frequency division multiplexing was published in the mid-1960s [2,3]. Some early development is traced back to the 1950s [4]. A U.S. patent was filed and issued in January, 1970 [5]. In a classical parallel data system, the total signal frequency band is divided into N nonoverlapping frequency subchannels. Each subchannel is modulated with a separate symbol and then the N subchannels are frequency-multiplexed. It seems good to avoid spectral overlap of channels to eliminate interchannel interference. However, this leads to inefficient use of the available spectrum. To cope with the inefficiency, the ideas proposed from the mid-1960s were to use parallel data and FDM with overlapping subchannels, in which each carrying a signaling rate b is spaced $b/2$ apart in frequency to avoid the use of high-speed equalization and to combat impulsive noise and multipath distortion, as well as to fully use the available bandwidth. Figure 10.3 illustrates the difference between the conventional nonoverlapping multicarrier technique and the overlapping multicarrier modulation technique. As shown in Figure 10.3, by using the overlapping multicarrier modulation technique, we save almost 50% of bandwidth. However, to realize the overlapping multicarrier technique we need to reduce crosstalk between subcarriers, which means that we keep the orthogonality of the carrier between subfrequency channels.

The word orthogonal indicates that there is a precise mathematical relationship between the frequencies of the carriers in the system. In a normal frequency division multiplex system, many carriers are spaced apart in such a way that the signals are received using conventional filters and demodulators. In such receivers, guard bands are introduced between the different carriers and in the frequency domain which results in a lowering of spectrum efficiency.

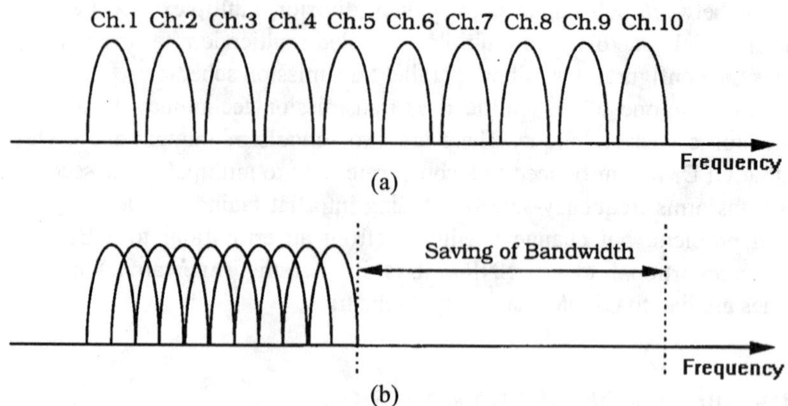

Figure 10.3 Concept of OFDM signal. (a) Conventional multicarrier technique, and (b) orthogonal multicarrier modulation technique.

However, it is possible to arrange the carriers in an OFDM signal so that the sidebands of the individual carriers overlap and the signals are still received without adjacent carrier interference. To do this the carriers must be mathematically orthogonal. The receiver acts as a bank of demodulators, translating each carrier down to DC, with the resulting signal integrated over a symbol period to recover the raw data. If the other carriers all beat down the frequencies which, in the time domain, have a whole number of cycles in the symbol period T_s, then the integration process results in zero contribution from all these other carriers. Thus, the carriers are linearly independent (i.e., orthogonal) if the carrier spacing is a multiple of $1/T_s$.

Mathematically, suppose we have a set of signals Ψ, where Ψ_p is the pth element in the set. The signals are orthogonal if

$$\int_a^b \psi_p(t)\psi_q^*(t)dt = \begin{cases} K & \text{for } p = q \\ 0 & \text{for } p \neq q \end{cases} \tag{10.1}$$

where * indicates the complex conjugate. A fairly simple mathematical proof exists that the series sin (mx) for $m = 1, 2, \ldots$ is orthogonal over the interval $-\pi$ to π. Much of transform theory makes use of orthogonal series, although they are by no means the only example.

Much of the research focuses on the high efficient multicarrier transmission scheme based on "orthogonal frequency" carriers. In 1971, Weinstein and Ebert [6] applied DFT to parallel data transmission system as part of the modulation and demodulation process. Using DFT to modulate and demodulate parallel data, the individual spectra are sinc functions and are not band-limited.

Figure 10.4 shows the spectrum of the individual data of the subchannel. The OFDM signal, multiplexed in the individual spectra at the space $b/2$ when the transmission speed of each subcarrier is b, is shown in Figure 10.4. Figure 10.4 shows that at the center frequency of each subcarrier, there are no cross-talks from other channels. Therefore, if we use DFT at the receiver and calculate correlation values with the center of frequency of each subcarrier, we recover the transmitted data with no cross-talk. In addition, using the DFT-based multicarrier technique, frequency division multiplex is achieved not by bandpass filtering but by baseband processing.

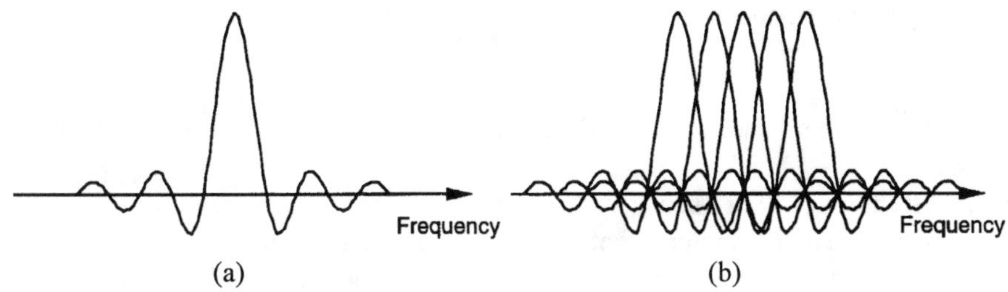

Figure 10.4 Spectra of (a) an OFDM subchannel, and (b) an OFDM signal.

Moreover, to eliminate the banks of subcarrier oscillators and coherent demodulators required by frequency division multiplex, completely digital implementations could be built around special-purpose hardware performing the FFT. Recent advances in very large scale integration (VLSI) technology make high-speed, large-size FFT chips commercially affordable. Using this method, both transmitter and receiver are implemented using efficient FFT techniques that reduce the number of operations from N^2 in DFT down to $N\log N$ [7].

In the 1960s, the OFDM technique was used in several high-frequency military systems such as KINEPLEX [4], ANDEFT [8], and KATHRYN [9]. For example, the variable rate data modem in KATHRYN was built for high-frequency radio and up to 34 parallel low-rate channels of PSK modulation were generated by a frequency multiplexed set of subchannels. Orthogonal frequency assignment was used with a channel spacing of 82 Hz to provide guard time between successive signaling elements.

In the 1980s, OFDM was studied for high-speed modems, digital mobile communications, and high-density recording. One of the systems realized the OFDM techniques for multiplexed QAM using DFT [10], and by using pilot tone, stabilizing carrier and clock frequency control and implementing trellis coding are also implemented [11]. Moreover, various-speed modems were developed for telephone networks [12].

In the 1990s, OFDM was exploited for wideband data communications over mobile radio FM channels, high-bit-rate digital subscriber lines (HDSL; 1.6 Mbps), asymmetric digital subscriber lines (ADSL; 1.536 Mbps), very-high-speed digital subscriber lines (VHDSL; 100 Mbps), digital audio broadcasting (DAB), and high definition television (HDTV) terrestrial broadcasting [13–18].

The OFDM transmission scheme has many advantages:

- More efficient use of the available bandwidth since the subchannels are overlapping.
- Spreading out the fade over many symbols. This effectively randomizes the burst errors caused by the Rayleigh fading, so that instead of several adjacent symbols being completely destroyed, many symbols are only slightly distorted.
- The symbol period is increased and thus the sensitivity of the system to delay spread is reduced.

On the other hand, the OFDM transmission scheme has some disadvantages:

- The technique is very sensitive to carrier frequency offset caused by the jitters of carrier wave and Doppler effect caused by a moving mobile terminal.
- The technique is one of multiamplitude and frequency modulation; therefore, the OFDM signal is contaminated by nonlinear distortion of transmitter power amplifier.
- At the receiver, it is very difficult to decide the start point of FFT when the OFDM signal arrives at the receiver. In other words, it is difficult to synchronize the OFDM signal.

These disadvantages become very important research tools when designing a whole system. In the next section, we introduce the configuration of the OFDM transmission scheme.

10.4 CONFIGURATION OF OFDM TRANSMISSION SYSTEM

This section presents a description of OFDM transmitter and receiver. The principle of guard interval and its effect on the system are also discussed.

10.4.1 Configuration of the Transmitter

Figure 10.5(a) illustrates the configuration of the OFDM transmitter. In the transmitter, the transmitted high-speed data are first converted to the parallel data of N subchannels. Then transmission data of each are modulated by QAM. Consider a quadrature-modulated data sequence of N channel (d_0, d_1, d_2,…,d_n), where each d_n is a complex number $d_n = a_n + jb_n$ in the case of QPSK, a_n, $b_n = \pm 1$ and for 16 QAM a_n,

$b_n = \pm 1, \pm 3$. Then, these modulated data are fed into the inverse discrete fourier transform (IDFT) circuit and the OFDM signal is generated. The transmitted data are given by

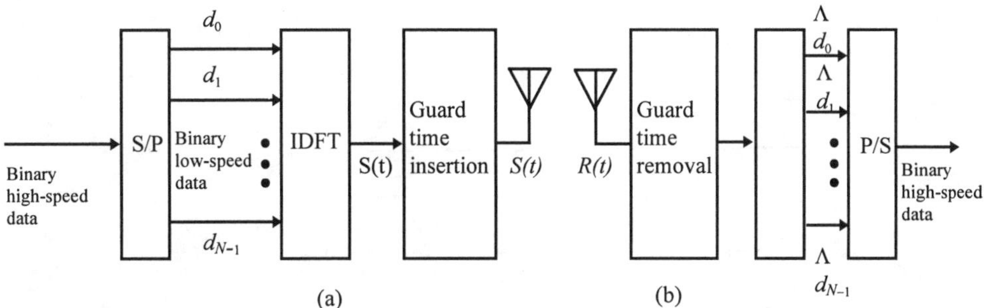

Figure 10.5 Configurations of OFDM: (a) transmitter and (b) receiver .

$$s(t) = \sum_{k=-\infty}^{\infty} \sum_{i=0}^{N-1} d_i(k) \exp\left(j2\pi\ f_i(t-kT_s)\right) f(t-kT_s)$$

$$= \sum_{k=-\infty}^{\infty} \sum_{i=0}^{N-1} \left[a_i(k) + jb_i(k)\right] \left[\cos\left(2\pi\ f_i(t-kT_s)\right) + j\sin\left(2\pi\ f_i(t-kTs)\right)\right] f(t-kT_s)$$

$$= \sum_{k=-\infty}^{\infty} \sum_{i=0}^{N-1} \left[a_i(k)\cos\left(2\pi\ f_i(t-kT_s)\right) - b_i(k)\sin\left(2\pi\ f_i(t-kTs)\right)\right] f(t-kT_s)$$

$$+ j\sum_{k=-\infty}^{\infty} \sum_{i=0}^{N-1} \left[a_i(k)\sin\left(2\pi\ f_i(t-kT_s)\right) + b_i(k)\cos\left(2\pi\ f_i(t-kTs)\right)\right] f(t-kT_s) \tag{10.2}$$

where T_s is the symbol duration of the OFDM signal and f_i (i = 0, 1, 2,...) is the frequency of ith subcarrier and given by

$$f_i = f_0 + \frac{i}{T_s} \tag{10.3}$$

Moreover, $f(t)$ is the pulse waveform of each symbol defined as

$$f(t) = \begin{cases} 1 & (0 \le t \le T_s) \\ 0 & (\text{otherwise}) \end{cases} \tag{10.4}$$

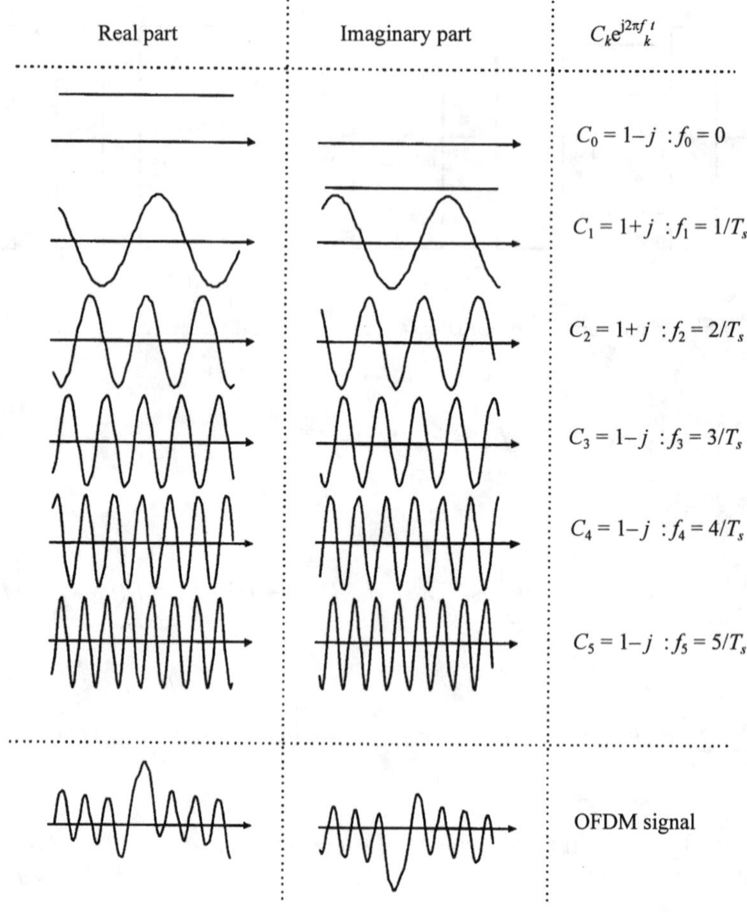

Figure 10.6 Waveform of OFDM-QPSK signal.

Figure 10.6 illustrates the waveform of the real and imaginary parts of the OFDM signal in each subchannel when defined d_i and f_i ($i = 0, 1, 2, \ldots, N-1$). As shown in Figure 10.6, the OFDM signal includes many carrier signals that have their own frequency. After that, the OFDM signal is fed to a guard time insertion circuit to reduce ISI .

Guard interval is explained as follows. The orthogonality of subchannels in OFDM is maintained, and individual subchannels are completely separated by the DFT at the receiver when there are no ISI and intercarrier interference (ICI) introduced by transmission channel distortion. In practice, these conditions cannot be obtained. Since the spectra of an OFDM signal is not strictly band-limited (sin *c*(*f*) function), linear distortion such as multipath causes each subchannel to spread energy into the adjacent channels and consequently cause ISI. A simple solution is to increase symbol duration or the number of carriers so that distortion becomes insignificant. However, this method is difficult to implement in terms of carrier stability, Doppler shift, and FFT size and latency.

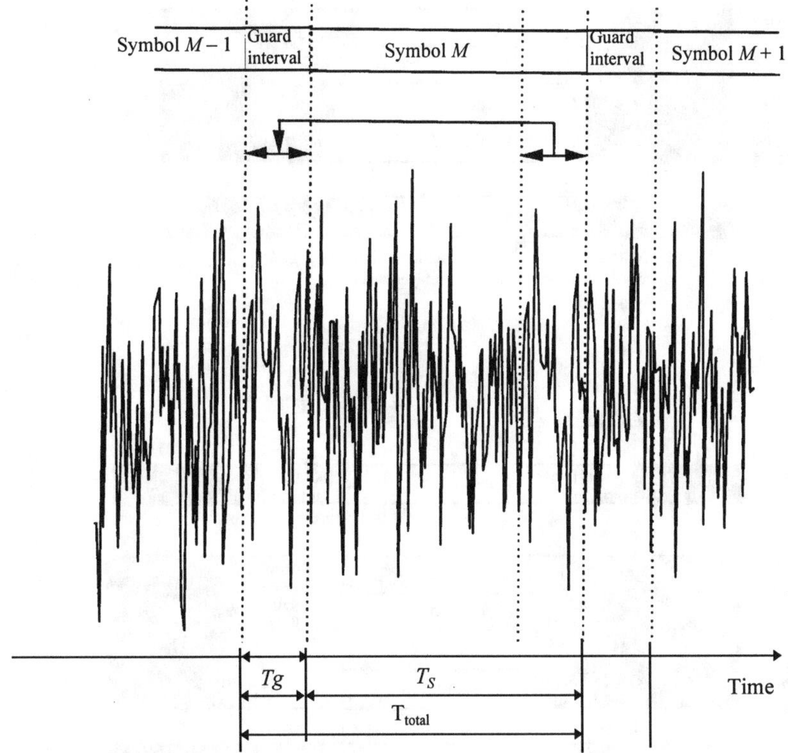

Figure 10.7 Principle of guard interval.

One way to prevent ISI is to create a cyclically extended guard interval, where each OFDM symbol is preceded by a periodic extension of the signal itself. The total symbol duration is $T_{\text{total}} = T_g + T_s$, where T_g is the guard interval. Figure 10.7 illustrates an example of the guard interval. Each symbol is made up of two parts. The whole signal is contained in the active symbol, the last part of which is also repeated at the start of the symbol and is called the guard interval. When the guard interval is longer than the channel impulse response or the multipath delay, ISI is eliminated. However, ICI, or inband fading, still exists. The ratio of the guard interval to useful symbol duration is application-dependent. Since the insertion of the guard interval will reduce data throughput, T_g is usually less than $T_s/4$.

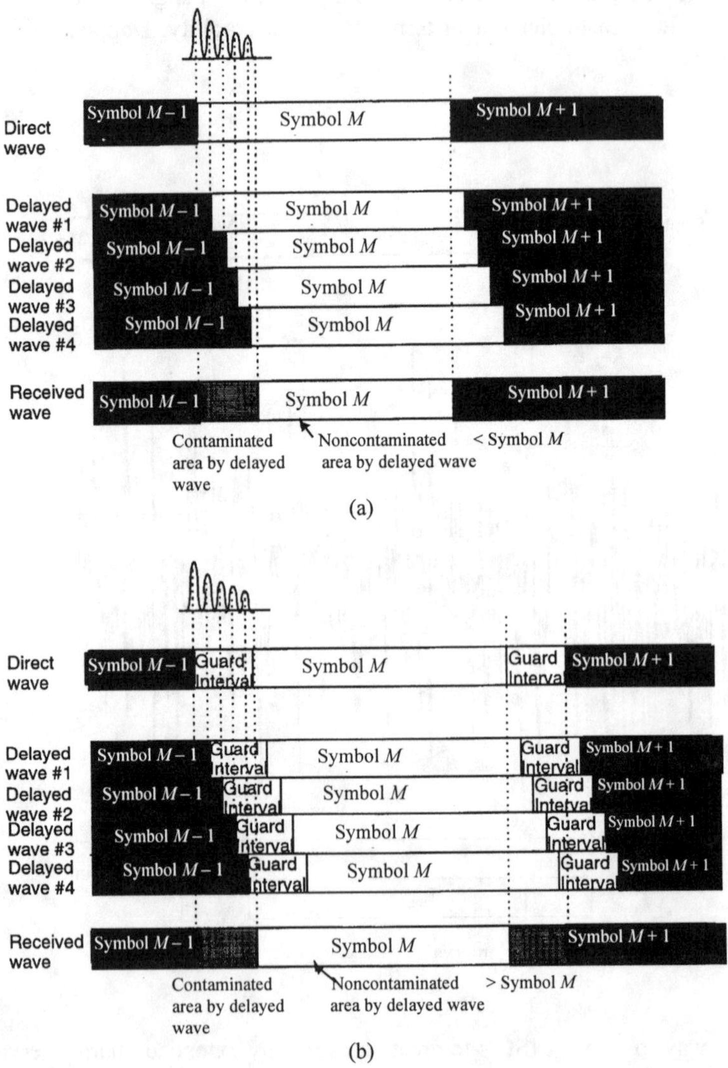

Figure 10.8 Effect of guard interval insertion: (a) without, and (b) with guard interval.

Figure 10.8 illustrates the effect of guard interval for a multipath-fading channel. As shown in Figure 10.8, if we do not insert the guard interval, the transmitted data are contaminated by the delayed waves and ISI occurs. On the other hand, by using the guard interval, the interference caused by all delayed waves are absorbed.

After guard interval insertion circuit, the OFDM signal is given as follows:

$$s'(t) = \sum_{k=-\infty}^{\infty} \sum_{i=0}^{N-1} d_i(k) \exp\left[j2\pi f_i (t - kT_{\text{total}})\right] f'(t - kT_{\text{total}}) \tag{10.5}$$

where $f'(t)$ is the modified pulse waveform of each symbol defined as

$$f'(t) = \begin{cases} 1 & \left(-T_g \le t \le T_s\right) \\ 0 & \left(t < -T_g, t > T_s\right) \end{cases} \tag{10.6}$$

After that, the OFDM signal transmits to the receiver; however the transmitted data $s'(t)$ are contaminated by multipath-fading and AWGN. In the receiver, the received signal is given by

$$r(t) = \int_0^{\infty} h(\tau;t) s(t - \tau) d\tau + n(t) \tag{10.7}$$

where $h(\tau;t)$ is the impulse response of the radio channel at time t, and $n(t)$ is the complex AWGN.

10.4.2 Configuration of the Receiver

Figure 10.5(b) illustrates a block diagram of the OFDM receiver. In the receiver, the received signal $r(t)$ is filtered by a bandpass filter, which is assumed to have enough passband to introduce negligible distortion on to the signal. The signal is then applied to an orthogonal detector where the signal is downconverted into an intermediate frequency (IF) band, and applied to a DFT circuit to obtain Fourier coefficients of the signals in observation periods $[iT_{\text{total}}, iT_{\text{total}} + T_s]$. The output $\hat{d}_i(k)$ of DFT circuit of ith OFDM subchannel is given by

$$\hat{d}_i(k) = \frac{1}{T_s} \int_{kT_{\text{total}}}^{T_s + kT_{\text{total}}} r(t) \exp\left(-j2\pi f_i (t - kT_{\text{total}})\right) dt \tag{10.8}$$

If we estimate the characteristics of delayed waves $\hat{h}_i(k)$ in a multipath-fading environment, we equalize these received data. The equalized data are given by

$$\hat{\hat{d}}_i(k) = \frac{\hat{h}_i^*(k)}{\hat{h}_i(k)\hat{h}_i^*(k)} \hat{d}_i(k) \tag{10.9}$$

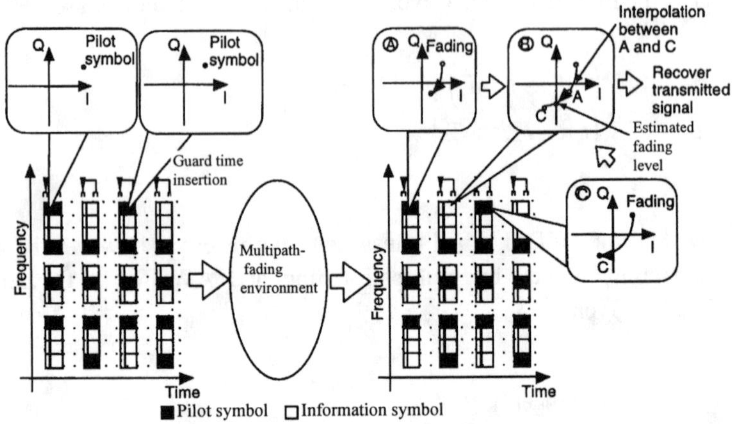

Figure 10.9 Pilot-symbol-aided OFDM system.

It is important to estimate the characteristics of delayed waves $\hat{h}_i(k)$. One of the channel characteristics estimation method is the pilot symbols insertion technique [18], in which pilot symbols are inserted at decided intervals as shown in Figure 10.9. At the receiver we estimate the channel characteristics with every pilot symbol. However, while the estimated information is very effective for the data of the pilot-inserted OFDM subchannels, it is not sufficient for the nonpilot-inserted OFDM subchannels. To estimate the channel characteristics in the nonpilot-inserted OFDM subchannels, we use the interpolation and extrapolation of the estimated channel characteristics of the pilot channel.

10.5 BEP

In this section, we present one of advantages of OFDM, which is that it has robustness for multipath-fading in comparison with the nonparallel transmission system.

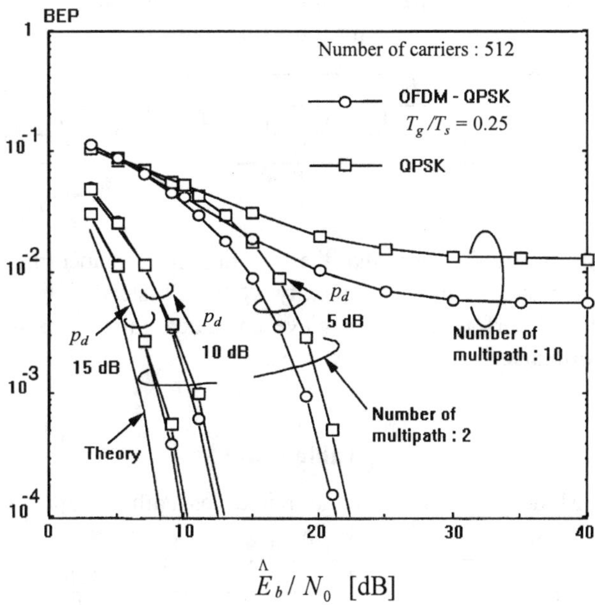

Figure 10.10 BEP performance of the OFDM transmission scheme.

Figure 10.10 shows the BEP performance of single-carrier QPSK and OFDM-QPSK transmission schemes as a function of the ratio of energy per bit and noise per density (E_b/N_0), under the multipath environment without any kinds of channel coding and equalizing techniques. For the multipath environment, we selected a two-path model that includes a direct and indirect wave and a ten-path model. These two channel models are used to evaluate performance results. In Tables 10.1 and 10.2, the configurations of the multipath model are described. In these tables, we use three parameters; namely, the normalized position of delayed wave d_w, the power ratio of direct wave and indirect wave p_d, and the phase rotation factor of delayed wave at the receiver ϕ_d. As shown in Figure 10.10, in the case of the two-path model, as p_d decreases the difference of BEP performance between QPSK and OFDM-QPSK increases. This is because the decrease of p_d means that the signal level of delayed wave increases; therefore, by using guard interval, OFDM-QPSK avoids the influence of delayed waves. Moreover, in the case of the ten-multipath channel, by using OFDM-QPSK, we obtain about a 3-dB better BEP performance than normal QPSK transmission. We conclude that the OFDM transmission scheme has robustness for a multipath-fading environment.

Table 10.1

Channel Model Using the Simulation in a Two-path Multipath Channel

	d_w	p_d [dB]	ϕ_d [deg]
Direct wave	0	0	0
Delayed wave	0.25	5, 10, 15	45

It is worth mentioning here that the uncoded performance improvement is not very impressive for such a high complexity system. Instead, it is recommended to investigate the coded performance to achieve enhanced gain. In particular, we would expect huge performance gains for the ten-tap case.

Table 10.2

Channel Model Using the Simulation in a Ten-Path Multipath Channel

	d_w	p_d [dB]	ϕ_d [deg]
Direct wave	0	0	0
Delayed wave no. 1	0.0049	5.3	311
Delayed wave no. 2	0.010	19.0	241
Delayed wave no. 3	0.023	26.0	317
Delayed wave no. 4	0.031	28.0	339
Delayed wave no. 5	0.034	30.9	268
Delayed wave no. 6	0.040	34.7	83
Delayed wave no. 7	0.047	29.9	344
Delayed wave no. 8	0.065	29.4	204
Delayed wave no. 9	0.076	33.6	268
Delayed wave no. 10	0.079	32.8	211

10.6 CONCLUSIONS

OFDM has long been studied and implemented to combat transmission channel impairments. Its applications have been extended from high-frequency radio communications to telephone networks, digital audio broadcasting, and digital television terrestrial broadcasting. The advantages of OFDM, especially in the multipath propagation, interference, and fading environment, make the technology a promising alternative in digital communications, including mobile multimedia. Current research and development of OFDM around the world will certainly provide us with valuable findings in theory and implementation.

The OFDM transmission scheme offers several research themes; for example,

- Channel estimation for multipath-fading environment;
- Synchronization of the OFDM signal;
- Countermeasure nonlinear distortion;
- Reduce peak-to-average-ratio of the OFDM signal.

We must take these research topics into account when designing novel OFDM transmission schemes.

REFERENCES

[1] R. Prasad, "Wireless broadband communication systems," *IEEE Comm. Mag.*, Vol. 25, p.18, January 1993.

[2] R.W. Chang, "Synthesis of band limited orthogonal signals for multichannel data transmission," *Bell Syst. Tech. J.*, Vol. 45, pp. 1775–1796, December 1996.

[3] B.R. Salzberg, "Performance of an efficient parallel data transmission system," *IEEE Trans. Comm.*, Vol. COM-15, pp. 805–813, December 1967.

[4] R.R. Mosier and R.G. Clabaugh, "Kineplex, a bandwidth efficient binary transmission system," *AIEE Trans.*, Vol. 76, pp. 723–728, January 1958.

[5] "Orthogonal frequency division multiplexing," U.S. Patent No. 3, 488,4555, filed November 14, 1966, issued January 6, 1970.

[6] S.B. Weinstein and P.M. Ebert, "Data transmission by frequency division multiplexing using the discrete Fourier transform," *IEEE Trans. Comm.*, Vol. COM-19, pp. 628–634, October 1971.

[7] W.Y. Zou and Y. Wu, "COFDM: an overview." *IEEE Trans. Broadc.*, Vol. 41, No. 1, pp. 1–8, March 1995.

[8] G.C. Porter, "Error distribution and diversity performance of a frequency differential PSK HF modem," *IEEE Trans. Comm.*, Vol., COM-16, pp. 567–575, August 1968.

[9] M.S. Zimmerman and A.L. Kirsch, "The AN/GSC-10 (KATHRYN) variable rate data modem for HF radio," *IEEE Trans. Comm.,* Vol., COM-15, pp. 197–205, April 1967.

[10] B. Hirosaki, "An orthogonally multiplexed QAM system using the discrete Fourier transform," *IEEE Trans. Comm.,* Vol., COM-29, pp. 982–989, July 1981.

[11] B. Hirosaki, "A 19.2 kbits voice band data modem based on orthogonality multiplexed QAM techniques," *Proc. of IEEE ICC'85,* pp. 21.1.1–5, 1985.

[12] W.E. Keasler and D.L. Bitzer, "High speed modem suitable for operating with a switched network," U.S. Patent No. 4,206,320, June 1980.

[13] P.S. Chow, J.C. Tu and J.M. Cioffi, "Performance evaluation of a multichannel transceiver system for ADSL and VHDSL services," *IEEE J. Selected Area,* Vol., SAC-9, No. 6, pp. 909–919, August 1991.

[14] P.S. Chow, J.C. Tu, and J.M. Cioffi, "A discrete multitone transceiver system for HDSL applications," *IEEE J. Selected Areas in Comm.,* Vol. SAC-9, No. 6, pp. 909–919, August 1991.

[15] R.V. Paiement, "Evaluation of single carrier and multicarrier modulation techniques for digital ATV terrestrial broadcasting," *CRC Report,* No. CRC-RP-004, Ottawa, Canada, December 1994.

[16] H. Sari, G. Karma, and I. Jeanclaude, "Transmission techniques for digital terrestrial TV broadcasting," *IEEE Comm. Mag.,* Vol.33, pp. 100–109, February 1995.

[17] A.V. Oppenheim and R.W. Schaffer, *Discrete-time Signal Processing*, Prentice-Hall International, ISBN 0-13-216771-9, 1989.

[18] S. Hara, M. Mouri, M. Okada, and N. Morinaga, "Transmission performance analysis of multi-carrier modulation in frequency selective fast Rayleigh fading channel," in *Wireless Personal Communications*, Kluwer Academic Publishers, Vol.2, pp. 335–356, 1996.

Chapter 11
Multicarrier CDMA

11.1 INTRODUCTION

Recently, the CDMA technique has been considered as a candidate to support multimedia services in mobile radio communications [1], because it has its own capabilities to cope with the asynchronous nature of multimedia data traffic to provide higher capacity over conventional access techniques such as TDMA and FDMA, and to combat the hostile channel frequency selectivity.

On the other hand, the multicarrier modulation scheme, also called OFDM, has drawn a lot of attention in the field of radio communications, mainly because of the need to transmit high data rate in a mobile environment that makes a highly hostile radio channel. To combat the problem, OFDM is seen as a solution, as discussed in Chapter 10.

It was in 1993, an epoch of CDMA application, that three types of new multiple access schemes based on a combination of code division and OFDM techniques were proposed, such as multicarrier (MC-) CDMA, multicarrier DS-CDMA, and multitone (MT-) CDMA. These schemes were developed by different researchers: MC-CDMA by N. Yee, J-P. Linnartz, and G. Fettweis [2], K. Fazel and L. Papke [3], and A. Chouly, A. Brajal, and S. Jourdan [4]; multicarrier DS-CDMA by V. DaSilva and E.S. Sousa [5]; and MT-CDMA by L. Vandendorpe [6]. These signals are easily transmitted and received using FFT device without increasing the transmitter and receiver complexities, and have the attractive feature of high spectral efficiency due to minimally densely subcarrier spacing.

This chapter reviews the three types of multicarrier CDMA schemes and discusses their advantages and disadvantages in terms of transmitter and receiver structures, spectral efficiency, and downlink BEP performance in a frequency-selective, slow Rayleigh-fading channel.

Section 11.2 presents the channel model. The multicarrier CDMA schemes are categorized mainly into two groups. The first spreads the original data stream using a given spreading code, and then modulates a different subcarrier with each chip (in a sense, the spreading operation in the frequency domain) [2 – 4]. The other spreads the serial-to-parallel (S/P) converted data streams using a given spreading code, and then modulates a different subcarrier with each of the data stream (the spreading operation in the time domain) [5,6], similar to a normal DS-CDMA scheme. There are two schemes corresponding to the second category, multicarrier DS-CDMA and (MT-) CDMA, whereas there is only one scheme, known as MC-CDMA, in the first category. Channel model is discussed in Section 11.2. Sections 11.3, 11.4, 11.5, and 11.6 present MC-CDMA in detail where multicarrier DS-CDMA and MT-CDMA are discussed briefly in Section 11.7. Section 11.8 compares BEP of the three systems. To have a fair comparison, performance of the three systems is evaluated considering the same spectral bandwidth and the same channel characteristics. Note that system performance is evaluated only for a downlink channel. Conclusions are given in Section 11.9.

11.2 CHANNEL MODEL

As a frequency-selective, fast Rayleigh-fading channel, we assume a wide-sense, stationary uncorrelated scattering (WSSUS) channel [7] with L received paths in the complex equivalent low-pass, time-variant impulse response:

$$h(\tau;t) = \sum_{l=1}^{L} g_l(t)\delta(\tau - \tau_l) \tag{11.1}$$

where t and τ are the time and the delay, respectively, $\delta(t)$ is the Dirac delta function $g_l(t)$ is the complex envelope of the signal received on the lth path that is a complex Gaussian random process with zero mean and variance σ_l^2, and τ_l is the propagation delay for the lth path.

Figure 11.1 shows the corresponding multipath delay profile given by (we assume there is no signal whose propagation delay exceeds the symbol duration)

$$\phi_c(\tau) = \frac{1}{2}E\left[h^*(\tau;t)h(\tau;t)\right] = \sum_{l=1}^{L}\sigma_l^2\delta(\tau - \tau_l) \tag{11.2}$$

where $E[.]$ is the expectation.

The rms delay spread (τ_{RMS}) is calculated as

$$\tau_{\text{RMS}} = \sqrt{\frac{\sum_{l=1}^{L} \sigma_l^2 \tau_l^2}{\sum_{l=1}^{L} \sigma_l^2} - \tau_{AV}^2} \tag{11.3}$$

$$\tau_{\text{AV}} = \frac{\sum_{l=1}^{L} \sigma_l^2 \tau_l}{\sum_{l=1}^{L} \sigma_l^2} \tag{11.4}$$

On the other hand, the channel time selectivity is characterized by the time ACF:

$$\rho_l(\Delta t) = \frac{1}{2} E\left[g_l(t+\Delta t) g_l^*(t)\right] \tag{11.5}$$

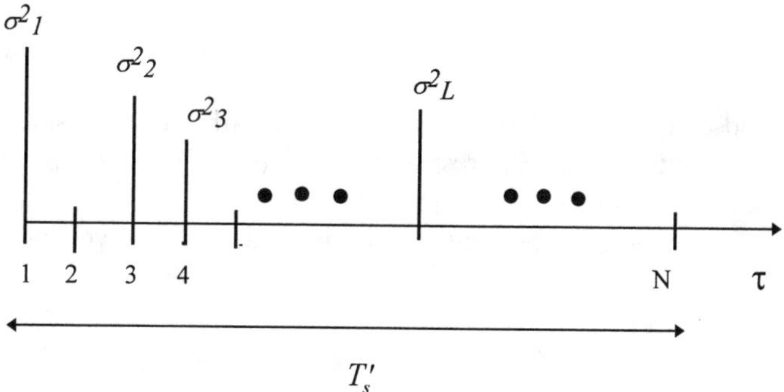

Figure 11.1 Multipath delay profile. Here, T_s' : symbol duration (at subcarrier), and N: the number of subcarriers (DFT points).

Assuming that an omnidirectional monopole antenna is used at the receiver, $\rho_l(\Delta t)$ is given by [8]

$$\rho_l(\Delta t) = \sigma_l^2 J_0\left(2\pi \ f_D \Delta t\right) \approx 1 - \left(\pi \ f_D \Delta t\right)^2 \quad \left(f_D \Delta t << 1\right) \tag{11.6}$$

where $J_0(.)$ and f_D are the zero-order Bessel function of the first kind and the maximum Doppler frequency, respectively.

11.3 DS-CDMA AND MC-CDMA SYSTEMS

11.3.1 DS-CDMA System

DS-CDMA transmitter spreads the original signal using a given spreading code in the time domain. The capability of suppressing multiple access interference (MAI) is determined by the cross-correlation characteristic of the spreading codes. Also, a frequency-selective-fading channel is characterized by the superimposition of several signals with different delays in the time domain, as shown in Section 11.2. Therefore, the capability of distinguishing one component from other components in the composite received signal (time resolution) is determined by the autocorrelation characteristic of the spreading codes.

Figure 11.2(a) shows the DS-CDMA transmitter for the jth user with BPSK modulation/coherent demodulation (CBPSK) scheme. The complex equivalent low-pass transmitted signal is written as

$$s_j(t) = \sum_{i=-\infty}^{+\infty} \sum_{k=0}^{K_{DS}-1} b_j(i) c_j(k) p_c\left(t - kT_c - iT_s\right) \tag{11.7}$$

where $b_j(i)$ and $c_j(k)$ are the ith information and the kth bit of the spreading code with length K_{DS} and chip duration T_C, respectively, T_S (= $1/R$) is the symbol duration (R is the symbol rate), and $p_C(t)$ is the chip pulse waveform.

For instance, when a rectangular pulse is used, $p_C(t)$ is given by

$$p_c(t) = \begin{cases} 1 & \left(0 \le t \le T_c\right) \\ 0 & \text{(otherwise)} \end{cases} \tag{11.8}$$

The transmitted signal for total J users is written as

$$s(t) = \sum_{j=1}^{J} s_j(t) \tag{11.9}$$

DS-CDMA Rake receiver contains multiple correlators, each synchronized to a different resolvable path in the received composite signal [9].

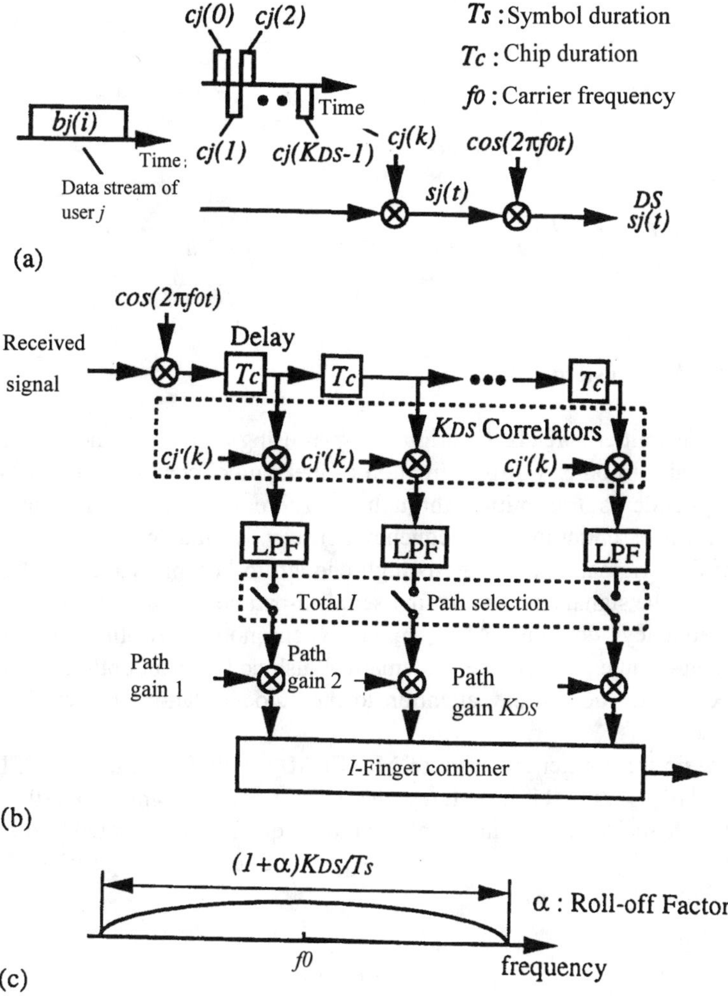

Figure 11.2 DS-CDMA system: (a) transmitter, (b) Rake receiver, and (c) power spectrum of its transmitted signal.

Figure 11.2(b) shows the I-finger DS-CDMA Rake receiver for the j' th user. The received signal through the frequency-selective, Rayleigh-fading channel given by (11.1) is written as

$$r(t) = \int_{-\infty}^{+\infty} s(t-\tau)*h(\tau;t)d\tau + n(t) = \sum_{l=1}^{L} r_l(t) + n(t) \tag{11.10}$$

$$r_l(t) = s(t-\tau_l)g_l(t) \tag{11.11}$$

The decision variable at $t = iT_S$ is written as

$$D_{j'} = \sum_{v=1}^{I} g_v(iT_s)\frac{1}{T_s}\int_{iT_s+\tau_v}^{iT_s+\tau_v+T_s}\sum_{k=0}^{K_{DS}-1} c_{j'}(k)p_c(t-uT_c-iT_s-\tau_v)r(t)dt \tag{11.12}$$

11.3.2 MC-CDMA System

MC-CDMA transmitter spreads the original signal using a given spreading code in the frequency domain. In other words, a fraction of the symbol corresponding to a chip of the spreading code is transmitted through a different subcarrier. For multicarrier transmission, it is essential to have frequency-nonselective fading over each subcarrier. Therefore, if the original symbol rate is high enough to become subject to frequency-selective fading, the signal needs to be first serial-to-parallel converted before spreading over the frequency domain. Also, in a synchronous downlink mobile radio communication channel, we use the Hadamard Walsh codes as an optimum orthogonal code set, because we do not pay attention to the autocorrelation characteristic of the spreading codes.

The basic transmitter structure of MC-CDMA is similar to that of OFDM in a DAB system [10]. Figure 11.3(a) shows the MC-CDMA transmitter for the *j*th user with CBPSK scheme, where the input information sequence is converted into P parallel data sequences $[a_{j,0}(i), a_{j,1}(i), \ldots, a_{j,P-1}(i)]$. The complex equivalent low-pass transmitted signal is written as

$$s_j(t) = \sum_{i=-\infty}^{+\infty}\sum_{p=0}^{P-1}\sum_{m=0}^{K_{MC}-1} a_{j,p}(i)d_j(m)p_s(t-iT_s')e^{j2\pi(Pm+p)\Delta f'(t-iT_s')} \tag{11.13}$$

$$\Delta f' = \frac{1}{(T_s'-\Delta)} \tag{11.14}$$

$$T_s' = PT_s \tag{11.15}$$

where $\{d_j(0), d_j(1), \ldots, d_j(K_{MC}-1)\}$ is the Hadamard Walsh code for the *j*th user (the length is K_{MC}), $\Delta f'$ is the minimum subcarrier separation (the subcarrier separation for

$a_{j,p}(i)$ is $\Delta f = P/\left(T_s' - \Delta\right)$, and the total number of subcarriers is $N = P \times K_{MC}$), T_s' is the symbol duration at subcarrier, Δ is the guard interval, and $p_s(t)$ is the rectangular symbol pulse waveform defined as

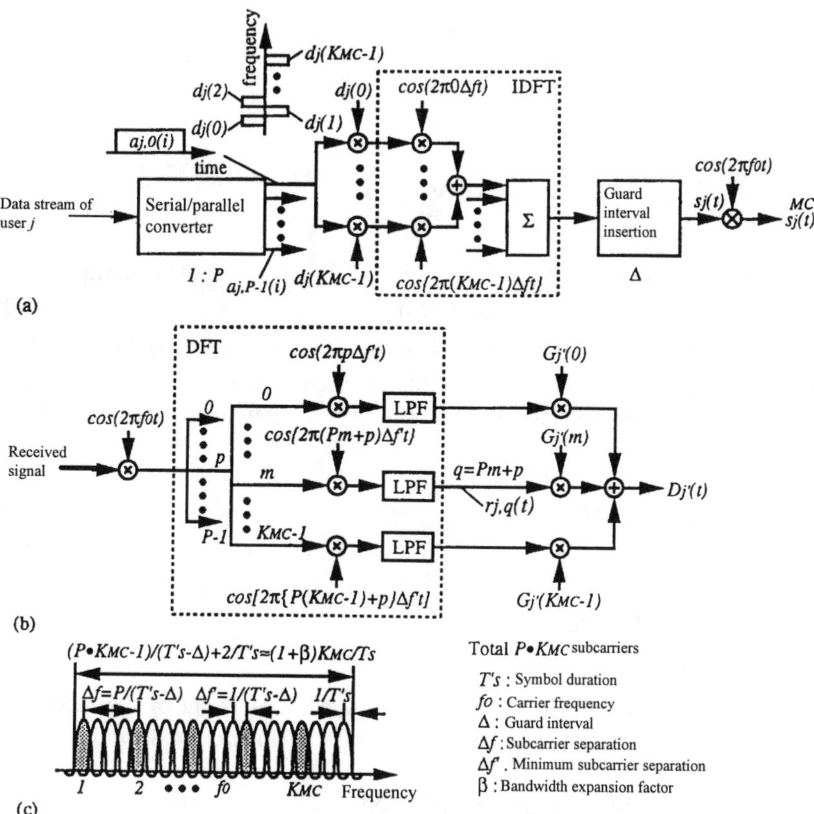

Figure 11.3 MC-CDMA system: (a) transmitter, (b) receiver, and (c) power spectrum of its transmitted signal.

$$p_s(t) = \begin{cases} 1 & \left(-\Delta \le t \le T_s' - \Delta\right) \\ 0 & \text{(otherwise)} \end{cases} \tag{11.16}$$

The transmitted signal for total J users is written as

$$s(t) = \sum_{j=1}^{J} s_j(t) \tag{11.17}$$

The received signal is written as

$$r(t) = \int_{-\infty}^{+\infty} s(t-\tau) * h(\tau;t)d\tau + n(t)$$

$$= \sum_{i=-\infty}^{+\infty}\sum_{p=0}^{P-1}\sum_{m=0}^{K_{MC}-1}\sum_{j=1}^{J} z_{m,p}(t)a_{j,p}(i)d_m^j p_s(t-iT_s')e^{j2\pi(Pm+p)\Delta f't} + n(t) \tag{11.18}$$

where $z_{m,p}(t)$ is the received complex envelop at the $(mP + p)$-th subcarrier.

Figure 11.3(b) shows the MC-CDMA receiver for the j'th user, where the mth subcarrier component for the received data $a_{j,p}(i)$ is multiplied by the gain $G_{j'}(m)$ to combine the energy of the received signal scattered in the frequency domain. The decision variable is given by (we omit the subscription p without loss of generality)

$$D_{j'}(t = iT_s) = \sum_{m=0}^{K_{MC}-1} G_{j'}(m)y(m) \tag{11.19}$$

$$y(m) = \sum_{j=1}^{J} z_m(iT_s)a_j d_m^j + n_m(iT_s) \tag{11.20}$$

where $y(m)$ and $n_m(iT_s)$ are the complex baseband component of the received signal after downconversion and the complex additive Gaussian noise at the mth subcarrier at $t = iT_s$, respectively.

Now, we discuss the following four combining strategies.

Orthogonality Restoring Combining (ORC)
Choosing the gain $G_{j'}(m)$ as

$$G_{j'}(m) = d_{j'}(m)z_m^* / |z_m|^2 \tag{11.21}$$

the receiver can eliminate the MAI perfectly [4]:

$$\hat{a}^{j'} = a^{j'} + \sum_{m=1}^{K_{MC}} d_{j'}(m)z_m^* / |z_m|^2 n_m \tag{11.22}$$

However, in (11.22), low-level subcarriers tend to be multiplied by the high gains, and the noise components are amplified at weaker subcarriers. This noise amplification effect degrades BEP performance.

Equal Gain Combining (EGC)
The gain for the equal gain combining is given by [2]

$$G_{j'}(m) = d_{j'}(m) z_m^{j'*} / \left| z_m^{j'} \right| \tag{11.23}$$

Maximum Ratio Combining (MRC)
The gain for the maximum ratio combining is given by [2]

$$G_{j'}(m) = d_{j'}(m) z_m^{j'*} \tag{11.24}$$

In the case of a single user, the maximum ratio combining method minimizes BEP.

Minimum Mean Square Error Combining (MMSEC)
Based on the MMSE criterion, the error must be orthogonal to all the baseband components of the received subcarriers:

$$E\left[\left(a^{j'} - \hat{a}^{j'}\right) y(m')^*\right] = 0 \qquad \left(m' = 1,\ 2, \ldots, K_{MC}\right) \tag{11.25}$$

$G_{j'}(m)$ is given by [4]

$$G_{j'}(m) = d_{j'}(m) z_m^{j'*} / \left(\sum_{j=1}^{J} \left| z_m^j \right|^2 + \sigma_n^2 \right) \tag{11.26}$$

Note that for small $|z_m|$, the gain becomes small to avoid the excessive noise amplification, while for large $|z_m|$, it is in proportion to the inverse of the subcarrier envelop $z_m^* / |z_m|^2$ to recover orthogonality among users [4].

Advantages and Disadvantages
For DS-CDMA system, BEP performance depends on how many fingers the Rake receiver employs. Usually a 1-, 2-, 3-, or 4-finger Rake receiver is used depending on hardware limitation. Therefore, if the received signal is composed of more paths than the number of Rake fingers, the receiver misses a part of its energy. Also, when the Nyquist filters are introduced in the transmitter and receiver for baseband pulse shaping, the Rake receiver may wrongly combine paths. This is because noise causing

distortion in autocorrelation characteristics often results in the wrong correlation. In single-user detection such as a Rake combiner, BEP is determined by MAI and self-interference (SI) due to the imperfect autocorrelation characteristic of spreading codes, and therefore, the system capacity is limited by them, while multiuser detection (MUD) improves the system capacity because the receiver jointly detects the signals in order to mitigate the nonorthogonal properties [11]. However, it is difficult for the DS-CDMA receivers to employ all the energy of the received signal scattered in the time domain.

We do not imagine introducing OFDM signaling in to a CDMA scheme, because both are respectively robust to adverse channel frequency selectivity. Also, OFDM has severe disadvantages such as difficulty in subcarrier synchronization and sensitivity to nonlinear amplification and frequency offset, which result from the fact that it is composed of a lot of subcarriers with their overlapping power spectra and it exhibits a nonconstant nature in its envelop. The OFDM drawback may be a burden to the CDMA scheme. However, the MC-CDMA receiver combines the received signal scattered in the frequency domain. The energy always exists where it should be. We believe that it is the main advantage of the MC-CDMA scheme over other schemes. However, through a frequency selective fading channel, all the subcarriers have different amplitude levels and different phase shifts (although they have high correlations among them), which results in distortion of the orthogonality among users.

The required frequency bandwidth (main-lobe) of DS-CDMA signal for the rectangular pulse waveform is given by

$$B_{\mathrm{DS}} = 2K_{\mathrm{DS}} / T_s \tag{11.27}$$

and for the Nyquist pulse waveform with roll-off factor of α (see Figure 11.2(c)),

$$B_{\mathrm{DS}} = (1+\alpha)K_{\mathrm{DS}} / T_s \qquad (0 \le \alpha \le 1.0) \tag{11.28}$$

On the other hand, for MC-CDMA signal (see Figure 11.3(c)),

$$\begin{aligned} B_{\mathrm{MC}} &= \left(PK_{\mathrm{MC}} - 1\right) / \left(T_s' - \Delta\right) + 2 / T_s' \\ &\approx K_{\mathrm{MC}} / T_s / (1 - \Delta / P) \\ &= \left(1+\beta\right)K_{\mathrm{MC}} / T_s \end{aligned} \tag{11.29}$$

$$\beta = \Delta / P \qquad (0 \le \alpha \le 1.0) \tag{11.30}$$

where β is the bandwidth expansion factor associated with the guard interval insertion. From (11.28) and (11.30), we conclude that DS-CDMA and MC-CDMA systems require almost the same frequency bandwidth.

11.4 MC-CDMA SYSTEM DESIGN

To determine the number of subcarriers and the length of guard interval, we derive the ACF of the received signal.

The jth received signal through the frequency selective fast Rayleigh fading channel is given by

$$r_j(t) = \int_{-\infty}^{+\infty} s_j(t-\tau)h(\tau;t)d\tau + n(t) \tag{11.31}$$

where $n(t)$ is a complex additive Gaussian noise.

$\{\tau_l\}$ is classified as follows:

$$\begin{array}{ll} 0 \le \tau_l \le \Delta & (l = 1,\ldots,L_1) \\ \Delta < \tau_l < T_s & (l = L_1+1,\ldots,L_1+L_2(=L)) \end{array} \tag{11.32}$$

The Fourier coefficient of the $q = Pm + p$th subcarrier at $t = iT_s'$ is given by Figure 11.3(b)

$$r_{j,q}(iT_s') = \frac{1}{T_s' - \Delta} \int_{iT_s'}^{iT_s'+T_s'-\Delta} r_j(t) e^{-j2\pi q\Delta f'(t-iT_s')} dt \tag{11.33}$$

The normalized ACF of the qth subcarrier between $t = iT_s'$ and $t = (i-1)T_s'$ for the jth user is written as (see Appendix 11A)

$$R_{j,q}(\Delta, N, R', f_{D,}\tau_{\text{RMS}}) = \frac{E\left[r_{j,q}(iT_s')r_{j,q}^*((i-1)T_s')\right]}{E\left[r_{j,q}(iT_s')r_{j,q}^*(iT_s')\right]}$$

$$= \frac{\sigma_{S1}^2}{\sigma_{S2}^2 + \sigma_I^2 + \sigma_n^2} \tag{11.34}$$

$$\sigma_{S1}^2 = \sum_{l=1}^{L_1} \sigma_l^2 \left\{ 1 - \left(\frac{\pi \ f_D}{R'} \right)^2 \left(\frac{(N - \Delta R')^2}{6} + N^2 \right) \right\}$$

$$+ \sum_{l=L_1+1}^{L_1+L_2} \sigma_l^2 \frac{(N - \tau_l R')^2}{(N - \Delta R')^2} \left\{ 1 - \left(\frac{\pi \ f_D}{R'} \right)^2 \left(\frac{(N - \tau_l R')^2}{6} + N^2 \right) \right\} \tag{11.35}$$

$$\sigma_{S2}^2 = \sum_{l=1}^{L_1} \sigma_l^2 \left\{ 1 - \left(\frac{\pi \quad f_D}{R'} \right)^2 \frac{(N - \Delta R')^2}{6} \right\}$$

$$+ \sum_{l=L_1+1}^{L_1+L_2} \sigma_l^2 \frac{\left(N - \tau_l R'\right)^2}{\left(N - \Delta R'\right)^2} \left\{ 1 - \left(\frac{\pi \quad f_D}{R'} \right)^2 \frac{\left(N - \tau_l R'\right)^2}{6} \right\}$$

$$+ \sum_{l=L_1+1}^{L_1+L_2} \sigma_l^2 \frac{\left(-\Delta R' + \tau_l R'\right)^2}{\left(N - \Delta R'\right)^2} \left\{ 1 - \left(\frac{\pi \quad f_D}{R'} \right)^2 \frac{\left(-\Delta R' + \tau_l R'\right)^2}{6} \right\} \tag{11.36}$$

$$\sigma_I^2 = \sum_{l=1}^{L_1} \sum_{k=0,k\neq q}^{N-1} \sigma_l^2 \frac{\left(\frac{\pi \quad f_D}{R'} \right)^2 (N - \Delta R')^2}{2\pi^2 (k-q)^2}$$

$$+ \sum_{l=L_1+1}^{L_1+L_2} \sum_{k=0,k\neq q}^{N-1} \sigma_l^2 \left\{ \frac{\left(\frac{\pi \quad f_D}{R'} \right)^2 \left(N - \tau_l R'\right)^2}{2\pi^2 (k-q)^2} \cos\left(\frac{2\pi(k-q)(N - \tau_l R')}{N - \Delta R'} \right) \right\}$$

$$+ \frac{\left(\frac{\pi \quad f_D}{R'} \right)^2 (N - \Delta R')\left(N - \tau_l R'\right)}{\pi^3 (k-q)^3} \sin\left(\frac{2\pi(k-q)(N - \tau_l R')}{N - \Delta R'} \right)$$

$$+ \left(\frac{1}{2\pi^2 (k-q)^2} + \frac{3\left(\frac{\pi f_D}{R'} \right)^2 (N - \Delta R')^2}{4\pi^4 (k-q)^4} \right) \left(1 - \cos\left(\frac{2\pi(k-q)(N - \tau_l R')}{N - \Delta R'} \right) \right)$$

$$+\frac{\left(\frac{\pi f_D}{R'}\right)^2\left(-\Delta R'+\tau_l R'\right)^2}{2\pi^2(k-q)^2}\cos\left(\frac{2\pi(k-q)(-\Delta R'+\tau_l R')}{N-\Delta R'}\right)$$

$$+\frac{\left(\frac{\pi f_D}{R'}\right)^2(N-\Delta R')\left(-\Delta R'+\tau_l R'\right)}{\pi^3(k-q)^3}\sin\left(\frac{2\pi(k-q)(-\Delta R'+\tau_l R')}{N-\Delta R'}\right)$$

$$+\left(\frac{1}{2\pi^2(k-q)^2}+\frac{3\left(\frac{\pi f_D}{R'}\right)^2(N-\Delta R')^2}{4\pi^4(k-q)^4}\right)\left(1-\cos\left(\frac{2\pi(k-q)\left(-\Delta R'+\tau_l R'\right)}{N-\Delta R'}\right)\right)$$

(11.37)

where $R'(=K_{\mathrm{MC}}R)$ is the chip rate.

In the OFDM transmission scheme, generally, when the transmission rate (R' in the case of the MC-CDMA scheme) is given, the transmission performance becomes more sensitive to the time selectivity as the number of subcarriers (N) increases because the wider symbol duration is less robust to the random FM noise, while it becomes poor as N decreases because the wider power spectrum of each subcarrier is less robust to the frequency selectivity (see Figure 11.4). On the other hand, the transmission performance becomes poor as the length of guard interval (Δ) increases because the signal transmission in the guard duration introduces the power loss, while it becomes more sensitive, to the frequency selectivity as Δ decreases because the shorter guard duration is less robust to the delay spread (see Figure 11.5). Therefore, for given R (R'), f_D, and τ_{RMS}, there exists an optimal value to minimize BEP in both N and Δ [12].

In the MC-CDMA scheme, N_{opt} and Δ_{opt} maximizes the ACF given by (11.34) to (11.37), because it means a measure to show how much the received signal is distorted in the time-frequency, selective fading channel (how we can place the signal on the time-frequency plane, so that it suffers from minimum distortion):

$$\left[N_{\mathrm{opt}},\Delta_{\mathrm{opt}}\right]=\arg\left\{\max R_j\left(N,\Delta|R',f_D,\tau_{\mathrm{RMS}}\right)\right\} \tag{11.38}$$

Therefore, with (11.38), we determine two parameters, N_{opt} and Δ_{opt}.

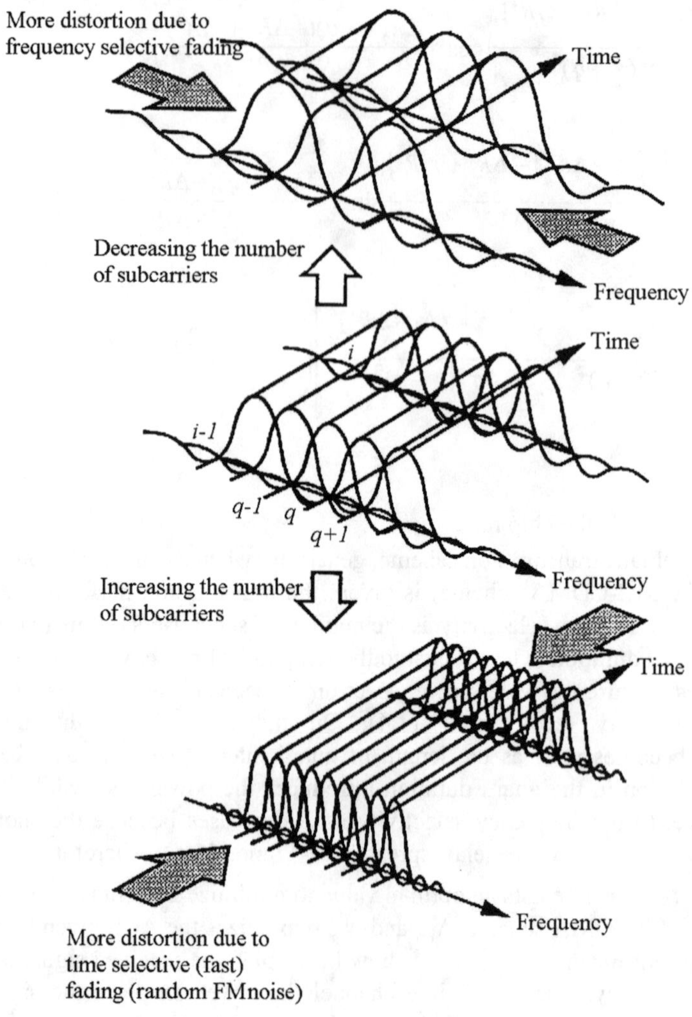

Figure 11.4 Optimum value of the number of subcarriers (N_{opt}).

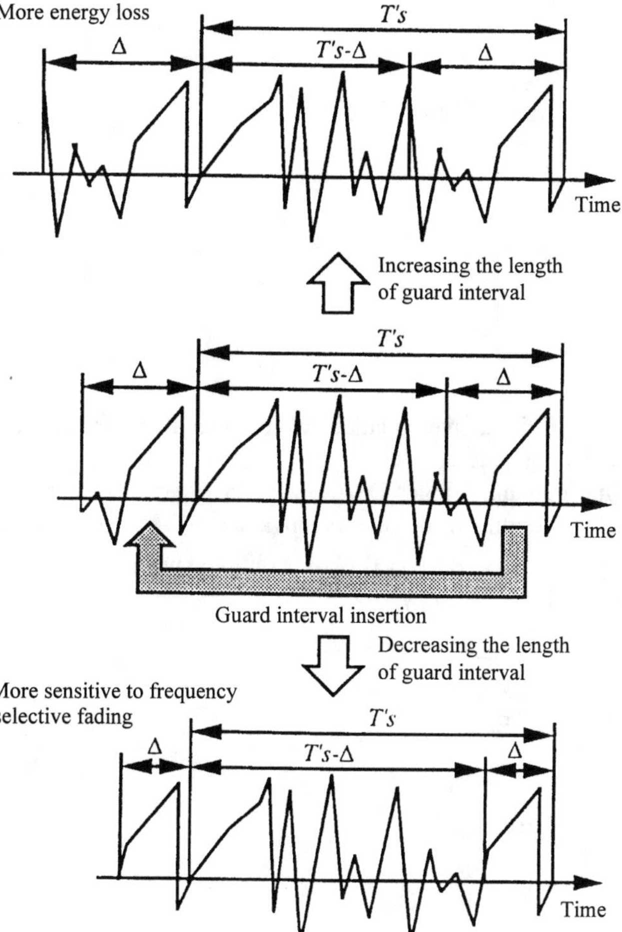

Figure 11.5 Optimum value of the length of guard interval (Δ_{opt}).

11.5 BEP LOWER BOUND

11.5.1 DS-CDMA System

Defining $\boldsymbol{r}_t$ as the received signal vector, the time domain covariance matrix $\boldsymbol{M}_t$ is given by

$$\boldsymbol{r}_t = \left[r_1, r_2, \ldots r_L\right]^T \tag{11.39}$$

$$\boldsymbol{M}_t = \frac{1}{2} E\,[\,\boldsymbol{r}_t \,.\, \boldsymbol{r}_t^T]$$

$$= \begin{bmatrix} \sigma_1^2 & 0 & . & . & . & 0 \\ 0 & \sigma_2^2 & & & & . \\ . & & . & & & . \\ . & & & . & & . \\ . & & & & . & 0 \\ 0 & . & . & . & 0 & \sigma_L^2 \end{bmatrix} \tag{11.40}$$

where T is the transpose. In the above equation, we assume a perfect autocorrelation characteristic for the spreading codes.

The BEP of time domain I-finger DS-CDMA Rake receiver in the case of a single user is uniquely determined by the eigenvalues of $\boldsymbol{M}_t$ (in this case, the eigenvalues are clearly $\sigma_1^2, \sigma_2^2, \ldots, \sigma_L^2$) [13]. For example, when $\sigma_l^2 \left(l = 1, \ldots, L\right)$ are different from each other, BEP is given by

$$\mathrm{BEP_{DS}} = \sum_{l=1}^{I} w_l \frac{1}{2}\left\{1 - \sqrt{\frac{\sigma_l^2 / \sigma_n^2}{1 + \sigma_l^2 / \sigma_n^2}}\right\} \tag{11.41}$$

$$w_l = \frac{1}{\prod_{v=1, v \neq l}^{I} \left(1 - \frac{\sigma_v^2}{\sigma_l^2}\right)} \tag{11.42}$$

$$\sigma_t^2 = \sum_{l=1}^{L} \sigma_l^2 \tag{11.43}$$

where σ_t^2 is the total power of the received signal.

Also, when $\sigma_l^2 \left(l = 1, \ldots, L\right)$ are all the same $\left(= \sigma_s^2\right)$ [7],

$$\mathrm{BEP_{DS}} = \left(\frac{1 - \mu_{\mathrm{DS}}}{2}\right)^I \sum_{l=0}^{I-1} \binom{I - 1 + l}{l} \left(\frac{1 + \mu_{\mathrm{DS}}}{2}\right)^l$$

$$\mu_{\mathrm{DS}} = \sqrt{\frac{\sigma_s^2 / \sigma_n^2}{1 + \sigma_s^2 / \sigma_n^2}} \tag{11.44}$$

Note that the L-finger Rake receiver achieves the minimum BEP (the BEP lower bound) [4].

11.5.2 MC-CDMA System

In the case of a single user, the frequency domain MC-CDMA Rake receiver based on the maximum ratio combining method achieves the minimum BEP (the BEP lower bound) [7].

Defining $\boldsymbol{r}_f$ as the received signal vector, the frequency domain covariance matrix $\boldsymbol{M}_f$ is given by

$$\boldsymbol{r}_f = \left[z_1, z_2, \ldots, z_{K_{MC}}\right]^T \tag{11.45}$$

$$\boldsymbol{M}_f = \frac{1}{2} E\left[\boldsymbol{r}_f \cdot \boldsymbol{r}_f^T\right] = \{m_f^{a,b}\}$$

$$m_f^{a,b} = \Phi_C\left((a-b)\Delta f\right) \tag{11.46}$$

where $m_f^{a,b}$ is the a - b element of $\boldsymbol{M}_f$, and $\Phi_C\left(\Delta f\right)$ is the spaced frequency correlation function defined as the Fourier transform of the multipath delay profile:

$$\Phi_C\left(\Delta f\right) = \int_{-\infty}^{+\infty} \phi_c(\tau) e^{-j2\pi\Delta f\tau} d\tau \tag{11.47}$$

Defining $\lambda_1, \lambda_2, \ldots, \lambda_{K_{MC}}$ as the nonzero eigenvalues of $\boldsymbol{M}_f$, BEP is given by a form similar to (11.41) or (11.44) [13]. For example, when $\lambda_m\left(m = 1, \ldots, K_{\mathrm{MC}}\right)$ are different from each other,

$$\mathrm{BEP}_{\mathrm{MC}} = \sum_{m=1}^{K_{\mathrm{MC}}} v_m \frac{1}{2}\left\{1 - \sqrt{\frac{\lambda_m / N_0}{1 + \lambda_m / N_0}}\right\} \tag{11.48}$$

$$v_m = \frac{1}{\prod_{u=1, u \neq m}^{K_{MC}} \left(1 - \frac{\lambda_u}{\lambda_m}\right)} \tag{11.49}$$

Also, when $\lambda_m\left(m = 1, \ldots, K_{MC}\right)$ are all the same (= λ),

$$\text{BEP}_{\text{MC}} = \left(\frac{1-\mu_{\text{MC}}}{2}\right)^{K_{\text{MC}}} \sum_{m=0}^{K_{\text{MC}}-1} \binom{M-1+m}{m} \left(\frac{1+\mu_{\text{MC}}}{2}\right)^{m}$$

$$\mu_{\text{MC}} = \sqrt{\frac{\lambda / N_0}{1+\lambda / N_0}} \tag{11.50}$$

11.5.3 BEP Lower Bound Equivalence

When there are L paths in the symbol duration at subcarrier T_s' in the multipath delay profile shown in Figure 11.1, we obtain the following $N \times N$ time domain covariance matrix with time resolution of T_s'/N:

$$\mathbf{M}_t' = \begin{bmatrix} \sigma_1^2 & 0 & \cdots & \cdots & \cdots & 0 \\ 0 & 0 & & & & \cdots \\ \cdots & & \sigma_1^2 & & & \cdots \\ \cdots & & & \sigma_3^2 & & \cdots \\ \cdots & & & & \cdots & 0 \\ 0 & \cdots & \cdots & \cdots & 0 & \cdots \end{bmatrix} \tag{11.51}$$

where the nonzero eigenvalues of $\mathbf{M}_t'$ are $\sigma_1^2, \sigma_2^2, \sigma_3^2, \ldots, \sigma_L^2$.

The corresponding $N \times N$ frequency domain covariance matrix with frequency resolution of $1/T_S$ is given by

$$\boldsymbol{M}'_f = \boldsymbol{W}\boldsymbol{M}'_t\boldsymbol{W}^* \tag{11.52}$$

where $\boldsymbol{W}$ is the $N \times N$ DFT matrix given by $\boldsymbol{W} = \{w^{i,j}\}$

$$w^{i,j} = e^{j2\pi\frac{ij}{N}} \tag{11.53}$$

We define $\mathbf{r}_1$ as the eigenvector corresponding to the eigenvalue σ_l^2:

$$\mathbf{M}_t'\mathbf{r}_1 = \sigma_l^2\ \mathbf{r}_1 \qquad (l = 1, 2, \ldots, L) \tag{11.54}$$

Also, we define $\mathbf{z}_1$ as

$$\mathbf{z}_1 = \boldsymbol{W}\mathbf{r}_1 \qquad (l = 1, 2, \ldots, L) \tag{11.55}$$

Now we can theoretically prove that the frequency domain covariance matrix has all the same eigenvalues as the time domain covariance matrix as follows:

$$\boldsymbol{M}'_f \mathbf{z}_1 = \boldsymbol{W}\boldsymbol{M}'_t\boldsymbol{W}^* \,.\, \boldsymbol{W}\mathbf{r}_1 = \boldsymbol{W}\boldsymbol{M}'_t\mathbf{r}_1 = \boldsymbol{W}\sigma_l^2\,\mathbf{r}_1 = \sigma_l^2\,\boldsymbol{W}\mathbf{r}_1 = \sigma_l^2\,\mathbf{z}_1 \tag{11.56}$$

The above equation clearly shows that the nonzero eigenvalues of $\boldsymbol{M}'_f$ are $\sigma_1^2, \sigma_2^2, \sigma_3^2, \ldots, \sigma_L^2$. Therefore, as long as we use the same frequency selective fading channel, the BEP lower bound of the MC-CDMA system is the same as that of the DS-CDMA system. Also, the assumption of independent fading characteristic at each subcarrier implies a frequency selective fading at each subcarrier, because it requires independent N paths uniformly scattered in the symbol duration at subcarrier.

11.6 NUMERICAL RESULTS

As mentioned in Section 11.3, the MC-CDMA system has some severe disadvantages associated with OFDM signaling, which are hostile to the uplink application (from mobile users to a central base station). Therefore, in this section we discuss the BEP of MC-CDMA system in a downlink channel and compare it with that of DS-CDMA system [14–16]. In this sense, we refer to them as MC-CDM and DS-CDM, respectively.
To demonstrate the numerical results, we assume:

- Two-path multipath delay profile with τ_{RMS} = 20 nsec, where the first and second paths have the same power;
- Doppler power spectrum with maximum Doppler frequency (f_D) = 10 Hz;
- Transmission rate (R) = 3 Msps (BPSK format);
- Walsh Hadamard codes with K_{MC} = 32 for the MC-CDMA system;
- Gold codes with K_{DS} = 31 for the DS-CDMA system.

In addition, we assume a perfect (sub-) carrier synchronization.

Figure 11.6 shows the optimal values in the number of subcarriers N and the length of guard interval Δ versus the rms delay spread σ_{RMS} as a function of the Doppler frequency f_D, where Δ, σ_{RMS} and f_D are normalized by R'. For given f_D and R', both N_{opt} and Δ_{opt} increases as σ_{RMS} increases. For the above parameters, we obtain $f_D / R' \approx 10^{-7}$ and $\tau_{\text{RMS}} R' = 1.92$, so we choose $\left[N_{\text{opt}}, \Delta_{\text{opt}} / T'_s\right] = [256, 0.015]$ as an optimal set. It means that the original data sequence is first converted into 8 parallel sequences (P = 8), and then each sequence is mapped onto 32 subcarriers, and the length of the guard interval is negligibly short, as compared with the symbol duration at subcarrier.

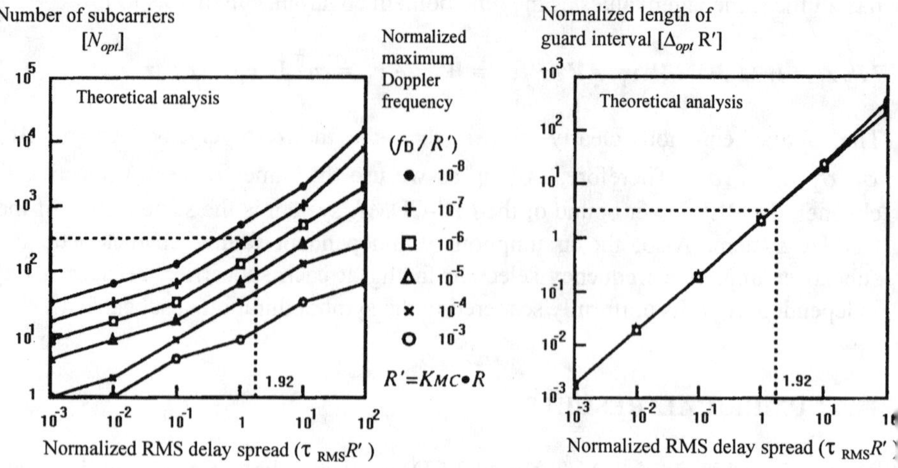

Figure 11.6 (a) Optimum number of subcarriers, and (b) optimum length of guard interval.

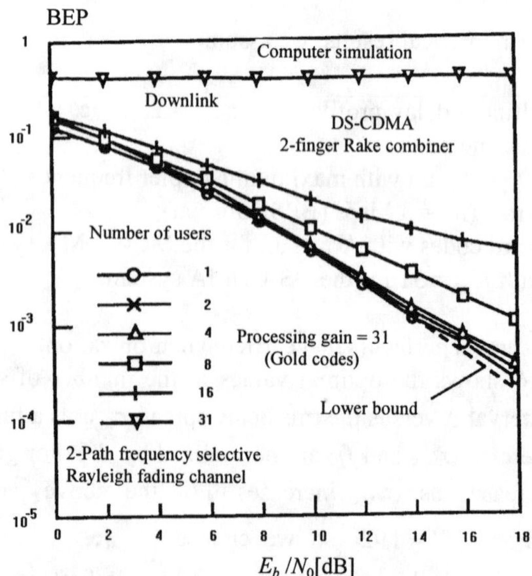

Figure 11.7 BEP of DS-CDMA system with 2-finger Rake combiner.

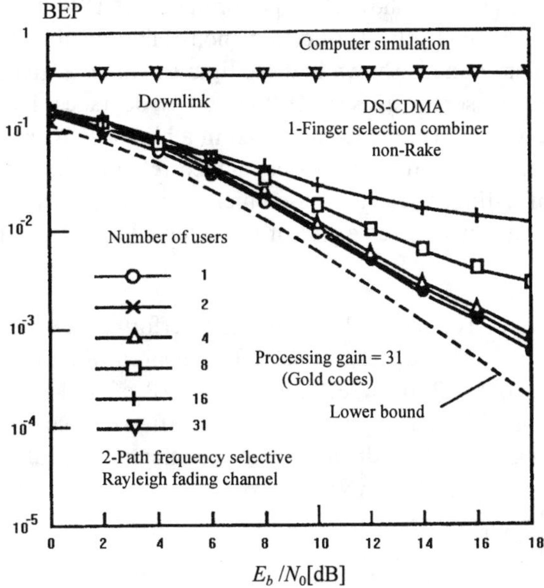

Figure 11.8 BEP of DS-CDMA system with 1-finger selection (non-Rake) combiner.

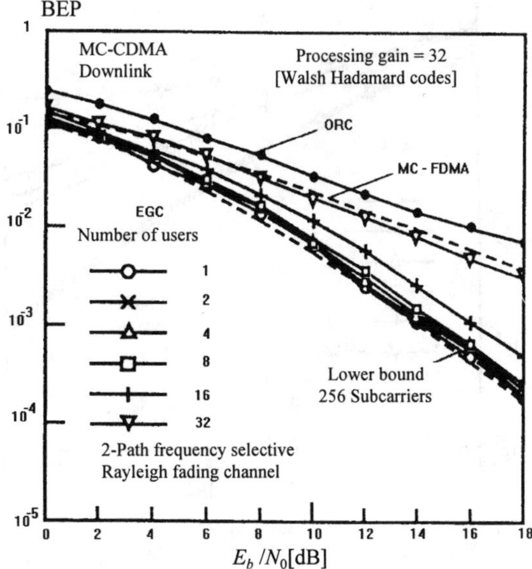

Figure 11.9 BEP of MC-CDMA system with EGC.

Figures 11.7 and 11.8 show BEP performance of DS-CDMA system with 2-finger Rake combiner and 1- finger selection (non-Rake) combiner, respectively. The theoretical BEP lower bound is shown in these figures, which is given by (11.44) with $I = 2$. As the number of users increases, BEP gradually degrades. Even in the case of 1 user, BEP is a little worse than the lower bound in a higher E_b/N_0 region because of the self-interference resulting from the imperfect autocorrelation characteristic of the Gold codes. The BEP of 1-finger selection combiner is worse than that of 2-finger Rake combiner because it always misses a part of the received signal energy scattered in the time domain.

Figures 11.9, 11.10, and 11.11 show BEP performance of MC-CDMA system with EGC, MRC, and MMSEC, respectively. The theoretical BEP lower bound is also given by (11.44) with $I = 2$.In these figures, the BEP of MC-FDMA system is also shown, which supports 32 users at most, assigning different 8 subcarriers to each user. This system obtains no frequency diversity effect, so the theoretical BEP is given by [7]

$$BEP_{\text{MC-FDMA}} = \frac{1}{2}\left(1 - \sqrt{\frac{E_b / N_0}{1 + E_b / N_0}}\right) \tag{11.57}$$

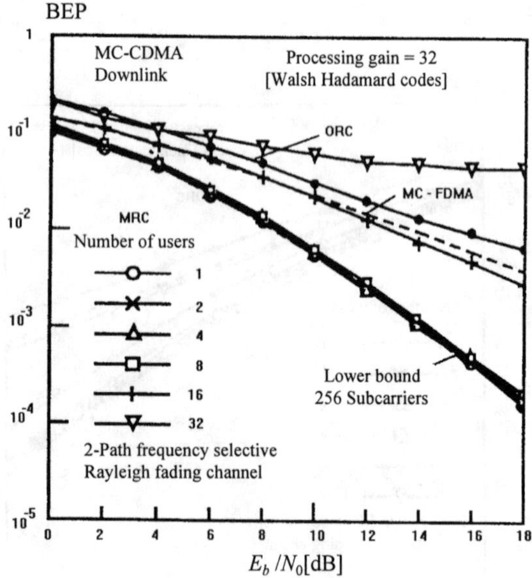

Figure 11.10 BEP of MC-CDMA system with MRC.

The performance of MC-CDMA with ORC is independent of the number of users; however, it is worse than that of the MC-FDMA system. Therefore, we do not employ the ORC even when we can estimate the channel condition perfectly.

The MC-CDMA system with MRC performs better when the number of users is less than 8. For more users, however, the performance abruptly becomes worse. This is because the interference resulting from distorted code orthogonality is multiplied in the combining process. On the other hand, as the interference is not multiplied in the EGC, it performs better than the MRC with of 16 and 32 users.

The MMSEC performs best among four combining strategies, although it requires the number of total active users and the noise power, in addition to information on the channel condition.

Figure 11.12 shows the BEP comparison between the DS-CDMA system with 2-finger Rake combiner and the MC-CDMA system with MMSEC. It is evident from Figure 11.12 that the MC-CDMA system achieves better performance than the DS-CDMA system, although it requires a little complexity in the receiver structure, such as subcarrier synchronization.

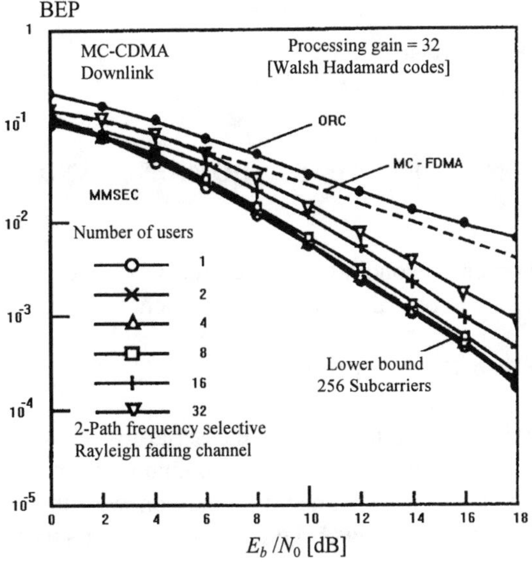

Figure 11.11 BEP of MC-CDMA system with MMSEC.

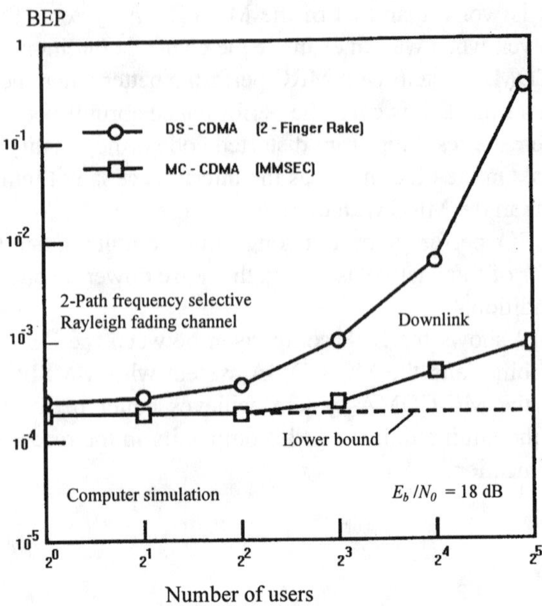

Figure 11.12 BEP comparison.

11.7 COMBINATION OF TIME DOMAIN AND MULTICARRIER MODULATION

There are two schemes corresponding to this category. In both schemes, when setting the number of subcarriers to one, they become equivalent to DS-CDMA schemes.

11.7.1 Multicarrier DS-CDMA Scheme

Multicarrier DS-CDMA transmitter spreads the serial-to-parallel(S/P)-converted data streams using a given spreading code in the time domain so that the resulting spectrum of each subcarrier satisfies the orthogonality condition with the minimum frequency separation [5]. This scheme can lower the data rate in each subcarrier so that a large chip time makes it easier to synchronize the spreading sequences. The multicarrier DS-CDMA scheme is originally proposed for a uplink communication channel, because this characteristic is effective for the establishment of a quasi-synchronous channel.

Figure 11.13(a) and (b) show the multicarrier DS-CDMA transmitter of the jth user and the power spectrum of the transmitted signal, respectively, where G_{MD} denotes the processing gain, N_c the number of subcarriers, and $C^j(t) = \left[C_1^j C_2^j \ldots C_{\text{MD}}^j\right.$ the spreading code of the jth user.

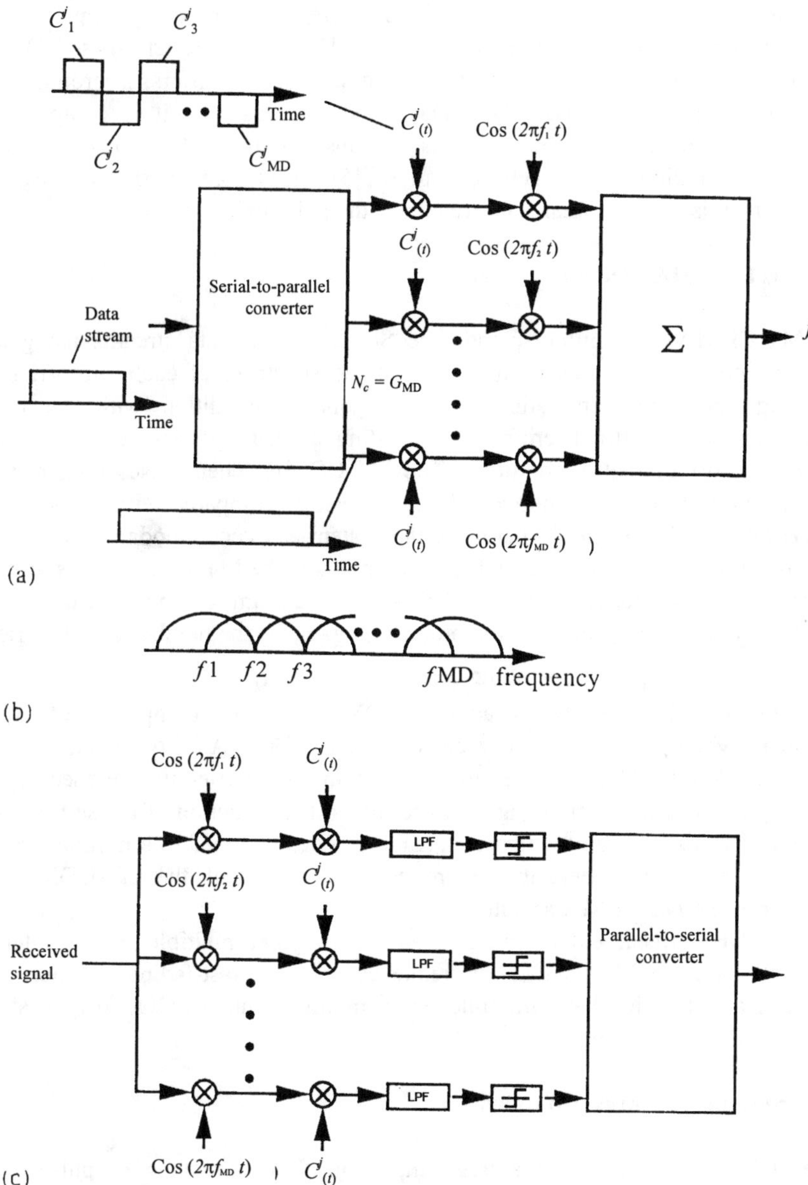

Figure 11.13 Multicarrier DS-CDMA scheme: (a) transmitter, (b) power spectrum of its transmitted signal, and (c) receiver.

In [17], a multicarrier-based DS-CDMA scheme with a larger subcarrier separation is proposed in order to yield both frequency diversity improvement and narrowband interference suppression. Also, in [18], a multicarrier-based DS-CDMA scheme, which transmits the same data using several subcarriers, is proposed. Figure 11.13(c) shows a multicarrier DS-CDMA receiver. Usually, it is composed of N_C normal coherent (non-Rake) receivers, because it is crucial to have frequency nonselective fading over each subcarrier [5]. Therefore, with no FEC among subcarriers, this scheme obtains no frequency diversity gain.

11.7.2 MT-CDMA Scheme

The MT-CDMA transmitter spreads the S/P-converted data streams using a given spreading code in the time domain so that the spectrum of each subcarrier, before spreading operation, can satisfy the orthogonality condition with the minimum frequency separation [6]. Therefore, the resulting spectrum of each subcarrier no longer satisfies the orthogonality condition. The MT-CDMA scheme uses longer spreading codes in proportion to the number of subcarriers, as compared with a normal (single carrier) DS-CDMA scheme. Therefore, the system can accommodate more users than the DS-CDMA scheme. Figure 11.14(a) and (b) show the MT-CDMA transmitter of the jth user for CBPSK scheme and the power spectrum of the transmitted signal, respectively, where G_{MT} denotes the processing gain, N_c the number of subcarriers, and $C^j(t) = \left[C_1^j C_2^j \ldots C_{\text{MT}}^j\right]$ the spreading code of the jth user.

Figure 11.14(c) shows an MT-CDMA receiver composed of N_C Rake combiners, which has an optimum receiver structure for an AWGN channel [6].

The MT-CDMA scheme suffers from intersubcarrier interference, while the capability to use longer spreading codes results in the reduction of the self interference (SI) and MAI, as compared with the spreading codes assigned to a normal DS-CDMA scheme. In a channel where this improvement is dominant, the MT-CDMA scheme outperforms the DS-CDMA scheme.

A DFE, linear equalizer (LE), and linear joint multiple access interference canceler/equalizer (JEIC) suited for frequency selective fast-fading channels are also proposed, all of which have multiple input, multiple output (MIMO) type structures based on MMSE criterion.

11.7.3 System Features Comparison

Table 11.1 shows the system features comparison. When a rectangular pulse is used in the DS-CDMA scheme, similar to other schemes, the required bandwidths of MC-CDMA and multicarrier DS-CDMA schemes are almost half as wide as that of DS-CDMA scheme, and the MT-CDMA scheme has almost the same bandwidth as the DS-CDMA scheme. However, when the Nyquist filter with a small roll-off factor is used in the DS-CDMA scheme, the required bandwidths of MC-CDMA and multicarrier DS-CDMA schemes become comparable with that of the DS-CDMA scheme.

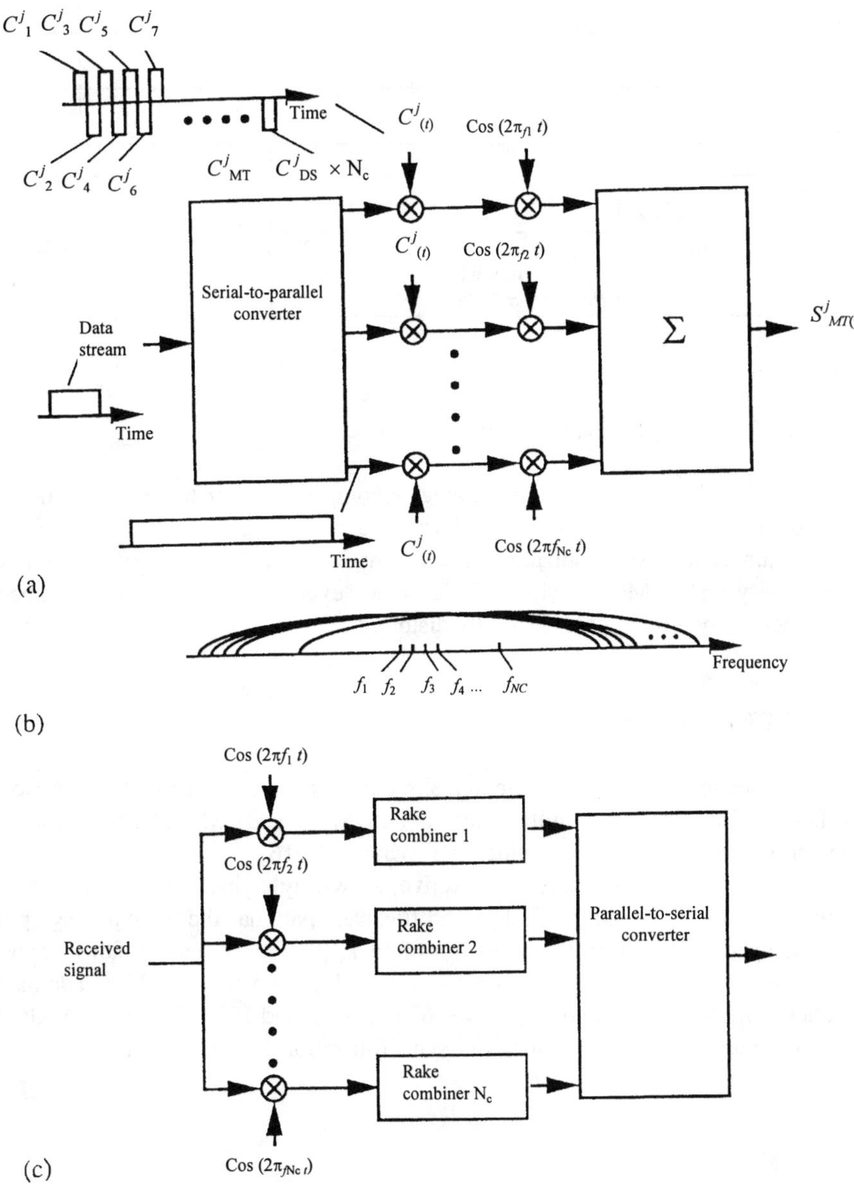

Figure 11.14 MT-CDMA scheme: (a) transmitter, (b) power spectrum of its transmitted signal and (c) receiver.

Table 11.1

System Features Comparison

	DS-CDMA	MC-CDMA	Multicarrier DS-CDMA	MT-CDMA
Symbol duration at subcarrier	T_S	T_S	$N_C T_S$	$N_C T_S$
Number of subcarriers	(1)	N_C	N_C	N_C
Processing gain	G_{DS}	$G_{MC} \approx G_{DS}$	$G_{MD} = G_{DS}$	$G_{MT} = N_C G_{DS}$
Chip duration	T_S / G_{DS}		$N_C T_S / G_{MD}$	$N_C T_S / G_{MT}$
Subcarrier separation		$1/T_S$	$G_{DS} / (N_C T_S)$	$1 / (N_C T_S)$
Required bandwidth	G_{DS} / T_S Nyquist filter with roll-off factor ≈ 0	$(N_C + 1)/ N_C \, . G_{MC} / T_S$	$(N_C + 1)/ N_C \, . G_{MD} / T_S$	$(N_C - 1 + 2G_{MT}) / (N_C T_S)$

11.7.4 Detection Strategies Comparison

Table 11.2 shows the detection strategies comparison. For the MC-CDMA scheme, a lot of strategies have been proposed and analyzed for the downlink channel because of easy subcarrier synchronization and estimation. Multiuser detection techniques are necessary for the MC-CDMA uplink channel, even without near/far effect, because the orthogonality among users is totally distorted.

11.8 BEP COMPARISON

In this section, we show computer simulation results on BEP performance of DS-CDMA, MC-CDMA, multicarrier DS-CDMA, and MT-CDMA schemes in a synchronous downlink communication channel [15,16].

We assume a frequency-selective, slow-Rayleigh-fading channel, where there are two paths in the multipath delay profile, each path has the same average power, and the delayed path uniformly ranges from 0 to T_s $(0, T_s)$. To make a fair comparison, we assume the processing gains $G_{DS} = 31$ (Gold codes), $G_{MC} = 32$ (Hadamard Walsh codes), $G_{MD} = 31$ (Gold codes), $G_{MT} = 63$ $(N_C = 2)$, and 127 $(N_C = 4)$ (Gold codes). Also, we assume a perfect subcarrier synchronization and subcarrier state estimation.

Table 11.2

Detection Strategies Comparison

Access Scheme	Detection Strategy	UPlink	Downlink
DS-CDMA	Rake combiner	o	o
	MUD	o	
MC-CDMA	ORC		o
	CE		o
	EGC	o	o
	MRC	o	o
	MMSEC		o
	ML MUD		o
	EGC-EGC MUD		o
	ORC-MRC MUD		o
	CE-MLD MUD		o
	DIC		o
	MMSEIC		o
Multicarrier DS-CDMA	Non-Rake Receiver	o	
MT-CDMA	Rake combiner	o	
	DFE	o	
	LE	o	
	JEIC	o	

Given a frequency-selective fading channel, we calculate the frequency correlation function defined as the Fourier transform of the multipath delay profile, which determines the frequency correlations among subcarriers. It is well known that BEP performance depends on the covariance matrix of the channel; in other words, it is uniquely determined by the eigenvalues. Therefore, BEP of the DS-CDMA scheme is determined by the eigenvalues of the time domain covariance matrix, while that of the MC-CDMA scheme is determined by the eigenvalues of the frequency domain covariance matrix. We showed that the frequency domain covariance matrix has all the same eigenvalues as the time domain covariance matrix [14]. Therefore, given a frequency selective channel, the best performance of DS-CDMA scheme (for a single user with a perfect autocorrelation characteristic of the spreading code) is the same as that of MC-CDMA scheme (for a single user). It also implies that we cannot assume an independent characteristic at each subcarrier even if we employ an ideal frequency interleaving, and that, considering the FFT operation, the assumption results in a frequency selective fading at each subcarrier.

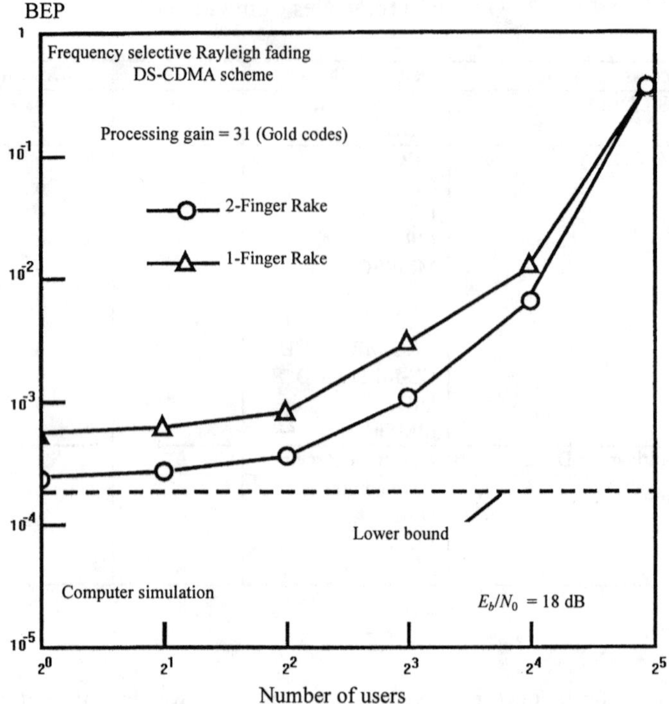

Figure 11.15 BEP performance of the DS-CDMA scheme.

BEP lower bound is given by (the 1 × 2-branch maximum ratio combiner) [7].

$$BEP_{\text{LowerBound}} = \left(\frac{1-\mu}{2}\right)^2 \sum_{l=0}^{1} \binom{l+1}{l} \left(\frac{1+\mu}{2}\right)^l \quad \text{with } \mu = \sqrt{\frac{E_b / N_0}{2 + E_b / N_0}} \tag{11.58}$$

Figure 11.15 shows the BEP of DS-CDMA scheme for 1-finger non-Rake (selection combiner) and 2-finger Rake receiver and the lower bound because of the SI, and as the number of users increases, BEP gradually degrades because of the MAI. The 1-finger non-Rake receiver, which selects a larger path, always misses a part of the received signal energy; therefore, the performance is worse than that of the 2-finger receiver.

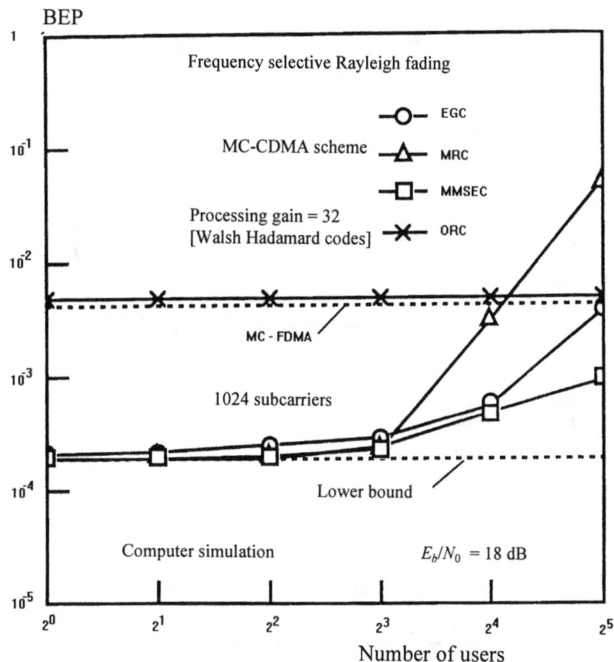

Figure 11.16 BEP performance of the MC-CDMA scheme.

Figure 11.16 shows BEP of the MC-CDMA scheme for ORC, EGC, MRC, and MMSEC, respectively, where 1024 subcarriers and 1% guard interval have frequency nonselective fading over each subcarrier [14]. Also shown in this figure is the performance of the MC-FDMA scheme, where 1024 subcarriers are employed as well and a different set of 32 subcarriers is assigned to a different user. BEP performance of the MC-CDMA scheme is given by $BEP_{\text{MC-FDMA}} = \frac{1}{2}\left(1 - \sqrt{\frac{E_b / N_0}{1 + E_b / N_0}}\right)$ [7], which is identical to the BEP expression for a frequency-nonselective, slow-Rayleigh-fading channel. As compared with the MC-FDMA scheme, the MC-CDMA scheme with ORC performs worse. Therefore, we should not use the ORC even if we can perfectly estimate the subcarrier state information. The MRC minimizes BEP for one user. However, as the number of users increases, BEP rapidly degrades. The EGC, which requires only subcarrier synchronization, keeps a good BEP performance; further, the MMSEC performs better than the EGC.

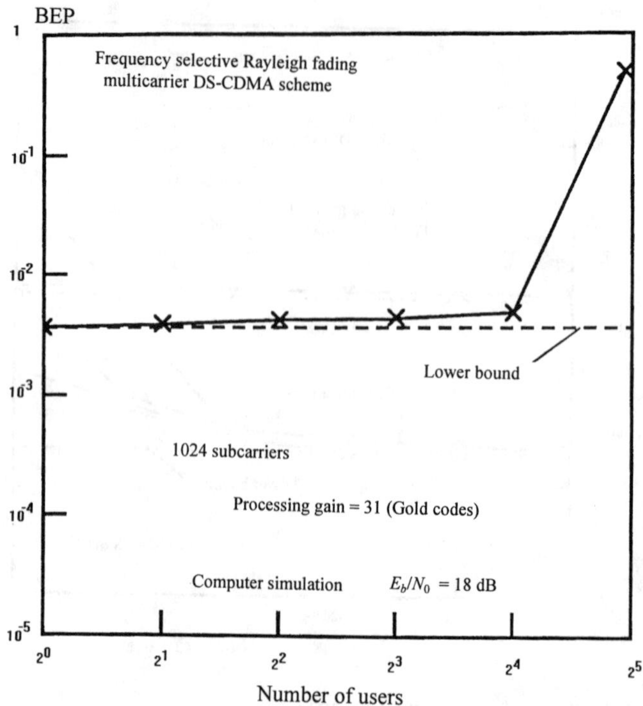

Figure 11.17 BEP performance of the multicarrier DS-CDMA scheme.

Figure 11.17 shows BEP of the multicarrier DS-CDMA scheme, where 1024 subcarriers and 1% guard interval are introduced as well. When we do not take account of FEC and interleaving techniques, BEP performance is lower-bounded by the same expression for the MC-FDMA scheme, because it cannot gain frequency and time diversity effect.

Figure 11.18 shows BEP of the MT-CDMA scheme with 2-finger Rake receiver for $N_c = 2$ and $N_c = 4$, respectively. The MT-CDMA scheme $N_c = 4$ performs better than if $N_c = 2$, because in the channel used in this computer simulation, the intersubcarrier interference, which increases in proportion to the number of subcarriers, affects the detection process less, as compared with the BEP improvement effect due to the use of longer spreading codes.

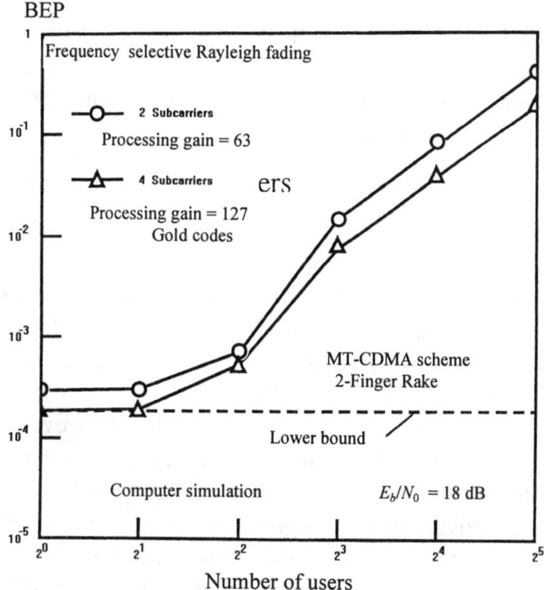

Figure 11.18 BEP performance of the MT-CDMA scheme.

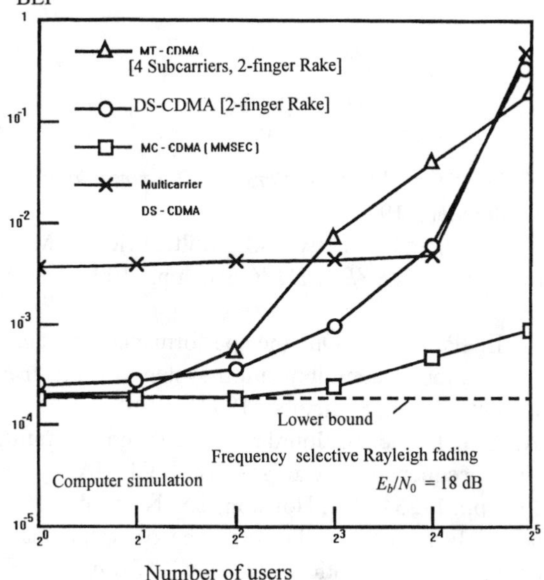

Figure 11.19 BEP comparison.

Figure 11.19 shows BEP comparison of the DS-CDMA scheme with 2-finger Rake receiver, MC-CDMA scheme with MMSEC, multicarrier DS-CDMA scheme, and 2-subcarrier MT-CDMA scheme with 2-finger Rake receiver. It is evident from Figure 11.19 that the MC-CDMA scheme with MMSEC outperforms all other schemes, although it requires the estimation of subcarrier state information and noise power, and knowledge of the number of active users.

11.9 CONCLUSIONS

This chapter reviewed the multi-carrier-based CDMA schemes such as MC-CDMA, multicarrier DS-CDMA, and MT-CDMA, and discussed their advantages and disadvantages with a normal DS-CDMA scheme. Computer simulation results show that the MC-CDMA scheme, with MMSEC, is a promising protocol in 2-path frequency-selective, slow-Rayleigh-fading channel. However, more detailed discussions and analyses using different multipath delay profiles are required.

Always employ all the received signal energy scattered in the time domain; the MC-CDMA system effectively combines all the received signal energy scattered in the frequency domain. In a 2-path, frequency-selective, slow-Rayleigh-fading channel, the MC-CDMA system with the combining method based on MMSEC achieves better performance than the DS-CDMA system with 2-finger Rake combiner, although it requires the estimation of subcarrier condition and noise power, and knowledge of the number of active users.

REFERENCES

[1] R. Prasad, *CDMA for Wireless Personal Communications,* Norwood, MA: Artech House Publishers, 1996.

[2] N. Yee, J.-P. Linnartz, and G. Fettweis, "Multicarrier CDMA in indoor wireless radio networks," *Proc. IEEE PIMRC'93*, pp. 109–113, Yokohama, Japan, September 1993.

[3] K. Fazel and L. Papke, "On the performance of convolutionally-coded CDMA/OFDM for mobile communication system," *Proc. IEEE PIMRC'93*, pp. 468–472, Yokohama, Japan, September 1993.

[4] A. Chauly, A. Brajal, and S. Jourdan, "Orthogonal multicarrier techniques applied to direct sequence spread spectrum CDMA systems," *Proc. IEEE GLOBECOM'93,* pp. 1723–1728, Houston, TX, November 1993.

[5] V.M. DaSilva and E.S. Sousa, "Performance of orthogonal CDMA codes for quasi-synchronous communication systems," *Proc. IEEE ICUPC'93*, pp. 995–999, Ottawa, Canada, October 1993.

[6] L. Vandendorpe, "Multitone direct sequence CDMA system in an indoor wireless environment," *Proc. IEEE First Symposium of Communications and Vehicular Technology in the Benelux*, pp. 4.1.1–4.1.8, Delft, The Netherlands, October 1993.

[7] J.G. Proakis, *Digital Communications*, 3rd Edition, McGraw Hill, 1995.

[8] W.C. Jakes, Jr., *Microwave Mobile Communications*, New York, John Wiley & Sons, 1974.

[9] M.J. Feuerstein and T.S. Rappaport, *Wireless Personal Communications*, Kluwer Academic Publishers, 1993.

[10] P. Dambacher, *Digital Broadcasting*, The Institution of Electrical Engineers, 1996.

[11] A. Duel-Hallen, J. Holtzman, and Z. Zvonar, "Multiuser detection for CDMA system," *IEEE Personal Communications*, Vol.2, No. 2, pp. 46–58, April 1995.

[12] S. Hara, M. Mouri, M. Okada, and N. Morinaga, "Transmission performance analysis of multi-carrier modulation in frequency selective fast Rayleigh fading channel," *Wireless Personal Communications*, Vol. 2, No. 4, pp. 335–356, 1995/1996.

[13] P. Monsen, "Digital transmission performance on fading dispersive diversity channels," *IEEE Trans. Commun.*, Vol. COM-21, pp. 33–39, January 1973.

[14] S. Hara, T.-H. Lee, and R. Prasad, "BER comparison of DS-CDMA and MC-CDMA for frequency selective fading channels," *Proc. 7th Tyrrhenian International Workshop on Digital Communications*, pp. 3–14, Viareggio, Italy, September 1995.

[15] S. Hara and R. Prasad, "DS-CDMA, MC-CDMA and MT-CDMA for mobile multi-media communications," *Proc. IEEE VTC'96*, pp. 1106–1110, Atlanta, GA, April 1996.

[16] R. Prasad and S. Hara, "An overview of multi-carrier CDMA," *Proc. 4th IEEE International Symposium on Spread Spectrum Techniques and Applications (ISSSTA'96)*, pp. 107–114, Mainz, Germany, 22–25, September 1996.

[17] S. Kondo and L.B. Milstein, "Performance of multicarrier DS-CDMA systems," *IEEE Trans. Comm.*, Vol. 44, No. 2, pp. 238–246, February 1996.

[18] E. Sourour and M. Nakagawa, "Performance of orthogonal multicarrier CDMA in a multipath fading channel," *IEEE Trans. Comm.*, Vol. 44, No. 3, pp. 356–367, March 1996.

APPENDIX 11A

Equation (11.15) is written as

$$s_j(t) = \sum_{i=-\infty}^{+\infty}\sum_{q=0}^{N-1} c_{j,q}(i) p_s\left(t - iT_s'\right) e^{j2\pi\Delta f/Pq(t-T_s')} \tag{11A.1}$$

$$c_{j,q=Pm+p}(i) = a_{j,p}(i) d_j(m) \tag{11A.2}$$

where $c_{j,q}(i)$ has the following property:

$$E\left[c_{j,q}(i) c_{j,q'}(i)\right] = \begin{cases} 1 & (q = q') \\ 0 & (q \neq q') \end{cases} \tag{11A.3}$$

Substituting (11.1), (11.31) and (11.58) into (11.33) leads

$$\begin{aligned} r_{j,q}\left(iT_s'\right) = & \left\{ \sum_{l=1}^{L_1} \frac{1}{T_s' - \Delta} \int_{iT_s'}^{iT_s'+T_s'-\Delta} g_l(t) dt c_{j,q}(i) e^{-j2\pi q\Delta f'\tau_l} \right. \\ & \left. + \sum_{l=L_1+1}^{L_1+L_2} \frac{1}{T_s' - \Delta} \int_{i'T_s-\Delta+\tau_l}^{i'T_s+T_s'-\Delta} g_l(t) dt c_{j,q}(i) e^{-j2\pi q\Delta f'\tau_l} \right\} \\ & + \left\{ \sum_{l=1}^{L_1} \sum_{k=0,k\neq q}^{N-1} \frac{1}{T_s' - \Delta} \int_{iT_s'}^{iT_s'+T_s'-\Delta} g_l(t) e^{j2\pi(k-q)\Delta f'(t-iT_s')} dt c_{j,k}(i) e^{-j2\pi k\Delta f'\tau_l} \right. \\ & + \sum_{l=L_1+1}^{L_1+L_2} \sum_{k=0,k\neq q}^{N-1} \frac{1}{T_s' - \Delta} \int_{iT_s'-\Delta+\tau_l}^{iT_s'+T_s'-\Delta} g_l(t) e^{j2\pi(k-q)\Delta f'(t-iT_s')} dt c_{j,k}(i) e^{-j2\pi k\Delta f'\tau_l} \\ & \left. + \sum_{l=L_1+1}^{L_1+L_2} \sum_{k=0,k\neq q}^{N-1} \frac{1}{T_s' - \Delta} \int_{iT_s'}^{iT_s'-\Delta+\tau_l} g_l(t) e^{j2\pi(k-q)\Delta f'(t-iT_s')} dt c_{j,k}(i-1) e^{-j2\pi k\Delta f'(\tau_l - T_s')} \right\} \\ & + \sum_{l=L_1+1}^{L_1+L_2} \frac{1}{T_s' - \Delta} \int_{iT_s'}^{iT_s'-\Delta+\tau_l} g_l(t) dt c_{j,q}(i-1) e^{-j2\pi q\Delta f'(\tau_l - T_s')} + n_q(i) \end{aligned} \tag{11A.4}$$

where the first term of (11A.4) represents the desired signal component, the second and third terms represent ICI and ISI components, respectively, and $n_q(i)$ is the Gaussian random noise component with zero mean and variance σ_n^2.

With (11A.4), $E\left[r_{j,q}(iT_s')r_{j,q}^*\left((i-1)T_s'\right)\right]$ and $E\left[r_{j,q}(iT_s')r_{j,q}^*(iT_s')\right]$ in (11.34) are written as

$$E\left[r_{j,q}(iT_s')r_{j,q}^*\left((i-1)T_s'\right)\right] = \sum_{l=1}^{L_1}\frac{1}{\left(T_s'-\Delta\right)^2}\int_0^{T_s'-\Delta}\int_{-T_s'}^{-\Delta}R_l(x-y)dxdy$$

$$+\sum_{l=L_1+1}^{L_1+L_2}\frac{1}{\left(T_s'-\Delta\right)^2}\int_{-\Delta+\tau_l}^{T_s'-\Delta}\int_{-T_s'-\Delta+\tau_l}^{-\Delta}R_l(x-y)dxdy \tag{11A.5}$$

$$E\left[r_{j,q}(iT_s')r_{j,q}^*(iT_s')\right] = \sum_{l=1}^{L_1}\frac{1}{\left(T_s'-\Delta\right)^2}\int_0^{T_s'-\Delta}\int_0^{T_s'-\Delta}R_l(x-y)dxdy$$

$$+\sum_{l=L_1+1}^{L_1+L_2}\frac{1}{\left(T_s'-\Delta\right)^2}\int_{-\Delta+\tau_l}^{T_s'-\Delta}\int_{-\Delta+\tau_l}^{T_s'-\Delta}R_l(x-y)dxdy$$

$$+\sum_{l=1}^{L_1}\sum_{k=0,k\neq q}^{N-1}\frac{1}{\left(T_s'-\Delta\right)^2}\int_0^{T_s'-\Delta}\int_0^{T_s'-\Delta}R_l(x-y)e^{j2\pi(k-q)\Delta f'(x-y)}dxdy$$

$$+\sum_{l=L_1+1}^{L_1+L_2}\sum_{k=0,k\neq q}^{N-1}\frac{1}{\left(T_s'-\Delta\right)^2}\int_{-\Delta+\tau_l}^{T_s'-\Delta}\int_{-\Delta+\tau_l}^{T_s'-\Delta}R_l(x-y)e^{j2\pi(k-q)\Delta f'(x-y)}dxdy$$

$$+\sum_{l=L_1+1}^{L_1+L_2}\sum_{k=0}^{N-1}\frac{1}{\left(T_s'-\Delta\right)^2}\int_0^{-\Delta+\tau_l}\int_0^{-\Delta+\tau_l}R_l(x-y)e^{j2\pi(k-q)\Delta f'(x-y)}dxdy+\sigma_n^2$$

(11A.6)

Taking account of $N = T_s'K_{\text{MC}}R$, substituting (11.6) into (11A.5) and (11A.6) leads (11.35), (11.36), and (11.37).

Chapter 12
Antenna Diversity

12.1 INTRODUCTION

Thorough knowledge of antenna theory is necessary for developing a wireless communication system. The wireless environment is very hostile in nature, typified by a multiple scattering process and the absence of direct LOS paths between the base and the mobile. The conventional free-space antenna pattern is thus greatly modified, and antenna designs have to be tailored to the statistical nature of the environment [1].

Since 1927, diversity schemes have helped in enhancing the performance of a wireless system. There are several diversity techniques: space diversity, polarization diversity, angle diversity, frequency diversity, and time diversity. Several combining techniques, such as selective, switched, maximal ratio, and equal-gain, also exist [1, 2].

Section 12.1 briefly discusses the role of the antenna in mobile communications. Sections 12.2 and 12.3 present several diversity and combining techniques. Section 12.4 focuses as the performance of switched and selection diversity in a faded radio environment. Finally, conclusions are given in Section 12.5.

12.2 ANTENNA

Antennas play a significant role in the development of any wireless communication system. Antennas are the outermost components in the transmission chain. Their behavior depends on the interaction between the antenna and the radio environment. This causes the network evaluation of the antenna system to be stochastic in nature. The primary task of the antenna is to convey information from transmitter to receiver. However, it is also of equal importance for the antenna to suppress or repair media impairments [3, 4].

There are three major areas to consider when using antennas for compensating for the impairments of a mobile radio channel:

1. Noise suppression through antenna gain and effectivity;
2. Temporal fading and interference suppression through multiport antenna diversity;
3. Spatially selective interference suppression through adaptive antenna diversity.

12.2.1 BS Antennas

First, it is important to decide on a choice of BS antennas. In general, BS antennas are mastmounted, omnidirectional, or sector antennas with respect to the azimuthal plan. In recent years, smart antennas have been suggested to suppress interference via spatial filtering in a space division multiple access (SDMA) scheme. The effectiveness of such antenna arrangements are closely linked with the directional properties of the radio environment as sensed at the BS site.

A hardware antenna testbed based on a DECT radio interference and digital baseband beamforming was developed by TSUMANI [5]. The antenna array consists of 8 dual-polarized wideband patch antennas. Each element gives approximately 6 dB gain and is configurable to offer a variety of array geometrics (linear, nonlinear, planar, circular) and element spacings (up to λ). Broadband BS arrays and antenna elements for mobile communications are summarized in [6].

The algorithms used in the TSUNAMI tests are grouped into:

1. Beam-oriented (SDMA, grid of beams, beam jitter);
2. Signal combination (selection, maximal ratio combining, etc.).

12.2.2 Antennas for Portables

Generally, most antennas for portables are external wire antennas, typically halfwave dipoles but also quarterwaves and partly helical types to make them smaller and better matched. The halfwave dipole antenna is a good candidate for portables with respect to antenna performance, but there are practical problems with external antennas. In the COST 231 project, several documents reported on internal antennas. There are two main goals for the design of integrated antennas. One goal is to obtain high antenna performance due to limited power available, and the other is to reduce absorption in the human head. These goals are not necessarily mutually exclusive, which was demonstrated by an antenna design in [7].

12.3 DIVERSITY AND COMBINING TECHNIQUES

A method to combat the effects of multipath fading is to obtain not just one, but several versions of a signal at the receiver. This principle is known as *diversity*. The probability that all of these signals are simultaneously in a fade is much lower than the probability that just one signal is in a fade. Naturally, this is only true if these signals are not strongly correlated. For partially correlated signals, there still is some diversity gain, but it is less than that for uncorrelated signals.

Once several versions of a signal, or *branches*, are obtained, they are combined into one "optimum" signal based on some performance measure such as CIR or BEP.

12.3.1 Diversity Techniques

There are several ways of obtaining more than one version of a signal at the receiver. The most important diversity techniques are described below.

Space or Antenna Diversity

With space or antenna diversity, two or more spatially separated antennas are used at the receiver. The distance between the antennas must be large enough to ensure that signals received by different antennas are uncorrelated; a spacing of at least half a wavelength is required [1]. A distinction is made between the following two types:

1. *Microdiversity*, where the distance between antennas is in the order of only a few wavelengths;
2. *Macrodiversity*, where the distance between antennas is at least 10 wavelengths.

In practice, this usually means that with microdiversity all antennas are on the same station, and with macrodiversity all antennas are on physically separated stations. A combination of both types is often used to combat both multipath fading (microdiversity) and shadowing (macrodiversity).

Frequency Diversity

With frequency diversity, the same signal is transmitted simultaneously on two or more different carrier frequencies. The spacing between carrier frequencies must be large enough to ensure that they exhibit uncorrelated fading. A measure for the spacing of the carriers is the coherence bandwidth B_C, defined as the rms width of the spaced frequency channel correlation function. An approximation for the coherence bandwidth is the reciprocal of rms delay spread τ_d (i.e., $B_C \approx 1/\tau_d$). For carrier spacings of more than several times the coherence bandwidth, the different branches are uncorrelated [1].

Time Diversity

With time diversity, the same signal is transmitted two or more distinct times. Signals are only uncorrelated if the channel is time-variant (exhibits Doppler spread), and if the speed of variation is such that the channel characteristics change sufficiently during the time interval between two transmissions. A measure for the minimum required time between two transmissions is the coherence time T_C, defined as the rms width of the spaced-time channel correlation function. Analogously to the relation between coherence bandwidth and rms delay spread, the coherence time is approximated by the reciprocal of rms Doppler spread σ_D (i.e., $T \approx 1/\sigma_D$).

Polarization Diversity

With polarization diversity, the receiver uses two separate and orthogonally polarized antennas. Reflection of the transmitted signal of objects causes decorrelation of both polarization directions because the reflection coefficient for both polarizations is usually different. This means that they exhibit different phase shifts and attenuations after a reflection. A sufficient number of random reflections ensures uncorrelated polarizations at the receiver. Therefore, this method cannot be used if the received signal mainly consists of a strong LOS component.

Angle Diversity

With angle diversity, one omnidirectional antenna is split up into two or more directional antennas, each covering a limited angle (sector). As an example: 6 antennas, each covering only a 60 deg angle, supply 6 branches. The difference in path length between signals reaching the receiver at different angles is usually significant enough to assume they are uncorrelated. This method relies on the presence of signals coming from different directions; in other words, there must be a number of different transmission paths.

Combinations of the above-mentioned diversity techniques are also possible, and increase the number of branches even more. However, the amount of improvement in fading characteristics achieved by adding branches decreases as the number of branches increases [1].

12.3.2 Combining Techniques

Once several versions of a signal are obtained, they are combined into one optimum signal. Two kinds of combining exist: predetection and postdetection combining. The former combines signals at RF level, the latter at baseband level. For combining schemes that add signals, this is an important difference since predetection signals are usually out of phase. Several different performance measures are used for combining. In the following combining schemes, the received SNR is used as a measure for combining. Note that although this gives optimal results in theory, it turns out to be very complicated, if not impossible, to determine the SNR in practice, so usually the total received power (signal and noise) is used instead.

In general, a received radio signal $r(t)$ consists of the different received copies

$$r_i(t) = \beta_i(t)s[t - \tau_i(t)] + n_i(t) \tag{12.1}$$

where $s(t)$ is the transmitted signal, $\beta_i(t)$ is a time-dependent attenuation factor, $\tau_i(t)$ is a time-dependent delay factor ,and $n_i(t)$ is a noise term.

Now suppose that there are n branches available, then the general *linear combining function* can be written as

$$r(t) = \alpha_1 r_1(t) + \alpha_2 r_2(t) + \ldots + \alpha_n r_n(t) = \sum_{i=1}^{n} \alpha_i r_i(t) \tag{12.2}$$

Depending on the combining method, the α_i may vary according to the statistics of the $r_i(t)$, so (12.2) is actually a "quasi-linear" combining function. The most important combining techniques are described below.

Switched Antenna Diversity

With switched antenna diversity, at any given time only one of the input branches is chosen as the output signal $r(t)$. A selector device scans all branches until finding one with an SNR above a certain preset value, then chooses this branch as the output signal. The selector remains idle until the SNR of the chosen branch drops below the threshold value and the process of scanning starts again (Figure 12.1). In practice, the received power of the chosen branch is measured and a simple comparator acts as switch. Seeing as only one signal is chosen at any time, there is no difference between predetection or postdetection combining of this type.

The advantage of switched diversity is:

- Only one receiver is required to sequentially scan all the different input branches.

The disadvantages are:

- There is a time delay; the "best" input branch cannot be chosen instantaneously;
- The output SNR is only as good as the SNR of one (selected) branch.

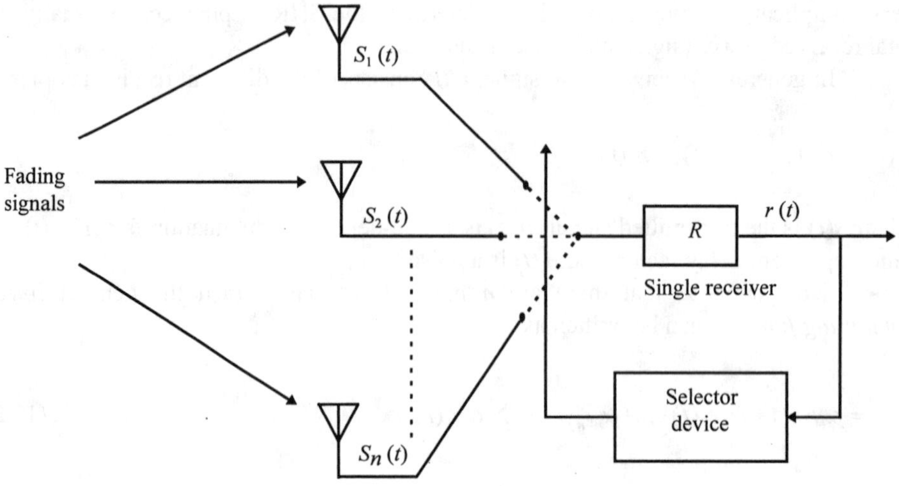

Figure 12.1 Switched antenna diversity.

Selection Diversity

At any given time only one of the α_i in (12.2) differs from zero, namely the signal having the highest SNR. The branches are not scanned sequentially, but they are simultaneously compared with each other and the optimum branch is chosen (Figure 12.2). Again, there is no difference between predetection and postdetection combining. The advantage of selection diversity is:

- The optimum branch is always instantaneously chosen.

The disadvantages are:

- *n* separate receivers are required;
- As with switched diversity, the output SNR is only as good as the SNR of the selected branch.

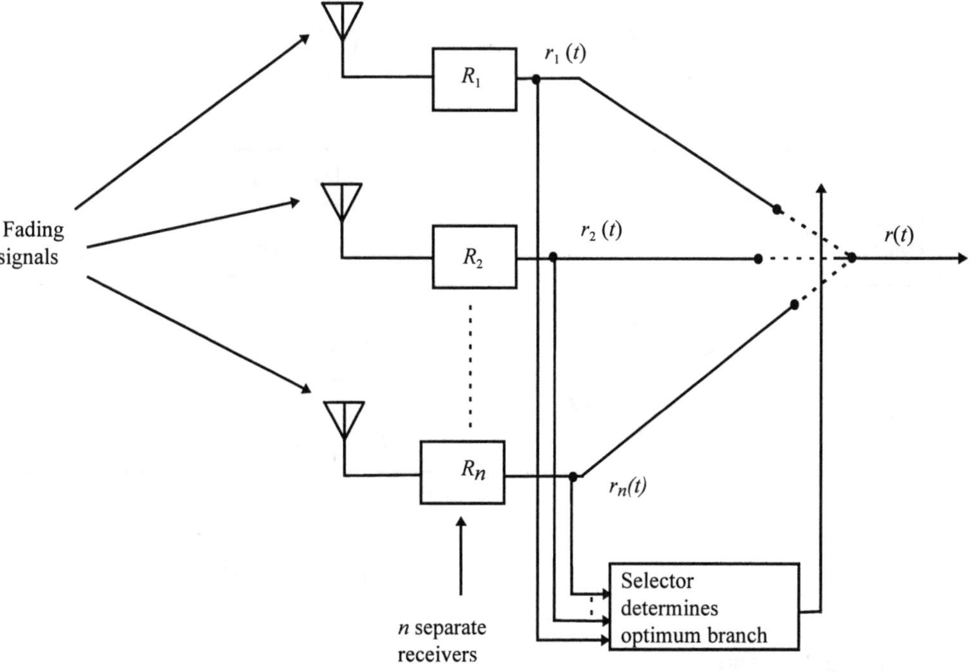

Figure 12.2 Selection diversity.

MRC

The α_i are chosen proportional to the ratio $\sqrt{<s_i^2>/<n_i^2>}$, the rms value of the signal over the mean noise power (Figure 12.3). MRC works only under the assumption that the phases of all branches are known so that coherent summation is possible.
The advantage of MRC is:

- This method yields the highest output SNR of all combining methods [1].

The disadvantages are:

- For the weighed summation, the SNR and the phase of each branch have to be known;
- *n* separate receivers, phase equalizers, and gain controls are required.

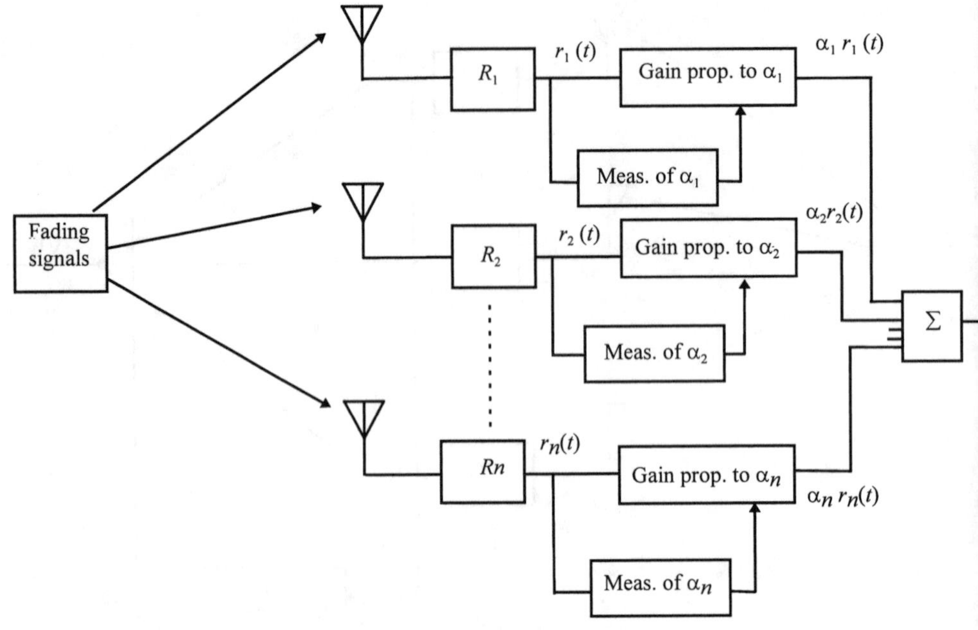

Figure 12.3 MRC.

EGC

In MRC (Figure 12.3), only all α_i equal to unity, are chosen (i.e., the output signal is simply the sum of all input signals). The advantage is that there is no need to measure the SNR of each branch. The disadvantage is that the output SNR is less than that for MRC.

Table 12.1 summarizes some of the above-mentioned points for the different combining methods in terms of complexity, performance, and cost.

Table 12.1

Qualification of Combining Methods According to Performance, Complexity, and Cost

Combining methods	Performance	Complexity	Cost
MRC	++	−−	−−
EGC	+	+/−	−
Selection diversity	+/−	+	+
Switched diversity	−	++	++

12.4 DIVERSITY IN A FREQUENCY NONSELECTIVE FADING CHANNEL

This section focuses on the performance of switched diversity in a faded radio environment. Both methods are compared with the theoretically superior method of EGC. Although MRC gives the highest diversity gain of the combining methods described in Section 12.3, it is omitted here because the extra costs and complexity, as compared to EGC, are not compensated by the extra gain achieved. The performance of the different diversity schemes is determined for frequency nonselective Rayleigh and Rician fading, for both independent and correlated branches. The effects of (the absence of) correlation between successive time samples in the case of switched diversity, as well as the effects of different power levels on different branches are also examined.

12.4.1 Switched Diversity and Selection Diversity Output Signals

In this section, we show what the output signals after diversity look like in the case of switched and selection diversity. Section 12.3.1 already describes the functioning of both diversity methods. Only two-branch diversity is considered here.

With *selection diversity*, a branch consists of an antenna plus a receiver. On the basis of a certain signal quality measure, the "best" branch at any instant is chosen as the output branch, so the signal received on that branch becomes the output signal. Due to the fact that each branch has its own separate receiver, there is no delay in the selection of the optimum branch. In this chapter, the instantaneous received signal power is chosen as the selection criterion. The resulting output signal for faded branches is shown in Figure 12.4(a).

With *switched diversity*, a branch consists of just an antenna. A switch can connect only one branch at a time to a single receiver. As with selection diversity, instantaneous received signal power on a branch is chosen as the switching criterion; the switching algorithm is as follows: the power on the chosen (first) branch is compared with a certain present threshold value A. If the power on this branch drops below A, the second branch is chosen. If the power on the second branch is above A, it becomes the output branch. If however, the power on the second branch is also below A, two different follow-up strategies can be discerned:

1. A switch back to the first branch is made, which leads to a period of continuous switching between branches until the power on one of them exceeds A. This strategy is called *switch-and-examine* [8].
2. The second branch stays the chosen branch to avoid needless switching; a strategy called *switch-and-stay*. A switch to the other branch is only made if power on the chosen branch crosses A in negative direction. The difference in the output signal between both strategies, for faded branches, can be seen in Figure 12.4(b) and (c).

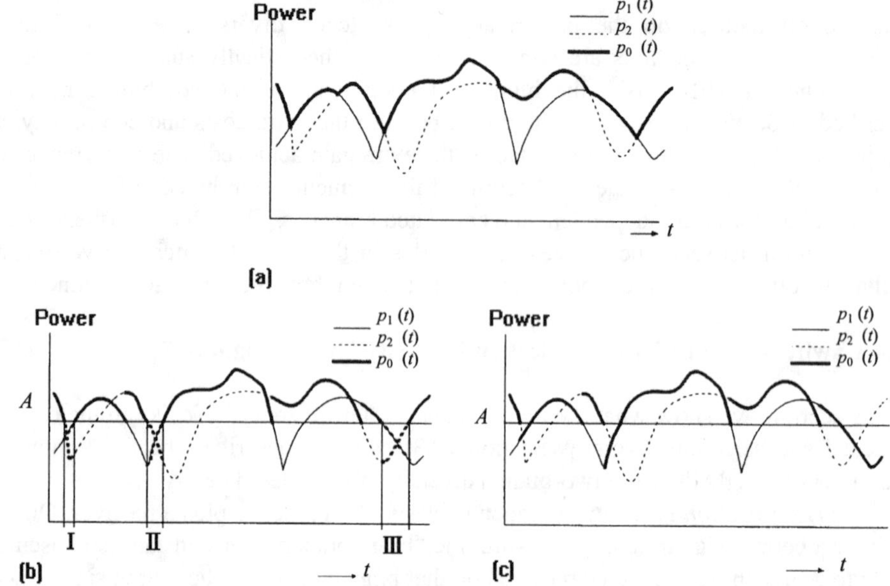

Figure 12.4 Power on both branches $p_1(t)$, $p_2(t)$, and output power $p_0(t)$ for: (a) selection diversity, (b) switch-and-examine diversity, and (c) switch-and-stay diversity. A is the threshold value for switched diversity. In (b) during the time intervals I, II, and III, the dotted line indicates rapid switching between branches.

In this chapter, all calculations involving switched diversity are based on the switch-and-examine strategy.

To keep handhelds in a DECT system small in size and weight, an option is to equip the handheld with just one antenna and the BS with two antennas. Then the base station thus has transmitter diversity on the downlink and receiver diversity on the uplink. The handheld also profits from this diversity function, assuming reciprocity of the channel. The channel (or branch) chosen by the BS for reception most likely also gives the best reception quality at the handheld if the base station uses it for transmission. However, this is only valid if the time correlation between the upward and downward channels is high, so channel statistics have not changed. This correlation factor is determined by the fading frequency and the length of a frame [9]. If the product of these factors is too large, the upward and downward channels become uncorrelated and there is no diversity gain for the handheld.

12.4.2 Two-Branch Diversity

Independently Faded, Equally Powered Branches

In this section, we assume two independently Rayleigh or Rician faded branches. We also assume that average power on both branches is equal and that successive time samples are totally correlated (i.e., branch fading statistics are time-independent). The instantaneous received power on both branches is represented by the random variables p_1 and p_2. The output power after the diversity operation is denoted by $p_0 = c(p_1, p_2)$, where $c(.)$ is the relevant combining function.

First we give the pdfs of the received power on each branch. The pdf of the instantaneous signal power is exponentially distributed corresponding to Rayleigh fading

$$f_{\text{Ray}, p_i}(p_i) = \frac{1}{\sigma_i^2} \exp\left(-\frac{p_i}{\sigma_i^2}\right), \qquad p_i \geq 0 \tag{12.3}$$

where σ_i^2 is the average received power, i = 1 or 2.
The pdf for Rician fading is

$$f_{\text{Ray}, p_i}(p_i) = \frac{1}{\sigma_i^2} \exp\left(-\frac{2p_i + s^2}{2\sigma_i^2}\right) I_0\left(\frac{\sqrt{2p_i}\; s}{\sigma_i^2}\right), \qquad p_i \geq 0 \tag{12.4}$$

where σ_i^2 is the average scattered component power, $s^2/2$ is the average dominant component power, and $I_0(.)$ is the modified Bessel function of the first kind and zero order.

The μth order modified Bessel function of the first kind is given by the infinite series [10]

$$I_\mu(x) = \sum_{k=0}^{\infty} \frac{\left(x/2\right)^{\mu+2k}}{k!\Gamma\left(\mu + k + 1)\right)} \tag{12.5}$$

where $\Gamma(.)$ denotes the gamma function.

An important factor in the case of Rician fading is the Rice factor K, defined as the ratio of average dominant component power and average scattered component power

$$K \triangleq \frac{s^2}{2\sigma_i^2} \tag{12.6}$$

Next, we determine the cdf of the output power p_0 for the different combining methods. A functional description of these methods is given in Section 12.3.2. We assume equal average power on both branches, so $\sigma_i^2 = \sigma^2$.

EGC

Omitting MRC, EGC gives the best theoretical performance of the combining techniques treated in the previous section. The combining function is simply $p_0 = c(p_1, p_2) = p_1 + p_2$. Summing random variables means convolving their pdfs. For Rayleigh fading this means convolving $f_{\text{Ray},p1}(p_1)$ and $f_{\text{Ray},p2}(p_2)$, with $n = 2$, this results in

$$f_{Ray,p_0}(p_0) = \frac{p_0}{\sigma^4}\exp\left(-\frac{p_0}{\sigma^2}\right), \qquad p_0 \geq 0 \tag{12.7}$$

The cdf of p_o is found by integrating (12.7)

$$f_{\text{Ray},p_0}(p) = \int_0^p f_{\text{Ray},p0}(p_0)dp_0 = 1 - \left(\frac{p}{\sigma^2} + 1\right)\exp\left(-\frac{p}{\sigma^2}\right), \qquad p \geq 0 \tag{12.8}$$

Convolving $f_{\text{Ric},p1}(p_1)$ and $f_{\text{Ric},p2}(p_2)$, with $n = 2$, results in the sum pdf for Rician fading

$$f_{\text{Ric},p_0}(p_0) = \frac{1}{\sigma^2}\left(\frac{2p_0}{v}\right)^{1/2} \exp\left(-\frac{2p_0 + v}{2\sigma^2}\right) I_1\left(\frac{\sqrt{2p_0 v}}{\sigma^2}\right), \qquad p_0 \geq 0 \tag{12.9}$$

where $v = 2s^2$ is the sum of dominant component powers. The cdf of p_0 for Rician fading is found by integrating (12.9)

$$f_{\text{Ric},p_0}(p) = \int_0^p \frac{1}{\sigma^2}\left(\frac{2p_0}{v}\right)^{1/2} \exp\left(-\frac{2p_0 + v}{2\sigma^2}\right) I_1\left(\frac{\sqrt{2p_0 v}}{\sigma^2}\right) dp_0 \tag{12.10}$$

for which no closed expression has been found.

Selection Diversity

It is apparent from Section 12.3.2 that for selection diversity the combining function is the nonlinear function $p_0 = c(p_1, p_2) = \max\,(p_1 . p_2)$. This leads to the following probability that p_0 exceeds a certain value of p

$$Pr\{p_0 > p\} = Pr\{p_1 > p | p_1 > p_2\} Pr\{p_1 > p_2\} + Pr\{p_2 > p | p_2 > p_1\} Pr\{p_2 > p_1\}$$

$$= Pr\{p_1 > p \wedge p_1 > p_2\} + Pr\{p_2 > p \wedge p_2 > p_1\} \tag{12.11}$$

Now, $F_{p0}(p) = 1 - Pr\{p_0 > p\}$ which, according to [11], is equal to

$$F_{\text{Ray},p_0}(p) = F_{p_1}(p)\, F_{p_2}(p) \tag{12.12}$$

For Rayleigh faded branches this is

$$F_{\text{Ray},p_0}(p) = 1 - Pr\{p_0 > p\} = F_{\text{Ray},p_1}(p)\, F_{\text{Ray},p_2}(p)$$

$$= 1 - 2\exp\left(-\frac{p}{\sigma^2}\right) + \exp\left(-\frac{2p}{\sigma^2}\right), \qquad p \geq 0 \tag{12.13}$$

Substituting $x = (2\,p_i)^{1/2}/\sigma_i$ in (12.4), the separate cdfs of p_1 and p_2 for Rician faded branches are given by

$$F_{Ric,p_i}(p) = 1 - \int_p^{\infty} F_{Ric,p_i}(p_i)dp_i = 1 - \int_{\frac{\sqrt{2p_i}}{\sigma_i}}^{\infty} x\exp\left(-\frac{x^2 + (s/\sigma_i)^2}{2}\right) I_0\left(x\frac{s}{\sigma_i}\right)dx$$

$$= 1 - Q(\sqrt{2}K, \frac{\sqrt{2p_i}}{\sigma_i}) \tag{12.14}$$

where $Q(.,.)$ is the Marcum's Q-function, defined [10] as

$$Q(a,b) \,\Delta \int_b^{\infty} x\exp\left(-\frac{x^2 + a^2}{2}\right) I_o(ax)dx,$$

$$= \exp\left(-\frac{a^2 + b^2}{2}\right) \sum_{k=0}^{\infty} \left(\frac{a}{b}\right)^k I_k(ab), \qquad b > a > 0 \tag{12.15}$$

and K is the Rice factor given by (12.6). The cdf of p_0 for Rician faded branches is found by squaring (12.14).

Switched Diversity

Since only one receiver is used, the optimum branch cannot be determined instantaneously. Therefore, the combining function not only depends on the current value of p_1 and p_2 but also on the values of p_1 and p_2 at the previous sampling instant. If we denote the value of p_i at sampling instant t_k as $p_i(k)$, the combining function for switch-and-examine is written as

$$p_0 = c(p_1, p_2) = \begin{cases} p_1(k) & \left[\begin{array}{l} \text{Branch 1 selected} \wedge p_1(k-1) > A \\ \vee \\ \text{Branch 2 selected} \wedge p_2(k-1) < A \end{array} \right. \\ p_2(k) & \left[\begin{array}{l} \text{Branch 2 selected} \wedge p_2(k-1) > A \\ \vee \\ \text{Branch 1 selected} \wedge p_1(k-1) < A \end{array} \right. \end{cases} \tag{12.16}$$

where A stands for the switching threshold. The probability that $p_i(k-1)$ is equal to A is infinitely small and therefore does not appear in (12.16). In terms of probabilities, (12.16) is written as

$$\begin{aligned} Pr\{p_0 > p\} = &(Pr\{p_1(k) > p \wedge p_1(k-1) > A\} \\ &+ Pr\{p_2(k) > p \wedge p_1(k-1) < A\})Pr\{1 \text{ selected}\} \\ &+ (Pr\{p_1(k) > p \wedge p_2(k-1) < A\} \\ &+ Pr\{p_2(k) > p \wedge p_2(k-1) > A\})Pr\{2 \text{ selected}\} \end{aligned} \tag{12.17}$$

Because of the assumption that the average received power in both branches is equal ($\sigma_i^2 = \sigma^2$ in (12.3) and (12.4)) and they have the same statistics, the probability of either branch 1 or 2 being selected is equal, (i.e., $Pr\{1 \text{ selected}\} = Pr\{2 \text{ selected}\} = ½$). Given this and the fact that p_1 and p_2 are interchangeable, (12.17) is reduced to

$$Pr\{p_0 > p\} = Pr\{p_1(k) > p \wedge p_1(k-1) > A\} + Pr\{p_2(k) > p \wedge p_1(k-1) < A\} \tag{12.18}$$

If successive time samples have a correlation of 1 so they have equal statistics, the cdf of p_0 is

$$F_{p_0}(p) = 1 - \int_{-\infty}^{\infty} \int_{\max(p,A)}^{\infty} f_{p_1}(p_1) f_{p_2}(p_2) dp_1\, dp_2 - \int_{p}^{\infty} \int_{-\infty}^{A} f_{p_1}(p_1) f_{p_2}(p_2) dp_1\, dp_2 \tag{12.19}$$

For independent Rayleigh faded branches, this results in

$$F_{\text{Ray},p_0}(p) = \begin{cases} 1 - \exp\left(-\frac{A}{\sigma^2}\right) - \exp\left(-\frac{p}{\sigma^2}\right) + \exp\left(-\frac{p+A}{\sigma^2}\right) & 0 < p < A \\ 1 - 2\exp\left(-\frac{p}{\sigma^2}\right) + \exp\left(-\frac{p+A}{\sigma^2}\right) & p > A \end{cases} \tag{12.20}$$

$$F_{\text{Ric},p_0}(p) = \begin{cases} 1 - Q\left(\sqrt{2K}, \frac{\sqrt{2A}}{\sigma}\right) - Q\left(\sqrt{2K}, \frac{\sqrt{2p}}{\sigma}\right)\left[1 - Q\sqrt{2K}, \frac{\sqrt{2A}}{\sigma}\right] & 0 < p < A \\ 1 - Q\left(\sqrt{2K}, \frac{\sqrt{2p}}{\sigma}\right)\left[2 - Q\left(\sqrt{2K}, \frac{\sqrt{2A}}{\sigma}\right)\right] & p > A \end{cases} \tag{12.21}$$

Using the equations derived for the different cdfs, the results shown in Figures 12.5 to 12.8 are calculated. Figures 12.5 and 12.6 show the results for Rayleigh fading; 12.7 and 12.8 for Rician fading. In Figure 12.5, curves are plotted for EGC, selection diversity, switched diversity, and no diversity.

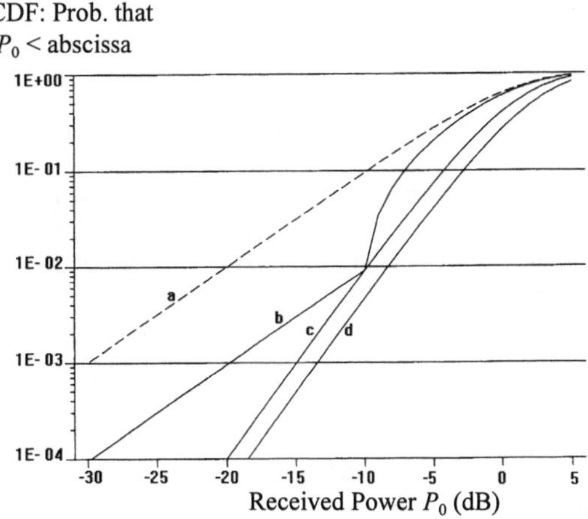

Figure 12.5 Cumulative density functions of power after diversity $p_0(t)$ for twobranch diversity and uncorrelated Rayleigh faded branches with equal average power. Curves are shown for (a) no diversity, (b) switched diversity, with a threshold $A = -10$ dB, (c) selection diversity, and (d) EGC. For all curves $\sigma = 0$ dB was taken.

From Figure 12.5, we see that, as expected, EGC performs best, followed by selection diversity and switched diversity. The difference between selection diversity and EGC is not so large as might be expected. The switched diversity curve actually consists of two parts; one below $p_0 = -10$ dB and one above this value, equal to the threshold A. The part below A is parallel to the no diversity curve. The reason for this is that if p_0 is below the threshold, the power on both branches must be below the threshold, so there is constant switching between branches. This has the same effect as with a random branch and therefore performance is no better than for no diversity. Above the threshold, the curve moves rapidly from the selection diversity curve to the no diversity curve. The most diversity gain with switched diversity is therefore achieved near the switching threshold A.

Figure 12.6 shows cdfs of $p_0(t)$ for switched diversity and Rayleigh fading for different values of the threshold A.

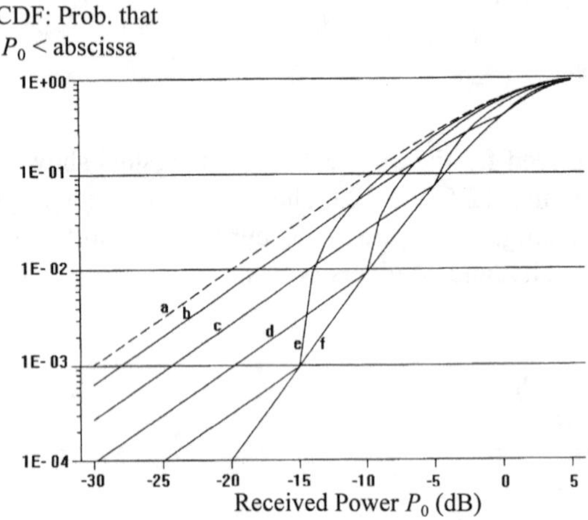

Figure 12.6 Cumulative density functions of power after diversity $p_0(t)$ for two-branch switched diversity and uncorrelated Rayleigh faded branches with equal average power, for different values of the threshold A. Curves are shown for (a) no diversity, (b) A = 0 dB, (c) $A = -5$ dB, (d) $A = -10$ dB, (e) $A = -15$ dB, and (f) selection diversity. For all curves $\sigma = 0$ dB was taken.

If A is very low, the probability that the power on a chosen branch stays above A increases, so the probability of a switch decreases. This implies that the behavior of $p_0(t)$ increasingly resembles its behavior for no diversity. Figure 12.6 shows this effect: the lower the A, the longer the curve follows the no diversity curve. The curve for selection diversity is also shown to emphasize that all switch-points (where $p_0 = A$) coincide with this curve. This checks out with (12.20). If we insert $p = A$, it reduces to (12.13), the equation for selection diversity. Figure 12.6 also shows that for selection diversity, diversity gain increases monotonously for decreasing probabilities. For switched diversity this is not the case. Diversity gain increases for decreasing probabilities until the threshold is reached and gain settles at a constant value. The higher the threshold A is set, the faster the increase in gain and the lower the final constant gain becomes. This effect is shown in Table 12.2, which lists values of diversity gain for different probabilities

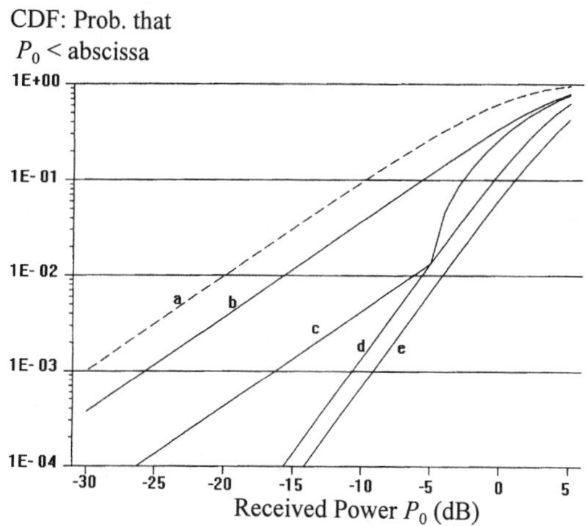

Figure 12.7 Cumulative density functions of power after diversity $p_0(t)$ for two-branch diversity and uncorrelated Rician faded branches with an equal Rice factor K. Curves are plotted for: (a) no diversity, Rayleigh fading, (b) no diversity, Rician fading, (c) switched-diversity with a threshold $A = -5$ dB, (d) selection diversity, and (e) EGC. For all Rice curves $K = 0$ dB was taken.

The results for Rician fading do not differ very much from those for Rayleigh fading shown in Figures 12.5 and 12.6. Figure 12.7 shows the cdfs for Rician fading with a Rice factor K of 0 dB, so the dominant signal component and the scattered signal components are equal in strength. As in Figure 12.5, curves are plotted for EGC, selection diversity, switched diversity, and no diversity. The curve for Rayleigh fading/no diversity is also plotted. Apart from a certain offset caused by the higher average received power due to the dominant component, the curves show a similar pattern to those shown in Figure 12.5 for Rayleigh fading, so the same observation holds.

Table 12.2

Diversity Gain (dBs) for Selection and Switched Diversity, for Two-Branch Diversity and Uncorrelated Rayleigh Faded Branches With Equal Statistics.

Probability	Selection Diversity Gain (dB)	Switched Diversity Gain (dB)		
		$A = 0$ dB	$A = -5$ dB	$A = -10$ dB
1	0	0	0	0
0.8	1.3	1	0.3	0.1
0.5	2.3	2.2	1.2	0.5
0.3	3.3	2.5	2	0.8
0.1	5.3	2	5	2.5
0.08	5.7	2	5.7	3
0.05	6.7	2	5.8	4.3
0.03	7.8	2	5.8	5.8
0.01	10	2	5.8	9.8
0.008	10.5	2	5.8	10.2

A is the threshold value.

Figure 12.8 shows the diversity curves for three different values of the Rice factor *K*. Logically, increasing *K* decreases the probability that p_0 is below a certain power level. Note that the curves for switched diversity again coincide with those for selection diversity at the threshold level, which was set at – 5 dB in this plot. The influence of *K* on diversity gain may not be obvious from the figure. For selection diversity, increasing *K* results in a decreasing gain for the probabilities of interest (down to approx. 1E-4). If the probability decreases further, the situation reverses and gain increases for an increasing *K*. This also holds for EGC, since we have shown that there is a constant gain difference between selection diversity and EGC. For switched diversity, two situations exist. For high probabilities where neither of the cdf curves is below the threshold, gain decreases when *K* is increased. For low probabilities where both cdf curves are below the threshold, gain increases when *K* is increased. Between these two situations there is a transition region, where one cdf curve has passed the threshold while the other has not.

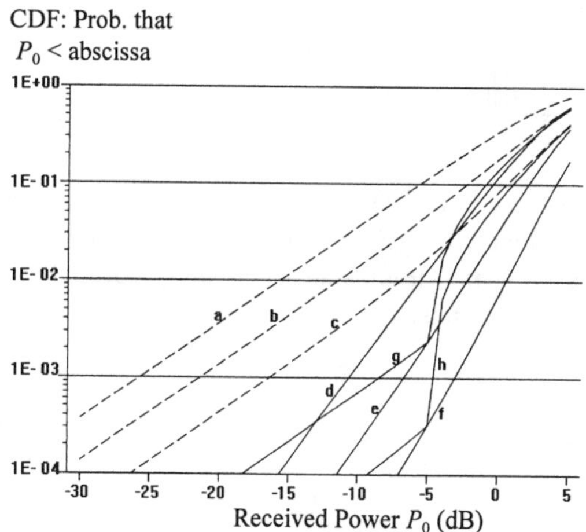

Figure 12.8 Cumulative density functions of power after diversity $p_0(t)$ for two-branch diversity and uncorrelated Rician faded branches with an equal Rice factor K. Curves are plotted for no diversity: (a) K = 0 dB, (b) K = 3 dB, (c) K = 5 dB, selection diversity, (d) K = 0 dB, (e) K = 3 dB, (f) K = 5 dB, switched diversity, (g) K = 3 dB, and (h) K = 5 dB. For the switched diversity curves, $A = -5$ dB was taken.

Independently Faded, Unequally Powered Branches

Diversity gain depends on several conditions, such as the space correlation between antennas, time correlation between succeeding time slots (if the switching decision is delayed), and branch statistics. With respect to this last point, if both branches have equal statistics, they each have an equal chance of being selected, which obviously results in the highest diversity gain. If the average power on both branches is unequal, so $\sigma_1^2 \neq \sigma_2^2$ or $K_1 \neq K_2$, the branch with the highest power is favored most of the time so the advantage of diversity is reduced. This section looks at the amount of reduction for unequal power levels on both branches.

For EGC the pdfs of instantaneous received power on both branches is summed and the cdf of power after diversity $p_0(t)$ for unequally powered Rayleigh faded branches ($\sigma_1^2 > \sigma_2^2$) is easily determined as

$$F_{\text{Ray},p_0} = 1 - \frac{1}{\sigma_1^2 - \sigma_2^2}\left(\sigma_1^2 \exp\left(-\frac{p}{\sigma_1^2}\right) - \sigma_2^2 \exp\left(-\frac{p}{\sigma_2^2}\right)\right) \qquad p \geq 0 \tag{12.22}$$

The cdf of $p_0(t)$ for *selection diversity* is determined by (12.12) (i.e., the product of the individual cdfs of each branch). For Rayleigh faded branches with unequal average power this results in

$$F_{\text{Ray},p_0}(p) = \left(1 - \exp\left[-\frac{p}{\sigma_1^2}\right]\right)\left(1 - \exp\left[-\frac{p}{\sigma_2^2}\right]\right) \qquad p \geq 0 \tag{12.23}$$

For *switched diversity* we need (12.17). Since we assume unequal branch statistics, we can no longer take $Pr\{1 \text{ selected}\} = Pr\{2 \text{ selected}\}$, so (12.18) is not valid here. If we still assume that successive time samples are completely correlated (and branches fade independently), (12.17) is written as

$$Pr\{p_0 > p\} = \left[\int_{\max(p,A)}^{\infty} f_{p_1}(p_1)dp_1 + \int_{p}^{\infty} f_{p_2}(p_2)\,dp_2 \int_{-\infty}^{A} f_{p_1}(p_1)\,dp_1\right] Pr\{1 \text{ selected}\}$$

$$+\left[\int_{\max(p,A)}^{\infty} f_{p_2}(p_2)dp_2 + \int_{p}^{\infty} f_{p_1}(p_1)\,dp_1 \int_{-\infty}^{A} f_{p_2}(p_2)\,dp_2\right] Pr\{2 \text{ selected}\} \tag{12.24}$$

which results in (12.19) if $f_{p_1}(p_1) = f_{p_2}(p_2)$. To compute the branch selection probabilities, we use the following Markov chain

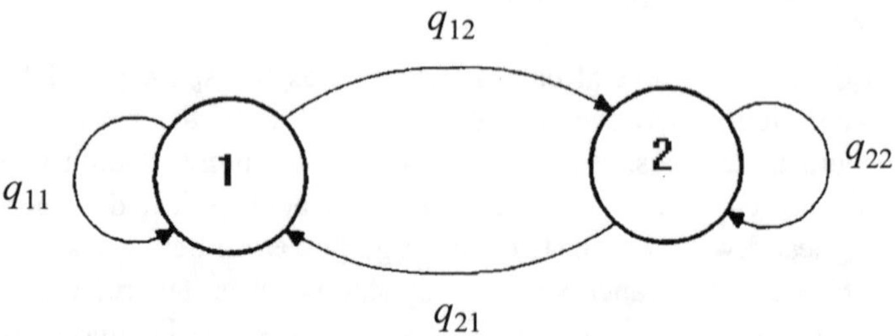

with two states; state 1, branch-one selected; and state 2, branch-two selected. The state transition probabilities q_{ij} are equal to the probability that power on a branch is either above or below the threshold A, that is,

$$q_{11} = \int_A^\infty f_{\mathrm{Ray},p_1}(p_1)dp_1 = \exp(-A/\sigma_1^2); \qquad q_{12} = 1 - q_{11}$$

(12.25)

$$q_{22} = \int_A^\infty f_{\mathrm{Ray},p_2}(p_2)dp_2 = \exp(-A/\sigma_2^2); \qquad q_{21} = 1 - q_{22}$$

The steady state solution of this Markov chain gives the branch selection probabilities as

$$Pr\{1\text{-selected}\} = \frac{1}{1+(q_{12}/q_{21})} = q$$

(12.26)

$$Pr\{2\text{-selected}\} = \frac{(q_{12}/q_{21})}{1+(q_{12}/q_{21})} = 1-q$$

Using this, (12.24) is written as
$Pr\{p_0 > p\}$

$$= \begin{cases} \left(1-\exp(-A/\sigma_1^2)\right)\exp(-p/\sigma_2^2)q + \left(1-\exp(-A/\sigma_2^2)\right)\exp(-p/\sigma_1^2)(1-q), & p > 0 \\ \quad + \exp(-p/\sigma_1^2)q + \exp(-p/\sigma_2^2)(1-q), & p > A \\ \quad + \exp(-A/\sigma_1^2)q + \exp(-A/\sigma_2^2)(1-q), & 0 < p < A \end{cases}$$

(12.27)

which reduces to (12.20) if $\sigma_1^2 = \sigma_2^2 = \sigma^2$, so $q = \frac{1}{2}$
Figure 12.9 shows the cdf of power after diversity $p_0(t)$ for uncorrelated Rayleigh faded branches for EGC and selection diversity. Different situations are compared: equal average power, a 6- and 12-dB power difference between both branches. Figure 12.10 shows the curves for switched diversity, for a 0-, 3-, 6-, and 12-dB power difference and a threshold value A of – 15 dB.

Figure 12.9 shows that for a large range of probabilities for both EGC and selection diversity, diversity gain is reduced by approximately the same amount as the power difference. Although diversity gain is decreased, performance is never worse than for the no-diversity case, contrary to the performance of switched diversity.

Figure 12.10 shows a large reduction of diversity gain below the threshold A for switched diversity. For a probability region where several cdf curves are below the threshold, diversity gain (in decibels) is decreased by approximately twice the power difference (in decibels). For a 12-dB power difference this means that the performance is even worse than for the no-diversity case. This is caused by the choice made for σ_i^2 in the calculations. For no diversity, σ = 0 dB was taken. For diversity, one branch was also taken 0 dB, the other – 3, – 6, or – 12 dB. With switched diversity, the best branch is not always chosen (contrary to selection diversity), so the branch having the lowest average power of the two has a finite chance of being chosen some of the time. Of course, performance on average will then be worse than for the no-diversity case.

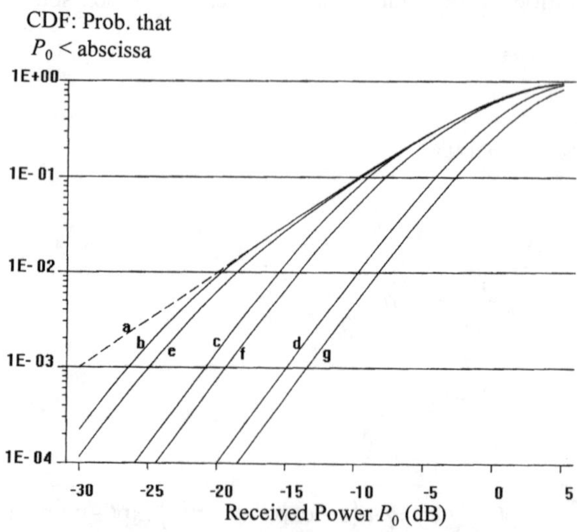

Figure 12.9 Cumulative density functions of power after diversity $p_0(t)$ for two-branch diversity and unequally powered, uncorrelated Rayleigh faded branches. Curves are plotted for (a) no diversity, selection diversity with a power difference between branches of (b) 12 dB, (c) 6 dB, (d) 0 dB, EGC with a power difference between branches of (e) 12 dB, (f) 6 dB, and (g) 0 dB.

Figure 12.11 shows what happens to the cdf for switched diversity for a power difference between branches if the sum of the power on both branches is kept equal to the power on a single Rayleigh faded branch. The sum power was chosen to be 0 dB; the power difference ranged from 3 to 12 dB. We see in Figure 12.10 that the curves for $p_0 > A$ stay relatively close. For $p_0 > A$, the difference is larger, but not as large as in Figure 12.10. The cdf for a 3-dB difference is almost equal to the cdf for no-power difference.

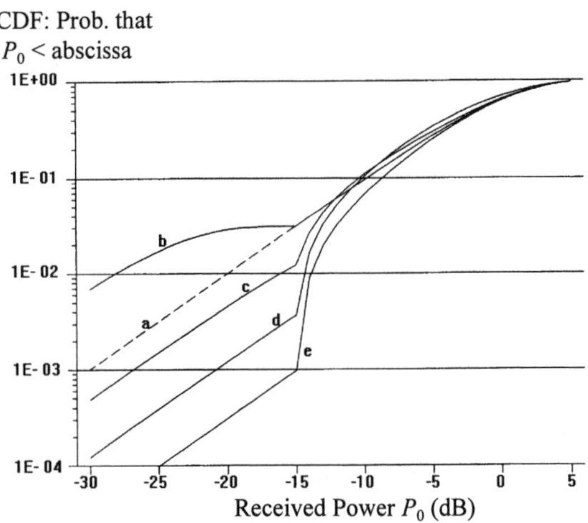

Figure 12.10 Cumulative density function of power after diversity $p_0(t)$ for two-branch diversity and unequally powered, uncorrelated Rayleigh faded branches. Curves are plotted for (a) no diversity, switched diversity with a power difference between branches of (b) 12 dB, (c) 6 dB, (d) 3 dB, and (e) 0 dB. The threshold value A is set at – 15 dB.

In the case of Rician fading with branches having different K factors, the cdf for EGC is very hard to determine, since we need to sum the pdfs. For this reason, we only look at selection and switched diversity in combination with Rician fading.

The cdf of power after diversity $p_0(t)$ for *selection diversity* and Rician faded branches with an unequal K factor, using (12.12) and (12.14), is $F_{\text{Ric},p_0}(p)$

$$\left(1-Q\left[\sqrt{2K_1},\frac{\sqrt{2p}}{\sigma_1}\right]\right)\left(1-Q\left[\sqrt{2K_2},\frac{\sqrt{2p}}{\sigma_2}\right]\right) \qquad p \geq 0 \tag{12.28}$$

For *switched diversity*, the state transition probabilities of the Markov chain used to calculate the branch selection probabilities as compared with Rayleigh fading, are changed to (compare (12.25))

$$\begin{aligned} q_{11} &= \int_A^\infty f_{\text{Ric},p_1}(p_1)\,dp_1 = Q\left(\sqrt{2K_1},\frac{\sqrt{2A}}{\sigma_1}\right); \qquad q_{12} = 1-q_{11} \\ q_{22} &= \int_A^\infty f_{\text{Ric},p_2}(p_2)\,dp_2 = Q\left(\sqrt{2K_2},\frac{\sqrt{2A}}{\sigma_2}\right); \qquad q_{21} = 1-q_{22} \end{aligned} \tag{12.29}$$

Inserting (12.29) in (12.26) results in the branch selection probabilities $Pr\{1\text{-selected}\} = q$ and $Pr\{2\text{-selected}\} = 1 - q$ which, combined with (12.24), give the following expression for the probability $p_0(t) > p$ for Rician faded branches with an unequal K factor

$$Pr\{p_0 > p\} = \begin{cases} Q_2(p)\left(1 - Q_1(A)\right)q + Q_1(p)\left(1 - Q_2(A)\right)(1 - q), & p \geq 0, \\ +Q_1(A)q + Q_2(A)(1 - q), & 0 < p < A, \\ +Q_1(p)q + Q_2(p)(1 - q), & p > A, \end{cases} \tag{12.30}$$

where $Q_i(x)$ represents $Q\left(\sqrt{(2K)}, \sqrt{(2x)} / \sigma_i\right.$, the Marcum's Q-function of (12.15). Figure 12.12 shows the cdfs for selection and switched diversity.

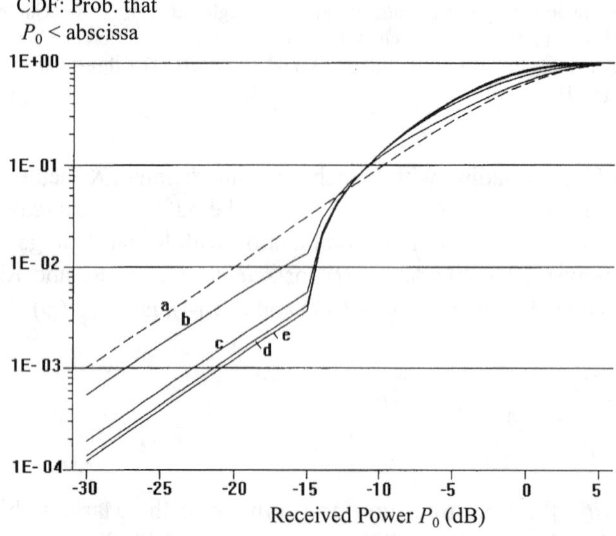

Figure 12.11 Cumulative density functions of power after diversity $p_0(t)$ for two-branch diversity and unequally powered, uncorrelated Rayleigh faded branches. Curves are plotted for (a) no diversity, switched diversity with a power difference between branches of (b) 12 dB, (c) 6 dB, (d) 3 dB, and (e) 0 dB. The sum power of both branches is kept equal to 0 dB, the same value as the power for the no-diversity curve. The threshold value A is set at – 15 dB.

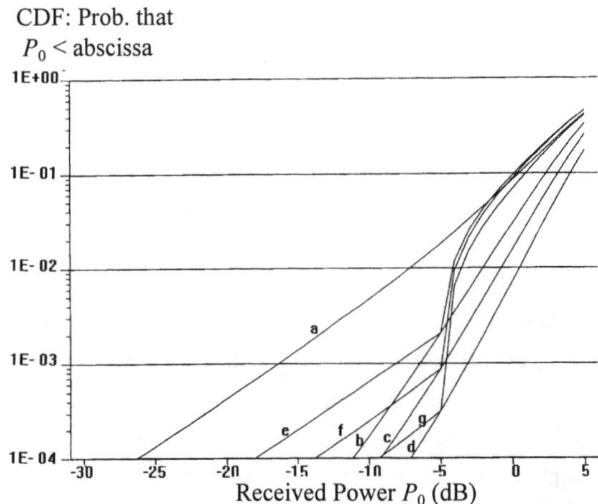

Figure 12.12 Cumulative density functions of power after diversity $p_0(t)$ for two-branch diversity and unequally powered, uncorrelated Rician faded branches. Curves are plotted for (a) no diversity, $K = 5$ dB, selection diversity with a difference between K factors of (b) 5 dB, (c) 2 dB, (d) 0 dB, switched diversity with a difference between K factors of (e) 5 dB, (f) 2 dB, and (g) 0 dB.

For a probability of 1E – 3, the loss of diversity gain for selection diversity is approximately equal to the difference in K factors, For a decreasing probability, this loss increases slowly since the cdf curves divert. As with Rayleigh fading, for switched diversity the loss of diversity gain strongly depends on whether $p_0(t) > A$ or $p_0(t) < A$. Above the threshold, loss is almost negligible; below the threshold it increases. As an example, for probabilities where both switched diversity cdf curves are below the threshold, diversity gain is reduced by approximately 4 dB if the difference in K factors increases from 0 to 2 dB.

Rayleigh Faded Branches, Correlated in Space

In this section we look at the effect of signal correlation between both antennas. The correlation between signals on both antennas increases if the distance between these antennas relative to the signal's wavelength becomes too small. In [12] we show that the correlation factor strongly depends on the presence of local scatterers around the receiver as well as on the angle of an incoming signal relative to the imaginary line connecting the antennas. Dependent of these factors, [12] shows that to achieve a correlation factor of 0.7 or less, the minimum antenna spacing ranges from a few wavelengths up to 70 wavelengths. For DECT carrier frequencies of 1880 to1900 MHz, the wavelength is approximately 16 cm, which would lead to a worst-case antenna separation of several meters.

To examine the effect of signal correlation we introduce the correlation factor ρ_S, defined as

$$\rho_S = \frac{\langle p_1 p_2\rangle - \langle p_1\rangle \langle p_2\rangle}{\sigma_1 \sigma_2} \tag{12.31}$$

where $\langle \, . \, \rangle$ denotes statistical averaging. If p_1 and p_2 are independent, $\langle p_1 p_2\rangle = \langle p_1\rangle \langle p_2\rangle$ so $\rho_s = 0$ (i.e., p_1 and p_2 are uncorrelated and diversity gain is the highest). For $\rho_S = 1$ (i.e., totally correlated branches), all diversity gain is lost. It is interesting to see what happens to the diversity gain for a correlation factor ρ_s that lies between these two extremes. In this section, we assume $\sigma_i^2 = \sigma^2 = 0$ dB for $i = 1, 2$.

We start with the joint pdf of two Rayleigh distributed random variables a_1 and a_2, which is given as [9]

$$f_{\text{Ray},a_1 a_2}(a_1,a_2) = \frac{a_1 a_2}{\sigma^4(1-\rho_s^2)} I_0\left(\frac{|\rho_s| a_1 a_2}{(1-\rho_s^2)\sigma^2}\right)\exp\left(-\frac{a_1^2 + a_2^2}{2(1-\rho_s^2)\sigma^2}\right), \qquad a_i \geq 0, i = 0,1 \tag{12.32}$$

which reduces to the product of two separate Rayleigh pdfs if $\rho_S = 0$. Transforming (12.32) with $p_i = a_i^2/2$ $(i = 0, 1)$ results in the joint pdf for the power of a_1 and a_2

$$F_{\text{Ray},p_1 p_2}(p_1, p_2) = \frac{1}{\sigma^4(1-\rho_s^2)} I_0\left(\frac{2|\rho_s|\sqrt{p_1 p_2}}{(1-\rho_s^2)\sigma^2}\right)\exp\left(-\frac{p_1 + p_2}{(1-\rho_s^2)\sigma^2}\right), \qquad p_i \geq 0, i = 0,1 \tag{12.33}$$

Given this joint power pdf, we determine the cdfs of power after diversity $p_o(t)$ for the different diversity techniques. Since $Pr\{p_o > p\} = Pr\{p_1 + p_2 > p\} = Pr\{p_1 > p - p_2\}$, the cdf for EGC is given by

$$F_{\text{Ray},p_0} = 1 - \int_0^{\infty}\int_{(p-p_2)}^{\infty} f_{\text{Ray},p_1 p_2}(p_1,p_2)dp_1 dp_2 \tag{12.34}$$

Using (12.11) and the fact that p_1 and p_2 in (12.33) are interchangeable, the cdf for *selection diversity* for correlated branches is given by

$$F_{\text{Ray},p_0}(p) = 1 - \left(\int_p^{\infty}\int_0^{p_1} f_{\text{Ray},p_1 p_2}(p_1,p_2)\, dp_2 dp_1 + \int_p^{\infty}\int_0^{p_2} f_{\text{Ray},p_1 p_2}(p_1,p_2)\, dp_1 dp_2 \right)$$

$$= \int_0^{p}\int_0^{p} f_{\text{Ray},p_1 p_2}(p_1,p_2) dp_2\, dp_1 \qquad (12.35)$$

For *switched diversity*, under the same assumptions as in independently faded equally powered branches (12.19) changed to

$$F_{\text{Ray},p_0}(p) = 1 - \int_{-\infty}^{\infty} \int_{\max(p,A)}^{\infty} f_{\text{Ray},p_1,p_2}(p_1,p_2)\, dp_1 dp_2 - \int_p^{\infty}\int_{-\infty}^{A} f_{\text{Ray},p_1 p_2}(p_1,p_2) dp_1 dp_2$$

(12.36)

$$= \begin{cases} \int_0^{p}\int_0^{A} f_{Ray,p_1,p_2}(p_1,p_2) dp_1 dp_2, & o < p < A \\ \int_0^{p}\int_0^{p} f_{Ray,p_1,p_2}(p_1 p_2) dp_1 dp_2 + \int_p^{\infty}\int_A^{p} f_{Ray,p_1,p_2}(p_1,p_2) dp_1 dp_2, & p > A \end{cases}$$

No closed expression is found for (12.34) to (12.36), so they have been solved numerically. The results are in Figure 12.13 for selection diversity and Figure 12.14 for switched diversity.

From Figure 12.13 it is apparent that even for a very high correlation factor (ρ_S = 0.95) diversity gain is still significant. The loss of diversity gain decreases rapidly for a decreasing correlation factor. Take $\rho_s = 0.7$ as an example; for this value the loss of diversity gain for a large range of probabilities is in the order of only 1 dB. The loss for even lower correlation factors is negligible. Figure 12.14 shows a similar behavior for switched diversity. For $p_0 > A$, almost no diversity gain is lost up to $\rho_s = 0.7$. For $p_0 < A$, diversity gain is decreased more than for selection diversity, but loss is still not very significant up to high values of ρ_S. For $\rho_S = 0.7$, diversity gain is decreased by approximately 2.5 dB.

To compare the performance of selection diversity with the performance of switched diversity, Table 12.3 lists some values of diversity gain for different probabilities, different correlations factors, and different threshold values A. Noted that, although the calculations in this section were all done for an average power σ_i^2 of 0 dB, values for the diversity gain in Table 12.3 are also valid for other values of σ_i^2, since all curves in Figures 12.13 and 12.14 are shifted by the same amount to the left or right if average power is changed. Of course, if σ_i^2 is changed, the values for A must be changed accordingly for the table to remain valid.

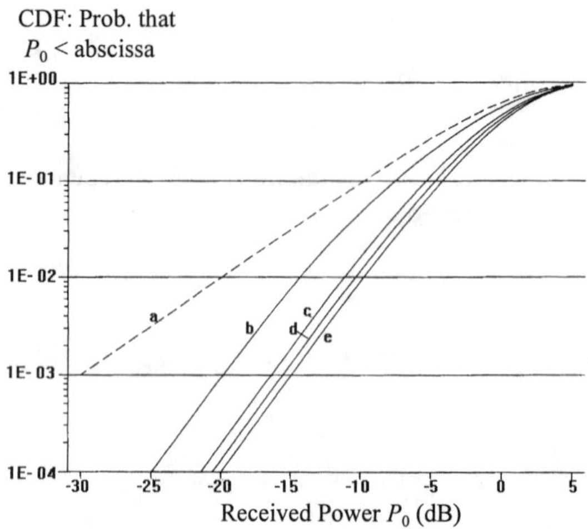

Figure 12.13 Cumulative density functions of power after diversity $p_0(t)$ for two-branch selection diversity and correlated Rayleigh faded branches. Curves are plotted for a correlation factor ρ_s of (a) 1 (no diversity), (b) 0.95, (c) 0.7, (d) 0.5, and (e) 0 (uncorrelated, optimum diversity gain).

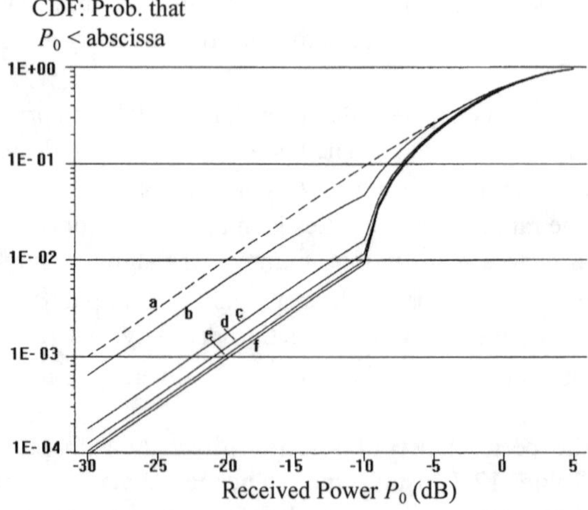

Figure 12.14 Cumulative density functions of power after diversity $p_0(t)$ for two-branch switched diversity and correlated Rayleigh faded branches. Curves are plotted for a correlation factor ρ_s of (a) 1 (no diversity), (b) 0.95, (c) 0.7, (d) 0.5, (e) 0.3, and (f) 0 (uncorrelated, optimum diversity gain). For all curves, $A = -10$ dB was taken.

Time Correlation in a Switched Diversity Scheme

The previous section treated space correlation between antennas. In this section, we look at a different type of correlation, namely time correlation between succeeding time samples. The value of the stochastic variable p_i at instant t_k is denoted by either $p_i(k)$ or $p_{ik}(t)$ in this section.

As was mentioned before, if only the BS in a DECT system is equipped with more than one antenna, the handheld still profits from this antenna diversity if the BS uses the same antenna for downlink transmission as it used for receiving on the uplink, assuming reciprocity of the channel[1]. In a TDMA/TDD system such as DECT, however, there is a time gap of half a frame length between the uplink and downlink time slots. During this time, the channel characteristics may change, and the correlation between these characteristics at the time of the uplink and downlink slots, determine the effectiveness of diversity on the downlink.

For space correlation, a low correlation factor ρ_s is desired to achieve the highest diversity gain. For time correlation this is exactly opposite. If we introduce the time correlation factor ρ_t, defined in a similar fashion as ρ_s as

$$\rho_t = \frac{\langle p_i(k)\, p_i(k-1)\rangle - \langle p_i(k)\rangle^2}{\sigma^2} \tag{12.37}$$

we can state that diversity gain is the highest if channel characteristics do not change at all, that is, $\langle p_i(k)p_i(k-1)\rangle = \langle p_i(k)\rangle^2$ so $\langle p_i(k)p_i(k-1)\rangle - \langle p_i(k)\rangle^2 = \sigma^2$ and $\rho_t = 1$. At the other extreme, if $\rho_t = 0$, succeeding time samples are completely uncorrelated and all diversity gain is lost. To illustrate this, we take the equation for $Pr\{p_o > p\}$ for switched diversity (12.18) and take $\rho_t = 0$, which results in

$$\begin{aligned} Pr\{p_o > p\} &= Pr\{p_1(k) > p\}Pr\{p_1(k-1) > A\} + Pr\{p_2(k) > p\}Pr\{p_1(k-1) < A\} \\ &= Pr\{p_1(k) > p\} \end{aligned} \tag{12.38}$$

where p_1 can be replaced by p_2. Whichever is chosen, all diversity gain is lost.

---

1 By definition, reciprocity means that channel characteristics are exactly in both transmission directions. Certain (periodic) structures and anisotropic media in a real indoor radio environment cause channel characteristics to be direction-sensitive. In that case, the assumption of reciprocity is not valid.

Table 12.3

Diversity Gain (decibels) for Selection Diversity and Switched Diversity, for Different Probabilities, Different Correlation Factors ρ_S, and Different Threshold Values A

Probability	ρ_S	Selection Diversity Gain (dB)	Switched Diversity Gain (dB)		
			A= – 5dB	– 10dB	– 15dB
0.1	0	5.8	5.4	2.8	1.2
	0.5	5.2	5.1	2.8	1.2
	0.7	4.6	4.3	2.7	1.1
	0.95	2.4	0.8	1.6	0.8
0.01	0	10.2	5.8	10.1	6.1
	0.5	9.7	4.8	9.2	6.1
	0.7	8.9	3.5	7.8	6
	0.95	5.8	0.2	2.1	5.3
0.001	0	15	5.7	10.3	15
	0.5	14.4	4.7	9	13.9
	0.7	13.6	3.4	7.5	12.2
	0.95	10	0.1	2	5.7

The value of ρ_t is determined by two factors: the time between the succeeding samples $k - 1$ and k (time between uplink and downlink slots) T, and the fading frequency or Doppler shift, respectively. In [1], the correlation factor for two Rayleigh faded signals is given as

$$\rho_t = J_0(2\pi f_D T) \tag{12.39}$$

where $f_D = 2\pi f_C v/c$ is the maximum Doppler shift, f_C is the carrier frequency and v is the receiver speed. For a DECT system, T is the time between the end of the uplink and the beginning of the downlink slot, which is approximately 5 meters per second (mps) (half the frame length). Some values of ρ_t for different speeds v and $T = 5$ mps are given in Table 12.4.

Table 12.4 shows that ρ_t decreases rapidly for an increasing speed. Normal walking speed is about 1 ms, a value for which ρ_t has already decreased to 0.65.

If we assume that p_1 and p_2 are independent and have an identical distribution, we can still use (12.18) to determine the cdf of p_0 for switched diversity. To solve (12.18), we need to know the joint probability density function of $p_1(k)$ and $p_1(k-1)$. This pdf is equal to (12.33), with ρ_s replaced by ρ_t and p_1 and p_2 replaced by $p_1(k)$ and $p_1(k-1)$, respectively. Since we assume that p_1 and p_2 are independent, the cdf of p_0 is given by

$$F_{\text{Ray},p_0k}(p) = 1 - \int_p^{\infty}\int_A^{\infty} f_{\text{Ray},p_{1k},p_{1k-1}^1}(p_{1k}, p_{1k-1})dp_{1k}dp_{1k-1}$$

$$-\int_0^A f_{\text{Ray},p_{1k}}(p_{1k})dp_{1k}\int_p^{\infty} f_{\text{Ray},p_{2k}}(p_{2k})dp_{2k} \tag{12.40}$$

where $f_{\text{Ray},p_{1k},p_{1k-1}}(\,.\,)$ is similar to (12.33) and $f_{\text{Ray},p_{1,2k}}(\,.\,)$ are similar to (12.3). The first integral in (12.40) is solved numerically, and the last two are simply solved analytically.

Table 12.4

Values of the Correlation Factor ρ_t for Different Receiver Speeds v and Corresponding Doppler Shifts f_D

v (mps)	f_D (Hz)	$\lvert\rho_t\rvert$
0.5	≅ 20	0.91
1.0	≅ 40	0.65
1.5	≅ 60	0.30
2.0	≅ 80	0.05

Figure 12.15 shows the cdf of p_0 for $\rho_t = 0$ (no diversity), 0.65, 0.9, 0.95, and 1 (optimum diversity). This figure shows the detrimental effect of incomplete time correlation. Even for a high correlation factor of 0.95, diversity gain is reduced considerably. For ρ_t = 0.65, almost all diversity gain is lost. This indicates that the switched diversity scheme as is examined here probably will not function very well when the receiver is moving around.

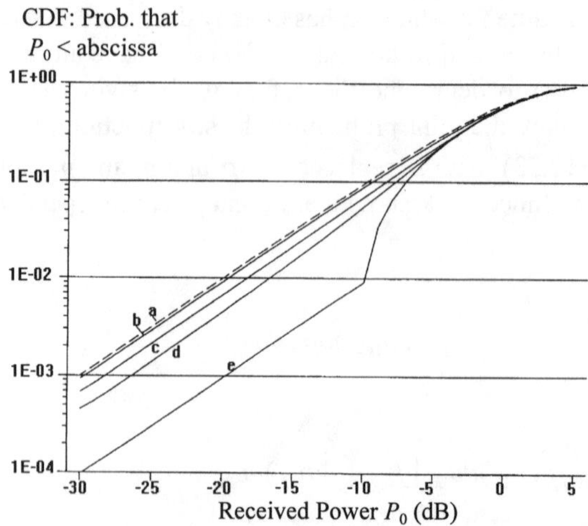

Figure 12.15 Cumulative density functions of power after diversity $p_0(t)$ for two-branch switched diversity and uncorrelated Rayleigh faded branches with equal average power. Curves are drawn for different values of the time correlation ρ_t between successive samples k and $k-1$: (a) $\rho_t = 0$ (uncorrelated, no diversity gain), (b) 0.65, (c) 0.9, (d) 0.95, and (e) 1 (correlated).

12.4.3 Multiple Branch Diversity

Section 12.4.2 considered two-branch diversity extensively. This section gives some results for the performance of EGC and selection diversity with more than two branches. Switched diversity is omitted because it does not profit from more than two branches. This is made clear by an example. Suppose that a switched diversity scheme used three branches, and a cyclic switching algorithm (...-1-2-3-1-2-...). If a switch is made from branch 1 to branch 2, the succeeding switch could just as well be back to branch 1 as forward to branch 3, since the delay in a practical system causes the time samples to be uncorrelated. Branch 3 thus does not contribute to diversity gain, even more branches are therefore also useless.

For EGC, each extra branch does contribute to diversity gain. The pdf of power after diversity $p_0(t)$ is found by summing all individual pdfs. If average power on each branch is equal, for Rayleigh fading this results in the gamma distribution.

$$f^{(n)}_{\mathrm{Ray},p_0}(p) = \frac{1}{\sigma^2} \frac{(p/\sigma^2)^{n-1}}{(n-1)!} \exp\left(-\frac{p}{\sigma^2}\right) \tag{12.41}$$

where n is the number of branches and σ^2 is the average received power on each branch.

To determine the cdf of $p_0(t)$, (12.41) is integrated. Repeated partial integration results in

$$F^n_{\text{Ray},p_0}(p) = \int_0^p f^{(n)}_{\text{Ray},p_0}(p_0)dp_o = 1 - \exp(-p/\sigma^2)\sum_{i=0}^{n-1}\frac{p^i}{(\sigma^2)^i i!} \tag{12.42}$$

If n goes to infinity, the sum in (12.42) is the series expansion for $\exp(p/\sigma^2)$, so then the cdf would go to 0 for all p. For Rice fading, the sum pdf is given by

$$f^{(n)}_{\text{Ric},p_0}(p) = \frac{1}{\sigma^2}\left(\frac{2p}{v}\right)^{\frac{n-1}{2}} \exp\left(-\frac{2p+v}{2\sigma^2}\right) \boldsymbol{I}_{n-1}\left(\frac{\sqrt{2pv}}{\sigma^2}\right) \tag{12.43}$$

This expression is integrated numerically to determine the cdf of p_0(t).

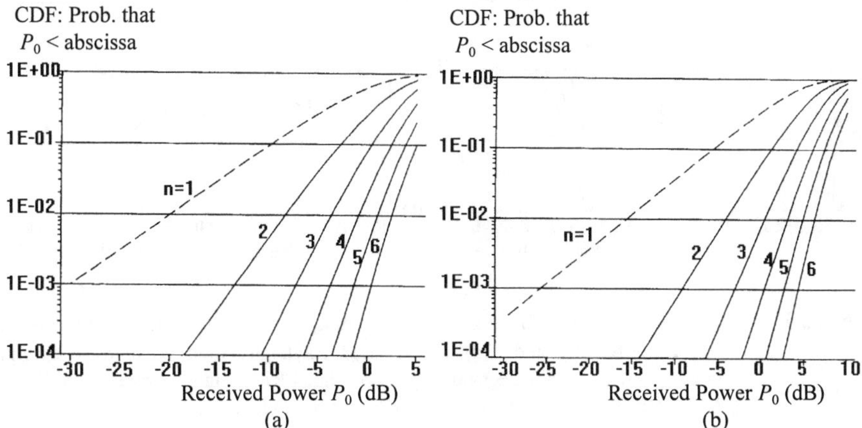

Figure 12.16 Cumulative density functions of power after diversity $p_0(t)$ for multiple branch equal gain combining and (a) uncorrelated Rayleigh faded branches with equal average power σ^2, and (b) uncorrelated Rician faded branches with equal Rice factor $K = 0$ dB. n indicates the number of branches.

The results for EGC up to $n = 6$ are shown in Figures 12.16 (a) and (b) for Rayleigh and Rice fading, respectively. Figure 12.16 (a) and (b) show that the increase in diversity gain for each extra branch decreases rapidly. From 5 to 6 branches, this increase is in the order of 1 to 2 dB for a large range of probabilities. Such a small gain generally does not justify the extra costs of adding an extra branch to a receiver.

The cdf of $p_0(t)$ for *selection diversity* and n branches is a simple extension of (12.12), namely the product of the n individual pdfs

$$F^{(n)}_{p_0}(p) = F_{p_1}(p).F_{p_2}(p). \dots .F_{p_n}(p) \qquad (12.44)$$

To obtain the cdf for Rayleigh faded branches, (12.3) is integrated and inserted in (11.44); for Rician faded branches (12.13), is inserted.

Figure 12.17 (a) and (b) show the results for Rayleigh and Rician fading, respectively. We see that for an increasing number of branches, diversity gain decreases a little more than for EGC. For example, from 5 to 6 branches the increase of diversity gain is less than 1 dB for a large range of probabilities.

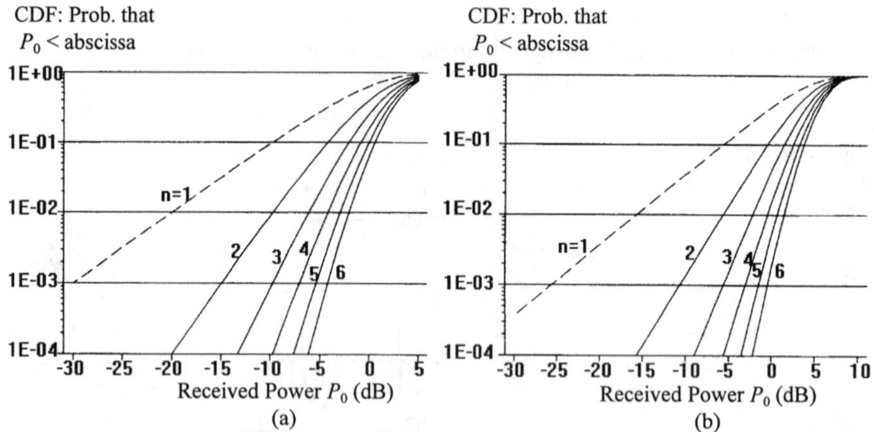

Figure 12.17 Cumulative density functions of power after diversity $p_0(t)$ for multiple branch selection diversity and (a) uncorrelated Rayleigh faded branches with equal average power $\sigma^2 = 0$ dB, and (b) uncorrelated Rician faded branches with equal Rice factor $K = 0$ dB. n indicates the number of branches.

The extra gain achieved by adding a branch is lost if there is a power difference between branches. Figure 12.18 shows this effect for selection diversity. The performance for 4, 5, and 6 equally-powered branches is compared with the situation of 6 branches, of which either 1, 2, or 3 have a 6-dB lower average power. This figure shows that all diversity gain of a branch is lost if that branch has a lower average power. The power level difference that results in a complete loss of diversity gain for a large range of probabilities depends on the number of branches. The larger the number of branches, the smaller the extra gain achieved by adding a branch, and the smaller the power level difference is to cancel the effect of the extra branch.

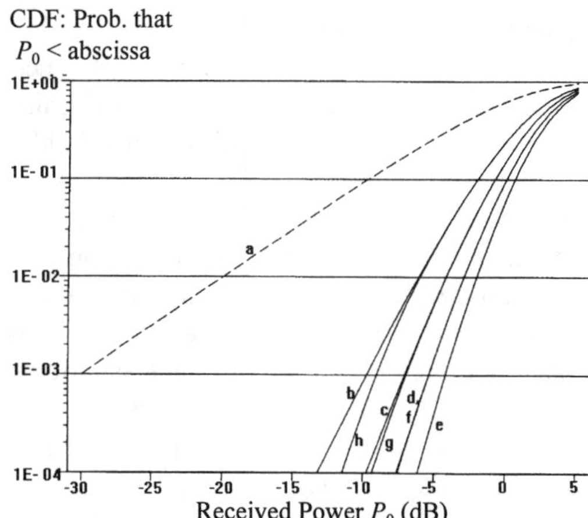

Figure 12.18 Cumulative density functions of power after diversity $p_0(t)$ for multiple branch selection diversity and uncorrelated Rayleigh faded branches. Curves are plotted for (a) no diversity, n equally-powered branches, where (b) $n = 3$, (c) $n = 4$, (d) $n = 5$, and (e) $n = 6$, and a total of 6 branches, of which n have a 6-dB lower average power, where (f) $n = 1$, (g) $n = 2$, and (h) $n = 3$.

12.5 CONCLUSIONS

The performance of a switched diversity scheme based on a received power is investigated and compared with the performance of selection diversity and EGC under the assumption of a frequency nonselective Rayleigh or Rician fading channel.

From the obtained results, the following conclusions are drawn (unless mentioned otherwise, these conclusions hold for both Rayleigh and Rice fading):

- Switched diversity performance is always inferior to the performance of the other diversity techniques.
- The largest diversity gain with switched diversity is achieved if average power on both branches is near the switching threshold. Performance is then nearly equal to the performance of selection diversity.
- An optimum switching threshold in general cannot be given. This depends on the exact channel statistics. Diversity gain for switched diversity is largely influenced by the choice of the threshold value. If the threshold value is very low, the probability that power on a chosen branch is above the threshold increases, which means that the probability of a switch decreases. The behavior of power after diversity will then increasingly resemble the behavior for no diversity. If power on both branches is below the threshold, diversity gain becomes a constant.

- The performance difference between selection diversity and EGC is not very large. The difference in diversity gain is never more than several decibels.
- In case of Rice fading, the Rice factor K has a nonlinear influence on diversity gain. In the cdf plots this means that for an increasing K, diversity gain decreases for high probabilities but increases for low probabilities. This holds for all diversity techniques.
- Different branch statistics degrade diversity performance for all diversity techniques. For selection diversity and EGC the reduction of diversity gain for Rayleigh fading is approximately equal to the difference in average power between branches. For switched diversity, performance can even get worse than for no diversity, due to the fact that there is always a finite possibility that the bad branch is selected. For the part of the cdf curve above the threshold, little diversity gain is lost; for the part below the threshold, approximately twice the difference in power level between branches is lost (in decibels).
- Antenna correlation does not degrade performance of any of the diversity techniques very much. Up to an antenna correlation factor 0.7, the loss of diversity gain is almost negligible (less than 2 dB for selection diversity and switched diversity). Above this value, the loss increases only very slowly.
- With switched diversity, performance is greatly reduced if the correlation between succeeding time samples is not 1. For example, for a very high correlation factor of 0.95 diversity gain for a large range of probabilities is reduced from 10 to 3 dB. For walking speeds, the correlation has dropped to a level where practically no diversity gain is left. This means that the examined switched diversity scheme probably does not function if the receiver is moving.
- For selection diversity and EGC, the extra gain achieved by adding a branch decreases rapidly if the number of branches increases. The gain of an extra branch is lost if average power on that branch is lower than the other branches.

REFERENCES

[1] W.C. Jakes, Jr. *Microwave Mobile Communications,* New York: John Wiley & Sons, 1993

[2] W.C.Y. Lee, *Mobile Communication Design Fundamentals,* New York: John Wiley & Sons, 1993.

[3] R. Prasad, "Overview of antennas in the development for future wireless personal communications," *Proc. Workshop on Large Antennas in Radio Astronomy, ESTEC,* Nordwijk, The Netherlands, pp.195–206, February 1996.

[4] P. Eggers (Ed.), Chapter 3 from COST 231 final report.

[5] T. Bull, M. Barret, and R. Amott, "Technology in smart antennas for universal advanced mobile infrastructure (TSUMANI R2108) Overview," *Proc. RACE Mobile Telecom. Summit,* Cascais, Portugal, pp. 88–97, November 22–24, 1995.

[6] Q. Garcia and F. Municiv, "Broadband base-station arrays and antenna elements for mobile communications," *Proc. EPMCC'95,* Bologna, Italy, pp. 56–61, November 28–30, 1995.

[7] G.F. Pedersen and J. Bach Andersen, "Integrated antenna for hand-held telephones with low absorption," *Proc. 44th IEEE VTC Conference,* Stockholm, Sweden, pp. 1537–1541, June 8–10, 1994.

[8] A.J. Rustako, Y.S. Yeh, and R.R. Murray, "Performance of feedback and switch space diversity 900 MHz FM mobile radio systems with Rayleigh fading," *IEEE Trans. on Comm., Vol. COM-21, No.11,* pp. 1257–1268, November 1973.

[9] K. Suwa and Y. Kondo, "Transmitter diversity characteristics in microcellular TDMA/TDD mobile radio," *Proceedings PIMRC'92,* Boston, 1992.

[10] J.G. Proakis, *Digital Communications,* 2nd edition, Singapore, McGraw-Hill, 1989.

[11] A. Papoulis, *Probability, Random Variables and Stochastic Processes,* McGraw-Hill, 1991.

[12] W.C.Y. Lee, "Effects on correlation between two mobile radio base station antennas," *IEEE Trans. on Comm.,* Vol. COM-21, No. 11, pp. 1214–1224, November 1973.

Chapter 13
Millimeter-Wave Communications

13.1 INTRODUCTION

Technology has advanced to the point where commercial exploitation of the millimeter wave can soon be a reality. The range of a millimeter wave is from 30 to 300 GHz [1]. This chapter considers only a 60-GHz frequency band because of its worldwide overlapping allocations for wireless broadband multimedia applications [2]. Millimeter waves from 59 to 62 and 62 to 63 GHz together with 65 to 66 GHz are under investigation in Europe for the cordless local area network and mobile broadband systems, respectively. In the United States, a 59 to 64 GHz frequency band is for general unlicensed applications and a 60-GHz band is considered for high-speed applications. In Japan, investigations are taking place in the 59 to 64 GHz frequency band for mobile broad band systems.

The 60-GHz frequency band combines the advantages of UHF and IR bands.

IR systems have the disadvantage that the LOS path must always be available between transmitter and receiver, because IR behaves like visible light. In the UHF region the spectrum is already crowded, so only small bandwidths remain, which puts restrictions on the maximum bit rate.

An alternative that combines the advantages of IR (enough free bandwidth, convenient antenna size) and UHF (good coverage) could be the millimetric wavelength band. The 20 to 60 GHz region is practically unused and allows for a large bandwidth and high bit rates (10 – 15 Mbps or more). As in the UHF region, the waves refract when touching a surface, so an LOS and multiple reflected paths can be received.

In a cellular communication system, it is important that cochannel interference is low so reuse distance can be small. For instance millimetric waves do not propagate easily through walls, which means the power is contained within existing physical boundaries such as rooms, floors, or buildings. Also, the signal is absorbed by rain or fog. Additionally, in the 60-GHz band, radio signals interact with the electrical dipole moment of oxygen molecules, causing absorption of the signal of about 14 dB/km [3]. This oxygen absorption produces a natural physical boundary in larger open sites that limits interference. Figure 13.1 presents the attenuation by atmospheric gases in the whole millimetric wavelength region [3]. We see that there are two kinds of attenuations; gaseous absorption, and attenuation caused by rain and fog.

However, millimetric waves also have their disadvantages: electronic components for the RF parts of the terminals are still not fully developed or far too expensive. Another drawback may be the possible health hazard of high frequencies. However, if sufficient low-transmitting power is used and users are shielded from the vicinity of the antenna, the hazard is minimal. It is worth mentioning that no definitive evidence has been shown to date documenting any hazards to the general public arising from prolonged exposure in the fields of less than 10 mw/cm^2 in the millimetric waves.

The study of millimeter-wave communication systems has drawn the attention of many researchers [4 – 33] because of the above advantages, and because millimeter waves also have the potential to support broadband service access, which is especially relevant due to the advent of B-ISDN.

This chapter presents some preliminary results on the performance of millimeter-wave communications in terms of CIP, fade duration, and BEP.

The rms delay spread is an important parameter in a wireless environment because it imposes a limit to the maximum bit rate that data can be signaled without ISI. Some typical values for the rms delay spread are given in Table 13.1 for the frequencies 1.7 and 60 GHz. Measurement results in [11] show that rms delay spread lies between 15 to 45 ns and 30 to 75 ns for small and large rooms, respectively. A worst-case rms delay spread value of 100 ns was observed.

Table 13.1

Probability Distribution for the rms Delay Spread [25]

Probability (%)	1.7 GHz	60 GHz
25	< 15 ns	< 2 ns
50	< 23 ns	< 7 ns
75	< 35 ns	< 8 ns
95	< 45 ns	<13 ns

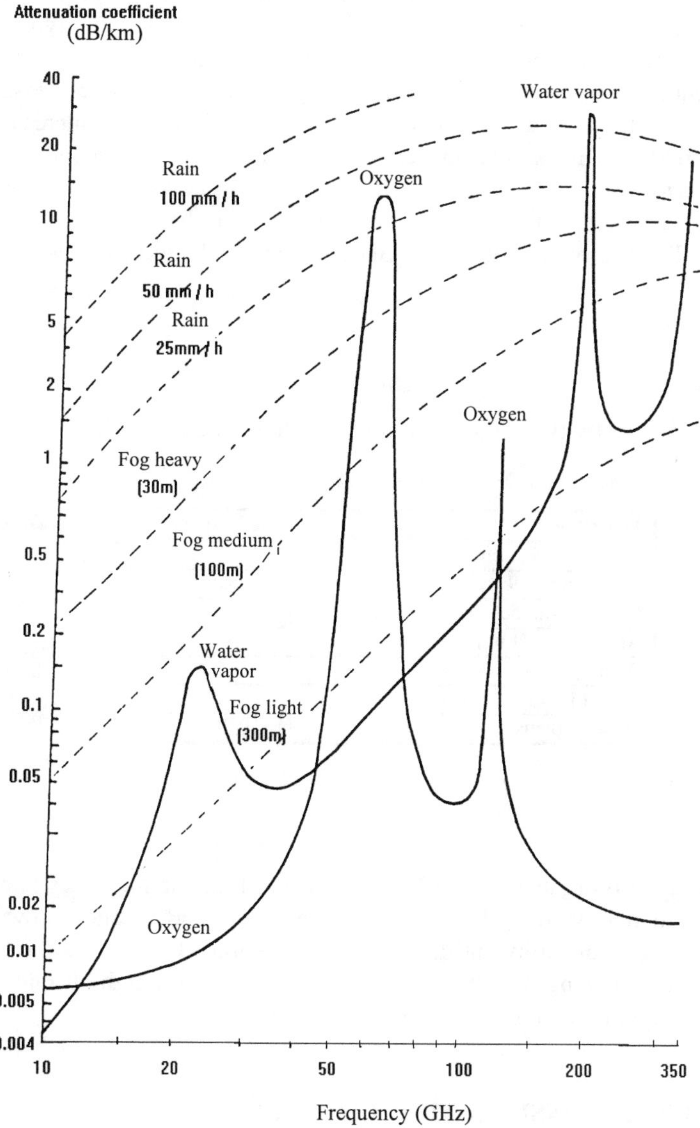

Figure 13.1 Attenuation due to oxygen, water vapor, rain, and fog as a function of frequency in gigahertz.

The effect of different building materials on the signal propagation is discussed in several papers. It appears that hard building materials, such as concrete and steel, give a high attenuation, while soft materials like sheetrock, glass, and gypsum give a much lower attenuation. For example, at 35 GHz, the attenuation of gypsum is 5 dB, while the attenuation of concrete is approximately 45 dB. From these measurements, we also conclude that for all considered materials the attenuation increases as the frequency increases. This knowledge can be used in designing an indoor wireless communication system, because we know that all buildings nowadays have concrete for floors and outside walls, which gives a good shielding, but have sheetrock or gypsum for internal walls. The reflectivity at 60 GHz of some building materials is shown in Table 13.2.

Table 13.2

Reflectivity at 60 GHz of Some Building Materials [32]

Material	Reflectivity (%)
Plasterboard (1 cm)	2
Chipboard (1.9 cm)	20
Concrete (autoclaved aerated blocks)	16
Aluminium (1 mm)	> 99
Wood (20 mm)	2
Glass (3 mm)	16
Homo sapiens	< 0.06

This chapter is organized as follows. In Section 13.2, the frequency nonselective channel is considered assuming low bit rates. The system performance is evaluated in terms of the outage probability and fade duration. Section 13.3 presents the frequency selective channel assuming high bit rate. The performance of a high bit rate system is evaluated by computing BEP using BPSK modulation.

13.2 FREQUENCY NONSELECTIVE CHANNEL

Outage probability and fade duration are evaluated for a cellular radio system at 60 GHz. The effect of cochannel interference on system performance is studied for a frequency nonselective channel modeled by an uncorrelated Rician fading environment. The influence of oxygen absorption at 60 GHz on performance is investigated.

13.2.1 Received Power Versus Distance

Because the area of every transmitted wave becomes greater with the considered distance, the power per unit area, and thus the received power, gets smaller at the considered distance point. Considering oxygen absorption due to the 60-GHz region of 14 dB/km [23], the (normalized) mean power P received from a terminal at a (normalized) distance ρ from the central receiver is given by

$$P = \frac{P(d)}{P(R_{\max})} = \frac{1}{\rho^m 10^{0.0014 R_{\max}(\rho-1)}} \tag{13.1}$$

where m is the path loss exponent and the normalized distance ρ is defined as:

$$\rho = \frac{d}{R_{\max}} \tag{13.2}$$

Here, $R_{\max}$ is the radius of the circular area (centered on the BS) in which the associated terminals are situated. Figure 13.2 shows the power decay for various values of $R_{\max}$ using $m = 2$. We see that due to the influence of oxygen absorption, the necessary transmitting power gets too high for large cell radii. The cell radius in the 60-GHz region therefore has an upper limit somewhere at about 2 km, depending on the acceptable transmitting power. The effect of path-loss law exponent on P is depicted in Figure 13.3 in the presence of oxygen absorption.

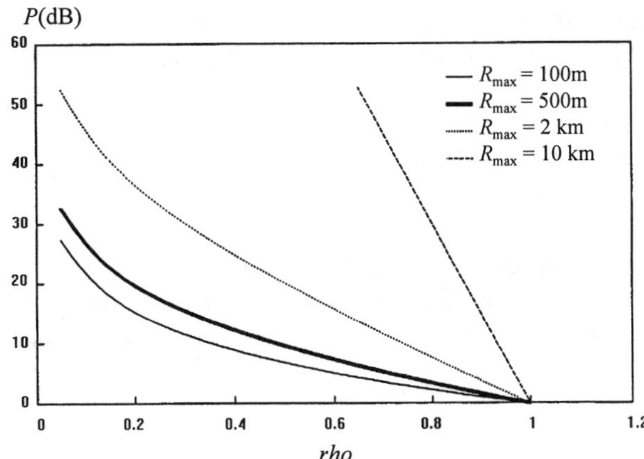

Figure 13.2 Normalized mean power (P) versus normalized distance ρ with $R_{\max}$ as a parameter for $m = 2$.

This power decay depends strongly on the environment. Every time the ray reflects on a surface, it gets attenuated depending on the reflection coefficient of the surface, which varies from 0.06% (Homo sapiens) to 99% (1-mm aluminium).

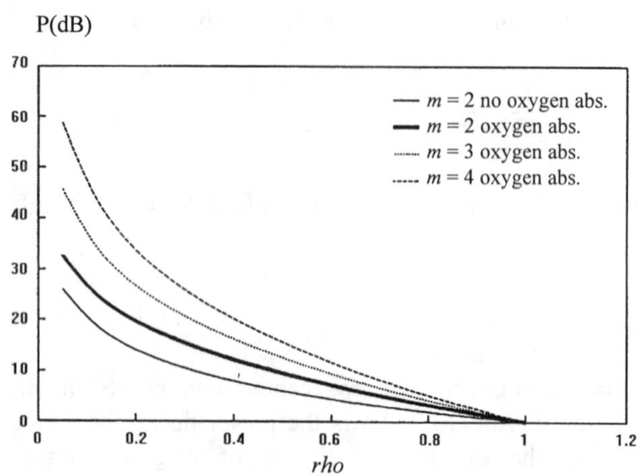

Figure 13.3 Normalized mean power (P) versus normalized distance ρ with m as a parameter for $R_{\max} = 500$m.

13.2.2 OUTAGE PROBABILITY

The conditional outage probability for desired and interfering signals having Rician pdf is evaluated by using the formula of conditional CIP given in Chapter 4 to 6:

$$F(\mathrm{CI}|n) = \int_0^\infty \left[1 - Q\left(\sqrt{2K_d}, \frac{u\sigma_u\sqrt{\alpha}}{\sigma_d}\right)\right] \frac{u^n}{(2nK_u)^{\frac{n-1}{2}}} e^{-\frac{u^2+2nK_u}{2}} I_{n-1}\left(u\sqrt{2nK_u}\right) du \tag{13.3}$$

where α is the specified cochannel protection ratio, K_d and K_u are the desired and interference Rician parameter, respectively, σ_d^2 and σ_u^2 are the average power of the reflected signal due to desired and interfering signals, respectively, and n is the number of interferers.
With

$$\frac{\sigma_d^2}{\sigma_u^2} = \frac{1+K_u}{1+K_d} \mathrm{SIR}_{\mathrm{LOS-LOS}} \tag{13.4}$$

here,

$$\text{SIR}_{\text{LOS-LOS}} = R_u^2 10^{0.0014(D-R)} \tag{13.5}$$

and R is the cell radius, D is the distance between the centers of the nearest neighboring cochannel cells, and R_u is defined as the (normalized) reuse distance [36] given by (3.4).

For desired signal Rician-distributed and interfering signal Rayleigh-distributed, $F(\text{CI}|n)$ is written as:

$$F(\text{CI}|n) = \int_0^\infty \left[1 - Q\left(\sqrt{2Kd}, \frac{u\sigma_u \sqrt{2\alpha}}{\sigma_d}\right)\right] \frac{2u^{2n-1}}{(n-1)!} e^{-u^2} du \tag{13.6}$$

with

$$\frac{\sigma_d^2}{\sigma_u^2} = \frac{1}{1+K_d} \text{SIR}_{\text{LOS-Reflected}} \tag{13.7}$$

here,

$$\text{SIR}_{\text{LOS-Reflected}} = KR_u^2 \left(1 + \frac{D}{g}\right)^2 10^{0.0014(D-R)} \tag{13.8}$$

and g is the turning point [37].
$F(\text{CI}|n)$ is derived for both desired and interfering signals Rayleigh distributed as:

$$F(\text{CI}|n) = 1 - \left(\frac{1}{\alpha\sigma_u^2 / \sigma_d^2 + 1}\right)^n \tag{13.9}$$

with

$$\frac{\sigma_d^2}{\sigma_u^2} = \text{SIR}_{\text{Reflected-Reflected}} \tag{13.10}$$

Here,

$$\text{SIR}_{\text{Reflected-Reflected}} = \frac{R_u^2 \left(1 + \frac{D}{g}\right)^2}{\left(1 + \frac{R}{g}\right)^2} 10^{0.0014(D-R)} \tag{13.11}$$

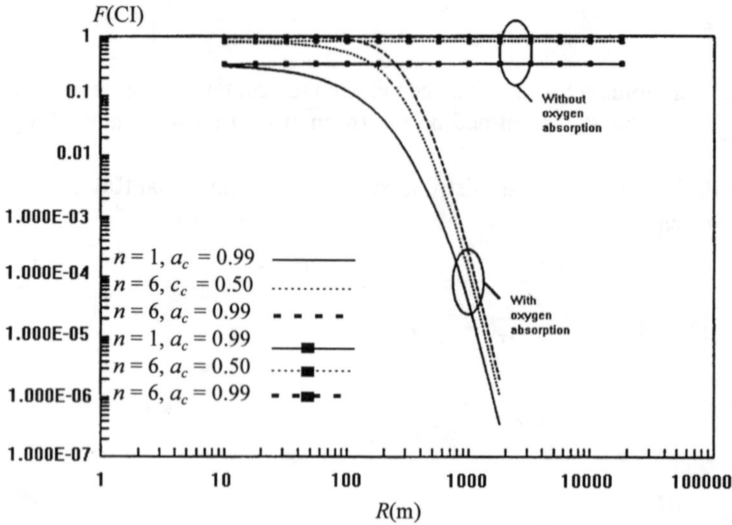

Figure 13.4 Outage probability [F(CI)] versus cell radius R for Rice-distributed, desired, and interfering signals with different number of interferers, $a_c = 0.5$, 0.99 erlang /channel and $\alpha = 8$ dB.

In Figure 13.4, the effect of oxygen absorption on the outage probability is shown for Rician desired and interfering signals for n and a_c as parameters. Figure 13.4 shows that the oxygen absorption enhances the performance of the cellular system. However, for small distances (<100m), the reduction in the outage probability is small and for large distances (>100m), the reduction in outage probability is significant as compared with the system without oxygen absorption.

Figure 13.5 shows the outage probability versus the reuse distance with oxygen absorption for Rician desired and interfering signals for R = 100m. The effect of carried traffic on the performance is also shown in Figure 13.5.

13.2.3 Fade Duration

The multipath fades result in burst errors in the transmitted data. To compute the length of burst errors, the average fade duration is computed. The level crossing rate N_R is defined as the expected rate at which the signal envelope crosses a specified signal level R^* in the positive direction [38]. The probability of the envelope passing through the value R^* per time interval t, $t + dt$ with positive slope is [39]

$$N_R = \int_0^\infty R_*' P(R_*, R_*') dR_*' \tag{13.12}$$

where $P(R^*, R^{*'})$ is the probability density of R^* and its time derivative $R^{*'}$. Rice [39] gives the expression for this pdf that reduces after simplification to

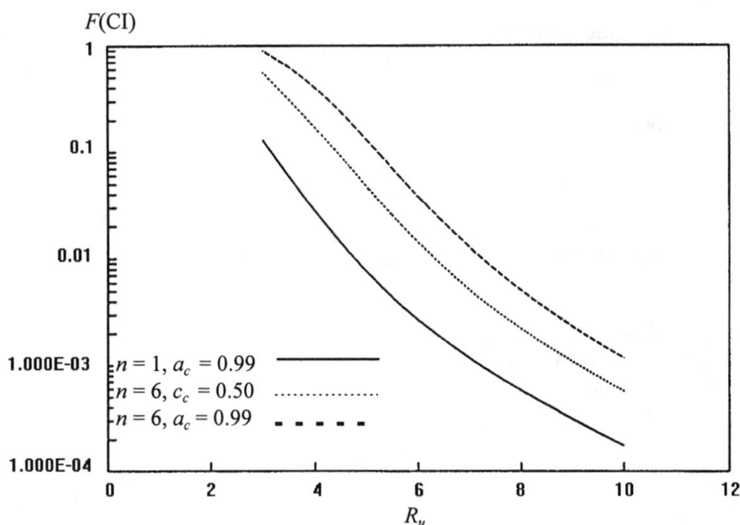

Figure 13.5 Outage probability [F(CI)] versus reuse distance (R_u) for Rice distribution desired and interfering signals with different numbers of interferers, a_c = 0.5, 0.99 erlang/channel, R = 100m, and α = 8 dB.

$$N_R = \sqrt{\frac{b_2}{2\pi}} p(R_*) \tag{13.13}$$

where $p(R_*)$ is the pdf of the signal envelope, which is derived from the signal power pdf. Here b_2 is defined as

$$b_2 = \sigma_r^2 (2\pi f_m)^2 \tag{13.14}$$

and the maximum Doppler frequency f_m as

$$f_m = \frac{v}{\lambda} = \frac{fv}{c} \tag{13.15}$$

The average fade duration is related to N_R:

$$\overline{T}_f = \frac{1}{N_R}\int_0^{R_*} p(r)d(r) \tag{13.16}$$

and the average nonfade duration

$$\overline{T}_n = \frac{1-\int_0^{R_*} p(r)d(r)}{N_R} \tag{13.17}$$

The integral is computed with

$$\int_0^{R_*} p(r)dr = \int_0^{R_*} xe^{-\frac{x^2+2K_d}{2}} I_0\left(x\sqrt{2K_d}\right) dx \tag{13.18}$$

where

$$R_* = \sigma_u = \frac{1}{\sqrt{\text{SINR}}} \tag{13.19}$$

with $\sigma_d^2 = 1$.

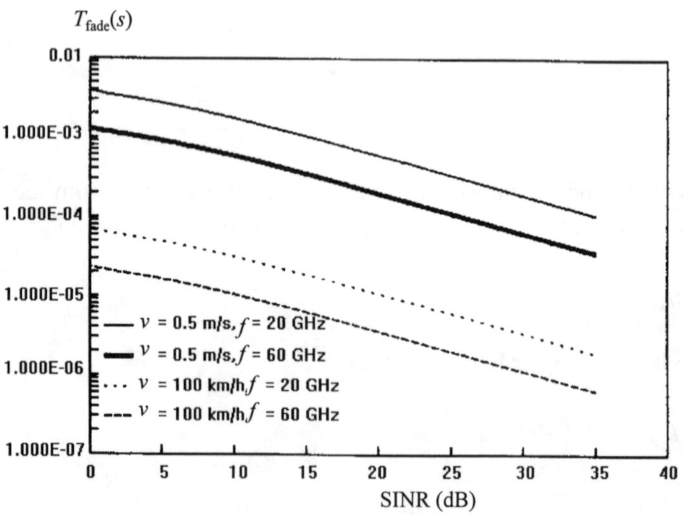

Figure 13.6 Average fade duration (T_{fade}) versus SINR for different receiver velocities and frequencies.

The influence of signal-to-interference-plus-noise ratio (SINR) on the average fade duration is computed. The results are shown in Figure 13.6. These results have general importance, since the propagation model is not yet used. The influence of receiver velocity is clear: a larger speed gives shorter fade durations.

When we introduce the propagation model, the influence of cell radius on fade duration is shown in Figure 13.7 with SNR as the parameter. We conclude that the SNR is dominant for cell radii larger than 50m to 100m.

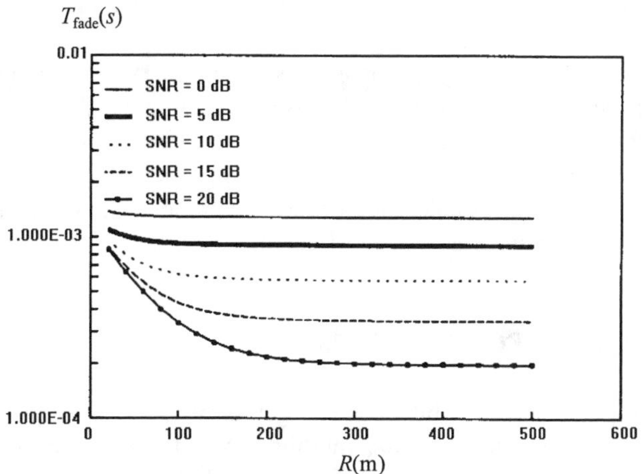

Figure 13.7 Average fade duration (T_{fade}) versus cell radius (R) for different values of the SNR (v = 0.5 m/s, f = 60 GHz, C = 7, ac = 0.5 erlang/channel).

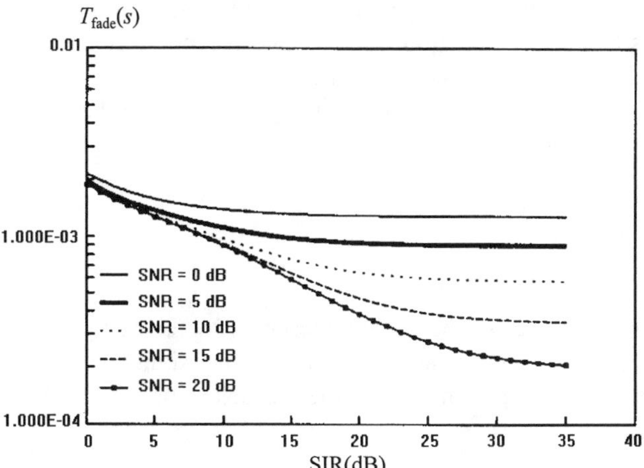

Figure 13.8 Average fade duration (T_{fade}) versus signal-to-interference ratio for different values of the SNR (v = 0.5 m/s, f = 60 GHz).

Measurements in [11] show signal propagation and the resulting signal-to-interference-ratio is very much dependent on the topology and used construction materials of the indoor environment. Therefore, average fade duration as a function of signal-to-interference ratio is plotted for SNR as the parameter in Figure 13.8.

13.2.4 Discussion

In the millimetric frequency region, oxygen absorption has a significant influence on system performance. The computed results show that the outage probability gets reduced due to oxygen absorption.

The reduction of the outage probability is quite remarkable at large distances (> 100m). However, larger than 2-km cell radius is not recommended because of the extra power attenuation at the rate of 14 dB/km at 60 GHz. At small distances (<100m), such as, for picocells, performance improvement is not so significant.

Using the calculation of fade duration, we recommend designing a proper error-correction coding scheme and studying its effect on performance.

13.3 FREQUENCY SELECTIVE CHANNEL

In a wireless computer network, bit rates can get as high as 10 Mbps or more. This means that the different echoes of the multipath faded signal are spread out over successive bits of the bitstream. Another difference with a mobile phone network is that the terminals are stationary instead of mobile. This means that the received signal is not a signal with fast fluctuating average power anymore, but has more or less constant average power. Due to the stationary character, the Rician or Rayleigh pdf is no longer appropriate: a sample is taken from all possible reception conditions so a different approach is used.

To investigate the performance of the system, we use the impulse response of the channel, which is written as

$$h(t) = \sum_{k=0}^{N} \beta_k e^{j\theta_k} \delta(t - \tau_k) \tag{13.20}$$

where LOS path 0 and the N echoes are represented by their amplitude β_k, phase shift θ_k and delay τ_k. It is convenient to describe the signal by some characteristic parameters. The most used parameters are the rms delay spread (T_{RMS}) where we use the discrete version

$$T_{\text{RMS}} = \sqrt{\frac{M_2}{M_0} - \left(\frac{M_1}{M_0}\right)^2} \tag{13.21}$$

where

$$M_0 = \sum_{k=0}^{N} \beta_k^2, \qquad M_1 = \sum_{k=0}^{N} \tau_k \beta_k^2, \qquad M_2 = \sum_{k=0}^{N} \tau_k^2 \beta_k^2 \tag{13.22}$$

and the Rician factor K, which is defined as

$$K = \frac{P_{\text{los}}}{P_{\text{reflected}}} = \frac{\beta_0^2}{\sum_{k=1}^{N} \beta_k^2} \tag{13.23}$$

which assumes that path 0 is the LOS path.

Further, the impulse response is characterized by its PDP type. The PDP type depends on the antenna used. When a (semi-) omnidirectional antenna is used, the PDP is roughly exponential.

The use of biconical horns [11] results in a slower decay in general (higher T_{rms}), and a PDP type called a flat-exponential PDP (F-PDP) in contrast with the exponential PDP (E-PDP).

Figures 13.9 and 13.10, respectively, show the E-PDP and F-PDP. As we see the F-PDP does not have a distinctive LOS component, therefore the Rician factor cannot be defined for the F-PDP. The T_{rms} parameter can still be computed.

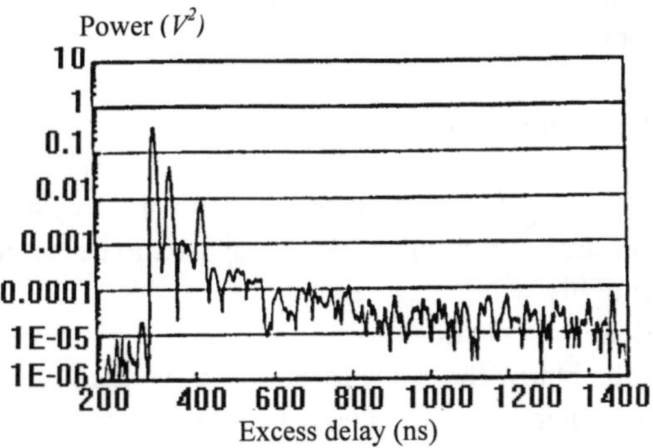

Figure 13.9 Example of a measurement of an E-PDP[17].

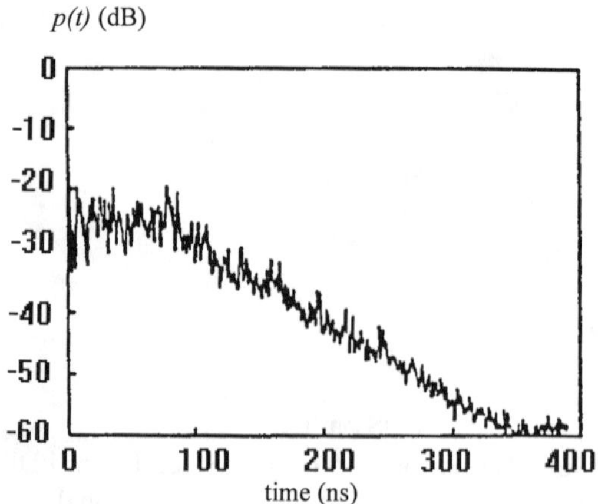

Figure 13.10 Example of a measured F-PDP[14].

13.3.1 BEP Calculations

To evaluate BEP under different circumstances it is possible to gather a large number of measurements and use these in simulations. But it is also possible to simulate several different impulse responses and use these for simulations of the receiver. The second method is chosen because of its simplicity and lack of enough measurement data.

A computer program was written to generate impulse responses according to the following specifications:

- **Approximate number of paths**
  The arrivals of the paths are assumed to be Poisson distributed [17]. The approximate number of paths is used to compute the average arrival rate of the Poisson distribution. In most simulations, the approximate number of paths is set at 25, which agrees roughly with measurements.
- **Minimum significant amplitude**
  When the power falls below the noise level, the paths are no longer significant. The program stops the generation of paths when the amplitude gets lower than the minimum level. This level can be, for example, the noise level. In the simulation, –40 dB relative to the LOS path is chosen.
- **PDP type**
  Considering the results of [14] and [17], we simulate E-PDP and F-PDP models. E-PDP and F-PDP models are shown in Figures 13.11 and 13.12, respectively.

To compare the two decay functions in the simulations, the total power is normalized at a value of $1 + 1/K$, which equals the total received power of the E-PDP. Reference [14] uses a biconical horn antenna to derive a model of the F-PDP. When an amplitude range of 40 dB is assumed, the average ratio of T_{max} and T_{flat} is about 6.6.

First, computations were done with one PDP and one set of phase values for different bit rates.

From Figure 13.13, the conclusion can be drawn that BEP does not always get worse when the bit rate rises. BEP is dependent on accidentally colliding echoes. Such a diagram does not give us information about the usefulness of a system in practice. This particular PDP may give good results but can never be reproduced. It would be useful to have a set of parameters that describe a channel, or even better, a channel type from which we derive BEP directly.

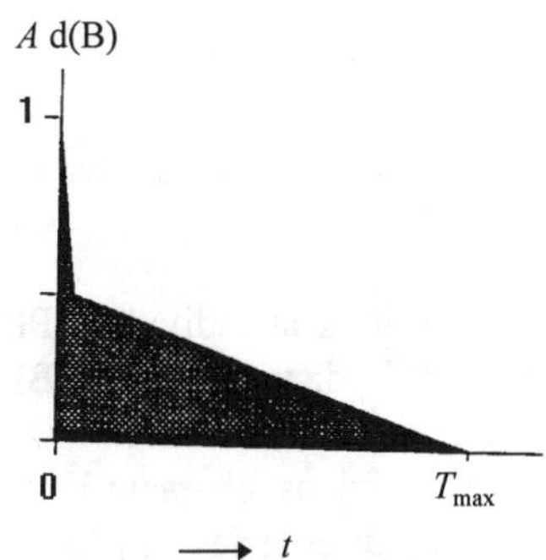

Figure 13.11 Diagram of the model used to simulate the E-PDP.

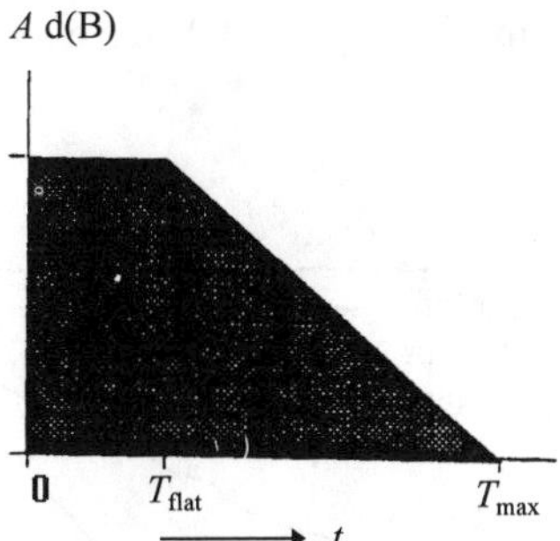

Figure 13.12 Diagram of the model used to simulate the F-PDP.

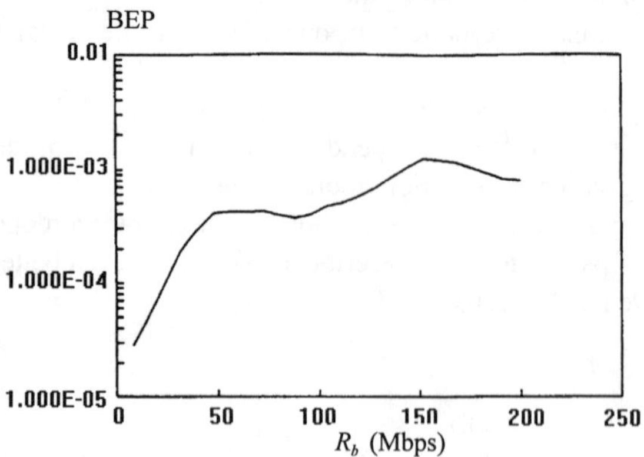

Figure 13.13 Example plot of the average BEP against data rate for a single PDP.

In order to do so we do not look at individual PDPs, but take a more general view by computing cumulative BEP distributions (CBDs) for normalized PDPs at different data rates. In this process we see which normalizing parameters are of relevant influence on BEP and thus are useful as characteristic parameters. The CBDs are computed by taking 100 different random series of phase values for a certain PDP and calculating 100 BEPs. CBD is composed through a histogram-making program.

Figures 13.14 and 13.15 show two CBDs of PDPs with the same K and T_{rms} parameters. We see that although the two PDPs are not identical, significant similarity is present. Because we are dealing with statistical samples, a complete identical CBD is not expected. From a series of experiments, it followed that when the Rice factor and T_{rms} are normalized for a certain PDP type, CBDs have a great resemblance (at least for E-PDP and F-PDP).

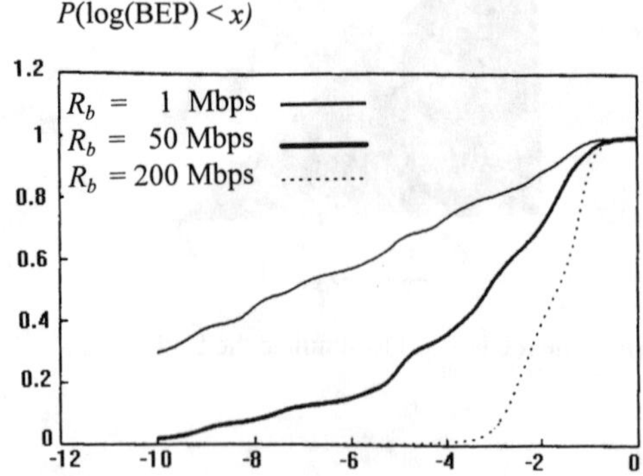

Figure 13.14 Example 1 of CBD for different data rates ($K = 1$ dB, $T_{rms} = 15$ ns, E-PDP).

Because an increase of bit rate equals a stretch in time of the PDP, we define the dimensionless normalized delay spread T_{norm} as

$$T_{\text{norm}} = T_{\text{rms}} R_b \qquad (13.24)$$

A realistic value for T_{rms} is15 ns, which gives a T_{norm} of 0.015 when the bit rate R_b is 1 Mbps.

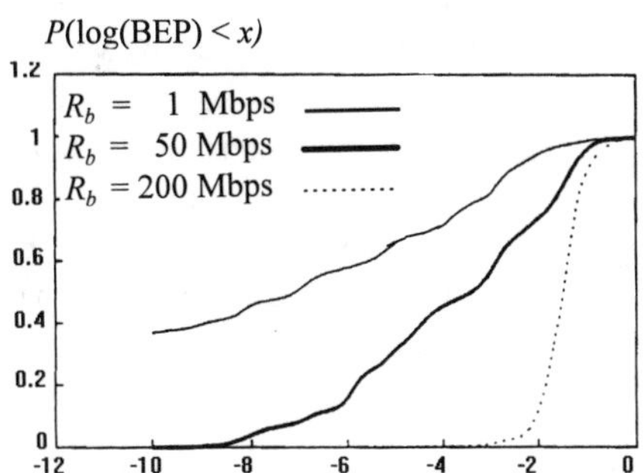

Figure 13.15 Example 2 of CBD for different data rates ($K = 1$ dB, $T_{\text{rms}} = 15$ ns, E-PDP).

13.3.2 The Influence of T_{rms} on BEP

When investigating the influence of T_{rms} on BEP, it is possible to try different values for T_{rms} and compute BEP for one data rate. But when the T_{norm} is used instead of T_{rms}, a generalized result is obtained.

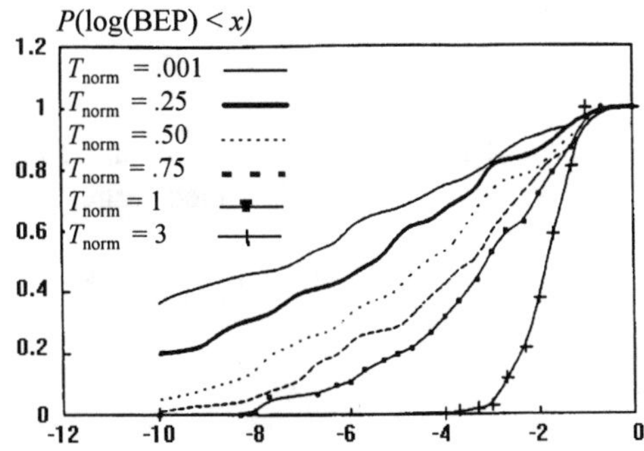

Figure 13.16 CBD for different T_{norm} values ($K = 1$ dB).

When we look at Figure 13.16, we see that when the data rate becomes higher, BEP increases, as expected. Further, we see that, when for instance, the criteria is used where 95% of BEP samples are better than 10^{-2}, a T_{norm} of 0.50 is acceptable. With a T_{rms} of 10 ns, this gives a bit rate of 50 Mbps.

13.3.3 Comparison of PDP Types

In previous sections, two types of PDPs are mentioned: E-PDP and F-PDP. We also stated that the E-PDP type is measured when (semi-) omnidirectional antennas are used. When biconical horns are used, which are more directional sensitive, F-PDP results. The reason to use the latter antenna type is that equal power is received by terminals in a large distance range from a BS. As well, about equal rms delay spreads are measured (15 – 98 ns, depending on the room characteristics [10]). The advantage is that terminals experience equal performance and access chances.

Both PDP types are modeled and used in simulations. For these simulations, the total received power and T_{rms} are equal to achieve a fair comparison. Figures 13.17 and 13.18 show CBDs for both E-PDP and F-PDP type, respectively.

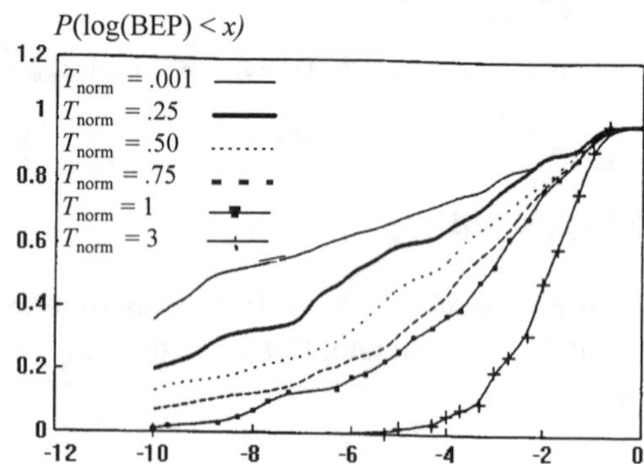

Figure 13.17 CBD of an E-PDP example for different T_{norm} ($K = 1$ dB).

It is clear from Figures 13.17 and 13.18 that performance in BEP is much worse for the F-PDP type. Note that T_{rms} values are typically much larger for F-PDPs than for E-PDPs [14, 17].

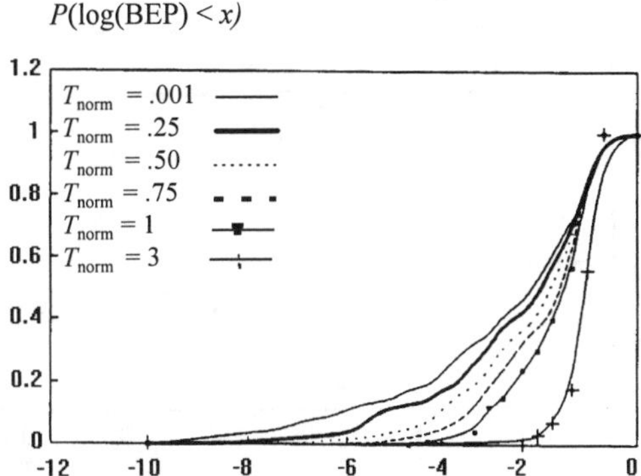

Figure 13.18 CBD of an F-PDP example for different T_{norm}.

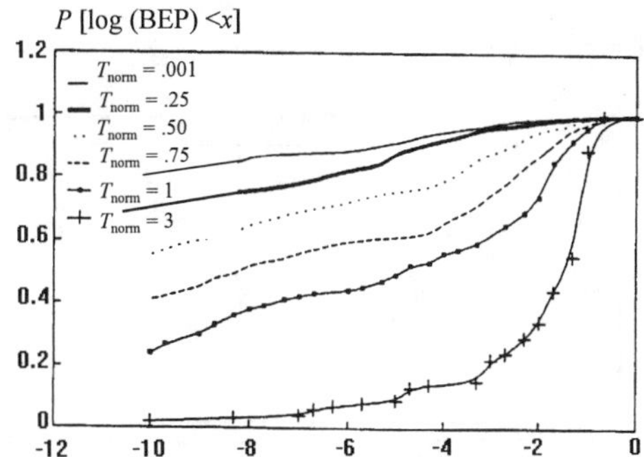

Figure 13.19 **CBD** of an F-PDP example for different T_{norm} with increased SNR (+ 6 dB).

In Figure 13.19, we see that performance improves drastically when the SNR is increased. When the SNR is increased to infinity, we compute the irreducible BEP: errors are only counted when caused by ISI.

In Figures 13.20 and 13.21, the irreducible bit error rate is shown for the E-PDP and F-PDP types. Surprisingly, there is little difference between the two types. F-PDP scores better for $T_{norm} \leq 1$, while E-PDP scores better for $T_{norm} = 3$; the ISI is about equal for the two types.

Apparently, the SNR of the strongest path dictates BEP and CBD. This path is the LOS path in the E-PDPs, but in the F-PDPs is one of the paths that have equal strength. Because power is divided over several of these paths, the strongest path is less strong than in the E-PDP case. Possibly, the fact that no path dictates the carrier recovery increases the loss of power.

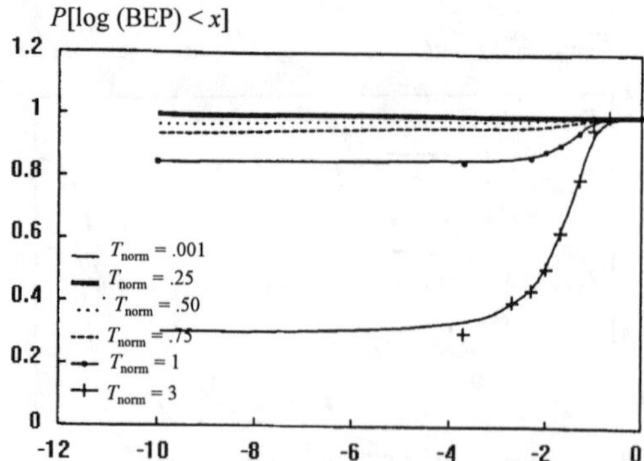

Figure 13.20 Cumulative distribution of irreducible errors of an E-PDP example for different T_{norm} (K= 1 dB, SNR = ∞).

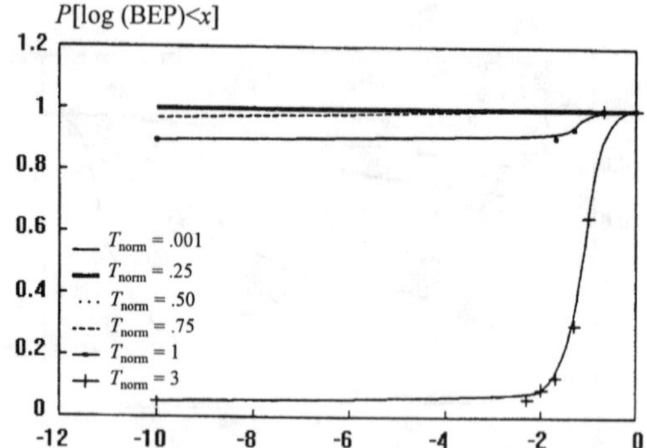

Figure 13.21 Cumulative distribution of irreducible errors of an F-PDP example for different T_{norm} (SNR = ∞).

This conclusion asks for a comparison of received power levels for both antenna types. In the literature [11, 17, 23] power levels are mentioned but are hard to compare due to the different antennas used. Received power levels for (semi-) omnidirectional antennas are higher than those of direction-sensitive antennas. In general, power level is limited due to the health hazard the radiation may pose, and by transmitter power consumption, which pose problems for portable equipment.

13.4 CONCLUSIONS

The following conclusions are drawn, based on the described research on the stationary frequency selective channel. The computed CBDs based on different PDPs with the same Rician factor, normalized delay spread, and SNR are alike. So, the performance of a system in terms of the CBD is computed when these parameters are known.

Irreducible errors of the omnidirectional and direction-sensitive antennas are about equal; the direction-sensitive antenna performs better for normalized delay spreads less than or equal to 1 and worse when it is equal to 3. The performance of the omnidirectional antenna is much better than the direction-sensitive antenna. In the second case, BEP is in 50% of the received signals worse than 10^{-2} for all delay spreads. The reason for the worse performance is that the SNR after demodulation is much lower than in the case of the omnidirectional antenna. In the literature, received power levels of the omnidirectional antenna systems are generally higher. In a single (omnidirectional) antenna system, in 90% of the cases BEP is better than 10^{-2} ($K = 1$ dB, $T_{\text{norm}} = 0.25$).

REFERENCES

[1] A.D. Olver, "Millimeter wave systems — past, present and future," *IEE Proc.*, Vol. 136, Pt. F, pp. 35–52, February 1989.

[2] R. Prasad and L.M. Correia, "An overview of wireless broadband multimedia communications," *Proc. 4th International communications* (*MoMuC'97*), Seoul, Korea, pp. 17–31, September 29 – October 2, 1997.

[3] L. Lake, "Millimetric waveband frequency planning," *IEE Electron. Commun. Eng. J.*, pp. 171–174, August 1991.

[4] R. Prasad, "Overview of wireless personal communications: microwave perspectives," *IEEE Comm. Mag*. Vol. 35, pp. 104–108, April 1997.

[5] R. Prasad, "An overview of millimeter waves for future personal communication system," *Proc. IEEE First Symposium on Communications and Vehicular Technology in the Benelux,* Delft, The Netherlands, pp. K3-1 – K3-8, October 1993.

[6] R. Prasad and L. Vandendorpe, "An overview of millimeter wave indoor wireless communication systems," *Proc. 2nd Int. Conf. on Universal Personal Communication*, Ottawa, Canada, pp. 885–889, October 1993.

[7] R. Prasad and B.C. van Lieshout, "Cochannel interference probability for micro- and pico- cellular systems at 60 GHz," *Electron. Lett.* Vol. 29, pp. 1909–1910, October 1993.

[8] R. Prasad and B. van Lieshout, "Performance of indoor and outdoor radio communication in millimeter wavelength," *Proc. PIMRC'93*, Yohahama, Japan, pp. 20–24, September 1993.

[9] P.F.M. Smulders, "Broadband wireless LANs: a feasibility study," Ph.D. Thesis, Eindhoven, University of Technology, The Netherlands, December 1995.

[10] P.F.M. Smulders and A.G. Wagemans, "Wideband indoor radio propagation measurements at 58 GHz," *Electron. Lett.*, Vol. 28, pp. 1270–1271, June 1992.

[11] P.F.M. Smulders and A.G. Wagemans, "Mm-wave biconical horn antennas for near uniform coverage in indoor picocells," *Electron. Lett.*, Vol. 28, pp. 679–681, March 1992.

[12] P.F.M. Smulders and A.G. Wagemans, "Wideband measurements of mm-wave indoor radio channels," *Proc. 3rd Int. Symp. on Personal, Indoor and Mobile Radio Commun.*, Boston, pp. 329–333, October 1992.

[13] P.F.M. Smulders and A.G. Wagemans, "Wideband measurements of mm-wave indoor radio channels," *Proc. 3rd Int. Symp. on Personal, Indoor and Mobile Radio Commun.*, Boston, pp. 329–333, October 1992.

[14] A.G. Wagemans, "Simulation of an indoor radio channel at mm-wave frequencies," Graduation Thesis, Eindhoven University of Technology, The Netherlands, Telecommunications Division, February 1992.

[15] W. Schafer, "Channel modeling for short range radio links at 60 GHz for mobile intervehicle communications," *Proc. 41st IEEE Veh. Technol. Conf.*, St. Louis, MO, pp. 314–319, May 1991.

[16] A.R. Threk and J.P. McGeehan, "Propagation and bit error measurements within buildings in the mm-wave band about 60 GHz," *IEEE ICC'88*, pp. 318–321, 1988.

[17] M. Bensebti, J.P. McGeehan, and M.A. Beach, "Indoor multipath propagation measurements and characterization at 60 GHz," *Proc. 21st European Microwave Conf.*, Stuttgart, Germany, September 1991.

[18] G. Allen and A. Hammonudeh, "Frequency diversity propagation measurements for an indoor 60 GHz mobile radio link," *Proc. IEE 7th Int. Conf. on Ant. and Prop.*, pp. 298–301, April 1991.

[19] H. Meinel, A. Plattner, and G. Reinhold, "A 60 GHz railway communication system," *IEEE JSAC*, pp. 615–622, September 1983.

[20] H. Meinel and B. Remhold, "Commercial and scientific applications of mm- and sub mm-waves," *Radio Electron. Eng.*, pp. 351–360, July–August 1979.

[21] J.J. Pucell, "Millimeter wavelength in path sources," *Radio Electron. Eng.*, pp. 347–350, July–August 1979.

[22] R. Davies, M. Bensebti, M.A. Beach, J.P. McGeehan, and D.C. Rickard, "Channel sound measurements at 60 GHz using wideband technique with particular reference to microcellular personal communications," *Proc. IEE 5th Int. Conf. in Mobile, Radio & Personal Communications.*

[23] N.D. Hawkins and R. Steel, "Path loss characteristics of 60 GHz transmissions," *Electron. Lett.*, Vol. 21, pp. 1054–1055, October 1985.

[24] A.H. Aghvami, I.D. Robertson, and S.A. Mohamed, "High bit rate indoor radio communications," *Proc. 3rd IEE Conf. on Telecommunications*, Edinburgh, pp. 101–106, 1991.

[25] R. Davies, M. Bensebti, M.A. Beach, and J.P. McGeehand, "Wireless propagation measurements in indoor multipath environment at 1.7 GHz and 60 GHz for small

cell systems," *Proc. 41st IEEE Veh. Technol. Conf.*, St. Louis, MO, pp. 589–593, May 1991.

[26] E.J. Violette, et. al., "Millimeter wave propagation at street in an urban environment," *IEEE Trans. on Geoscience & Remote Sensing*, Vol. 26, pp. 368–380, May 1988.

[27] L. Golding and A. Livne, "RALAN—A radio local area network for voice and data communications," *Proc. IEEE GLOBECOM'87*, pp. 1900–1908, 1987.

[28] S.E. Alexander and P. Pugliese, "Cordless communication within buildings: results of measurements at 900 MHz and 60 GHz," *Br. Telecom. Tech. J.*, Vol.1, pp. 99–105, July 1983.

[29] A.R. Tharek and J.P. McGeehan, "Indoor propagation and bit error rate measurements at 60 GHz using phase-locked oscillation," *Proc. IEEE Veh. Technol. Conf.*, pp.127–133, 1986.

[30] G.A. Kalivas, M. El. Tanany, and S.A. Mahmoud, "Millimeter wave channel measurements for indoor wireless communications," *Proc. IEEE Veh. Technol. Conf.*, pp. 609–612, 1992.

[31] V.R.M. Thyragarajan, H.M. Hafez, and D.D. Falconer, "Traffic study of TDMA-TDD in millimeter wave indoor wireless communications," *Proc. 3rd IEEE Symposium on Personal Indoor and Mobile Radio Communications,* Boston, pp. 566–571, October 1992.

[32] M.A.A. Melters, "Simulation of an indoor radio channel at mm wave frequencies," Graduation Thesis, Eindhoven University of Technology, The Netherlands, August 1990.

[33] S.W. Wales and D.C. Rickard, "Wideband propagation measurements of short range millimetric radio channels," *Electron. Commun. Eng. J.*, Vol. 5, pp. 249–254, August 1993.

[34] R. Prasad and A. Kegel, "Improved assessment of interference limits in cellular radio performance," *IEEE Trans. on Vehicular Technology*, Vol. 40, pp. 412–419, May 1991.

[35] R. Prasad, A. Kegel, and M.B. Loog, "Cochannel interference probability for picocellular system with multiple Rician faded interferers," *Electron. Lett.*, Vol. 28, pp. 2225–2226, November 1992.

[36] V.H. MacDonald, "The cellular concept," *Bell Syst. Tech. J.*, Vol. 58, pp. 15–41, January 1979.

[37] P. Harley, "Short distance measurements at 900 Hz and 1.8 GHz using low antenna heights for microcells," *IEEE J. on Selected Areas in Commun.*, Vol. SAC 7, pp. 5–10, January 1989.

[38] William C.Y. Lee, *Mobile Communications Design Fundamentals*, Indianapolis: Howard W. Sams & Co. 1986.

[39] S.O. Rice, "Properties of a sine wave plus noise," *Bell Syst. Tech. J.*, Vol. 27, pp. 109–157, 1948.

Chapter 14
Dynamic Channel Assignment and DECT

14.1 INTRODUCTION

The future of mobile communications depends on techniques of network planning and mobile radio equipment design that will enable efficient and economical use of the radio spectrum. It is also clear that the need for communication, especially mobile communication, is growing. To handle this need, optimum use should be made of the frequency band that is defined for mobile radio communications. This leads us to use the channels dynamically.

A practical radio system based on dynamic channel assignment algorithms is DECT. DECT is a standard developed by ETSI regarding indoor mobile systems. The standard provides high speech quality and flexibility for a variety of applications. The main attribute of the standard is actually the dynamic channel selection algorithm. Dynamic channel selection causes a great efficiency and flexibility in the system [1–13].

A mathematical model for the performance of dynamic channel selection prescribed by the DECT standard was developed in [1]. Based on this mathematical analyses, this chapter introduces the delay between the moment of gathering information by the portable parts and the moment of using this information in the study of dynamic channel selection. During this delay time, the state of system can change because of mobility of portable parts and birth and death of calls. Therefore, the collected information of one delay time before may not be valid at the moment of a call initiation. The main goal of this chapter is to clarify the effect of this delay on the performance of dynamic channel assignment using DECT. Section 14.2 presents a short description of different channel assignment methods and reviews the DECT concept. Section 14.3 shows the results of simulation and the effect of delay on the performance of DECT. The cause of this effect is also determined. Some interesting conclusions are also outlined at the end of this section. Finally, the conclusions of this study are given in Section 14.4.

14.2 CHANNEL ASSIGNMENT

The rapid development of handheld wireless terminals and the tremendous growth of the wireless/mobile users population, coupled with the bandwidth requirements of multimedia applications, requires efficient reuse of the scarce radio spectrum allocated to wireless/mobile communications. In this chapter, some of the proposals to make more efficient use of radio frequency band are described. The different algorithms namely, dynamic channel assignment, borrowing channel assignment, and hybrid channel assignment are compared with each other in different aspects such as blocking probability and forced call termination.

14.2.1 The Cellular Concept

The cellular concept was presented in Chapter 3; however, it is reviewed in this chapter for the benefit of the readers for a fast understanding. The basic elements of the cellular concept are *frequency reuse* and *cell splitting.* Frequency reuse refers to the use of radio channels on the same carrier frequency to cover different areas separated from each other by sufficient distances so that cochannel interference can be acceptable.

Through frequency reuse, a cellular mobile telephone system in one coverage area can handle a number of simultaneous calls largely exceeding the total number of allocated channel frequencies. The multiplier by which the available spectrum by means of frequency reuse exceeds depends on the number of cells. If the total allocation of C channels is partitioned into N sets, then each set will contain nominally $S = C/N$ channels; however, according to the traffic offered in each cell, some cell can have less channels than S and some other more. Early in the evolution of the cellular concept, system designers recognized that visualizing all calls as having the same shape helps to systematize the design and layout of the cellular systems. A cellular system can be designed with square or equilateral triangular cells, but for economic reasons, system designers adopted the regular hexagonal shape several years ago [2].

In the case of hexagonal cochannel cells to R— the cell radius— is sometimes called the *the cochannel reuse ratio.* This ratio is related to the number of cells per cluster, N, as follows [2]:

$$\frac{D}{R} = \sqrt{3N} \tag{14.1}$$

The minimum number of channel sets N required to serve the entire coverage area is

$$N = \frac{1}{3}\left(\frac{D}{R}\right)^2 \tag{14.2}$$

In a practical system, the choice of the number of cells per cluster is governed by cochannel interference considerations.

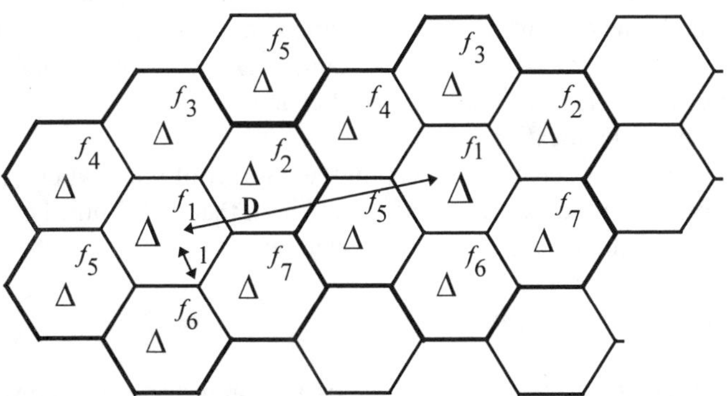

Figure 14.1 Idealized cell structure according to a hexagonal frequency repetition pattern with cluster 77777 of $N = 7$ cells.

In Figure 14.1, the meaning of reuse distance and cluster is illustrated. In reality the cluster, being composed of a group of contiguous hexagonal cells, cannot also be exactly hexagonal in shape. As is shown in Figure 14.1, each cluster is a large hexagonal that assumes to contain 7 small hexagonal cells. D is the distance between the centers of two large hexagons (clusters), thus the distance from f_1 to f_1.

14.2.2 Channel Allocation Schemes

A given radio spectrum or bandwidth is divided into a set of disjoint and noninterfering radio channels. To divide a given radio spectrum into such channels, many techniques such as frequency division, time division, and code division can be used. In frequency division, the spectrum is divided into disjoint frequency bands, whereas in time division, the channel separation is achieved by dividing the usage of the channel into disjoint time periods called time slots. In code division the channel separation is achieved by using different modulation codes. Combinations of these techniques are also possible.

Fixed Channel Assignment

In the fixed channel assignment strategy, a set of nominal channels is permanently allocated to each cell of a cluster for its exclusive use. Cells on the reuse distance use the same channels. Here a definite relationship is assumed between each cell and each channel, in accordance to cochannel reuse constraints [3–5].

In a simple fixed channel assignment strategy, the same number of nominal channels is allocated to each cell. This uniform channel distribution is efficient if the traffic distribution of the system is also uniform. In that case the overall average blocking probability of the mobile system is the same as the call blocking probability in a cell. Because traffic in cellular systems can be nonuniform, a uniform allocation of channels to cells may result in a high blocking in some cells, while the others might have a number of spare channels. This could result in poor channel utilization. It is therefore appropriate to tailor the number of channels in a cell to match the load in it by nonuniform channel allocation.

In a nonuniform channel allocation the number of nominal channels allocated to each cell depends on the expected traffic profile in that cell. Thus, heavily loaded cells are assigned more channels than lightly loaded ones.

Channel Borrowing Strategies

Channel borrowing is one of the strategies to compensate the temporal and spatial fluctuations of the traffic.

In a channel borrowing scheme, an acceptor cell that uses all its nominal channels borrows free channels from its neighboring calls (donors) to accommodate new calls. A channel is borrowed by a cell if the borrowed channel does not interfere with the existing calls. When a channel is borrowed, several other cells are prohibited from using it. This is called *channel locking.* To borrow a channel, there are different schemes, such as:

- *Borrow from the richest, simplest form* (*SBR*). In this scheme, channels that are candidates for borrowing are available channels nominally assigned to one of the adjacent cells has available channels for borrowing, the channel is borrowed from the cell with the largest number of available channels for the borrowing.
- *Basic algorithm* (*BA*). This is an improved version of the SBR strategy that takes channel locking into account when selecting a candidate channel for borrowing. This scheme tries to minimize the future call blocking probability in the cells most affected by the channel borrowing.

- *Basic algorithm with reassignment* (*BAR*). This scheme provides for the transfer of a call from a borrowed channel to a nominal channel whenever a nominal channel becomes available.
- *Borrow first available* (*BFA*). Instead of trying to optimize when borrowing, this algorithm selects the first candidate channel it finds.

In [5], simulation results showed a large variation in complexity of these algorithms depending on network load. For example, for a 20% increase in traffic, SBR requires 50%, and BA 100% more channel tests compared with BFA.

Hybrid Channel Borrowing Schemes

As mentioned earlier, fixed channel assignment performs better in heavy stationary traffic. Hybrid channel borrowing schemes are used to decrease the number of search steps for a borrowing channel and to adapt the channel assignment scheme to the traffic load. Some of the different hybrid channel schemes are described here [6].

- *Simple hybrid channel borrowing* (*SHCB*)
  In an SHCB scheme, the available channels in a cell are divided into two sets, one fixed channel set, and one set that is lent to the neighboring cells.
- *Borrowing with channel ordering* (*BCO*)
  In the BCO scheme, the fixed-to-borrowable channel ratio is dynamically varied according to changing traffic conditions. In the BCO strategy, all nominal channels are ordered such that the first channel has the highest priority for being assigned to the next local call, and the last channel is given the highest priority for being borrowed by the neighboring cells. In [7], the results of simulation of such an algorithm show that BCO performs better than SHCB and FCA.
- *Borrowing with directional channel locking* (*BDCL*)
  In the BCO strategy, a channel is borrowed if it is simultaneously free in three nearby cochannel cells; this requirement is too stringent and decreases the number of channels available for borrowing. In the BDCL strategy, channel locking in cochannel cells is restricted to those directions affected by the borrowing, and locked directions are specified for each locked channel.

Dynamic Channel Assignment

Due to short-term temporal and spatial variations of traffic in cellular systems, FCA schemes are not able to attain high channel efficiency. In contrast to FCA, there is no fixed relationship between channels and cells in dynamic channel assignment (DCA). In DCA, all channels are put in a pool and assigned dynamically to the radio cells as new calls arrive in the system. Because, in general, more than one channel might be available in the central pool to be assigned to a cell that requires a channel, some strategies are applied to select the assigned channel.

Based on information used for channel assignment, DCA strategies are classified either as *call-by-call DCA* or *adaptive DCA* schemes. In a call-by-call DCA, the channel assignment is based only on current channel usage conditions in the service area, while in adaptive DCA, the channel assignment is adaptively carried out using information on previous as well as present channel usage conditions.

DCA schemes are also divided into *centralized* and *distributed* schemes with respect to the type of control they employ.

Centralized DCA Schemes

In centralized DCA strategies, a channel is assigned to a call for temporary use by a centralized control. Some of these schemes are [6–8]:

- *Locally optimized dynamic assignment* (*LODA*). In the LODA strategy, the selected cost function is based on the future blocking probability in the vicinity of the cell in which a call is initiated.
- *Selection with maximum usage on the reuse ring* (*Ring*). In this scheme, a candidate channel is selected which is in use in the most cells in the cochannel set. If more than one channel has this maximum usage, an arbitrary selection among such channels is made to serve the call. If none is available, the selection is made based on the FA strategy.
- *Mean square* (*MSQ*), *nearest neighbor* (*NN*), *nearest neighbor plus one* (*NN*+1). The MSQ scheme selects the available channel that minimizes the mean square of the distance among the cells using the same channel (reuse distance). The NN strategy selects the available channel occupied in the nearest cell in distance $\geq D$ (reuse distance). The NN +1 scheme selects a channel occupied in the nearest cell within distance $\geq D+1$ or distance D if an available channel is not found in distance $D+1$.

Table 14.1 gives a comparison of these DCA schemes [8].

Distributed DCA Schemes

Several analysis and simulation results show that centralized DCA schemes can produce near-optimum channel production, but at the expense of a high centralization overhead. Distributed schemes are therefore more attractive for implementation in microcellular systems, due to the simplicity of the assignment algorithm in each base station or mobile station. In cell-based schemes, a channel is allocated to a call by the station at which the call is initiated. The difference with the centralized approach is that each base station keeps information about the current available channels in its vicinity.

Table 14.1
Comparison of Some Centralized DCA Schemes

Category	Results
Blocking probability	NN, MSQ, FA, NN+1 ⟶ less more
Forced termination rate[1]	NN+1, NN, MSQ, FA ⟶
Channel changing[2]	NN+1, NN, MSQ, FA ⟶
Carried traffic	NN, NN+1, Ring, MSQ, FA ⟶

The channel pattern information is updated by exchanging status information between base stations. Isolated DCA interference adaptation schemes are an approach that relies on signal strength measurements. In these schemes a base station uses only local information, without the need to communicate with any other base station in the network.

Thus, the system is self-organizing, and channels are placed or added everywhere as needed, to increase capacity or to improve radio coverage in a distributed fashion. These schemes allow real-time processing and maximal channel packing at the expense of increased CIP to ongoing calls in adjacent cells, which may lead to undesirable effects such as interruption, deadlock, and instability.

Some of the cell-based distributed DCA schemes are [8]:

---

1 Forced termination rate is a factor that defines how many of the crossing mobile users out of all crossings from one coverage area to another should terminate their calls because of the lack of available channel in the new coverage area.

2 If a mobile crosses a coverage area, it is possible to still use the same channel, or the mobile has to change the occupied channel. The less channel, change the better. If the call quality is not acceptable because of interference, an interruption occurs.

- *Local packing dynamic distributed channel assignment* (*LP-DDCA*).
Each base station assigns channels to calls using the augmented channel occupancy (ACO) matrix. Each base station has its own ACO, which contains necessary and sufficient information about the occupied channels by the base stations within the cochannel interference distance of that channel. For example, let M be the number of channels and k_i the number of neighboring cells to cell i within the cochannel interference distance. The ACO matrix as shown in Figure 14.2 has M columns and k_i +1 rows. An empty column indicates an idle channel assigned to cell i. When a call requests service from cell i, its base station uses the ACO matrix and assigns the first channel with an empty column. The content of the ACO table is updated by collecting channel occupancy information from interfering cells .

	Channel Numbers							
Base Station	1	2	3	5	6	7	...	M
I		X					...	
i_1	X			X			...	
i_2		X					...	
:	:	:	:	:	:	:	:	:
i_{ki}			X		X			

Figure 14.2 ACO matrix at base station I.

- *Moving direction* (*MD*). The MD strategy uses information about movement directions of the mobile units, to decrease both the forced call termination and channel changing. An available channel is selected among those assigned to mobile units that are elsewhere in the service area and are moving in the same direction as the mobile in question. The sets of mobiles moving in the same direction and assigned the same channel are formed. Thus, when a mobile of a set crosses a cell boundary, it is likely that mobiles within the same set already crossed the boundary of its cell to the next cell. In this manner, a mobile uses the same channel after handoff with higher probability. This lowers the probability of both changing channels and forced termination. The strategy is efficient in systems where mobiles move at nearly the same speed through the cells laid along a road or a highway, i.e., for one-dimensional microcellular systems.

Some of the isolated interference adaptation DCA schemes are [8]:

- *Sequential channel search* (*SCS*). In the simplest-case SCSs, all mobile/base station pairs examine channels in the same order and choose the first available with the acceptable CIR.
- *Minimum signal-to-noise interference-ratio* (*MSIR*). In MSIR, a base station searches for the channel with the minimum interference ratio in the uplink direction. MSIR has relatively lower interruption probability[3] than SCS, but more blocking probability. There is a trade-off between the goals of avoiding call blocking and avoiding interruption.
- *Dynamic channel selection.* DCS supposes that mobile stations are able to measure the amount of interference they experience in each channel. In DCS, each mobile station estimates the interference probability and selects the base station that minimizes its value. The interference probability is a function of a number of parameters, such as received signal power from the base station, the availability of channels, and cochannel interference.
- *Channel segregation.* Channel segregation is a self-organized DCA scheme. By scanning all channels, each cell selects a vacant channel with an acceptable cochannel interference level. The scanning order is formed independently for each cell in accordance with the probability of channel selectibility, $P(i)$, which is renewed by learning. The definition of $P(i)$ stands for an algorithm. This definition can be also adaptive to the traffic load. For every channel i in the system, each cell keeps the current value of $P(i)$. When a call request arrives at the base station, the base station with the highest value of $P(i)$ under observation is selected. Subsequently, the received power level of the selected channel is measured to determine whether the channel is used or not. If the measured power level is below (or above) a threshold level, the channel is determined to be idle (or busy). If the channel is idle, the base station starts communication using the channel and its priority is increased. If the channel is busy, the priority of the channel is decreased and the next highest priority channel tried. If all channels are busy, the call is blocked. The value of $P(i)$ and the update mechanism determine the performance of the algorithm.

Table 14.2 gives a comparison of different types of DCA schemes [8].

---

[3] If the quality of call is not acceptable because of interference, it will happen as an interruption.

Table 14.2
Comparison Between DCA Schemes

	Centralized DCA	Cell-based Control Distributed DCA	Measurement-Based Isolated DCA
Advantages	Near optimum channel allocation	Near optimum channel allocation	Suboptimum channel allocation Simple assignment algorithm Use of local information Minimum communication with other base stations Self organized Increase system capacity, efficiency, radio coverage Fast real time processing
Disadvantages	High centralized overhead	Extensive communication with other stations	Adaptive to traffic changes Under heavy traffic: Increased cochannel interference Increased interruption, deadlock probability and instability

A comparison between FCA and DCA is given in Table 14.3 [8].

Hybrid Channel Assignment

Hybrid channel assignment (HCA) schemes are a mixture of DCA and FCA. In HCA, the total number of channels available is divided in fixed and dynamic sets. The fixed set contains a number of nominal channels assigned to cells as in the FCA schemes, and in all cases, are preferred for use in their respective cells. The dynamic set of channels is shared by all users in the system to increase flexibility.

The ratio of dynamic and fixed channels is a significant parameter that defines the performance of the system. It would be interesting to find the optimum ratio in order to achieve better system performance. In general, the ratio of fixed and dynamic channels is a function of the traffic load and would vary over time according to offered load distribution estimations. The simulation results in [3] show that the performance of HCA is better than pure DCA or pure FCA; however, the performance of FCA in stationary heavy traffic load is better than DCA, but the performance of HCA, with reassignment of channels, is better than FCA. There are two reasons for this advantage: (1) HCAs have channels available that can be moved around to serve the normal fluctuations in the offered traffic, and (2) most of the channels are reused at the spaced minimum channel reuse distance by reassigning whenever possible calls using dynamic channels to fixed channels.

Table 14.3

Comparison Between FCA and DCA

FCA	DCA
• Performs better under stationary heavy traffic • Low flexibility in channel assignment • Maximum channel reusability • Sensitive to time and spatial changes • Not stable grade of service per cell in an interference cell group • High forced call termination probability • Suitable for large cell environment	• Performs better under low and moderate traffic • Flexible allocation of channels • Not always maximum channel reusability • Insensitive to time and time spatial changes • Stable grade of service per cell in an interference cell group • Low to moderate forced call termination probability • Suitable in micro-cellular environment
• Radio equipment covers all channels assigned to cell • Independent channel ⇒ control fully centralized to fully distributed • Low computational effort • Low set up delay • Low implementation complexity • Complex, labor intensive frequency planning • Low signaling load	• Radio equipment covers the temporary channels assigned to the cell • Control dependent on the scheme • High computational effort • Moderate to high call setup delay • Moderate to high implementation complexity • No frequency planning • Moderate to high signaling load

The effectiveness of dynamic channel reassignment in [3] illustrates that the blocking does not become significant (greater than 1%) until the traffic offered is over 50% greater than the traffic offered, which produce the same blocking for the FCA. It is also shown that the better-performing of dynamic channel reassignment systems is significant when the traffic offered is about 20% greater than the traffic offered for which the DCA produces the same blocking.

Other Schemes of DCA

Fuzzy Logic-Based DCA

From a decision-making point of view, conventional algorithms are global and crisp decision-makers. Decisions are TO or NOT TO assign a channel for an incoming call is a crisp logic evaluation over a set-up crisp system constraints. Complex logic operations are required to achieve the performance bound, resulting in increased cost and delay. In contrast, fuzzy distributed DCA (FD-DCA) algorithms are soft decision-makers and the decision-making is distributed. Every base station is responsible for making decisions of assigning channels for incoming calls. In [9] fuzzy logic is employed as the decision-making logic, and interference constraints are softened and treated as fuzzy sets characterized by membership functions. There are also two parameters defined: (1) *hotness* is defined as the ratio of offered traffic in a given cell to the total offered traffic in the system, and (2) *usability* is defined as a membership function that depends on the spectral distance between two channels. We also assume that at some special states of system, such as heavy traffic, some of the so-called soft constraints are violated at the expense of, for example, less voice quality.

On the arrival of a call request, the usability and hotness figures are obtained, then the decision to honor or block the call is determined by inferencing of a set of control rules stating the relationship of usability and hotness to the desirability to honor or block the call. The set of control rules used in [9] are:

*R*1: if usability LOW, then BLOCK
*R*2: if usability HIGH, then HONOR
*R*3: if usability MEDIUM AND *hotness* LOW, then BLOCK
*R*4: if usability MEDIUM AND *hotness* HIGH, then HONOR

Two membership functions, LOW and HIGH, are set for the hotness index, and three membership functions, LOW, MEDIUM, and HIGH are set for the usability index. The results of simulation in [9] show that the situation is quite different, especially in heavy traffic, if a softer rule is applied to permit base stations to assign channels more aggressively. The results show that when the traffic increase is large, the softening strategies establish a method to improve throughput performance without increase of computational complexities.

DCA Based on Hopfield Neural Networks

The best software-implemented algorithms for DCA suffer from high computational times that reduce the possibility of a practical implementation. In [10], a DCA based on an energy function whose minimization gives the optimal solution is described. Due to the particular formulation of such an energy function, the minimization can be performed by a Hopfield neural network for which a hardware implementation was recently proposed in the literature.

Two kinds of conditions are defined: (1) *hard condition*, such as simultaneous use of the same channel in two interfering cells; this condition cannot be violated, and (2) *soft condition*, which can be violated at the expense of a slight decrease of the performance of the allocation algorithm. The most important soft conditions in [10] are:

the *packing condition* and the *resonance condition*. With the packing condition, assignment solutions that tend to use the minimum number of channels to satisfy the global channel demand are preferred. The impact of this condition on the assignment is to favor channels already used in other cells without violating any hard condition. If more choices are possible, channels used in the nearest cells are taken into account. With the resonance condition, the algorithm in [10] tends to assign the same channels to the cells that belong to the same reuse scheme, obtained by jumping from one cell to another with steps of length exactly equal to the reuse distance.

The simulation results in [10] show that for an increase of 0% to 140% percent of traffic load in each cell, the performance of DCA based on Hopfield neural networks, because of a massively parallel computational structure, is much better than FCA, BCO, BDCL, and LODA.

14.2.3 DECT as a DCA-Based Mobile Radio System

The DECT standard is a European telecommunication standard (ETS). ETSI designed this standard to provide cordless communications for primarily voice traffic and to provide support for a range of data traffic requirements.

The standard is designed to support this wide range of applications at low cost. The DECT standard is targeted at the following applications:

- Residential-domestic cordless telephones;
- Public access services;
- Cordless business telephones (PBXs);
- Cordless data — local area networks (LANs);
- Evolutionary applications (extensions to existing structures as cellular radio and the local public networks).

Figure 14.3 shows the DECT network reference model.

General Description of the DECT Standard

DECT is based on a microcellular radio communication system that provides low power access at ranges up to a few hundred meters. The basic characteristics are [11]:

Frequency band	:	1880–1900 MHz;
Number of carriers	:	10;
Carrier spacing	:	1.728 MHz;
Peak transmit power	:	250mW ± 1 dB;
Multiplex method	:	TDMA, 24 time slots per frame on each carrier;
Frame length	:	10 ms;

Basic duplexing	:	TDD using 2 slots on the same RF carrier;
Gross bit rate	:	1.152 Mbps per carrier;
Net channel rates	:	32 kbps B-field (traffic) per slot,
		6.4 kbps A-field(control/signaling) per slot;
Modulation method	:	Gaussian minimum shift keying (GMSK).

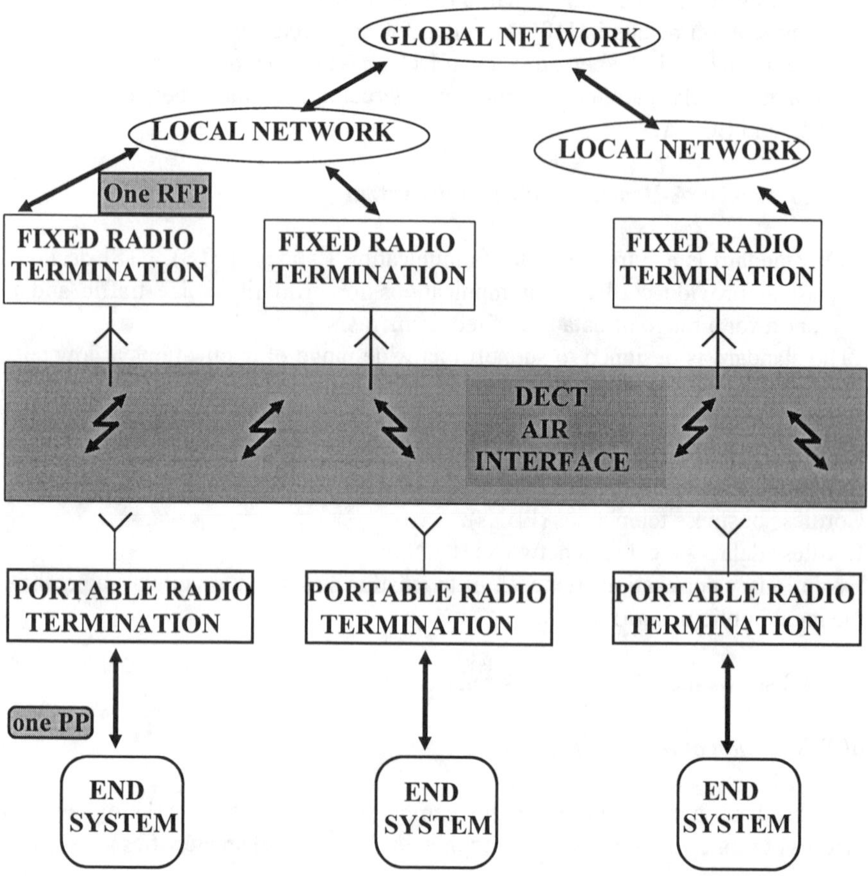

Figure 14.3 DECT network reference model. Here, RFP and PP stand for radio fixed part and portable part, respectively.

The two basic parts of the mobile system are the mobile handsets and the fixed base stations. We use the term portable part (PP) for mobile telephones (handhelds, portables, wireless telephones, etc.) and the term radio fixed parts (RFP) for fixed base stations(see Figure 14.3).

A connection is provided by transmitting bursts of data in the defined time slots. These are used to provide simplex or duplex communications. Duplex operation uses a pair of evenly spaced slots; one for transmitting and one for receiving.

The simplest duplex service uses a single pair of time slots to provide a 32-kbps digital information channel capable of carrying coded speech or other low-rate digital data. Higher data rates are achieved by using more time slots in the TDMA structure, and a lower data rate is achieved by using half-slot data bursts.

The protocols in DECT are designed to support uncoordinated system installation, even where the system coexists in the same physical location. Efficient sharing of the radio spectrum (of the physical channels) is achieved using a mechanism for selection of channels prior to their use: this is called dynamic channel selection.

The procedure is based on measurements and offers a flexible capacity allocation. Physical channels are allocated decentralized, between each PP and the RFP that offers the best quality. The RFPs offer a number of time and frequency multiplexed channels to portable that comes into reach. In DECT, 10 carriers are used, divided each into 24 time slots. Time division is used, so 120 physical duplex channels result.

The RFP has only a single transceiver; it cannot use two channels within the same time slot. Dynamic channel selection thus makes all channels available for use in all cells, but at a given point in time, there are limitations on the use of those channels. The RFPs cannot cope with more than 12 users at a time. In some implementations, the receiver changes the frequency during the neighboring time slots, and in other implementations, the PP cannot see the neighboring time slot. No information on received signal strength is obtained. It does not choose any of the neighboring channels to set up a call.

In addition, the DECT protocols provide two internal mechanisms to support rapid handover of calls in progress: *intercell* and *intracell.*

The intercell handover mechanism is used to switch a call in progress from one cell to another cell, and the intracell handover mechanism is used to switch a call in progress from one physical channel of one cell to another physical channel of the same cell.

These handover mechanisms allow a high quality of service to be maintained where the mobility of the portable part requires transparent re-connection to another fixed part, or when a new physical channel is required in response to a disturbances in the radio environment.

DECT Layered Structure

A structure of four layers is used for the signaling protocols as shown in Figure 14.4.

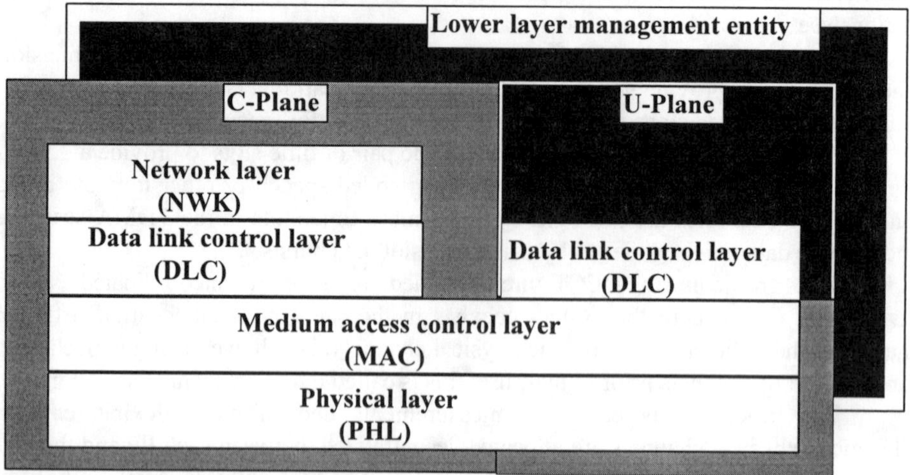

Figure 14.4 The DECT layered structure. Here C-plane and U-plane stand for control plane and user plane, respectively.

Physical Layer (PHL)

The physical layer divides the radio spectrum into physical channels. This division occurs in two fixed dimensions: frequency and time.

The frequency and time divisions uses TDMA operation on multiple RF carriers. Ten carriers are provided in the frequency band of 1.880 to 1.900 MHz. The ETS also provides for possible extensions of the band for the future demand. On each carrier the TDMA structure defines 24 time slots in a 10-ms frame, where each time slot is used to transmit one self-contained packet of data. Each transmitted packet contains a synchronization field, together with control information, service information, and error control.

Each fixed part radio end point operates according to a local timing reference and the physical layer is then responsible for transmitting packets of data under direct control of the MAC layer.

MAC Layer

The MAC layer performs two main functions. First, it selects physical channels and then establishes and releases connections on those channels. Second, it multiplexes and demultiplexes control information, together with higher layer information and error control information, into slot-sized packets. These functions are used to provide three independent services: a broadcast service, a connection-oriented service, and a connectionless service.

The broadcast service is a special DECT feature: it multiplexes a range of broadcast informations into a reserved field (the A-field), and this field appears a part of all transmissions. The broadcast service is always transmitted in every cell (even in the absence of user traffic) on at least on physical channel. This "beacon" transmission allows PPs to quickly identify all RFPs within range, select one, and lock into it without requiring any portable transmission.

DLC Layer

The DLC layer is concerned with the provision of very reliable data links to the network layer. Many of the imperfections of the radio transmissions are already removed by the efforts of the MAC layer. The DLC layer is designed to work closely with the MAC layer to provide higher levels of data integrity than are provided by the MAC layer alone.

The DECT layered model separates into two planes of operation at the DLC layer: the control plane and the user plane.

The C-plane is common to all applications, and provides very reliable links for the transmission of internal control signaling and limited quantities of user information traffic. Full error control is provided with a balanced link access protocol called LAPC.

The U-plane provides a family of alternative services, when each service is optimized to the particular need of a specific type of services. The simplest service is the transparent unprotected service used for speech transmission. Other services support circuit mode and packet mode data transmission, with varying levels of protection.

Network Layer

The network layer is the main signaling layer of the protocol. It operates using an exchange of messages between peer entities. The basic set of messages support the establishment, maintenance, and release of calls. Additional messages support a range of extended capabilities.

The basic call control provides a circuit switched service selected from one of the range of DLC options. Other network layer services are supplementary services, connection oriented messages service, connectionless messages service, and mobility management. These services are arranged as independent entities, and a particular application is realized using more than one.

Lower-Layer Management Entity

The lower-layer management entity contains defined procedures that concern more than one layer. Most of these procedures have only local significance, and they are defined in general terms to allow for alternative implementations.

Power Considerations

The following definitions concerning the power transmission concept of DECT are important to know for the simulation coming in the next chapters.

A reference radio endpoint in DECT is an equipment defined by the standard protocols and corresponding both the PPs and the RFPs in subject to power. This reference endpoint should maintain a transmit power of 24 dBm (=250 mW) ± 1 dB during the transmission of a physical packet.

The DECT standard defines the receiver sensitivity for signal strength measurements between –93 and –33 dBm. Values out of this range are considered as "≤ –93 dBm" or "≥ –33 dBm," respectively. Receiver resolution should be at least 6 dB.

The DECT standard also prescribes a minimum acceptable quality of performance. In the presence of severe interference, the receiver must be below a maximum BEP. To meet this condition, the ratio of desired and undesired signals should be more than 21 dB (a fade margin of 10 dB included).

Channel Selection in DECT

The DECT standard uses an isolated dynamic channel selection scheme based on measurement. This offers a flexible capacity allocation, because all channels are available to all RFPs and PPs. The only consideration for using a channel is an acceptable carrier power and interference power. The minimum acceptable CIR is 21 dB.

The RFPs in DECT offer 10 carriers. Each carrier is divided in 24 time slots. So, 120 duplex channels are available.

The RFP has only a single transceiver, so it uses only one frequency at each time. Because there are only 12 duplex time slots, DECT uses only 12 duplex channels simultaneously. Figure 14.5 shows a schematic of this situation.

14.2.4 The Influence of Delay in Channel Assignment Algorithms

Some different kinds of channel assignment algorithms are discussed above. We have also seen what kind of channel assignment is used in DECT. DCA algorithms are easy; however, research in this area goes back to the decade of 1960 to 1970. There are a lot of problems that make implementation in large-scale radio networks not yet possible.

Some of these problems are fast-frequency switching between different carriers in different time slots, and fast searching to find a proper channel between all the channels of the network and assigning it to an incoming call request. Optimal control on the traffic in the whole network (decentralized or centralized) to have as acceptable quality with the least blocking probability, power considerations, call interrupts, and so forth should also be considered.

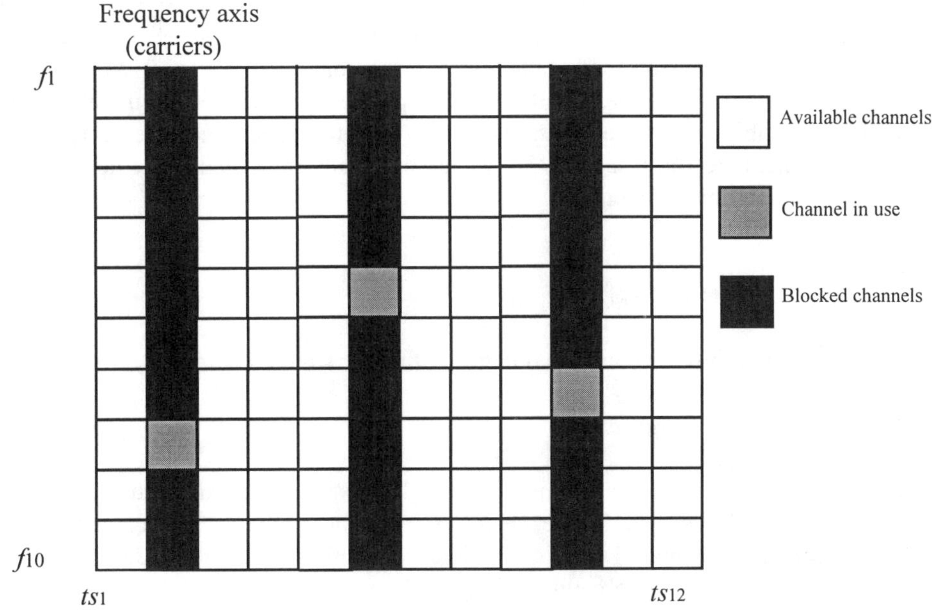

Figure 14.5 The (downlink) channels offered at a base station. Three channels are already occupied.

In this chapter, we also pay attention to another aspect of DCA algorithms, delay.

When a call request is initiated by a PP or when a handover should be done, the DCA algorithm begins to search for a proper channel to answer this demand. This search action costs time. This time duration is called "delay in the DCA algorithm" and is a combination of different smaller delays such as delay because of measurement, delay because of execution of the algorithm to choose the best channel concerning the measurement data, and delay because of internal interconnections and adding the control and signaling information to assign the found channel to the relevant PP.

Due to this delay between the moment of a call request and the moment of assigning a channel to that call request, a lot of changes can occur in the state of network. The amount of these changes depends obviously on the traffic in the network.

There is no clear idea about the exact amount of this delay in DCA algorithms and unfortunately, no article could be found in the literature.

The simulation program is flexible to choose different amounts of delay. It is possible to choose a fixed delay or a variable delay. Using a variable delay is much closer to reality.

There is a look-up table in each PP. When a call request is initiated, the algorithm looks for a proper channel by means of this look-up table. PPs are always busy with measurements and gathering information about the received power on different channels of the nearest RFP. Based on these measurements, the elements of the look-up table are updated each 160 msec. Therefore, some measurements are very new and others are 160 msec old. So, we are especially interested in the delays from 0 to 1 multiframe time (160 msec).

14.3 SIMULATION RESULTS

The following points are considered in generating the simulation results.

A. Area Size and Configuration

The shape of area studied is assumed square. The hexagonal cell size is used. Each RFP is assumed to be placed in the center of each hexagonal cell occupied with an omnidirectional antenna. Diversity is used to combat multipath fading. RFPs are assumed hanging from the ceiling of the building at such a height that the distance between an RFP and any PP carried by a person should always be larger than 1m, where different radio characteristics apply. The signal propagation model becomes simpler while the assumption is quite realistic, keeping in mind an indoor environment.

B. RFPs and PPs

The RFPs use a combined frequency and time multiplex scheme. According to the DECT standard, ten carriers are used, and each one is divided into 24 time slots. That offers 120 full duplex channels. Each RFP is occupied with a single transceiver. Thus, it operates only in one frequency each time slot; therefore, a maximum number of 12 users are served simultaneously.

The level of transmitted power from an RFP, according to DECT specifications, is supposed to be stable during the whole period of the RFPs transmissions and equal to 24 dBm (250 mW). The characteristic also applies to the PPs. However, note that through simulation, only the RFP's transmissions (i.e., only downlink) are of interest.

Each PP has a single transceiver, so it scans only one frequency at a time. The frequencies are scanned randomly with an equal probability.

C. Traffic Generation Model

The traffic is assumed to be voice only. Each RFP handles 8 erlangs of traffic, while every PP generates 0.2 erlangs approximately. A Poisson process with a varying mean value is used to produce this traffic load. The basic concept is to determine the number of call arrivals in the system depending on the number of free (noncommunicating) PPs within the system.

D. Mobility Model

For each call request, a PP with a random position, velocity, and direction is generated. Regarding an indoor mobile system, we assume that a large number of PPs (70%) are not moving. The velocity of the PPs is a sample from a normal distribution with a mean value equal to 0.67 ms and a standard deviation equal to 0.33 ms.

The PPs move toward all the directions of the *X-Y* axes. They are supposed to move and be active only inside the region covered by the DECT communication system. To guarantee that, whenever a PP reaches the outer limits of the building with a direction toward the outer area, it is reflected back inside.

E. Signal Propagation

The model is developed to incorporate all three attenuation phenomena of the radio environment: large-scale path loss, shadowing, and multipath fading.

The received signal power (rsp) resulting from large-scale effects, the "area mean" rsp is used in this model:

$$\text{rsp}(d) = P_t - (\beta + 10\alpha \log d)\ \text{dBm}$$

where: P_t is the transmission power, α and β are the path-loss constants, and d is the distance from the RFP.

The value of α and β vary with the distance from an RFP; this is a way to simulate an indoor radio environment and incorporate features such as walls and floors in the model. Shadow fading is added by distributing the signal power lognormally around the area mean. Multipath fading is involved only with the addition of a 10-dB margin to the threshold of CIR.

F. Simulated System Parameters

The simulated system has a covering area size of 70m by 70m. The distance between each two RFPs is 20m. The traffic intensity per RFP is 8 erlangs. With such a configuration, the system will have 12 RFPs and 575 PPs.

In this section, the simulation results such as the number of available channels at setup, carrier power at setup, interference power at setup, and CIR setup, for different amounts of delay are shown. Attention is paid mostly to the results at setup. It is logical that delay in measurements causes only a call setup (the beginning step) to be unreliable. After a call is initiated, then the only effect of delay is handovers. We found that the delay plays a very significant role in handover handling. We saw that delay affects only the setup results, is no more complete, and therefore, at the end of this chapter the results of simulation of the average carrier power and interference power are shown. Some mathematical calculations also give a clear idea about the origin of the effect of delay.

The state of the system changes during the delay between the moment of measurement and the moment of assignment. This change of state can affect the mobile radio environment. The measured information of a delay time before may not be valid any more.

14.3.1 Channel Scanning Method

The channel scanning method is an important issue for the functioning of the DCS algorithm. Based on the information gathered through this procedure, a PP determines its performance. To set up a call, a PP should always find at least one channel among the ones it can scan. The chosen channel should meet some constraints about CIR to experience a reasonable call quality.

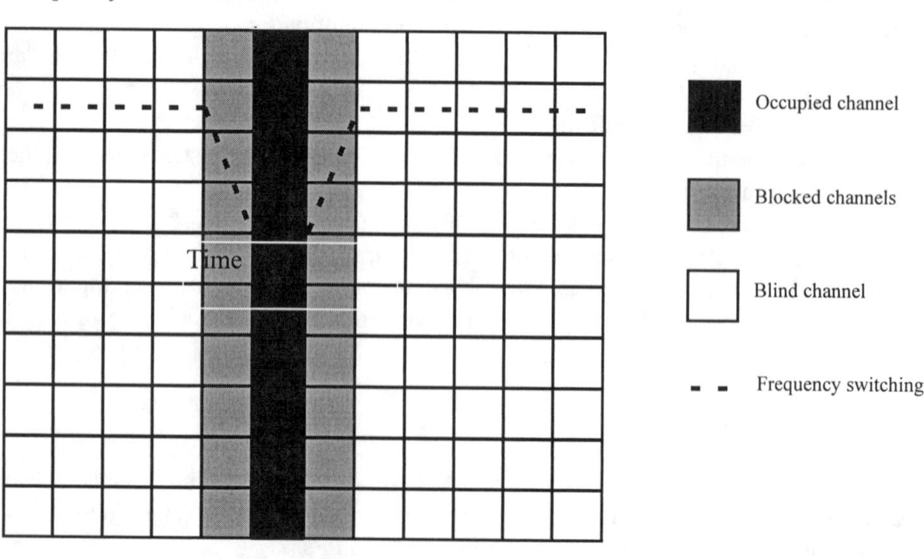

Figure 14.6 Blindness of neighboring time slots.

The PPs scan the channels once every multiframe. There are some limitations that reduce the total number of channels that a PP actually scans. The fact is that the transceiver of each PP is not able to operate in more than one frequency at the same time. Thus, it is impossible for the PP to scan all 120 channels that corresponds to one frame. Only 12 channels can be accessed (the number of time slots per frame) every frame, because only one frequency can be used at a time. Another limitation is that the radio endpoints of a PP cannot switch from scanning frequency to the "locked" or "used" frequency immediately. The time interval for switching is large enough to prevent any scanning of the two neighboring time slots. These are time slots exactly before and after the time slot corresponding to the locked channel. This fact is named "blindness" of the neighboring time slots. Thus, because of technical characteristics the portables are not able to scan more than nine channels plus the used or the locked channel. The lost channels are called blind channels and they cannot be chosen. This is shown in Figure 14.6.

14.3.2 Number of Free Channels at Setup

Considering a system with delay, the channels that are measured for the call request will be occupied after a delay time. Before occupation, these channels that are measured by the PPs for their later use are still available to the new call arrivals during one delay time later. Therefore, the number of free channels at setup for the new call arrivals increases with increasing delay. The PDF of that is shown in Figure 14.7.

It is remarkable to know that a great number of available channels at setup gives no guarantee for a successful call initiation. The number of available channels are only the number of nonoccupied channels concerning channel blindness. Thus, it is not the number of proper channels. Besides, at the moment of channel assignment, the chosen channel may be not good enough for a call initiation.

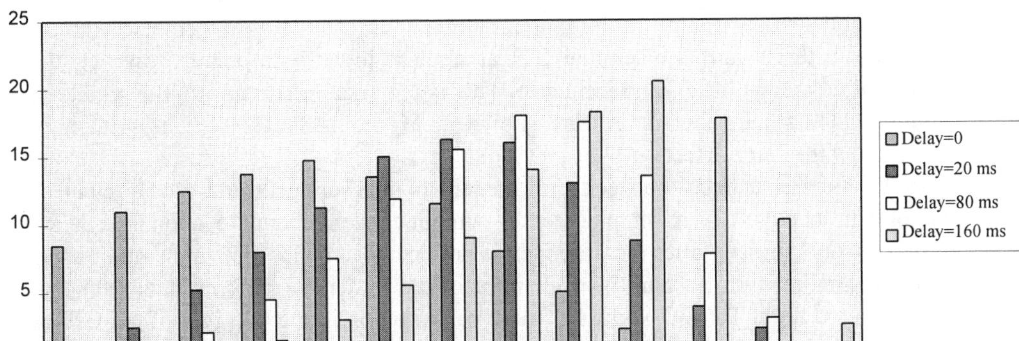

Figure 14.7 The PDF of the number of free channels at setup for different layers.

14.3.3 The Effect of Delay

CIR at setup is a good tool to take a better view on the effect of delay. Because it is the most important value that defines the ability of the beginning of a communication, CIR at setup does not have a prominent effect on the performance of the call in the whole calling time. After a call initiation, a lot of handovers may occur that are not only due to a bad call initiation, but also because of the total changes through the system.CIR for the whole call duration is of course a much better tool to judge the performance of the system.

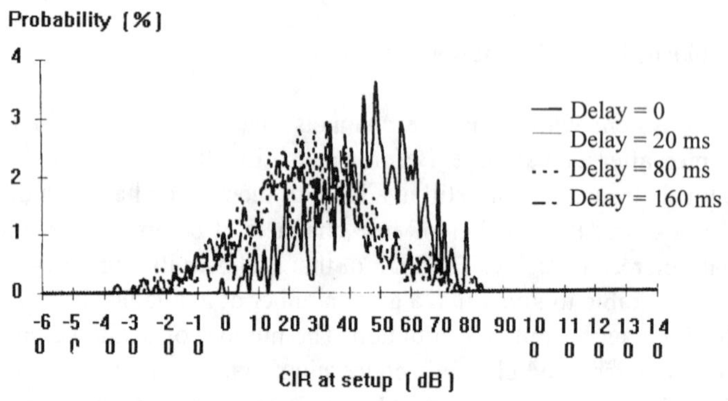

Figure 14.8 Carrier-to-interference at setup for delays (0–160 ms) as a parameter.

Figure 14.8 shows the PDF of carrier to interference at setup for different amounts of delay.

From this figure, we see that CIR at setup is lower with delay. However, the amount of delay has little or no influence. This is remarkable. Therefore, the results of simulation for some other delays are shown in Figure 14.9. This results with high delays show the same effect.

If there is a delay between the moment of measurement and the moment of assignment, then the CIR at setup decreases without concern about the amount of delay. Or, if a channel directly after measurement is not occupied and is free to the other users until assignment, the probability of changes in the system and its influence on that channel is too high. To find which has more participation in the decrease of the CIR at setup, the carrier or the interference, the carrier at setup and interference at setup should be checked separately.

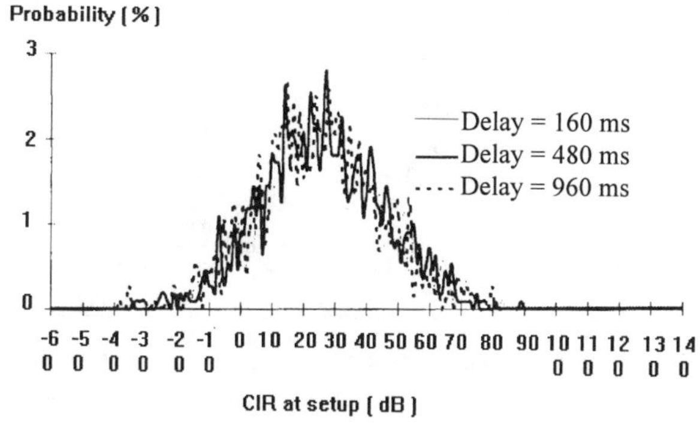

Figure 14.9 CIR at setup for some delays (160–960 ms) as a parameter.

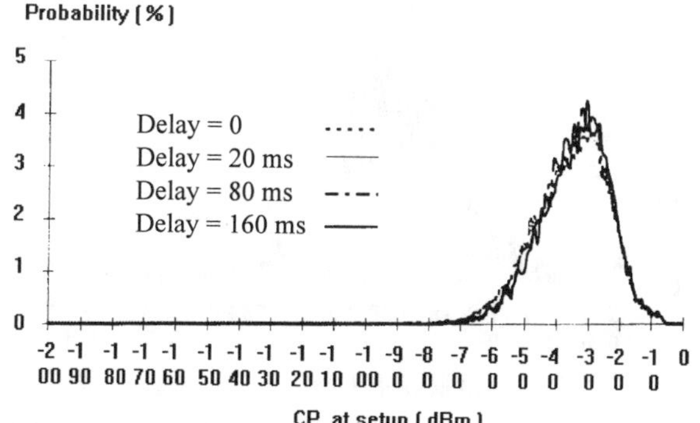

Figure 14.10 Carrier power at setup for different delays.

Figure 14.10 shows carrier at setup and Figure 14.11 shows interference at setup.

Figure 14.10 shows that delay has no remarkable effect on carrier power at setup, because carrier power depends only on the distance between the PP and the RFP. It means that during delay times, the position of a PP does not change so much that it cannot influence the carrier power. This is expected, keeping in mind a velocity (2σ) of 1.33 ms. To prove this hypothesis, setting the velocity of the PPs to zero shows that mobility of the PPs is not the reason of decreasing of CIR at setup. Figure 14.12 shows CIR at setup for the simulated mobile radio system with nonmoving PPs.

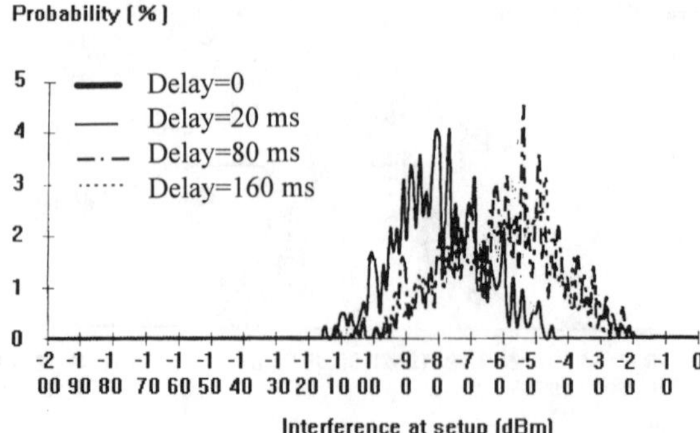

Figure 14.11 Interference power at setup for different delays.

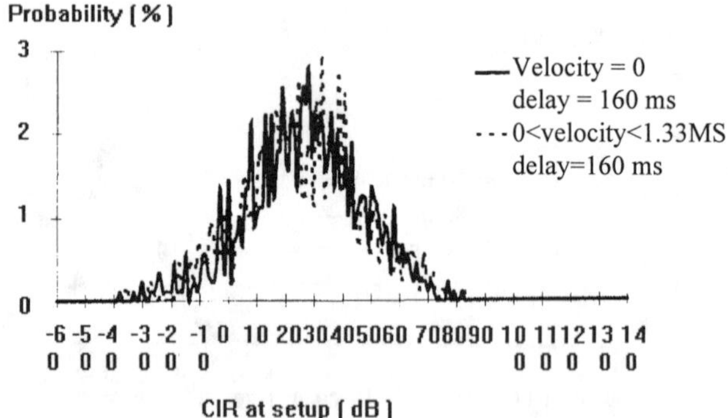

Figure 14.12 Carrier to interference at setup for moving and nonmoving PPs.

So, we are already sure that mobility of the PPs has no effect on decreasing of CIR with delay. As shown in Figure 14.11, the reason of decreasing of CIR at setup is actually the increasing of interference during delay between measurement and occupation of a channel. Now the question is, what is the cause of increasing interference? Is that the rate of birth and death of calls, handovers, or something else?

To find the reason, some information about number of call setups, number of call installations, number of handovers, and so forth is gathered in Table 14.4.

Table 14.4

Results of Simulation for Different Delays

Delay in measurement (ms)	0	80	160
Number of PPs	575	575	575
Number of RFPs	12	12	12
Area length (m)	70	70	70
Cell diameter (m)	20	20	20
Real time of simulation (hours)	5	5	5
Average received power at setup is (dBm)	–35.420	–34.758	–34.110
Average number of free channels	40.454	72.845	81.780
Blocking probability	0.000%	0.000%	0.000%
Grade of service	0.000%	28.152%	41.596%
Average of attempts for call setup	1.000	1.029	1.043
Average of attempts for handover	1.000	1.000	1.000
Probability of communication without interference	98%	65%	59%
Number of call interrupted during setup	0	318	394
Number of bad connections	0	262	323
Call installations	17279	10978	9078
Calls terminated successfully	17183	10945	9048
Intercell handovers per call	14.770	11.073	9.552
Forced intercell handovers per call	0.017	0.033	0.016
Intracell handovers per call	0.141	9.742	11.682
Handover rate (handovers per call)	14.928	20.847	21.250
Average of users per channel	0.7955	0.5055	0.4165
Probability that a channel is free	21%	59%	68%

Some definitions in Table 14.4 should be described:

- *Grade of service.* A tool to determine the performance of the system and is calculated as follows
  GoS = {number of blocked calls + 10 × (number of interrupted calls)}/{number of call attempts}
  It is clear that a larger GoS stands for a worse efficiency.
- *Number of calls interrupted during setup.* This is the number of calls that try to initiate with the same channel from the same RFP. Because of delay between measurement and assignment moments, it is possible that a channel measured by a PP is also chosen by another PP.
- *Number of bad connections.* This is the number of call initiation (channel assignments) with a CIR < 21 dB. During measurement, the CIR should be greater or equal to 21 dB for choosing a channel. However, because of delay, the state of the system can change and has influence on the measured information and

especially on the interference. Therefore, it is possible to have a CIR less than 21 dB at the moment of assignment.

Based on the data from Table 14.4, we can do some calculations:
During 5-hour simulation, we have:

$$\text{Number of call installations per second} = \frac{\text{number of call installations}}{\text{simulation time}}$$

0.96 (delay = 0), 0.61 (delay = 80 ms), 0.50 (delay = 160 ms).

It is very interesting that in spite of increasing the number of available channels shown in Figure 14.7, the number of successful call setups is decreasing. It means that there are a lot of free channels in the system that cannot be used by the users and that decreases the efficiency of the whole system. As shown in the last row of Table 14.4, the probability that a channel is free increases with increasing delay. Being free of a channel does not mean that it is a good channel for a call initiation. Average attempts for a call setup increases with increasing delay. Thus for a call setup, a PP tries more and more and it increases the load on the system.

Until now, we saw the effect of delay on some of the parameters of the DECT system. In the next part, we try to clarify the origin of the effect.

14.3.4 Handovers

There are two causes to make a handover in the simulated DECT system:

1. The received CIR is less than the acceptable CIR (CIR threshold = 21 dB);
2. The received CIR is enough but there is an RFP that offers a carrier power 10 dBm more than the currently used RFP and this RFP has a free channel available that offers a CIR ≥ 21 dB.

Using Table 14.4, the rate of these handovers are calculated. Cause one is related to the number of intracell[4] handovers per call and forced[5] handovers per call. Cause two is related to the number of intercell[6] handovers.

---

[4] If CIR < 21 dB and PP finds a channel with a CIR ≥ 21dB from the same RFP to which it is connected.
[5] If CIR <21 dB and PP finds a channel with a CIR ≥ 21 dB from another RFP than the one to which it is connected.
[6] If CIR ≥ 21 dB, but PP finds a RFP that offers a carrier power 10 dB more than the one to which it is connected and that RFP has a free channel available with CIR ≥ 21 dB.

Number of intracell handovers per second =

$$\frac{\text{(intracell handovers per call)} \times \text{(number of call installations)}}{\text{simulation time}}$$

The same formula is used for the calculation of the other kinds of handovers. The results are gathered in Table 14.5.

Table 14.5

Results of Handover Calculations

Delay (ms)	0	80	160
Number of intracell handovers per second	0.135	5.94	5.89
Number of forced intercell handovers per second	0.016	0.020	0.008
Number of intercell handovers per second	14.18	6.75	4.82
Number of total handovers per second	14.33	12.71	10.71

The rate of handover is much higher than the rate of call setups. The rate of call setups is maximum 0.96 per second for delay = 0. Comparing the number of call arrivals and the number of handovers, we conclude that the large number of handovers causes the state of system change during delay.

Forced intercell and intracell handovers should be done to have a CIR threshold above 21 dB. We can do nothing about these kind of handovers. But unforced intercell handovers are done to allow best performance in spite of having an acceptable CIR. In this policy, we make some changes.

As mentioned above in handover cause two, intercell handovers are done when a PP finds an RFP that offers a carrier power of 10 dB more than the RFP to which it is connected. The difference between these carrier powers is called "carrier-diff." The participation of this kind of handovers in the total rate of handovers is high. Enlarging the value of 10 dB (carrier-diff) produces the graphs in Figure 14.13.

Figure 14.13 shows that with increasing carrier-diff, the CIR with delay will approximate the CIR without delay (delay = 0). In Figures 14.6 to 14.12, the value of carrier-diff is always 10 dB according to the DECT standard. The reason why the CIR with a delay of 160 ms approximates the CIR with delay = 0 is actually the decrease of the number of handovers due to enlarging the carrier threshold. The number of handovers in this case is shown in Table 14.6.

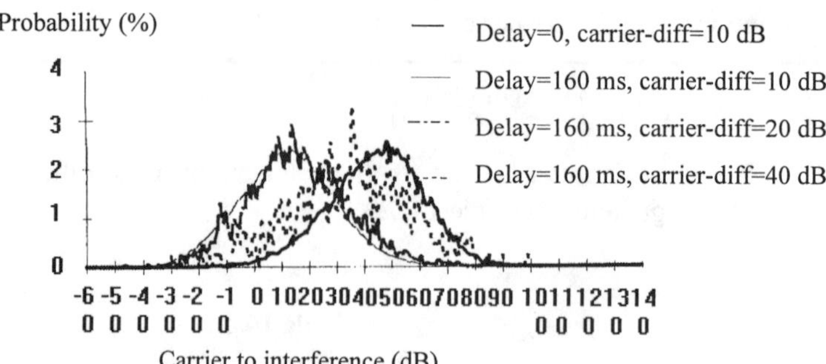

Figure 14.13 PDF of CIR of different amount of carrier differences for intercell handovers.

It is remarkable to see that decreasing the number of intercell handovers by changing the carrier-diff also influences the other handovers: they also decrease.

These are very interesting results, because the performance of the system and its efficiency is much better in the case of not doing intercell handover. It compensates for the bad effect of delay.

Considering Table 14.6, the number of call installations and the average number of users per channel increases with increasing carrier-diff for intercell handovers. Besides, the probability that a channel is completely free and the number of call interruptions during call setup decreases. These are all indications of increasing efficiency of the system.

Table 14.6

Handover Rate for Different Amounts of Carrier Threshold (Delay = 160 ms)

Carrier difference (dB)	10 (DECT Standard)	20	40
Intercell handover per call	9.552	3.994	0.164
Intracell handover per call	11.682	5.806	0.006
Forced intercell handover per call	0.016	0.020	0.371
Handover rate per call	11.682	9.771	0.542
Call installations	9078	11556	15145
Calls terminated successfully	9048	11515	15068
Number of calls interrupted during setup	394	300	118
Probability that a channel is free	68%	53%	31%
Average number of users per channel	0.4165	0.5338	0.7008

CONCLUSIONS

The goal of the simulation model presented in this chapter was the investigation of the effect of delay between measurement's moment and assignment's moment on the performance of DCA algorithms. DECT is the only practical radio indoor system based on an isolated DCA algorithm.

After finding the effect of delay (reaching the main goal), we were curious to find the reason for those effects; in other words, which aspects of the DECT standard is more sensible to the delay. Since that moment, the investigation got more and more interesting and we got some unexpected and elegant conclusions.

As a general result, delay damages the performance of a DCA algorithm. This is expected and proved through the simulation. At first glance at the results, some contradictions appeared. For example, the results such as:

1. *The number of available channels increases with increasing delay*;
2. *The number of call installations decreases with increasing delay.*

However, the higher number of free channels guarantee no extra proper channel availability. The number of available channels increases because they cannot be used. This fact is also confirmed by the facts :

1. *The number of channels that are completely free increases with increasing delay;*
2. *The average number of users per channel decreases with increasing delay.*

The results shown in Section 14.3, and finding the effect of delay appear quite satisfactory. The expectations that we had in the beginning was that delay should have a bad effect on the performance of DECT. The reason was searched for mostly in the mobility of the PPs during delay. It is now clear that for an indoor system, the delay has a bad effect but the cause is not the mobility nor the birth and death of the calls during delay.

It is remarkable that we can compensate the bad effect of delay by means of doing handover if and only if it is really necessary (see the different kinds of handover in Section 14.3.4). As shown in the last part of Section 14.3, not doing any unnecessary handovers makes the efficiency of the system much better. In this case, having a better CIR improves the performance of the system.

REFERENCES

[1] J.B. Punt, D. Sparreboom, F. Brouwer, and R. Prasad, "Mathematical analysis of dynamic channel selection in indoor mobile wireless communication system," accepted for publication in *IEEE Trans. on Vehicular Technology*.

[2] V.H. Macdonald, "The cellular concept," *Bell Syst. Tech. J.*, Vol. 58, No.1, pp.15–41, January 1979.

[3] D.C. Cox, "Increasing channel occupancy in large scale mobile radio systems dynamic channel reassignment," *IEEE Trans. on Vehicular Technology.*, Vol. VT-22, No.4, pp. 218–222, November 1973.

[4] W.C. Jakes, *Microwave Mobile Communication*, John Wiley & Sons, 1974.

[5] L. G. Anderson, "A simulation study of dynamic channel assignment in a high capacity mobile telecommunication systems," *IEEE Trans. on Vehicular Technology* Vol. VT-22 No. 4, pp. 210–217, November 1973.

[6] M. Zang, "Comparisons of channel assignment strategies in cellular mobile telephone systems," *IEEE Trans. on Vehicular Technology*, Vol. 38, No.4, pp.211–215, November 1989.

[7] S.M. Elnoubi, "A new frequency channel assignment algorithm in high capacity mobile communication systems," *IEEE Trans. on Vehicular Technology*, Vol. VT-31, No. 3, pp. 125–131, August 1982.

[8] I. Katzela and M. Naghshineh, "Channel assignment schemes for cellular mobile telecommunication systems: a comprehensive survey," *IEEE Personal Comm.*, pp. 10–31, June 1996.

[9] M. Abdul-Haleem, "Fuzzy logic based dynamic channel assignment," *Proc. ICCS*, Singapore, pp. 773–777, 1994.

[10] E. Del Re, "A dynamic channel allocation technique based on Hopfield neural networks," *IEEE Trans. on Vehicular Technology*, Vol. VT-45, No. 1, pp. 26–31, February 1996.

[11] ETSI Technical Report, "Radio equipment and systems digital European cordless telecommunications (DECT) reference document," *ETROIS*, March 1991.

[12] B. Bout, D. Sparreboom, F. Brouwer, and R. Prasad, "A mathematical model for dynamic channel selection in conformity with the DECT standards," *Proc. PIMRC'93*, Yokohama, Japan, pp. 547–551, September 1993.

[13] C.-L.I, "Distributed dynamic channel allocation algorithms in microcells under light traffic loading," *Wireless Personal Communications, An Int. J.*, Vol. 1, No. 3, pp. 155–163, 1995.

Chapter 15
Air Interface Multiple Access Schemes for FPLMTS/IMT-2000/UMTS

15.1 INTRODUCTION

Since the work started in the standardization bodies ITU TG8/1 committee for FPLMTS/IMT-2000 in 1986 and SMG5 subtechnical committee for UMTS in 1991, the third-generation activities in the field of wireless personal communications have formed an umbrella for advanced radio system development. In recent years, standardization activities for FPLMTS/IMT-2000/UMTS have accelerated towards concrete specifications. According to present plans, after the selection of UMTS Radio Access Concept in January 1998 by the main system parameters will be frozen by the end of 1998. Moreover, the FPLMTS/IMT-2000 radio transmission techniques evaluation process has started in ITU. Submissions of candidate technologies are expected by October 1998 at the latest and ITU selection process will be completed by early 1999.

To support the standardization activities, several research programs throughout the world have developed third-generation air interface concepts. In addition, laboratory and field trials have been performed. In this chapter, we focus on those air interface research activities that currently are considered as candidates for UMTS and FPLMTS/IMT-2000 air interface standards [1– 61].

IMT-2000 Study Committee in Association of Radio Industries and Business (ARIB) was established in April 1993 to coordinate the Japanese research and development activities for the FPLMTS system. In October 1994, it established the

Radio Transmission Special Group for radio transmission technical studies and production of draft specification for FPLMTS [16]. The Special Group consists of two ad hoc groups: CDMA and TDMA [17]. Originally, 13 different wideband CDMA/frequency division duplex (FDD) radio interfaces were presented to the FPLMTS Study Committee. In the beginning of 1995, they were merged into three CDMA/FDD proposals, Core A, B, and C, and into one TDD proposal. At the end of 1996, the four schemes were further combined into a single proposal where the main parameters are from Core A. In the following we focus on the Core A proposal, which is presented in more detail in [18–22]. For TDMA, there were seven proposals for pedestrian environment, six for office environment, and another six for vehicular environment. From these proposals the group compiled a single carrier TDMA system, multimode and multimedia TDMA. Since one of the TDMA proposals was based on multicarrier OFDM technology band division multiple access (BDMA) proposal, it was decided to continue studies of this technology [16].

In Europe, the Research of Advanced Communication Technologies in Europe (RACE I) program, which was launched in 1988 and lasted until June 1992, started the third-generation research activities. Between 1992 and 1995, in the RACE II program, code division multiple testbed (CODIT) and advanced TDMA (ATDMA) projects developed air interface specifications and testbeds for UMTS radio access [1– 4]. In addition, several industry projects developed air interface concepts and trial systems for UMTS [5–9]. European Research Program ACTS was launched at the end of 1995 to support collaborative mobile research and development. Within ACTS, the project FRAMES was set up with an objective to define a proposal for the UMTS radio access system. The first target of FRAMES was to investigate hybrid multiple access technologies, and based on thorough evaluation of several candidate schemes, select the best combination as a basis for further detailed development of the UMTS radio access system. The comparison results are presented in [10–12]. Based on this comprehensive evaluation, a harmonized FRAMES multiple access (FMA) platform was designed consisting of two modes: FMA1 [13], a wideband TDMA with and without spreading, and FMA2 [14], a wideband CDMA [15]. Both FRAMES multiple access schemes can be interfaced with the GSM core network, offering a future proof system platform for UMTS/FPLMTS/IMT-2000.

The UMTS air interface selection process in ETSI was started in 1997 by grouping the submitted air interface concepts into five different concept groups. FMA2 and three wideband CDMA schemes from Japan were submitted into the Alpha group. The Beta group evaluated FMA1 spreading together with some other TDMA ideas. Two OFDM concepts were submitted into the Gamma group. The Delta group evaluated FMA1 with spreading, i.e., the TD-CDMA scheme. In the beginning another hybrid scheme, CTDMA, was also considered in the Delta group. The Epsilon group considered opportunity driven multiple access (ODMA), which is a relay technology in principle applicable to all multiple access schemes.

In the concept group evaluation process, the FMA2 proposal was considered together with the wideband CDMA schemes from Japan. At the same time, international coordination group (ICG) in ARIB had discussions on the harmonization of different CDMA proposals. These two efforts led to the harmonization of the parameters for ETSI and ARIB wideband CDMA schemes. The main parameters of the current scheme are based in the uplink on the FMA2 scheme and in the downlink on the ARIB wideband CDMA. Also, contributions from other proposals and parties have been incorporated to further enhance the concept. In January 1998, ETSI SMG agreed to base the UMTS FDD component on the harmonized WCDMA scheme and UMTS TDD component on TD/CDMA.

In the US, the Telecommunication Industry Association (TIA) technical committee TR45 is responsible for the standardization mobile and personal communications systems. During 1997, several wideband CDMA proposals were submitted to TR45.5 for the development of wideband cdmaOne scheme. A common characteristic of all of these schemes is backward compatibility to IS-95. In December 1997, TR45.5 agreed upon the basic framework for the Wideband cdmaOne.

In the beginning of 1997, within UWCC, global TDMA forum (GTF) established the high speed data (HSD) group to evaluate air interface candidates for the IS-136 evolution towards third-generation. Several schemes including TDMA, wideband CDMA and two OFDM schemes, were submitted to the HSD group. Based on these proposals, UWCC developed the UWC-136 concepts which were accepted by TR45.3 in February 1998. UWC-136 consists of an enhanced IS-136 30 kHz carrier, 200 kHz HSD carrier and 1.6 MHz wideband TDMA carrier for an indoor radio environment. The 200 kHz HSD carrier has the same parameters as the enhanced GSM carrier (EDGE-Enhanced Data Rates for GSM Evolution), currently a work item in the GSM phase 2 + standardization. The wideband TDMA is based on the FRAMES FMA1 without spreading.

In Korea, two wideband CDMA schemes have been developed and submitted to the Telecommunication Technology Association (TTA) for standardization. The Electronics and Telecommunications Research Institute (ETRI) has developed a wideband CDMA scheme (TTA I) whose parameters are similar to wideband cdmaOne. The development of a network asynchronous wideband CDMA scheme was started in 1994. This scheme has formed the basis for the TTA II wideband CDMA scheme.

In addition to the air interface concept developments described above, an enormous amount of research has been devoted to different component technologies for third-generation systems both in industry and in universities. However, the work is continuing towards detailed standards and thus the different concepts and parameters presented in this chapter are still subject to change.

The third-generation air interface standardization for schemes based on CDMA seems to focus on two main types of wideband CDMA: network asynchronous and synchronous. In network asynchronous schemes the base stations are not synchronized while in network synchronous schemes the base stations are synchronized to each other within few microseconds. As discussed, there are three network asynchronous CDMA proposals: WCDMA in ETSI, and ARIB and TTA II wideband CDMA in Korea have almost similar parameters. A network synchronous wideband CDMA scheme has been proposed by TR45.5 (Wideband cdmaOne) and is being considered by Korea (TTA I). In the following, we discuss the features of WCDMA and Wideband cdmaOne. The main features of WCDMA and Wideband cdmaOne are presented in Table 15.1.

The main differences of WCDMA and Wideband cdmaOne systems are chip rate, downlink channel structure, and network synchronization. Wideband cdmaOne uses a chip rate of 3.6864 Mcps for the 5 MHz band allocation with the direct spread downlink and a 1.2288 Mcps chip rate for the multicarrier down link. WCDMA uses direct spread with a chip rate of 4.096 Mcps. The multicarrier approach is motivated by a spectrum overlay of Wideband cdmaOne with existing IS-95 carriers. Similar to IS-95B, the spreading codes of Wideband cdmaOne are generated using different phase shifts of the same *M* sequence. This is possible due to the synchronous network operation. Since WCDMA has asynchronous network, different long codes rather than different phase shifts of the same code are used for the cell and user separation. The code structure further impacts how code synchronization, cell acquisition, and handover synchronization are performed.

Compared to the second generation CDMA, the following new capabilities characterize third generation wideband CDMA:

- Wider bandwidth and chip rate;
- Provision of multirate services;
- Packet data;
- Complex spreading;
- A coherent uplink using a user dedicated pilot;
- Additional pilot channel in the downlink for beamforming;
- Seamless interfrequency handover;
- Fast power control in the downlink;
- Optional multiuser detection.

As a compromise in the ETSI air interface selection, the TD-CDMA was adopted as the TDD solution. The role of CDMA and TD-CDMA is to multiplex the different channels within a time slot. The spreading ratio is small and thus, if more than few users are desired per frame, joint detection is needed to remove the intracell interference. Therefore, in a fully loaded system the spreading does not help against the interference from other cells. Also the slow power control results in large variations in the received signal levels and thus joint detection is needed. Since the joint detection is a mandatory feature, it is more critical compared with wideband CDMA. However, if the number of users is small, the complexity of joint detection may not be excessive.

This chapter is organized as follows. Wideband CDMA-based radio interface is discussed in Section 15.2. Sections 15.3 and 15.4 assess TDMA and hybrid schemes, respectively, and Section 15.5 presents OFDM-based air interface schemes. Section 15.6 covers TDD concepts and related problems. Conclusions are presented in Section 15.7. Before discussing different air interference, multiple access schemes and spectrum issues are introduced in Section 15.1.1.

Table 15.1

Parameters of WCDMA and Wideband cdmaOne[1]

	WCDMA	Wideband cdmaOne
Channel Bandwidth	5, 10, 20 MHz	1.25, 5, 10, 15, 20 MHz
Downlink RF channel structure	Direct spread	Direct spread or multicarrier
Chip rate	4.096/8.192/16.384 Mcps	1.2288/3.6864/7.3728/11.0593/ 14.7456 Mcps for direct spread $n \times 1.2288$ Mcps (n = 1,3,6,9,12) for multicarrier
Roll-off factor	0.22	Similar to IS-95
Frame length	10 ms/ 20 ms (optional)	20 ms for data and control/5 ms for control information on the fundamental channel
Spreading modulation	Balanced QPSK (downlink) Dual channel QPSK (uplink) Complex spreading circuit	Balanced QPSK (downlink) Dual channel QPSK (uplink) Complex spreading circuit
Data modulation	QPSK (downlink) BPSK (uplink)	QPSK (downlink) BPSK (uplink)
Coherent detection	User dedicated time multiplexed pilot (downlink and uplink), no common pilot in downlink	Pilot time multiplexed with PC and EIB (uplink) Common continuous pilot channel and auxiliary pilot (downlink)
Channel multiplexing in uplink	Control and pilot channel time multiplexed I&Q multiplexing for data and control channel	Control pilot, fundamental and supplemental code multiplexed I&Q multiplexing for data and control channel
Multirate	Variable spreading and multicode	Variable spreading and multicode
Spreading factors	4 - 256 (4.096 Mcps)	4 - 256 (3.6864 Mcps)
Power Control	Open and fast closed loop (1.6 kHz)	Open loop and fast closed loop (800Hz, higher rates under study)
Spreading (downlink)	Variable length orthogonal sequences for channel separation Gold sequence for cell and user separation	Variable length Walsh sequences for channel separation, M sequence 2^{15} (same sequence with time shift utilized in different cells, different sequence in I&Q channel
Spreading (uplink)	Variable length orthogonal sequences for channel separation, Gold sequence 2^{41} for user separation (different time shifts in I and Q channel, cycle 2^{16} 10 ms radio frames)	Variable length orthogonal sequences for channel separation, M sequence 2^{15} (same for all users different sequences in I&Q channels), M sequence $2^{41}-1$ for user separation (different time shifts for different users)
Handover	Soft handover Interfrequency handover	Soft handover Interfrequency handover

[1] Note that Wideband cdmaOne is now known as cdma2000.

15.1.1 Spectrum Issues [61]

Critical for the development of the new third-generation systems are the recommendations of spectrum allocation for FPLMTS/IMT-2000 systems. The present spectrum recommendations for third-generation systems are briefly analyzed below.

WARC'92 Spectrum Assignments for FPLMTS

The 1992 World Administration Radio Conference (WARC) of the ITU targeted 230 MHz, in the 2-GHz band (1885–2025 MHz and 2110–2200 MHz), on a worldwide basis for FPLMTS/IMT-2000, including both terrestrial and satellite components. The objective was identified as establishing, through the appropriate global standards and the coordinated assignment of spectrum by the various national and regional authorities, the future, truly ubiquitous personal mobile communications system, creating a seamless radio infrastructure capable of offering a wide range of services, in all radio environments, with the quality we have come to expect from fixed networks.

The availability of the spectrum assigned to FPLMTS/IMT-2000 and the strategies followed to ensure its availability when the time comes varies from region to region. Another aspect that is region-specific, is how that spectrum can, or cannot, be shared with other systems. The whole 230 MHz of spectrum identified by WARC'92 was reserved in Europe to third-generation UMTS technology, even if part of the band allocated for DECT falls into the FPLMTS band.

The issue of technology migration from second to third-generation via the use of spectrum in the FPLMTS/UMTS bands has been repeatedly raised in many forums. This may result in the spectrum being allocated, in some parts of the world, in an inefficient piecemeal fashion to evolved second-generation technologies and potentially many new narrow-application systems, thereby impeding the development of broadband mobile multimedia services.

Terminal, system, and network technology as researched within the EU-funded ACTS projects may alleviate to a large extent the complexity of the sharing of the spectrum between the second- and third-generation systems.

Spectrum for UMTS in Europe

The European Radio communications Office (ERO), recently prepared a draft decision that advocates a two-phase approach. Phase 1 aims to safeguard the spectrum already allocated to third-generation mobile communications systems, and phase 2 is an effort for allocation of additional spectrum, taking into account the ever-increasing market demand for multimedia mobile services being catered to by mobile communications networks. Some preliminary indications of a possible way forward are:

- The total amount of spectrum needed by third-generation mobile communications systems is still unclear. It is estimated that the requirements will be between 300 and 500 MHz. These figures assume a multimedia service provision environment with competing operators. It is further noted that the minimum bandwidth required for one public operator is in the order of 2×20 MHz, and that each country will have to accommodate at least three public networks.

- Further R&D is needed in technologies that will enable a more efficient use of spectrum (e.g., continuous dynamic channel selection, smart antennas, adaptive radio access, and coding).
- The possibility of using the mobile satellite systems (MSS) frequency bands for terrestrial indoor use under a frequency sharing scenario needs to be further studied.
- There is no immediate need (before year the 2000) for additional spectrum, but additional spectrum will become a necessity for the second phase of the UMTS introduction.

Figure 15.1 depicts the current usage of spectrum and planned UMTS bands in Europe.

15.2 CDMA-BASED SCHEMES

In this section, we analyze the similarities and differences of wideband CDMA schemes proposed for a third-generation system. We cover five different wideband CDMA designs: FRAMES FMA2, which is presented in more detail in [13–15], Core A [18–22] and two Korean air interfaces (ETRI [23] and KMT [2] [24]).

Most of the proposals utilize the following new capabilities that characterize the third-generation wideband CDMA:

- Fast power control in downlink;
- Coherent uplink;
- Seamless interfrequency handover (not in WB-IS-95 and Korean proposals);
- Optional multiuser detection;
- Under dedicated pilot symbols both in uplink and downlink (not in WB-IS-95 and Korean proposals);

All the other schemes except ETRI and WB IS-95 are asynchronous and thus do not need external synchronization, such as global positioning system (GPS). This is especially important for a low-cost base station.

In addition to the above schemes, CODIT proposed a wideband DS-CDMA air interface [2]. CODIT had three bandwidths: 1, 5, and 20 MHz. The main features of CODIT are coherent uplink, long-spreading codes facilitating asynchronous network, and support for interfrequency handover with compressed mode [1,2]. The idea of coherent uplink and compressed mode is used in later wideband CDMA proposals.

---

[2] Note that KMT is now known as SK Telecom.

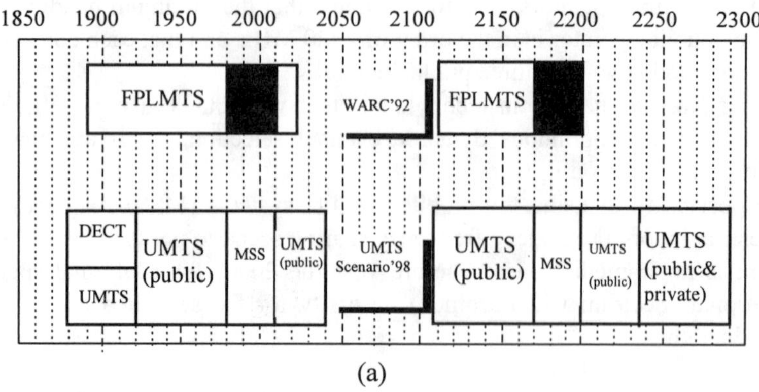

(a)

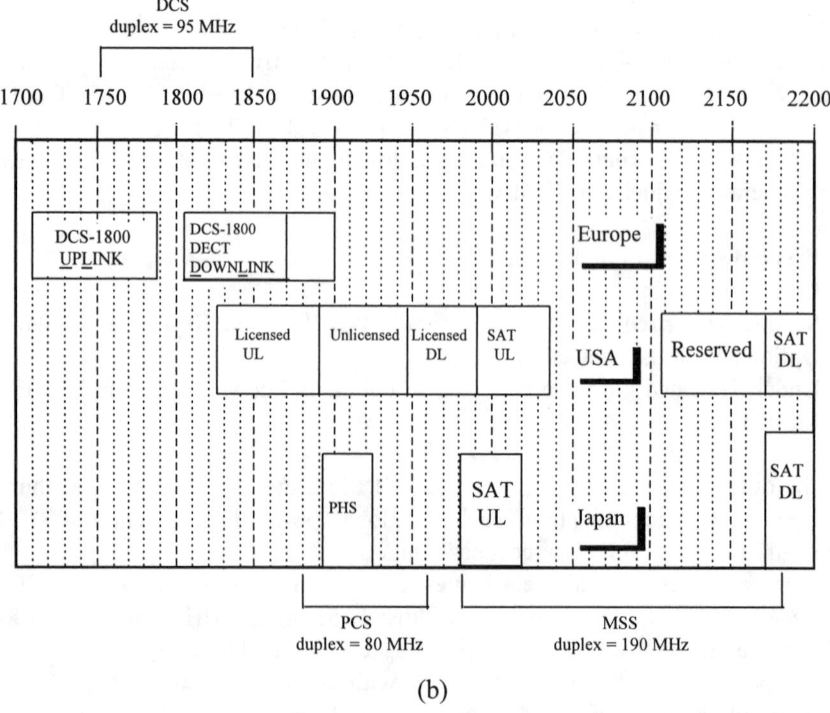

(b)

Figure 15.1 (a) FPLMTS/IMT-2000 and UMTS spectrum versus (b) current use.

15.2.1 Carrier Spacing and Chip Rate

For FMA2, a single RF bandwidth for all services was a target. In [7] it was shown by simulation that a 2-Mbps transmission rate is possible to achieve even with 6-MHz carrier spacing. However, better performance is achieved if a larger carrier spacing (e.g., 10 MHz) is used. Therefore, FMA2 employs two carrier spacings; 5 and 10 MHz [14].

The Core A also proposes several RF bandwidths: 1.25, 5, 10, and 20 MHz with chip rates of 1.024, 4.096, 8.192, and 16.384 Mchip/s, respectively [16]. However, earlier chip rates of 0.96, 3.84, 7.68, and 15.36 Mchip/s were also considered based on the testbed implementation [20]. Bandwidth of 1.25 MHz is employed up to 32 kbps, 5 MHz up to 384 kbps and higher bandwidths up to 2 Mbps.

WB IS-95 and the Korean CDMA proposals have 5-MHz bandwidth. The chip rates of the KMT (Note that now KMT is known as SK Telecom) and ETRI systems are 4.608 Mchip/s and 3.6864 Mchip/s, respectively, [23, 25–28].

15.2.2 Modulation and Detection

The data modulator maps the m incoming coded data bits into one of M (= 2^m) possible real or complex valued transmitted data symbols (i.e., data modulated). The M-ary data symbol is fed to the spreading circuit, where the resulting signal is filtered and mixed with the carrier signal (i.e., spreading modulation). Typical data modulation schemes are BPSK and QPSK. Spreading circuit can be binary, balanced quaternary, dual channel quaternary, or complex DS spreading. Typical spreading modulations are BPSK used with binary spreading circuit, QPSK, and O-QPSK used with quaternary spreading circuit. Roll-off factor depends on several factors such as power amplifier linearity and adjacent attenuation requirements. FMA 2 applies dual-channel QPSK spreading circuit [29] (i.e., symbol streams on I and Q channels are independent of each other).

FMA2, KMT, and ETRI schemes apply QPSK data modulation in both uplink and downlink. Spreading modulation for them is QPSK in the downlink and O-QPSK in the uplink. The data modulation of the Core A is QPSK both in uplink and downlink and the spreading modulation is QPSK in the uplink and BPSK in the downlink [22]. The BPSK spreading in this case means that the same code is used in in-phase and quaternary channels to save the number of orthogonal codes in the downlink. However, this leads to performance degradation due to crosstalk as a consequence of the phase error in the receiver. The roll-off factors of FMA2 and Core A are 0.2 and 0.22, respectively.

All schemes apply coherent detection both in uplink and downlink to enhance performance [30]. Further, FMA2 and Core A employ user-dedicated pilot symbols in both uplinks and downlinks to obtain the channel estimate for coherent detection. The user-dedicated downlink pilot symbol facilitates channel estimation with adaptive antennas. Channel estimate cannot be obtained from a common pilot proposed in [1], [5], and [10] since it experiences a different radio channel as the downlink beam for a specific user. FMA2 pilot symbols can be transmitted at variable rate from 0.5 to 2 kHz depending on the radio channel conditions. Core A transmits four consecutive symbols every 0.625 ms (i.e., with 1.6 kHz [16]). In both the KMT and ETRI schemes, continuous code multiplexed pilot facilitates coherent detection. In the downlink, the pilot signal is common to all users and in the uplink it is user-dedicated [27,28].

All schemes employ antenna diversity in the uplink. In addition, Core A employs antenna diversity in mobile station to improve the inferior downlink performance [20]. Other potential methods to improve the downlink performance are investigated in [31].

15.2.3 Spreading and Scrambling Codes

In the downlink, FMA2 employs short orthogonal codes to separate physical channels. Base-station-specific scrambling code is an extended Gold code of length 256. Also, the uplink data and control channel are separated by Walsh codes and the scrambling code is an extended VL-Kasami sequence of length 256.

Short codes ease the multiuser detection and allow use of orthogonal signal sets if the data symbol length is a multiple of power of two chips. The DS-CDMA air interface designs presented in [5] (MUD-CDMA) and [14] (FMA2) shows that the limitations of short codes pointed out by [1] can be circumvented by a smart system design. Flexibility is not sacrificed because unequal puncturing or repetition coding is used to obtain any user data rate regardless of the spreading factor. Intercell synchronization is avoided by using a special synchronization scheme during the soft handover [5]. Code planning is negligible because the number of VL-Kasami sequences, used in the uplink, is more than 1 million, and in the downlink same orthogonal code set can be reused in each cell and sector since it is overlaid by a cell-specific scrambling sequence. However, in certain cases, the use of short codes may lead to bad correlation properties, especially with very small spreading factors. In the uplink, multiuser detection is used to overcome this problem. In case the multiuser detection is not used (e.g., due to complexity reasons), adaptive code allocation is used to change the spreading code so that sufficiently good correlation properties are restored. As an alternative, an optional long-code scrambling for FMA2 is considered, where the long code is removed in cells that utilize multiuser detection.

Core A has a two-layered spreading code allocation [20]. Similar to FMA2, short codes are used for channel separation (i.e., all users are allocated the same set of orthogonal short spreading codes). In addition, cells are distinguished by different long spreading codes and hence a special arrangement is required for cell search [32]. In [16], tree structure Walsh codes are proposed as short codes. The generation of such codes is described in [33]. Earlier, short orthogonal Gold sequences and Gold sequences length of $2^{33}-1$ were proposed [20].

The KMT system uses short orthogonal codes of length 128 to separate the physical channels both in the uplink and downlink. Cells and users in the uplink are distinguished by long spreading codes of length 16,384 (2^{14}). If the uplink is not operated in synchronous mode, the long code of length 2^{32} is used to separate the users. Two pilots, cluster and cell pilot, are needed to avoid long synchronization time due to long codes. A cluster consists of several cells and under each cluster the long spreading sequence pilots are reused. Each cluster has a cluster pilot that is also a long spreading sequence. There are sixteen cluster pilots and the cluster can have 32 cell sequences. Thus, a maximum of 48 sequences are searched. A cluster pilot is transmitted by the center cell of a cluster only or by each cell. The former technique is suited for hierarchical cell system [24].

Like all the other systems, ETRI also uses short orthogonal Walsh codes to separate the physical channels. The length of the code is 256. In the downlink, all other physical channels except the signaling channel for each user are distinguished by Walsh codes and scrambled by a cell/sector-specific, 20-ms long PN code. Signaling channel for each user uses the same Walsh code as the data channel and is scrambled by a different PN sequence [23,28]. In the uplink, data channel, pilot channel, and signaling channel are distinguished by three different Walsh codes and scrambled by a common PN code [28]. Since the system employs the same PN codes with different phase offset, similar to the IS-95 standard, careful PN offset planning analog to frequency planning is needed [34].

15.2.4 Multirate

Multirate design is how different services with different QoS requirements are multiplexed together in a flexible and spectrum-efficient way. The provision of flexible data rates with different QoS requirements are divided into three subproblems: how to map different bit rates into the allocated bandwidth, how to provide desired QoS, and how to inform the receiver about characteristics of the received signal. The first problem concerns issues such as multicode transmission and variable spreading, the second problem concerns coding schemes, and the third problem deals with control channel multiplexing and coding.

FMA2 applies multicode transmission in the downlink and variable spreading gain in the uplink [12]. Variable spreading scheme eases the linearity requirements of power amplifier. There is also the option to apply multicode transmission in the uplink for the highest bit rates. Multicode transmission also facilitates control of QoS by power allocation instead of coding only. An arbitrary small granularity is achieved by a unequal repetition coding/puncturing scheme mapping the symbol stream after channel coding into the final, predefined symbol rate [35].

Originally, the Core A multirate scheme was based on multicode transmission for bit rates higher than 128 kbps and discontinuous transmission (on-off gating) for bit rates of 128 kbps or less [20]. However, this scheme leads into audible interference due to pulsed transmission (e.g., 8 kbps has to be gated on-off using 1.6-kHz frequency).

Due to these problems and the desire to relax power amplifier linearity requirements, the variable spreading scheme was introduced as a primary transmission scheme and multicode is employed only for the highest data rates [22].

In all schemes, the different services of a same user are time-multiplexed into the frame. Further, Core A and KMT time multiplex are also the control information. FMA2 and ETRI transmit the control information on a parallel code channel. In addition, FMA2 has a possibility to control the data rate on frame-by-frame basis. The FMA2 scheme is the most flexible and, as already said, facilitates arbitrary small granularity. The other scheme adjusts the bit rate in upper layers.

Since FMA2 transmits the FCH header within the same frame as the data, the receiver needs to buffer the traffic channel before the control channel is decoded. Since the spreading ratios are multiples of each other, it is possible to despread with the lowest spreading ratio and to buffer the subsymbols after despreading, instead of the samples before the despreading, as in CODIT, which has arbitrary spreading ratios.

The frame length of FMA2 and Core A is 10 ms, while the frame length of ETRI and KMT systems is 12 ms. The frame length is also the basic interleaving period. Further, FMA2 has also possibility for interframe interleaving. All schemes employ convolutional code for services with BER 10^{-3}, and concatenated convolutional code and RS code for services with BER 10^{-6}.

15.2.5 Power Control

All schemes use open loop and fast closed-loop power control in the uplink. In addition, FMA2 and Core A apply fast closed-loop power control also in the downlink. The power target is decided based on the required quality impacted by the channel quality and interference conditions. FMA2 transmits the power control information at a variable rate from 0.5 up to 2 kbps. The variable rate power control avoids unnecessary waste of capacity for power control signaling, especially for low bit rate services [30]. Variable rate transmission was also tested for the KMT system [25]. Power control parameters are given in Table 15.2.

Typically, SIR target for the power control is based on the frame error rate calculated from the Viterbi decoder output. The measured frame error rate is compared with the target frame error ratio (FER) and an SIR target is determined based on this. The target SIR is compared with the measured SIR, which, for example, Core A measures from the time multiplexed pilot.

The impact of the fast power control in the downlink is twofold. On one hand, it improves the performance against fading multipath channel, and on the other hand, it increases the multiuser interference variance within the cell, since orthogonality between users is not perfect due to multipath channel [35–37]. The net effect, however, is increased performance.

Table 15.2

Power Control Parameters

Uplink			
	Command Rate kHz	Dynamic Range	Step Size
FMA2	0.5 – 2	80 dB	1.0 dB
CoreA	1.6 kHz	80 dB	1.0 dB
ETRI	1.6 kHz	80 dB	0.5 dB
KMT	2 – 6 ms	Not given	0.5 – 2 dB
Downlink			
	Command Rate	Dynamic Range	Step Size
FMA2	0.5 – 2	20	1.0
CoreA			
ETRI	No details given, open-loop slow PC		
KMT	No details given		

15.2.6 Handover

All the proposals employ soft handover. All except ETRI also assume unsynchronized network, which is taken into account in the handover synchronization scheme. FMA2 employs short codes impacting the soft handover planning. However, using synchronization channel and adjustment of transmission time from a new base station, the seamless soft handover is carried out [5]. Core A employs long codes, and thus the mobile station can measure the timing uncertainty between the base stations involved in the soft handover.

15.2.7 Interfrequency Handover

For interfrequency handover there are two implementation alternatives: compressed mode/slotted mode and dual receiver. The dual receiver has an expensive receiver but a simple system operation. However, if there is already a diversity receiver (two receiver chains, not selection diversity), then it is possible to use the other receiver for measurements on other carriers [15]. Of course, diversity gain is lost during the measurement period. Compressed mode, on the other hand, has a simple receiver but a more complex system operation, and the receiver algorithms operate on burst nature during slotted mode (i.e., poorer performance and more control).

There exists a number of alternatives on how to implement the compressed mode. In [36], four different methods are presented: variable spreading factor, code rate increase, multicode, and higher-order modulation. Variable rate spreading and coding rate increment (puncturing) cause 1.5 to 2.5 dB loss in E_b/N_0 performance and in higher-order modulation even higher loss (5 dB). This is due to breaks in power control and less coding. A further drawback of variable spreading ratio is that simple terminals must be able to operate with different spreading ratios (e.g., speech terminal normally operates with a spreading ratio of 256, but now it must also operate with a spreading ratio of 128).

For FMA2, a dual receiver was proposed in [12]. However, compressed mode is also currently under study [15]. Both Core A and CODIT use compressed mode. Core A employs 16-QAM during the compressed mode operation [20]. The KMT and ETRI proposals do not consider interfrequency handover.

15.2.8 Multiuser Detection and Other Interference Reduction Methods

Even though multiuser detection is only a receiver technique, it impacts the overall system design (e.g., short codes ease the implementation of multiuser detection). Further, multiuser detection changes the relation between system load and coverage. Since multiuser detection reduces the intracell interference, cell size does not shrink as fast as without multiuser detection when the system load increases [38]. Due to complexity reasons, multiuser detection cannot be used in a similar way in the downlink than in the uplink. In addition, the terminal must only receive its own signal in contrast to the base station. Therefore, a simpler interference suppression scheme is applied in the terminal.

FMA2 employs an optional interference cancellation scheme in the uplink. In simulations, multistage scheme has been used. Performance results for this scheme are presented in [39–43]. For Core A, a pilot symbol assisted coherent multistage interference cancellation is proposed [21]. Downlink interference cancellation has been studied for the MUD-CDMA, scheme which was the basis for FMA2 [37]. The downlink multiuser detection improves performance and allows more accurate power control since it reduces the intracell interference [37]. Even though orthogonal codes are used in the downlink, there is gain from the multiuser detection since the multipath channel partially destroys the orthogonality [37].

To reduce the intracell interference, the KMT scheme time synchronizes all users in the uplink with accuracy of 1/8 chip [25]. This is done by measuring the timing in base station and signaling timing adjustment commands with a rate of 2 kbps to the mobile station. However, multipath still results to intracell interference and gain from the orthogonal uplink depends on the channel profile. In addition, the signaling traffic reduces the downlink capacity for each user by 2 kbps.

15.2.9 Packet Data

Packet access procedure in CDMA minimizes the interference to other users. If no connection is established, the initial access is not power controlled, thus we minimize time and information transmitted during this period. The FMA2 packet access scheme is similar to Core A except that it does not allow information transmission in the initial access burst. When there is nothing to send, FMA2 either cuts off the transmission or keeps only the physical connection transmitting power control and reference symbols. In the former case, a virtual connection (authentication and other information) is retained to quickly establish the link again in case of new transmission. Selection between these two alternatives is a trade-off between the resource spending for synchronization and power control information, and resources spent for random access.

Core A employs an adaptive mode, selective packet data transmission system where the initial access burst contains information on the data length and data volume [22]. Data are either transmitted on a common channel or a dedicated channel. Closed-loop power control is not possible for the common channel so only small volumes of information are transmitted on that channel. On the other hand, dedicated channels are inefficient since resources are tied up even when there are no data to transmit. The most appropriate mode is selected according to the traffic type. In case there is nothing to transmit on a dedicated channel, closed-loop power control is still applied [22].

15.3 TDMA-BASED SCHEMES

This section presents the similarities and differences of TDMA schemes; namely, FMA1 without spreading (wideband TDMA) [14], ATDMA [4], and MTDMA [16]. Further, we cover proposed GSM evolution using higher-level modulation [44]. Note that parameters for the enhanced GSM proposal should be seen as examples since detailed standardization is currently underway. References [45–50] present a modified ATDMA scheme, where the carrier bandwidth and frame structure are aligned with GSM. Evolution of GSM air interface is handled also in [44] and [51]. In the following, we focus on comparison of the schemes.

15.3.1 Modulation

Both ATDMA and FMA1 use the same linear modulation methods: B-O-QAM and Q-O-QAM. The modulation method of the MTDMA is QPSK and 16-QAM in picocell environments. For FMA1, coded modulation methods are under consideration for quaternary modulation. Linear modulation was selected since it is bandwidth efficient (i.e., it provides a high bit rate per bandwidth unit). The Q-O-QAM modulation is a flexible way of extending the chosen B-O-QAM modulation to higher bit rates without increasing the bandwidth requirement. Offset modulation is selected, since it reduces amplitude variation, which reduces the backoff requirement at the transmitter amplifier, giving more efficient transmission. In ATDMA, the GSM modulation scheme GMSK is used in long macrocells to extend the maximum range of macrocells. It uses a lower carrier bit rate (360 kbps) to maintain the same channel spacing as in the short macrocells [3]. For enhanced GSM, trellis-coded 16-QAM modulation is proposed [44].

The bandwidth efficiency for FMA1 is with B-O-QAM modulation 1.625 bits/s/Hz and for ATDMA is almost the same with a few percentages variation. Q-O-QAM has double the bandwidth efficiency (i.e., 3.25 bits/s/Hz). The bandwidth efficiency of MTDMA is 1.92 bits/s/Hz, which is considerably higher than for the two other schemes. The receiver filter for FMA1 and MTDMA is square-root, raised-cosine filter with a roll-off factor of 0.35.

15.3.2 Carrier Spacing and Symbol Rate

In ATDMA, the carrier spacing and symbol rate are adapted according to different cell types. The largest channel spacing is approximately 1107 MHz. This results in a maximum user bit rate of 1.16 Mbps using same coding rate as in FMA1. Thus, a user bit rate of 2 Mbps needs multicarrier transmission with two carriers. The carrier bit rate of 1800 kbps is motivated by the requirement of 100m range for a picocell environment. With the used propagation model, it was considered impossible to achieve a higher carrier bit rate within the given range requirements. However, in the modification of ATDMA for better compatibility with GSM, presented in [45], the carrier spacing was defined as 1.6 MHz in picocells providing 2 Mbps user bit rate. In microcells, the carrier spacing and user bit rate of ATDMA were halved into 800 kHz and 1 Mbps, respectively.

FMAl without spreading and MTDMA have both carrier spacing of 1.6 MHz and both can provide 2 Mbps user data peak rate. For them, the driving force is the provision of 2 Mbps transmission rate in all environments using a single RF bandwidth. The reasoning for the FMAI carrier spacing is as follows. For circuit-switched users, we dedicate the entire frame to one user to achieve at minimum a raw bit rate of 3.38 Mbps. To provide 2 Mbps user data rate, it is possible to use a punctured rate 1/2 convolutional coding. For packet switched users, 10 % of the raw data bits are reserved for the control information of the packet protocol. This reduces the available raw data rate. In packet mode, some raw bits are also reserved for the error detection (e.g., CRC). Then, for the error correction, rate 2/3 punctured convolutional code is used, for example.

The relation of channel spacing and carrier bit rate depends also on adjacent channel suppression requirements. GSM requires adjacent channel suppression of 18 dB, which means that commonly adjacent carriers cannot be used within the same cell and normally also not in neighboring cells. For UMTS a higher value is desired. The power amplifier linearity aspects also influence the adjacent channel interference, since power slip over can occur when power amplifiers are not sufficiently linear.

15.3.3 Burst and Frame Structures

Burst and frame structure are designed based on:

- Range and characteristics of services to be supported;

- Constraints based on multipath propagation and propagation environment;
- Constraints based on terminal hardware complexity.

The frame and slot structure of FMAl without spreading was optimized for microcell and picocell coverage for high bit rates up to 2 Mbps. As shown in [8], a range of approximately 1km was achieved for WB-TDMA, even though the carrier bit rate was 6.5 Mbps (carrier spacing of 2 MHz) compared with 5.2 Mbps (1.6 MHz) in FMAl. FMAl without spreading has two different slot sizes, 1/16 and 1/64. The larger slot reduces overhead for medium and high bit rate services. Frame length is selected to be the same as GSM since this supports an interleaving depth at least of four for services with a tight delay requirement. The frame and slot structure of FMA1 and ATDMA facilitate both frame-by-frame and slot-by-slot hopping. The latter is used for high bit rate services occupying several slots within a frame. Both FMAI and ATDMA use rate compatible punctured codes to achieve different coding rates in a flexible way. For FMA1 Turbo codes are also under study [46,47].

In ATDMA, the frame length and slot length design is based on transmission of speech service with a gross bit rate of 13 kbps in a macrocell environment. A minimum interleaving depth of four was expected to be required to achieve good performance. This led to specification of payload for burst of 66 bits with 10 bits for inburst signaling and a frame length of 5 ms. Further, when identifying the macrocell carrier bandwidth, a requirement of maximum user bit rate of 64 kbps was used in defining the number of slots required within a frame. The QoS requirement was thought to require rate 1/4 coding and thus led into 450 kbps and 18 slots/frame, taking into account one extra slot for measurement of other carriers. Then, microcell carrier bit rate was taken to be fourfold compared with macrocell, since delay spread is four times less. This resulted in 72 slots/frame. Picocell carrier bit rate was taken to be the same as in microcell due to expected peak power limitations to achieve the specified range of 100m. With the carrier bit rate of 1800 kbps, maximum user bit rate is 1.16 Mbps with Q-O-QAM taking into account the 40% overhead.

MTDMA is clearly targeted for data transmission. The frame length and interleaving depth is 10 ms, which is too long for a low-rate, low-delay speech transmission.

15. 3.4 Training Sequence Design

As depicted in Figure 15.2 (a typical training sequence is composed of L precursor symbols followed by P reference symbol. The training sequence is placed into the middle of the burst to get a better estimate of the channel over the whole burst. Length of the training sequence depends on the carrier symbol rate and the desired length of delay spread that needs to be estimated. The number of precursor symbols has to be at least as long as the length of the channel impulse response in symbols. After channel delay spread exceeds the specified limit, performance starts to degrade gradually, and hence the length depends also on the design paradigm: average or worst-case conditions. The number of required reference symbols in the training sequence depends on the length of the channel's impulse response, the required SNR, the expected maximum Doppler frequency shift, and the number of modulation levels. The number of reference symbols should not be too large so that the channel characteristics remain practically stable within the correlation window [3]. The training sequence has to exhibit good correlation properties. Typically, training sequences are optimized for good autocorrelation properties. However, FMA1 utilizes also interference cancellation and thus good cross-correlation properties are desirable for good performance [49]. For the FMAl nonspread data bursts, the longer sequence can handle about 7 μs of time dispersion and the shorter one 2.7 μs. Note that if the time dispersion is larger, the drop in performance is slow and depends on the power delay profile.

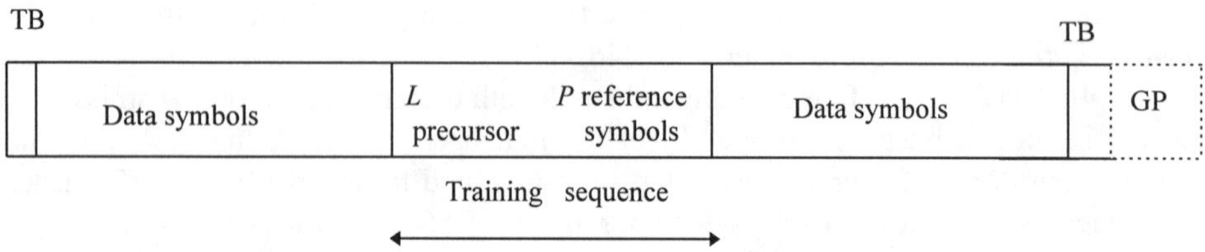

Figure 15.2 Training sequence (TB = tail bits, GP = guard period).

MTDMA training sequence is in the beginning of the burst. This facilitates direct processing of data without the need to buffer it as in the case of midample. On the other hand, performance might degrade for the last bits in the burst. In MTDMA, there is *S* precursor symbols that facilitate delay spread estimation up to 2.6 μs. However, the guard time limits the performance to 1.3 μs.

The guard bits provide a protection interval between the bursts for time alignment uncertainty, time dispersion, and power ramping, and the tail bits are used to assist the detection process. Faster ramping leaks more power into the adjacent channels and hence the power ramp up at the beginning of the bursts and power ramp down at the end of the burst must be smooth enough to avoid spreading of the output spectrum of the transmitter. According to calculations for GSM, about 3 bits ramp time would be sufficient [50,51]. Further, one symbol is needed for the timing alignment uncertainty. Thus, for FMA1 microcell subburst we have 5.5 symbols left for covering the length of the impulse response, which means 2.1 μs excess delay before the soft degradation in performance starts. For data burst/regular burst this is 2.3 μs. For ATDMA microcell we have 6.7 μs guard time; 3 bits are reserved for power ramping and 1 bit for timing alignment uncertainty, which leaves 4.5 μs for excess delay spread. MTDMA ramp time is 4 bits, which leaves 4 bits for channel impulse response (i.e., 1.3 μs). ATDMA combines two bursts into double bursts to reduce the overhead for higher bit rates. However, a slightly longer training sequence is needed to guarantee receiver performance at the burst ends. Contrary to ATDMA, FMA1 employs the guard times between bursts for user data to avoid use of two training sequence types.

15.3.5 Radio Resource Management

ATDMA transport chain has been designed around so-called transport modes, which is characterized by a certain configuration of a set consisting of modulation, error control code, and amount of radio resources to guarantee a given performance for a given level of signal-to-noise + interference. Link adaptation controls this configuration [3].

An improved packet access, PRMA++, has been developed for ATDMA [4]. For FMA1, an even more advanced scheme is under development that can multiplex both real time and nonreal time services in a flexible way.

Both ATDMA and FMA1 use dynamic channel for channel assignments.

15.4 HYBRID CDMA/TDMA

Several types of hybrid CDMA/FDMA are investigated in the literature [52]. FRAMES FMAl with spreading is based on a hybrid CDMA/TDMA concept, also called joint detection [53–55]. The role of CDMA is to multiplex the different channels within a timeslot. The spreading ratio is small and thus if more than few users are desired per frame, joint detection is needed to remove the intracell interference. Therefore, in a fully loaded system there is no benefit of the spreading against interference from other cells. Also the slow power control results in high variation in the received signals, and thus joint detection is needed. Since the joint detection is a mandatory feature, it is more critical compared with wideband CDMA. However, since the number of users is small, the complexity of joint detection is not excessive.

In the following, we explain the main differences between the FMAI without spreading, covered in Section 15.3, and FMA1 with spreading. Original parameters of joint detection CDMA, frame length of 6 ms and 12 slots per frame [53], have been aligned with FMAl without spreading to be backward compatible to GSM (i.e., 4.615 ms and 8 slots per frame, respectively). The main difference to FMAl without spreading is, of course, spreading. Short orthogonal spreading codes of length 16 are used. Spreading modulation method is GMSK and data modulation QPSK. The training sequence length of mode I is adapted to the expected channel impulse response length. There are bursts with two different training sequence lengths. The longer training sequence is suited for estimating the eight different uplink channel impulse responses of eight users within the same time slot with a time dispersion of up to about 15 μs. If the number of users is reduced, the tolerable time dispersion is increased almost proportionally. The shorter training sequence is suited for estimating the eight uplink channel impulse responses with a time dispersion of up to about 5.5 μs. Further, it is suited for estimating the downlink channel impulse response with a time dispersion of up to about 25 μs, independent of the number of active users, and the uplink channel impulse response with this same time dispersion in case all bursts within a slot are allocated to one and the same user.

15.5. OFDM-BASED SCHEMES

Introduction of OFDM into the cellular world is driven by two main benefits:

- Flexibility: each transceiver has access to all subcarriers within a cell layer;
- Easy equalization: OFDM symbols are longer than the maximum delay spread resulting into a flat fading channel that is easily equalized.

Also, the introduction of DAB based on OFDM and research of OFDM for high performance local area network (HIPERLAN) type II and wireless ATM have increased the interest in OFDM [56].

The main drawback of OFDM is the high peak to average power. This is especially severe for the mobile station and for long-range applications. Different encoding techniques have been investigated to overcome this problem. Furthermore, the possibility to access to all resources within the system bandwidth results in an equally complex receiver for all services, regardless of the bit rate. Of course, a partial FFT for only one OFDM block is possible for low bit rate services, but this would require an RF synthesizer for frequency hopping.

For UMTS and FPLMTS/IMT-2000, two OFDM air interface concepts are presented; BDMA and OFDM by Telia. BDMA concept has actually been proposed both in ETSI and ARIB FPLMTS Study Committee. More information from Telia's OFDM concept is found in [57–60]. The main difference between the Telia concept and BDMA is the detection method that impacts the overall design of the systems. Telia OFDM uses coherent detection and BDMA differentially coherent detection. Telia OFDM employs coherent detection due to two reasons: performance gain and opportunity to use arbitrary signal constellations such as 16-QAM [9].

15.5.1 Bandwidth

Both Telia OFDM and BDMA have a 5-MHz bandwidth. The reasoning for the bandwidth choice is that two times 60-MHz FDD spectra is allocated for FPLMTS/IMT-2000. So, assuming four operators and three cell layers results in a 5-MHz system bandwidth [9]. Each transceiver has access to the whole 5-MHz bandwidth, which is divided into subcarriers. This facilitates very high bit rates and flexible allocation of resources. Telia OFDM has within the 5-MHz bandwidth 1024 subcarrier, each at 5 kHz. BDMA has 800 subcarriers with bandwidth of 6.25 kHz [16].

15.5.2 Frame Design

The Telia OFDM frame length is 15 ms and the BDMA frame length is 5 ms. However, the total interleaving period for BDMA is 20 ms. BDMA frame consists of 4 subframes, and each are further divided into six timeslots. This frame design facilitates the hopping scheme for BDMA described below.

The OFDM block design principles are different for the schemes since Telia OFDM applies coherent detection and BDMA differential detection. The block size for Telia OFDM is 25 OFDMA carrier and 3 OFDM symbols. The time and frequency correlation function impacts the block shape. The shape is selected to reflect the peak of the time-frequency correlation function. However, since there is large variation in the radio channels, the shape is designed for the worst-case channels. For BDMA, the slot size is 24 OFDMA carriers and 1 OFDM symbol slot. In Telia OFDM, pilot symbols occupy 18.7% of a block. The downlink of Telia OFDM has smaller overhead than the uplink since a common pilot is utilized. However, this is not true for cheap speech terminals if only partial FFT is carried out to demodulate only a user's own block. In BDMA, one subcarrier is allocated to be used as reference for differential detection. The Telia scheme results into finer granularity, 3.9 kbps against 11.2 kbps, but seems to offer less diversity for low bit rates compared with BDMA.

For the purpose of coherent detection, the OFDM block has to be large enough to allow efficient use of channel correlation in estimation but the number of pilots should be minimal. In the uplink, blocks belonging to the same user should be kept together to utilize the same timing and to better maintain orthogonality. On the other hand, information should be transmitted on different places in the time-frequency space to utilize diversity. The placement of the blocks is a trade-off between these two contradicting requirements.

15.5.3 Coding, Interleaving, and Frequency Hopping

Frequency hopping together with coding, interleaving improves the performance through increased diversity. Telia OFDM hops on frame-by-frame basis; that is, every 15 ms (67 hops/s) being effective against long fading dips. BDMA hops on a burst-by-basis (800 hops/s) and thus better diversity is achieved.

Telia OFDM interleaves the symbols over all transmitted block within one frame. BDMA interleaves over 20 ms (i.e., 4 TDMA frames) and thus, also in this respect better diversity is achieved. Of course, the delay is also longer. The 20-ms interleaving delay might be too long with respect to the 30 ms minimum delay requirements of UMTS bearer services.

If the BDMA receiver is based on partial FFT and hopping synthesizers, the synthesizer settling times are very tight and are even tighter for multislot services. Therefore, full utilization of all six slots for one user might be difficult.

15.5.4 Handover

Handover principles are the same for both schemes. If the base stations are synchronized, handover within the same cell layer does not require any extra measures but measurements are performed during idle slots. On the other hand, if the base stations are not synchronized, the mobile station releases the existing link and switches to a new frequency band to perform measurements. Another possibility is to have a dual receiver [9].

15.5.5 Uplink Synchronization

The transmitted signals from all mobiles must reach the base station with a certain synchrony in order to maintain their orthogonality. Both time and frequency offsets are estimated. A number of studies were carried out to solve this critical aspect for OFDM [59].

15.5.6 Power Control

For Telia OFDM, the power control scheme is still open. In BDMA, a closed-loop power control based on quality measurements is applied [16]. The power control period is 1.2 ms and step size 1 dB. Also, open-loop power control is used. If the frequency hops are correlated, which is the case for slow-moving mobiles, the power control compensates fading and thus improves performance. For higher mobile speeds, it minimizes the transmission power but cannot follow the fast fading. However, the frequency hopping mitigates the impact of fast fading.

15.5.7 DCA

The Telia OFDM employs a distributed CIR measurement-based DCA scheme [58]. Channel rearrangement per frame basis is used to improve performance. The relation between frequency hoping (random vs. cyclic hopping) and DCA needs still further investigation.

15.6 TDD

The main discussion about UMTS air interface is around technologies for FDD. However, there are several reasons why TDD should be used. First of all, there will be most likely a dedicated frequency band for TDD within the identified UMTS frequency bands. Further, FDD requires exclusive paired bands and spectrum for such systems is therefore hard to find. On the other hand, TDD makes use of individual bands that do not need to be mirrored for the return path, and hence spectrum is more easily identified. With a proper design including powerful FEC, TDD can be used even in outdoor cells. It has been argued that the TDD guard interval would result in excessive overhead in large cells. However, in a cell with a range of 3 km, we would need a 20-μs guard time to prevent transmission and reception time slots to overlap, (i.e., an overhead of approximately 4% assuming frame length of 4.615 ms). When the propagation delay exceeds the guard period, soft degradation of performance occurs. Thus, UMTS TDD mode need not to be restricted into unlicensed indoor solutions but perhaps even most short-range UMTS should be TDD, even those used by the traditional cellular operators (e.g., for high-capacity microcells). The second reason for using TDD is the flexibility in radio resource allocation; that is, bandwidth is allocated by changing the number of time slots for uplink and downlink.

The asymmetric allocation of radio resources leads into two interference scenarios that impact the overall spectrum efficiency of TDD scheme:

- Asymmetric usage of TDD slots impacts the radio resource in neighboring cells;
- Asymmetric usage of TDD slots leads into blocking of slots in adjacent carriers within its own cell.

Figure 15.3 depicts the first scenario. Mobile station 2 (MS2) is transmitting at full power at the cell border. Since mobile station 1 (MSl) has different asymmetric slot allocation than MS2, its downlink slots received at the sensitivity limit are interfered by MS1 causing blocking. On the other hand, since base station 1 (BS1) has much higher effective isotropically radiated power (EIRP) than MS2, it interferes with BS2 receiving MS2. Hence, the radio resource algorithm needs to avoid this kind of situation.

In the second scenario, two mobiles are connected into the same cell but use different frequencies. The base station is receiving MS1 on the frequency f_1 using the same time slot it uses on frequency f_2 to transmit into MS2. As shown in Table 15.3, the transmission blocks the reception due to irreducible noise floor of the transmitter regardless of the frequency separation between f_1 and f_2.

Table 15.3

Adjacent Channel Interference Calculation

BTS transmission power for MS2 in downlink1W	30 dBm
Received power for MS1	−100 dBm
Adjacent channel attenuation due to irreducible noise floor	50 - 70 dB
Signal to adjacent channel interference ratio	– 60 - – 80 dB

The third scenario, where the above-described blocking effect exists, is an FDD system where if at any moment traffic in a cell is unbalanced between the uplink and downlink, then the spare capacity in the low-traffic direction is momentarily used for two-way operation (i.e., TDD).

DECT is a second-generation TDD system with carrier spacing of 1728 kHz and carrier bit rate of 1152 kbps. DECT frame length is 10 ms and each frame is divided into 24 slots. DECT provides bit rates up 512 kbps half duplex and 256 kbps full duplex. The fundamental difference between DECT and UMTS are the bit rate capabilities, operating point, and channel coding. Since the UMTS TDD mode has powerful channel coding, the required C/I and hence also reuse factor is smaller compared with DECT. Therefore, an assumption of a similar uplink and downlink interference situation does not hold anymore and the DECT dynamic channel selection is not suitable. Also, the performance of DECT in high-delay spread environments is not very good, which limits the outdoor cell range.

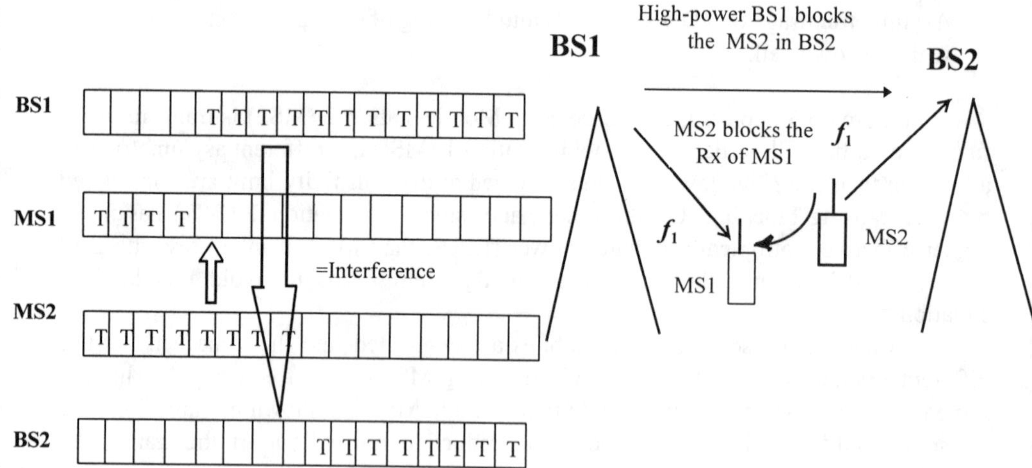

Figure 15.3 TDD interference scenario.

Both TDMA- and CDMA-based schemes have been proposed for TDD. Most aspects are common to TDMA- and CDMA-based air interfaces. However, in CDMA-based TDD systems, we need to change symmetry of all codes within one slot to prevent an interference situation where a high-power transmitter would block another receiver. Thus, TDMA-based solutions have higher flexibility. In FRAMES multiple access only FMA1 has a TDD option that can be used both with wideband TDMA without spreading and with spreading [13]. However, with the spreading option, the above-mentioned drawback of code allocation exists. CDMA/TDD has been proposed in [16].

15.7 CONCLUSIONS

We reviewed the third-generation, standardization-related radio access research activities. Further, a technical overview of the air interface designs for UMTS and FPLMTS/IMT-2000 covering wideband CDMA, TDMA, OFDM, and hybrid schemes were presented.

Asynchronous wideband CDMA has recently received a lot of attention in the research community. The most widely considered bandwidth seems to be 5 MHz providing bit rates up to 384 kbps. Given that Japanese and European approaches to the wideband CDMA are very similar, it is possible that a common air interface standard can be achieved.

For bit rates above the 384 kbps several alternatives are considered: wideband CDMA, wideband TDMA, hybrid CDMA/TDMA, and OFDM. In practice, wideband CDMA needs very large bandwidth to support 2 Mbps transmission. The critical question for OFDM and hybrid approaches is the maturity of the technology. Therefore, it seems that wideband TDMA would offer well-proven, high-performance technology for the high bit rate services up to 2 Mbps, especially for TDD mode, as a complement to 5-MHz wideband CDMA. Further, the TDMA-based UMTS proposals can be used to evolve GSM towards higher bit rates. The enhanced OSM using coded 16-QAM modulation can achieve a user data rate of over 300 kbps using a 200-kHz carrier bandwidth.

REFERENCES

[1] B. Andermo (Ed.), “UMTS code division testbed (CODIT),” *CODIT Final Review Report*, September 1995.

[2] A. Baier, U.-C. Fiebig, W. Granzow, W. Koch, P. Teder, and J. Thielecke, “Design study for a CDMA-based third generation mobile radio system,” *IEEE J. Selected Areas in Comm.*, Vol.12, No.4, pp. 733– 743, May 1994.

[3] A. Urie et al., "ATDMA system definition", ATDMA deliverable R2084/AMCFJPM2/DS1R/044lb1, January 1995.

[4] A. Urie et al., “An advanced TDMA mobile access system for UMTS,” *IEEE Personal Comm.*, Vol.2, No.1, pp. 38– 47, February 1995.

[5] T. Ojanpera et. al,, “Design of a 3rd generation multirate CDMA system with multiuser detection, MUD-CDMA,” *Proc. 1SSSTA’96 Conference*, Vol. 1, pp. 334–338., Mainz, Germany,1996.

[6] K. Pajukoski and J. Savusalo, “Wideband CDMA test system,” *Proc. PIMRC97*, Helsinki, Finland, September 1997.

[7] W. Tapani and H. Harri, “CDMA System for UMTS high bit rate services,” *Proc. of VTC97*, Vol., pp. 824–829, Phoenix, AZ, May 1997.

[8] E. Nikula and E. Malkamaki, "High bit rate services for UMTS using wideband TDMA carriers," *Proc. of ICUPC'96*, Vol. 2, pp. 562–566, Cambridge, MA, September/October 1996.

[9] FTSI SMG2, "Description of Telia's OFDM based proposal," *TD 180/97 ETSI SMG2*, May 1997.

[10] T. Ojanpera Tero et al., "FRAMES - Hybrid multiple access technology," *Proc. of 1555TA96 Conference*, Vol. 1, pp. 320–324, Mainz, Germany, 1996.

[11] T. Ojanpera et al., "A comparative study of hybrid multiple access schemes for UMTS," *Proc. of ACTS Mobile Summit Conference*, Vol. 1, pp.124–130, Granada, Spain, 1996.

[12] T. Ojanpera et. al, "Comparison of multiple access schemes for UMTS," *Proc. VTC97*, Vol. 2, pp. 490–494, Phoenix, AZ, May 1997.

[13] A. Klein, R. Pirhonen, J. Skold, and R. Suoranta, "FRAMES multiple access mode I—wideband TDMA with and without spreading," *Proc. PIMRC97*, Helsinki, Finland, September 1997.

[14] F. Ovesjo, E. Dahlman, T. Ojanpera, A. Toskala, and A. Klein, "FRAMES multiple access mode 2—wideband CDMA," *Proc. PIMRC97*, Helsinki, Finland, September 1997.

[15] T. Ojanpera, A. Klein, and P.O. Andersson, "FRAMES multiple access for UMTS," *IEE Colloquium on CDMA Techniques and Applications for Third Generation Mobile Systems*, London, May 1997.

[16] ARIB FPLMTS Study Committee, "Report on FPLMTS radio transmission technology SPECIAL GROUP, (round 2 activity report)," Draft v.E1.1, January 1997.

[17] A. Sasaki, "A perspective of third generation mobile systems in Japan," *IIR Conference Third Generation Mobile Systems*, The Route Towards UMTS, London, February 1997.

[18] F. Adachi et al., "Multimedia mobile radio access based on coherent DS-CDMA," *Proc. 2nd International Workshop on Mobile Multimedia Commun., A2.3*, Bristol University, UK, April 1995.

[19] K. Ohno et al., "Wideband coherent DS-CDMA," *Proc. IEEE VTC'95*, pp. 779–783, Chicago, IL, July 1995.

[20] T. Dohi et al., "Experiments on coherent multicode DS-CDMA," *Proc. IEEE VTC'96*, pp. 889–893, Atlanta, GA, 1996.

[21] F. Adachi et al., "Coherent DS-CIDMA: promising multiple access for wireless multimedia mobile communications," *Proc. IEEE ISSSTA'96*, pp. 351–358, Mainz, Germany, September 1996.

[22] S. Onoe et al., "Wideband-CDMA radio control techniques for third generation mobile communication systems," *Proc. VTC97*, Vol. 2, pp. 835–839, Phoenix, AZ, May 1997.

[23] S. Bang, et al., "Performance analysis of wideband CDMA system for FPLMTS," *Proc. VTC'97*, pp. 830–834, Phoenix, AZ, 1997.

[24] J.M. Koo, "Wideband CDMA technology for FPLMTS," *The 1st CDMA International Conference*, Seoul, Korea, November 1996.

[25] Y.-W. Park et al., "Radio characteristics of PCS using CDMA," *Proc. IEEE VTC'96*, pp.1661–1664, Atlanta GA, 1996.

[26] E.-K. Hong et al., "Radio interface design for CDMA-based PCS," *Proc. ICUPC'96*, pp. 365–368, 1996.

[27] J.M. Koo et al., "Implementation of prototype wideband CDMA system," *Proc. ICUPC'96*, pp. 797–800, 1996.

[28] H.-R. Park, "A third generation CDMA system for FPLMTS application," *The 1st CDMA International Conference*, Seoul, Korea, November 1996.

[29] R.L. Peterson, R.E. Ziemer, and D.E. Borth, *Introduction to Spread Spectrum Communications*, Prentice Hall, 1995.

[30] T. Ojanpera, P. Ranta, S. Hamalainen, and A. Lappetelainen, "Analysis of CDMA and TDMA for 3rd generation mobile radio systems," *Proc. VTC97*, Vol. 2, pp. 840–844, Phoenix, AZ, May 1997.

[31] S. Hamalainen, H. Holma, A. Toskala, and M. Laukkanen, "Analysis of CDMA downlink capacity enhancements," *Proc. PIMRC97*, Helsinki, Finland, September 1997.

[32] K. Higuchi et al., "Fast cell search algorithm in DS-CDMA mobile radio using long spreading codes," *Proc. VTC97*, Vol.3, pp. 1430–1434, Phoenix, AZ, May 1997.

[33] F. Adachi, M. Sawahashi, and K. Okawa, "Tree-structured generation of orthogonal spreading codes with different lengths for forward link of DS-CDMA mobile radio," *Electron. Lett.*, Vol. 33, No.1, pp. 27–28, January 1997.

[34] J. Yang et al., "PN offset planning in 15–95 based CDMA system," *Proc. VTC'97*, Vol.3, pp. 1435-1439, Phoenix, AZ, 1997.

[35] A. Hottinen and K. Pehkonen, "A flexible multirate CDMA concept with multiuser detection," *Proc. ISSSTA96 Conference*, Vol. 1 pp. 556–560, Mainz, Gennany,1996.

[36] M. Gustafsson, et. al, "Different compressed mode techniques for interfrequency measurements in a wide-band DS-CDMA system," *Proc. PIMRC'97,* Helsinki, Finland, 1997.

[37] R. Wichman and A. Hottinen, "Multiuser detection for downlink CDMA communications in multipath fading channels," *Proc. VTC97*, Vol. 2, pp. 572–576, Phoenix, AZ, May 1997.

[38] H. Holma, A. Toskala, and T. Ojanpera, "Cellular coverage analysis of wideband MUD-CDMA system," *proc. PIMRC97*, Helsinki, Finland, September 1997.

[39] A. Rottinen, H. Holma, and A. Toskala, "Performance of multistage multiuser detection in a fading multipath channel," *Proc. PIMRC'95*, pp. 960-964, Toronto, Canada, September 27–29, 1995.

[40] H. Holma, A. Toskala, and A. Hottinen, "Performance of CDMA multiuser detection with antenna diversity and closed loop power control," *Proc. VTC'96*, Atlanta, GA, April 1996.

[41] S. Hamalainen, H. Rolma, and A. Toskala, "Capacity evaluation of a cellular CDMA uplink with multiuser detection," *Proc. 1SSSTA96 Conference*, Vol. 1, pp. 556–560, Mainz, Germany,1996.

[42] R. Holma, A. Toskala, and A. Hottinen, "Performance of CDMA multiuser detection with antenna diversity and closed loop power control," *Proc. VTC'96*, Atlanta, GA, pp. 362–366, April/May 1996.

[43] A. Hottinen, R. Holma, and A. Toskala, "Multiuser detection for multirate CDMA communications," *Proc. ICC'96*, Dallas, TX, June 1996.

[44] J. Skold et. al, "Cellular evolution into wideband services," *Proc. VTC97*, Vol.2. pp. 485–489, Phoenix, AZ, May 1997.

[45] A. Urie, "Advanced GSM: a long term future scenario for GSM," *Proc. Telecom 95*, Vol. 2, pp.33–37, Geneva, Switzerland, October 1995.

[46] P. Jung, J. Plechinger, M. Doetsch, and F. Berens, "Pragmatic approach to rate compatible punctured turbo-codes for mobile radio applications," *6th International Conference on Advances in Communications and Control: Telecommunications/Signal Processing*, Grecotel Imperial, Corfu, Greece, June 23–27, 1997.

[47] P. Jung, J. Plechinger, M. Doetsch, and F. Berens, "Advances on the application of turbo-codes to data services in third generation mobile networks," *International Symposium on Turbo-Codes*, Brest, September 3–5,1997.

[48] P. Ranta, A. Lappetelainen, and Z-C Honkasalo, "Interference cancellation by joint detection in random frequency hopping TDMA networks," *Proc. ICUPC96 Conference*, Vol. 1, Boston, MA, pp. 428– 432, 1996.

[49] M. Pukkila and P. Ranta, "Simultaneous channel estimation for multiple co-channel signals in TDMA mobile systems," *Proc. IEEE Nordic Signal Processing Symposium (NORSIG'96)*, Helsinki, Finland, September 1996.

[50] F. Muratore and V. Palestioi, "Burst transients and channelization of a narrowband TDMA mobile radio system," *Proc. 38th IEEE Vehicular Technology Conference*, Philadelphia, PA, June15–17, 1988.

[51] H. Honkasalo, "The technical evolution of GSM," *Proc. Telecom 95*, Geneva, Switzerland, October 1995.

[52] R. Prasad, J.A.M. Nijhof, and H.I. Cikel, "Performance analysis of the hybrid TDMA/CDMA protocol for mobile multimedia communications," *Proc. ICC'97*, pp. 1063–1067, Montreal, Canada, June 1997.

[53] A. Klein and P.W. Baier, "Linear unbiased data estimation in mobile radio systems applying CDMA," *IEEE J. Selected Areas in Comm.*, Vol. SAC-11, pp. 1058–1066, 1993.

[54] M.M. Naflhan, P. Jung, A. Steil, and P.W. Baier, "On the effects of quantization, nonlinear amplification and band limitation in CDMA mobile radio systems using joint detection," *Proc. Fifth Annual International Conference on Wireless Communications* (*WIRELESS'93*), Calgary, Canada, pp. 173–186, 1993.

[55] P. Jung, J.J. Blanz, M.M. Nallhan, and P.W. Baier, "Simulation of the uplink of JD-CDMA mobile radio systems with coherent receiver antenna diversity,"

Wireless Personal Communications, An International Journal (Kluwer), Vol.1, pp. 61–89, 1994.

[56] J. Mikkonen and J. Kruys, "The magic WAND: a wireless ATM access system," *Proc. ACTS Mobile Summit Conference*, Vol. 2, pp. 535–542, Granada, Spain, 1996.

[57] B. Engstrom and C. Osterberg, "A system for test of multi access methods based on OFDM," *Proc. IEEE VTC'94*, Stockholm, Sweden, 1994.

[58] M. Ericson et al., "Evaluation of the mixed service ability for competitive third generation multiple access technologies," *Proc. VTC'97*, Phoenix, AZ, 1997.

[59] M. Wahlqvist, R. Larsson, and C. Osterberg, "Time synchronization in the uplink of an OFDM system," *Proc. IEEE VTC'96*, pp. 1569–1573, Atlanta, GA, 1996.

[60] R. Larsson, C. Osterberg, and M. Wahlqvist, "Mixed traffic in a multicarrier system," *Proc. IEEE VTC'96*, pp. 1259–1263, Atlanta, GA, 1996.

[61] J.S. da Silva, B. Arroyo-Ferna'dez, B. Barani, J. Pereira, and D. Ikonomou, "Mobile and personal communications: ACTS and beyond," in *Wireless Communications TDMA versus CDMA* (S.G. Glisic and P.A. Leppanen, Eds.) Kluwer Academic Publishers, pp. 379– 414, September 1997.

Chapter 16
WBMC

16.1 INTRODUCTION

A brief introduction of WBMCS, challenges in implementing WBMCS, and the worldwide activities that are trying to meet these challenges are presented in Section 1.3 [1–25].

Multimedia and computer communications are playing an increasing role in today's society, creating new challenges to those working in the development of telecommunications systems. Besides that, telecommunications is increasingly relied upon to establish links between or to nonwired terminals. Thus, the pressure for nonwired systems to cope with increasing data rates is enormous, and WBMCSs, those with data rates higher than 2Mbps, are emerging rapidly, even if at this moment applications for very high transmission rates do not exist.

Several WBMCSs are being considered for different users with different needs: they may accommodate data rates ranging between 2 and 155 Mbps; terminals can be mobile (moving while communicating) or portable (static while communicating), and moving speeds can be as high as that of a fast train; users may or may not, be allowed to use more than one channel if their application requires so; the system bandwidth may be fixed, or dynamically allocated according to the user's needs; communication between terminals may be done directly or must go through a base station; possibility of usage of ATM technology; and so on. Many other cases can be listed as making the difference between various perspectives of a WBMCS, but two major approaches are emerging: WLANs directed to communications between computers, from which HIPERLAN [8] and IEEE 802.11

[9] are examples, and MBS [10] intended as a cellular system providing full mobility to B-ISDN users.

The different requirements imposed by the various approaches to WBMCSs have consequences on system design and development. The trade-offs between maximum flexibility on one hand and complexity and cost on the other hand are always difficult to decide, since they have an impact not only on the deployment of a system, but also on its future evolution and market acceptance (GSM is a good example of a system foreseen to accommodate additional services and capacities to those initially offered, and the fact that operators are already implementing Phase 2+ is the proof of that fact). This means that many decisions must be made concerning the several WBMCSs that will appear in the market; for HIPERLAN, for example, those decisions have already been made, since the system will be commercialized in the very near future, but for other systems there are still many aspects to be decided. Of course this depends on what are the applications intended to be supported by the systems, and if these applications are targeted to the mass market or only to some niches. The former (from which mobile phones are a good example) will certainly include WLANs, since the expansion of personal computers will dictate this application as one of great success in WBMCSs; the latter will possibly have TV broadcasters among their users (to establish links between HDTV cameras and the central control room).

Not only market aspects are at stake in the development and deployment of WBMCSs, but many technical challenges are posed as well. The transmission of such high data rates via radio in a mobile environment creates additional difficulties, compared to the existing systems; these difficulties are augmented by the fact that frequencies higher than UHF are needed to support the corresponding bandwidths, thus pushing mobile technology challenges (size and weight among other things) to frequencies where these aspects were not much considered up to now. However, additional challenges are posed to those involved in the development of WBMCSs: in today's world, where consumers are in the habit of using a communications system that is available in different places (the roaming capability of GSM is an example, since users can make and receive phone calls in an increasing number of countries worldwide), or being able to exchange information between different systems (the exchange of files between different computer applications and systems is an example in this case), it does not make sense to consider systems for the future that offer a high data rate but that do not support these capabilities to some extent.

This chapter presents an overview of a WBMCSs. WBMCS, providing a data rate higher than 2 Mbps and up to 155 Mbps, is done by addressing some of the applications and services that are foreseen, as well as some of the technical challenges that need to be solved, and by referring to some safety considerations. After an introduction and a brief summary of the standardization issue, the need for high data rates is justified, and possible applications are listed and compared concerning user mobility and bandwidth. The issue of the antennas and batteries for these systems are also addressed. Finally, a wireless ATM-based MBS is discussed.

16.2 STANDARDIZATION AND FREQUENCY BANDS

There are five main forums for the standardization of wireless broadband services namely, ATM Forum, Internet Engineering Task Force (IETF), ITU, ETSI, and multimedia mobile access communications (MMAC), and a study group set up by the Japanese Ministry of Post and Telecommunications (MPT) [21].

It is expected that MMAC and the ATM Forum will standardize the new system by the end of 1998, whereas the ETSI will standardize the new system by the middle of 1999.

Figure 16.1 shows that 5 and 60 GHz are the commercially important frequency bands because of geographically wide spectrum allocations in Europe, USA and Japan for the wireless broadband multimedia communication networks. An analysis of the propagation aspects at the bands foreseen for WBMCS microwaves, millimeter, waves and infrared is presented in Chapter 13 and [26–29].

16.3 THE NEED FOR HIGH DATA RATES

Data rate is really what broadband is about. For example, HIPERLAN allows physical channel bit rates up to 23.5 Mbps, and in MBS it can go as high as 155 Mbps. Such high data rates impose large bandwidths, thus pushing carrier frequencies for values higher than the UHF band: HIPERLAN has frequencies allocated in the 5- and 17-GHz bands; MBS will occupy the 40- and 60-GHz bands, and even the infrared band is being considered for broadband WLANs. However, many people argue whether there is a need for such high-capacity systems, bearing in mind all the compression algorithms developed and the type of applications that do require tens of megabits per second. One can face this issue from another perspective.

The need for high-capacity systems is recognized by the "Visionary Group" [22], put together by the European Commission, to give a perspective of what should be the "hot topics" in the area of telecommunications for research in the next European programs (following RACE and ACTS). In this visionary perspective of the road to follow, in order to go along with the needs of society in the years to come as far as communications is concerned, capacity is one of the major issues to be developed due to the foreseen increase in demand for new services (especially those based on multimedia). Together with this, personal mobility will impose new challenges to the development of new personal and mobile communication systems.

A conclusion can be drawn from this: even if at a certain point in time it may look "academic" to develop a system for a capacity much higher than what seems reasonable (in the sense that there are no applications requiring such high capacity), it is worthwhile to do it, since almost certainly in the future (which may be not very far off) applications will come out that need those capacities and even more. The story of fiber optics is elucidative on that.

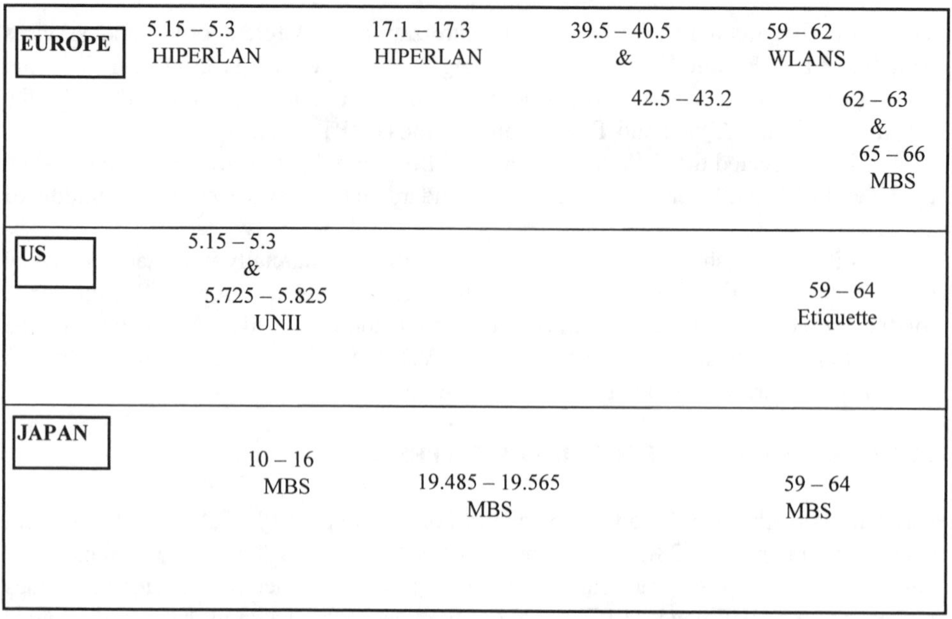

EUROPE	5.15 – 5.3 HIPERLAN	17.1 – 17.3 HIPERLAN	39.5 – 40.5 & 42.5 – 43.2	59 – 62 WLANS; 62 – 63 & 65 – 66 MBS
US	5.15 – 5.3 & 5.725 – 5.825 UNII			59 – 64 Etiquette
JAPAN	10 – 16 MBS	19.485 – 19.565 MBS		59 – 64 MBS

Figure 16.1 Frequency band in gegahertz for wireless broadband communications. UNII: unlicensed national information infrastructure.

16.4 SERVICES AND APPLICATIONS

The system concept of a WLAN like HIPERLAN and of a mobile broadband cellular system like MBS is totally different: they are directed to services and applications that differ in many aspects. A comparison of several systems, concerning two of the key features (mobility and data rate), is shown in Figure 16.2 [23], where it is clear that there is no competition between the two approaches.

The differences are more notorious when other parameters are compared (Table 16.1 [23]). One must be aware that HIPERLAN is already in the standardization phase at ETSI, while MBS is still in the development phase, and therefore these characteristics must be seen as goals and not as actual specifications; the intention of presenting this table is more for a comparison between the two types of systems, rather than a direct comparison between the two specific systems themselves.

The applications and services of the two systems are also different. HIPERLAN is mainly intended for communications between computers (thus being an extension of wired LANS); nevertheless, it can support real-time voice and image signals, and users are allowed some mobility and can have access to public networks. MBS services are wider in range. Figure 16.3 [24] shows several of the possible applications, according to their data rate and mobility; of course these applications are not exclusive of MBS, and many, if not all, are seen as possible applications of WBMSs in general. Some of these applications are already available for fixed and/or narrowband systems, thus WBMS represents only some extended upgrade in mobility or data rate; others are really specific of WBMS, and are only available through this type of system. Moreover, some of the applications require the existence of an operator, while others can be implemented on a private basis; the future, in the sense of market impositions, will dictate which ones will have success as key start services.

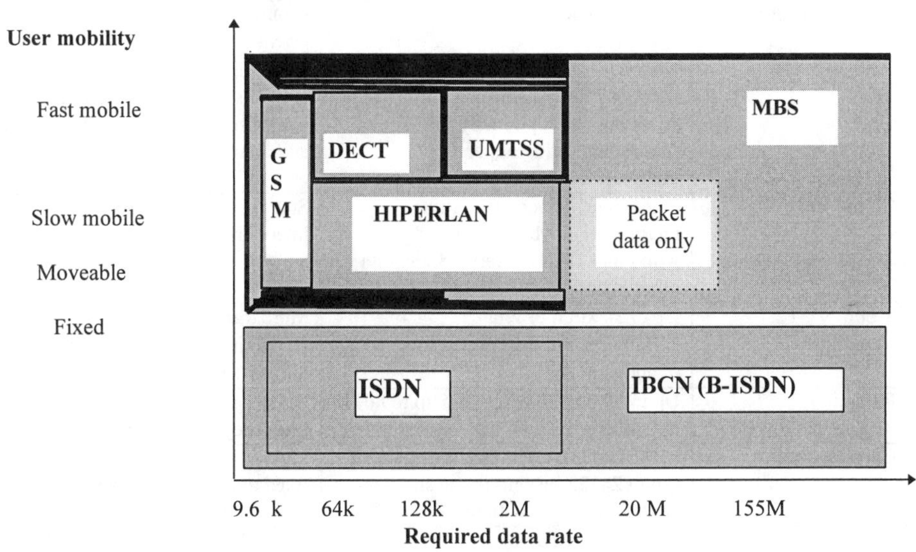

Figure 16.2 Comparison of mobility and data rates for several systems.

European R & D initiatives in mobile systems, and the corresponding efforts to standardize the upcoming systems, has lead to the definition of UMTS, which is considered a third-generation system.

Table 16.1
Comparison of HIPERLAN and MBS Characteristics

Parameter	HIPERLAN	MBS
Owner/ operator	Private system, owned and operated by the user	Public and private system
Objective	Extension or replacement of fixed LANS	Mobile or wireless extension or replacement of fixed B-ISDN
Applications	Primarily indoor and on-premises computer interconnection	Indoor and outdoor, all IBC services and applications
Service/data rates	MAC layer (bearer) service rates < 20 Mbps for asynchronous services 64 n kbps, up to 2.048 Mbps, for time-bounded services	ATM cell transfer capability: up to 155 Mbps
Communication	Connectionless	Connectionless and connection oriented
Infrastructure	No, only HIPERLAN nodes with functions for transmitting and receiving, and optionally for forwarding, bridging, and interworking	Yes, cellular system consisting of mobile stations, base stations comprising transceivers, controller, and interworking units
Configurations	Standalone, adhoc networking, integration or MAC level bridging with other LANs, interworking with other networks	Standalone, integration with B-ISDN, interworking with other networks
Mobility	Up to 36 km/h or 2 rad/s	More than 100 km/h
Coverage	Locally "unlimited" due to forwarding of active nodes	"Unlimited" due to cellular infrastructure and handover
Range	Up to 50m at 20 Mbps and up to 800m at 1 Mbps	Up to 1 km, depending on antenna and frequency
Channel access	FDMA/TDMA, variable data packets (up to 24322 bit); no frame structure, contention mode with priority	FDMA/TDMA, frame structure with fixed time slots and transmission bursts (356 symbols)
Frequency bands	5.15 – 5.30 GHz; 17.1 – 17.3 GHz	39.5 – 40.5 and 42.5 – 43.5 GHz; 62.0 – 63.0 and 65.0 – 66.0 GHz
Duplexing	1 frequency TDD	2 frequencies FDD, up to 4 carriers in parallel
Physical channel access	23.5 Mbps	40 to 160 Mbps
Modulation	GMSK	4 to 16 OQAM
Time scales	1996	2005

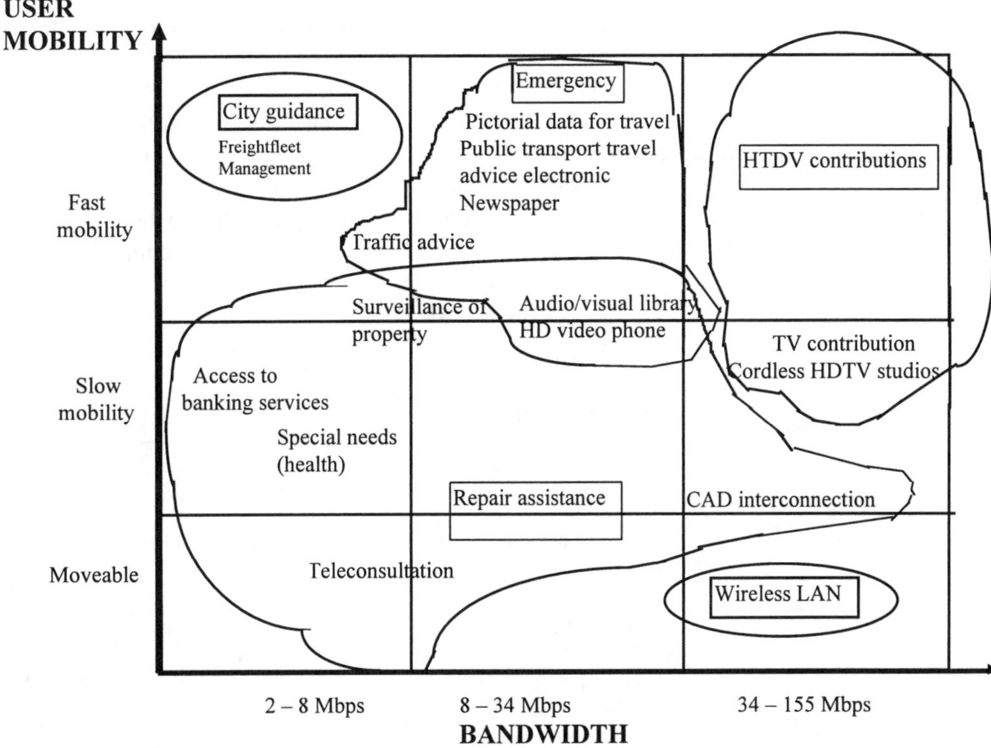

Figure 16.3 Possible WBMCS applications according to mobility and data rate.

However, there are differences between the foreseen features and applications of UMTS and MBS (Table 16.2) [25], the latter being capable of providing more advanced services, but with some possible restrictions on coverage.

16.5 ANTENNAS AND BATTERIES

Antennas and batteries play a key role in wireless systems: with the present capability of microelectronics to integrated circuits and to include signal processors into very small chips, antennas and batteries tend to impose the size and weight of mobile terminals. Of course the higher one goes in frequency, the less developed is the technology, and many problems are still found in size and weight at the millimeter-wave band, due to power consumption for example, but they are likely to be solved in the near future. The number of hours equipment based on batteries can work, or stand by, and the percentage of its weight corresponding to the battery is not a specific problem of WBMCSs. Laptop computers and cellular phones are the most common terminals relying on batteries these days. Huge investments on R&D have been and are being done on this technology, so that working time can be extended and weight can be reduced; a 100g battery corresponding to several hours of continuous work and a few days on standby are already on the market, but of course users want to have better numbers on this. Mobile multimedia terminals, certainly those to be used in some applications of WBMCSs, will be an extension of the current cellular phones. Therefore we foresee that the current problems associated with batteries will be transposed to the WBMCSs terminals; the same thing applies to laptop computers.

Table 16.2

Comparison Between UMTS and MBS

Features	UMTS	MBS
Spectrum allocation	2 GHz	40 GHz and/or 60 GHz
Bandwidth	Up to 2 Mbps	up to 155 Mbps
Environment	Indoor and outdoor(terrestrial and satellite)	Indoor and outdoor (terrestrial)
Coverage	Universal (except in remote zones)	
Service provision	Public and private	Public and private
Locality	Global	Service area
Mobility classes	Vehicle-mounted, portable, and hand-held	Portable, movable, and mobile
Personal mobility	User mobility using an identity module on any terminal; support of UPT	User mobility using smart cards or similar on any terminal; support of UPT
Interworking	PLMN, PSTN, ISDN and B-ISDN compatible	B-ISDN compatible (all functions for interworking with other systems will be provided by B-ISDN)
Applications		
Emergency services	Can be provided with severe restrictions on bandwidth	Offers additional services related to high-quality images
City guidance	Can be provided	Can be provided (offers high-quality full motion video)
High-definition TV	Not possible	Can be provided
High-definition video phone	Not possible	Can be provided
LAN interconnection	Not possible	Can be provided
Mobile office	Can be provided with severe restrictions on bandwidth	Offers additional services related to high-quality images
Interactive TV	Not possible	Can be provided
Radio extension to B-ISDN	Not possible	Can be provided
Telerobotics	Limited image quality	High-definition image quality

Antennas (size, type, technology, and so on) are not a specific problem of WBMCSs as well, but again they are very much related to the type of systems that will be made available for users. It does not make sense to impose restrictions (like having to point in a certain direction or avoiding someone to pass in between) on the type of mobility a mobile terminal can have. Even for portable terminals (like computers), these restrictions make no sense; hence, there are only two options as far as antennas are concerned: either an omnidirectional antenna (dipole type) or an adaptive array antenna (enabling the use of narrow beams) is used. Either way, patch antennas seem a very promising solution for general use in WBMCSs: for the frequency bands under concern, isolated patches or adaptive arrays (with many elements) can be made with a small size (e.g., a credit card), thus enabling the terminal not to be limited in size by the antenna system.

The role of antennas radiation patterns is not negligible when discussing the performance of a system and their influence on parameters associated with wave propagation. Although the trade-off between an omnidirectional and a narrowbeam antenna is not particular of WBMCS, it assumes particular importance at microwave and millimeter waves because of the characteristics of wave propagation at these bands. Using an omnidirectional antenna means a lower gain but also the possibility to receive signals from various directions without the requirement for knowing where the base station is, and allowing the received rays coming from reflections on the propagation scenario. On the other hand, the use of a very directive antenna provides a higher gain, but it has to be pointed at the base station and does not receive reflected waves coming from directions very different from the one to which it is pointed. A low value for the delay spread is desirable. It limits the maximum data rate that is transmitted, and even the use of equalizers cannot overcome totally the problem, for which narrowbeam antenna can be the solution; although it has the additional advantage of the high gain, which is important if the link budget does not present very large margins for fading, the need for pointing it can be very discouraging, if not a drawback when LOS does not exist. On the other hand, omnidirectional antennas lead to high values of delay spread, but they may ensure that the link still exists, relying on reflections if LOS is lost.

16.6 SAFETY CONSIDERATIONS

Up to a few years ago, the analysis of possible harmful effects of electromagnetic radiation on people was devoted mainly to power lines and radars, due to the huge power levels involved in those systems. Even when mobile phone systems appeared, there was no major concern, since the antennas were installed on the roofs of cars. With the development of personal communication systems, in which users carry mobile phones inside their coat pockets, and the antenna radiates a few centimeters from the head, the safety problem gained a great importance and a new perspective. Much research is found in the literature not only on the absorption of power inside the head, but also on the influence of the head on the antenna's radiation pattern and input impedance. However, these works have addressed only the frequency bands at use in today's systems; that is, up to 2 GHz (mainly on the 900- and 1800-MHz bands), and only very few references are made to systems working at higher frequencies, as it is the case of WBMCSs.

The problems associated with infrared are different from those posed by microwaves and millimeter waves. Eye safety, rather than power absorption inside the head, is the issue in these cases, since the eye acts as a filter to the electromagnetic radiation, allowing only light and near-frequency radiation to enter into it, and the amount of power absorption inside the human body is negligible. Exposure of the eye to high levels of infrared radiation may cause cataractlike diseases, and the maximum allowed transmitter power seems to limit the range to a few meters [29]. If this is the case, safety restrictions will pose severe limitations to the use of infrared in WBMCSs, as far as general applications are concerned. The question in this case is not that there are always problems during system operation (as in mobile phones, for example), but the damage that may be caused if someone looks at the transmitter during operation.

Microwaves and millimeter waves have no special effect on eyes; power absorption is really the problem. In the case of WLANs, antennas do not radiate very near (1 or 2 cm) the user as in the mobile phone case, thus enabling power limitations to be less restrictive; also, if mobile multimedia terminals are used like PDAs, the case will be similar to the previous one. But if terminals are used in the same form as mobile phones, then maximum transmitter powers have to be established, as it did for the current personal communication systems. The standards for safety levels have already been set in the United States and Europe, since the ones used for UHF extend up to 300 GHz (IEEE/ANSI and CENELEC recommendations are the references). Thus, it is left to researchers in this area to extend their work to higher frequencies, by evaluating SAR (the amount of power dissipated by unit of mass) levels inside the head (or other parts of the human body very near the radiating system), from which maximum transmitter powers will be established. However, this may not be as straightforward as it seems, since the calculation of SAR is usually done by solving integral or differential equations via numerical methods (method of moments or finite difference), which require models of the head made of small elements (cubes, for example) with dimensions of the order of a tenth of the wavelength. This already requires powerful computer resources (in memory and CPU time) for frequencies in the high UHF band, and may limit the possibility of analyzing frequencies much higher than UHF. On the other hand, the higher the frequency, the smaller the penetration of radio waves into the human body, hence making possible to have models of only some centimeters deep. This is really an area for further research.

16.7 ATM-BASED WIRELESS (MOBILE) BROADBAND MULTIMEDIA SYSTEMS

A recent study of MBS, ATM, and ATM-oriented MBS has drawn the attention of several researchers [30–39].

With enormous complexity of managing and operating the many different types of networks now in use, the door is open for finding a common platform — a network on which all established services can be supported and which will allow new services to be introduced without needing new networks on which to run them. The answer seems to be ATM. This is the technology being defined and standardized for B-ISDN. Thus, ATM, when adequately modified, is also an answer for the future mobile wireless broadband multimedia systems.

ATM is a packet-oriented transmission scheme. The transmission path of the packets of constant length, the so-called ATM cells, is established during connection setup between the two endpoints by assigning a virtual channel. At this time, the necessary resources are provided and the logical channels are assigned. All packets of a virtual channel are carried over the same path. The transmission capacity of the virtual channel are characterized by the parameters mean bit rate and peak bit rate during connection setup. ATM cells are generated according to the need of the data source. Thus, ATM is a very good method to meet the dynamic requirements of connections with variable data rates.

MBS is the interface between the fixed ATM net at the base-station side and mobile ATM net at the mobile-station side. Normally, the ATM net at the mobile-station side will only consist of one end system. For every end station, it is possible to operate several virtual channels with different data rates at the same time.

A conceptual view of the ATM-type broadband communications network is shown in Figure 16.4. The most important benefit of ATM is its flexibility; it is used for the new high bit rate services, which are either VBR or burst traffic. Several factors that tend to favor the use of ATM cell transport in MBS, are as follows:

- Flexible bandwidth allocation and service-type selection for a range of applications;
- Efficient multiplexing of traffic from bursty data/multimedia sources;
- End-to-end provisioning of broadband services over wired and wireless networks;
- Suitability of available ATM switching for intercell switching;
- Improved service reliability with packet-switching techniques;
- Ease of interfacing with wired B-ISDN systems.

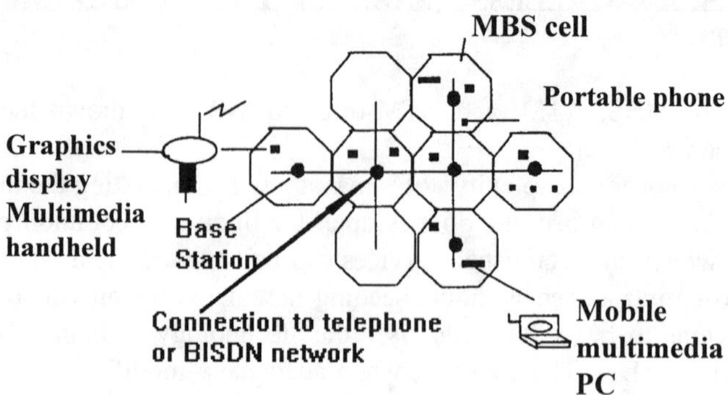

Figure 16.4 Conceptual view of MBS.

Taking the above points into consideration, adoption of ATM-compatible, fixed-length, cell-relay format for MBS is recommended. A possible ATM-compatible MBS approach is shown in Figure 16.5.

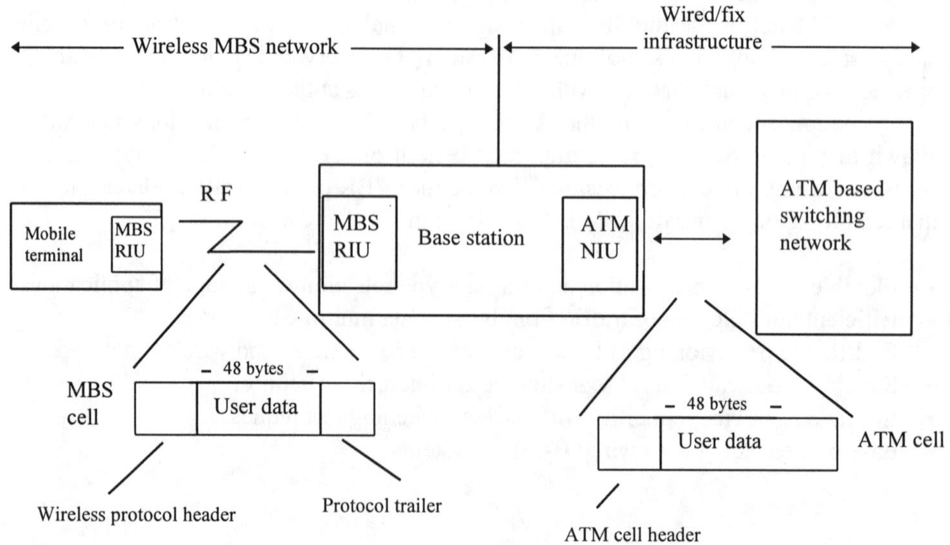

Figure 16.5 ATM-compatible MBS approach. Here, RIU and NIU stand for radio interface unit and network interface unit, respectively.

With this approach, the 48-byte ATM cell payload becomes the basic unit of data within the MBS network. Within MBS, specific protocol layers (e.g., data link and medium-access control layer) are added to the ATM payload as required, and replaced by ATM headers before entering the fixed network [37]. The use of ATM switching for intercell traffic also avoids the crucial problem of developing a new backbone network with sufficient throughput to support intercommunication among large numbers of small cells. ATM multiplexers are used to combine traffic from several base stations into a single ATM port.

For a seamless internetworking mechanism with the wired broadband network, it is vital to have the MBS protocol layering harmonized with the ATM stack. In Figure 16.6, a protocol reference model is shown. In this approach, new wireless channel-specific physical, medium-access control, and data link layers are added below the ATM network layer. This means that regular network layer and control services such as call set up, virtual channel identifier/virtual path identifier (VCI/VPI) addressing, cell prioritization and flow-control indication will continue to be used by mobile services. The baseline ATM network and signaling protocol are specified to specific mobility-related functions, such as address registration (roaming), broadcasting, handoff and so forth.

16.7.1 Fragments

As mentioned earlier, traffic is carried by ATM. Two concepts of ATM are important in this context, namely virtual paths and cells. Virtual paths are routes that traffic takes when going through an ATM network. Every element of a traffic stream follows this route. It would be natural to take these paths as the unit of communication between mobile and base stations. As suggested in [37], this is achieved by assigning a separate spreading code to each virtual path.

ATM traffic is divided into cells. These cells contain 48 octets or 384 bits of payload data, and a header of 5 octets. As in [37], we assume that the contents of the header are implied by the used spreading code, except for payload type and cell loss priority. These fields total 4 bits or about 1% of the total traffic flow, so they are not included in this discussion.

However small these ATM cells are, they may still be too large to be practically transmitted over a radio link. An important consideration is whether to break the cells into separate *fragments*, and how large these fragments should be. Assuming that a fragment is only accepted after it is received without any errors, it is well known that the probability p of receiving a fragment with size n bits without errors is found through

$$p = (1 - \text{BEP})^n$$

It is clear that the probability diminishes with increasing n. There is, of course, a limit to the number of fragments per cell. As every fragment needs to be acknowledged to correct errors, the use of a lot of fragments incurs a certain overhead [40].

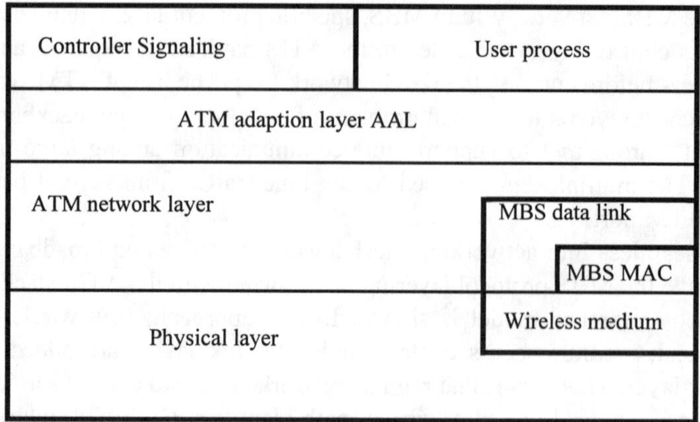

Figure 16.6 Relation of wireless network protocol layers.

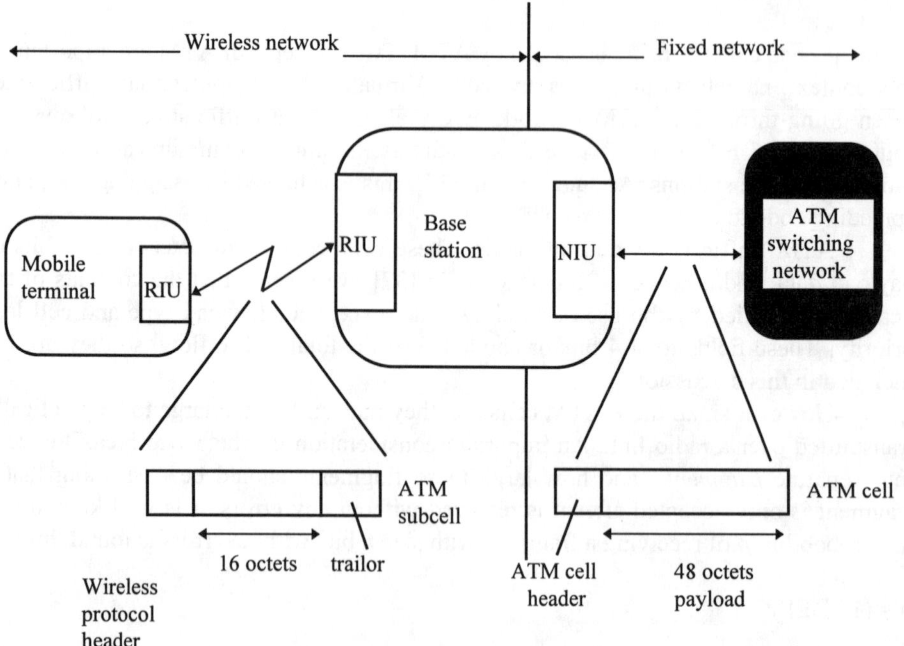

Figure 16.7 Conceptual view of an ATM-compatible air interface.

The payload of an ATM cell consists of 48 octets, meaning that without any segmentation, packet lengths of 384 bits would have to be transmitted over the mobile radio channel. Higher throughput levels over a fading mobile radio channel are achieved when using packet of smaller lengths. So a suitably integer submultiple of a cell is chosen as the basic unit of data over the wireless radio medium. Based on the system parameters, a submultiple cell size of 16 octets or 128 bits is used as an appropriate value.

The air interface is terminated directly at the base station, as illustrated in Figure 16.7. Also shown is the segmentation of an ATM cell into its three subcells and the addition of an extra wireless ATM header.

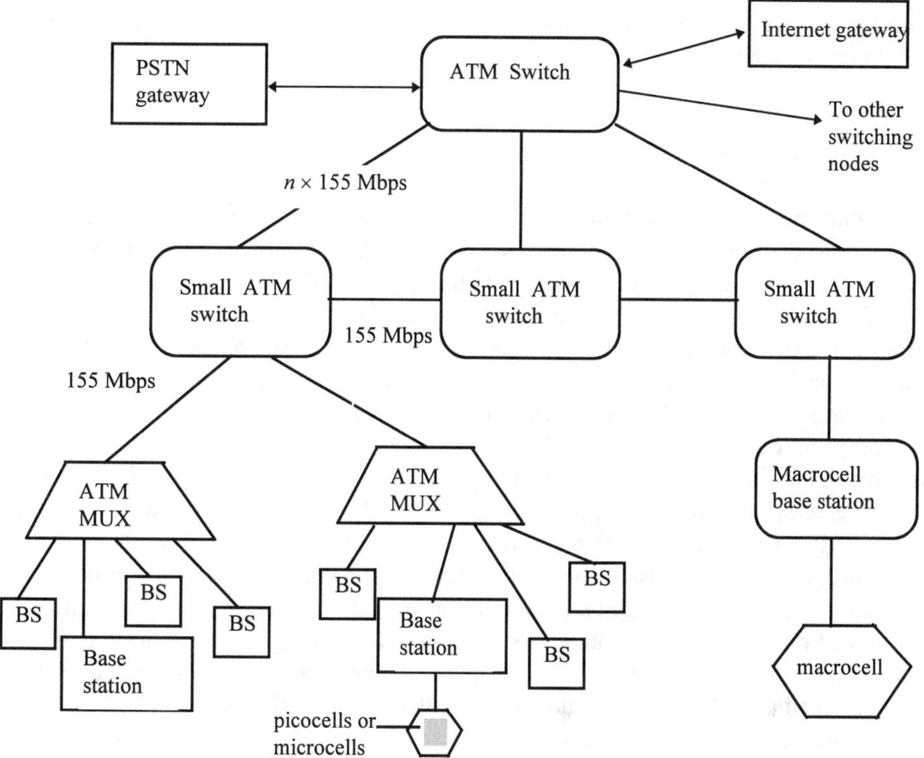

Figure 16.8 Organization of an ATM backbone switching network.

At the RIU, information is segmented into subcells of 16 octets and a wireless ATM-specific protocol header is added. At the receiving side, these subcells are reassembled for transmission over the fixed ATM network. This is done immediately at the base station or at the so-called base station interworking unit (BIU) via an intermediate (broadband) access network. Termination at the base station is the most straightforward method, but it makes the base stations more complex and expensive since additional processing has to be done. However, especially in picocellular and perhaps also microcellular environments and during the starting phase of ATM, it is advantageous to terminate the ATM air interface at the so-called BIUs. Several picocellular or microcellular base stations are interconnected to one BIU over either an ATM network or a non-ATM network. In the case of an ATM-based access network, the received subcell is encapsulated into a normal ATM cell with header, resulting in some spare space in the cell to be transmitted. It is also possible to transmit the received subcells over the particular (narrowband) access network as they are, and then let the BIU reassemble the subcells. The BIU then performs both the reassemble function of ATM subcells and further ATM switching and routing. Figure 16.8 shows the structure of a possible ATM backbone switching network for a wireless ATM-based mobile communications system.

16.8 CONCLUSIONS

This chapter presented some thoughts about current research on WBMCS, concerning technical issues as well as some aspects related to their deployment and to possible services and applications. Some requirements for WBMCS were listed, and it was justified that although it may seem that data rates as high as 155 Mbps are not needed at the present time, research will ensure that the necessary development has been done when the need arises. WLANs are one of the strongest candidates to use as an example of successful WBMCS, although other applications were referred to. Some aspects of wave propagation were addressed, like the average power decay (almost as in free space for the millimeter-wave band, and not very far from it for the microwave band), and the delay spread (which can be a limiting factor for high data rates, and can have median RMS values of 80 ns at 60 GHz); the problems related to modeling of wave propagation were also discussed, being geometrical optics approaches enough in many of the cases. The problem of battery lifetime was included, although it is not specific of WBMCSs. Some discussion on antennas was presented, mainly on the influence of the radiation pattern on system performance; patch antennas seem to be the solution for WBMCS terminals. Since the frequency bands that are, or will be, used by WBMCSs have started only recently to be considered for these type of applications, almost nothing is found in the literature concerning user safety, but this is an area that will attract the attention of researchers in the near future. An ATM-compatible MBS approach was also presented. ATM can be used for bit rates needed for MBS. For the use of ATM with MBS, an architecture of ATM cell was described.

REFERENCES

[1] R. Prasad, "Wireless broadband communication systems," *IEEE Comm. Mag.*, Vol. 35, pp. 18, January 1997.

[2] L.M. Correia and R. Prasad, "An overview of wireless broadband communications," *IEEE Comm. Mag.*, Vol. 35, pp. 28–33, January 1997.

[3] R. Prasad and L.M. Correia, "Wireless broadband multimedia communications," *International Wireless and Telecommunications Symposium*, Shah Alam, Malaysia, Vol. 2, pp. 55–70 May 14–16,1997.

[4] R. Prasad and L.M. Corriea, "An overview of wireless broadband multimedia communications," *Proc. MoMuC'97*, Seoul, Korea, pp. 17–31, September/October 1997.

[5] W. Honcharenko, J.P. Kruys, D.Y. Lee, and N.J. Shah, "Broadband wireless access," *IEEE Comm. Mag.*, Vol. 35, pp. 20–26, January 1997.

[6] M. Morinaga, M. Nakagawa, and R. Kohno, "New concepts and technologies for achieving highly reliable and high-capacity multimedia wireless communications systems," *IEEE Comm. Mag.*, Vol. 35, pp. 34–40, January 1997.

[7] M. Chelouche, S. Hethuin, and L. Ramel, "Digital wireless broadband corporate and private networks: RNET concepts and applications," *IEEE Comm. Mag.*, Vol. 35, pp. 42–51, January 1997.

[8] ETSI/RES — European Telecommunications Standards Institute I Radio Equipment and Systems, *HIPERLAN, Services and Facilities*, Sophia-Antipolis, France, December 1992.

[9] IEEE 802.11, *IEEE 802.11, Wireless Access Method and Physical Layer Specifications*, New York, NY, September 1994.

[10] L. Fernandes, "R2067 MBS - mobile broadband system," in *Proc. ICUPC'93 — 2nd IEEE International Conference on Universal Personal Communications*, Ottawa, Canada, October 1993.

[11] R. Prasad, "Research challenges in future wireless personal communications: microwave perspective," *Proc. 25th European Microwave Conference* (Keynote opening address), Bologna, Italy, pp.4–11, September 4–7, 1995.

[12] R. Prasad, "Overview of wireless personal communications: microwave perspectives," *IEEE Comm. Mag.*, Vol. 35, pp.104–108, April 1997.

[13] J. S. da Silva, B. Arroyo-Fernandez, B. Barani, J. Pereira, and D. Ikonomou, "Mobile and personal communications: ACTS and beyond," *PIMRC'97*, Helsinki, Finland, September 1997.

[14] *Wireless ATM Workshop*, Espoo, Finland, September 1996.

[15] J. Aldis and F. Bauchot, "Magic WAND project: radio subsystem design," *Wireless ATM Workshop*, Espoo, Finland, September 1996.

[16] M. Umehira, "AWA: ATM wireless access system," *Wireless ATM Workshop*, Espoo, Finland, September 1996.

[17] C. Ciotti, "ACTS MEDIAN— wireless broadband CPN/LAN for professional and residential multimedia applications," *Wireless ATM Workshop*, Espoo, Finland, September 1996.

[18] A.S. Krishnakumar, "Wireless ATM research in Bell Laboratories," *Wireless ATM Workshop*, Espoo, Finland, September 1996.

[19] L. French and D. Raychaudhuri, "WATMnet: a prototype wireless ATM system for multimedia personal communications," *Wireless ATM Workshop*, Espoo, Finland, September 1996.

[20] G. Wu, Y. Hase, K. Taira, and K. Iwasak, "A wireless ATM oriented MAC protocol for high-speed wireless LAN," *Proc. PIMRC'97,* Helsinki, Finland, pp. 198–203, September 1997.

[21] J. Kruys, "Standardization of wireless high speed premises data networks," *Wireless ATM Workshop*, Espoo, Finland, September 1996.

[22] European Commission, "Communications for society visionary research," European Commission, DG XIII/B, Brussels, Belgium, February 1997.

[23] C.-H. Rokitansky and M. Scheibenbogen (Eds.), "Updated version of system description document," *RACE Deliverable R2067/UA/WP215/DS/P/068.b1*, RACE Central Office, European Comrnission, Brussels, Belgium, December 1995.

[24] J.T. Zubryzclci and S.P. Ashby (Eds.), "Report on services and applications," *RACE Deliverable R2O67/BBC/WP122/DR/P/027.b1*, RACE Central Office, European Commission, Brussels, Belgium, May 1994.

[25] T.M. Gomes, "MBS and UMTS: the relationship between these systems," in *Proc. RACE Mobile Telecommunications Summit*, Cascais, Portugal, pp. 516–519, November 1995.

[26] P.F.M. Smulders, "Broadband wireless LANs: a feasibility study;" *Ph.D. Thesis*, Eindhoven Univ. of Technology, Eindhoven, The Netherlands, December 1995.

[27] L.M. Correia and P.O, Frances, "A propagation model for the estimation of the average received power in an outdoor environment at the millimeter wave band," in *Proc. VTC'94 — IEEE VTS 44th Vehicular Technology Conference*, Stockholm, Sweden, pp. 1785–1788 , July 1994.

[28] G. Lovnes, J. J. Reis, and R.H. Raekken, "Channel sounding measurements at 59 GHz in city streets," in *Proc. PIMRC'94 — 5th IEEE International Symposium on Personal, Indoor; and Mobile Radio Communications*, The Hague, The Netherlands, pp. 496–500, September 1994.

[29] J. J. Fernandes, P.A. Watson, and J.C. Neves, "Wireless LANs: physical properties of infrared systems vs Mmw systems," *IEEE Comm. Mag.,* Vol. 32, No.8, pp. 68–73, August 1994.

[30] A. R. Prasad, "Asynchronous trasfer mode based mobile broadband system," *Proc. IEEE 3rd Symposium on Comm. & Vech. Technology in Benelux*, Eindhoven, The Netherlands, pp.143–148, October 1995.

[31] R. Prasad and L. Vandendorpe, "An overview of millimeter wave indoor wireless communication system," *Proc. 2nd Int. Conf. on Universal Personal Communications*, pp. 885–889, Ottawa, Canada, 1993.

[32] R. Prasad and L. Vandendorpe, "Cost 231 project: performance evaluation of a millimetric-wave indoor wireless communications system," *RACE Mobile Telecommunications Workshop*, Metz, France, pp. 137–144, June16–18 1993.

[33] K. Zauner, "On the ATM road to broadband," *Telecom Report International*, No.3, pp. 26–29, 1993.

[34] Tubtiang, H.I. Kwon and O. Pujolle, "A simple ATM switching architecture for broadband-ISDN and its performance," *Modeling and Performance Evaluation of ATM Technology (C-I5)*, pp. 361–367, IFIP 1993.

[35] W. Stalings, *Advances in ISDN and Broadband ISDN*, IEEE Computer Society Press, 1992.

[36] I.M. Leslie, D.R. McAuley, and D.L. Tennenhouse, "ATM everywhere?" *IEEE Network*, pp. 40–46, March 1993.

[37] M.J. McTiffin, A.P. Hulbert, T.J. Ketseoglou, W. Heimsch, and O. Crisp, "Mobile access to an ATM network using a CDMA air interface," *IEEE J. Selected Areas in Comm.*, Vol. 12, No. 5, pp. 900–908, June 1994.

[38] D. Raychaudhri, "ATM-based transport architecture for multiservices wireless personal communication networks," *IEEE J. Selected Areas in Comm.*, Vol.12, No. 8, pp. 1401–1414, October 1994.

[39] R.R. Geij, "Mobile multimedia scenario using ATM and microcellular technologies," *Vehicular Technology*, Vol. 43, No. 3, pp. 699–703, August 1994.

[40] J.D. Bakker and R. Prasad, "Wireless multimedia communications using a CDMA-based ATM air interface," *Proc. IEEE ISSSTA'96*, Mainz, Germany, pp. 1128–1132, September 1996.

Chapter 17
OFDM-Based Wireless ATM Transmission System

17.1 INTRODUCTION

As mentioned in Chapter 10, OFDM has attracted attention as a high-rate transmission scheme in multipath channel [1–7]. In the OFDM systems, ISI is eliminated by inserting a guard interval between successively transmitted signals, which makes complicated equalization unnecessary. However, this system must be equipped with guard interval insertion and removal circuits. In addition, the OFDM technique is very sensitive to the carrier frequency offset between the transmitter and receiver. The frequency offset destroys the orthogonality among the subcarriers. Therefore, we must estimate both the amounts of frequency offset and propagation characteristics.

This chapter describes one configuration of the OFDM-based wireless ATM transmission system. This configuration includes the synchronization method of wireless ATM cell and channel and frequency offset estimation method to realize robust OFDM transmission.

This chapter is organized as follows: Section 17.2 describes the system configuration, numerical results are presented in Section 17.3, and conclusions are given in Section 17.7.

17.2 SYSTEM CONFIGURATION

This section presents the transmitter and receiver configurations of the proposed OFDM-based wireless ATM system.

17.2.1 Transmitter Configuration

Figure 17.1 shows a block diagram of the proposed OFDM-based wireless ATM cell transmitter. At the transmitter, the transmitted high-speed serial data are converted to $N1$ parallel data and fed to one OFDM-ATM cell-maker. This OFDM-ATM cell-maker consists of ATM cell header generator, ATM cell frame-maker and pilot PN (pseudorandom) noise sequence generator. In the ATM cell frame-maker, using header

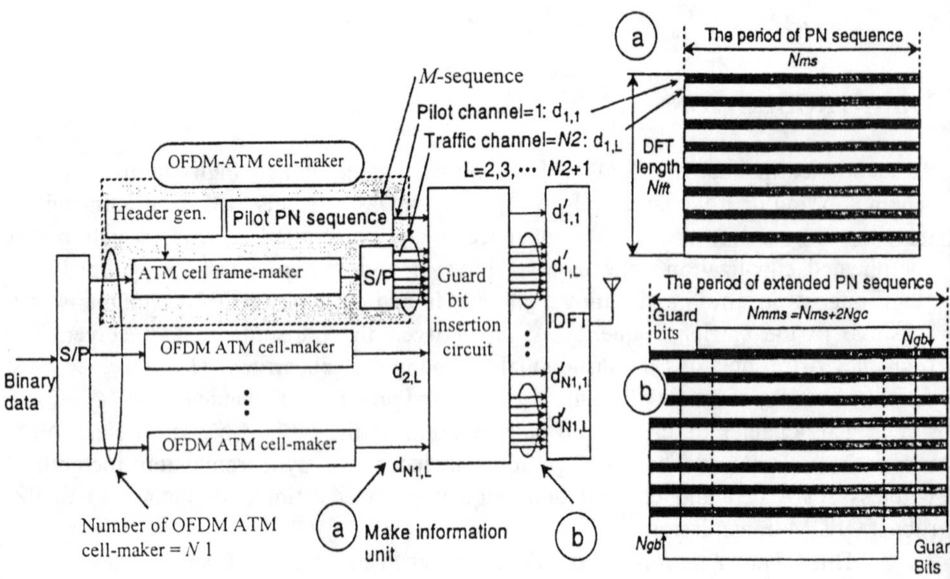

Figure 17.1 Configuration of the transmitter.

generator and input data sequence, 53 bytes (424 bits) of ATM cells are generated and coded to reduce ATM cell error rate. After that, these ATM cells are generated and coded to reduce ATM cell error rate. After that, these ATM cell data are converted to low-speed parallel data of $N2$ channels. At the same time, one fixed pattern pilot PN sequence with the period N_{ms} (in this chapter, M sequence) is inserted at every OFDM-ATM cell-maker. We set the period of PN sequence as one information unit time. As a consequence, one OFDM-ATM cell-maker generates $N2+1$ parallel data. This transmitter includes $N1$ OFDM-ATM cell-maker. Therefore, one information unit consists of $N1 \times (N2 + 1) \times N_{ms}$ bits. The data information unit is given by ($k = -\infty \ldots +\infty$)

$$d_{i,L}(k) = \begin{cases} dI_{i,L}(k) + j.dQ_{i,L}(k) & L = 2,3,\ldots \\ dp(k) + j.0 & L = 1 \end{cases} \tag{17.1}$$

where $d_{i,L}(k)$, $dI_{i.L}(k)$ and $dQ_{i,L}(k)$ are the complex, in-phase, and quadrature channel transmission data, which are allocated in Lth parallel channel of ith OFDM-ATM cell-maker, respectively.

In the case of $L = 1$, we insert a pilot M sequence $dp(k)$ with the period N_{ms} to only I channel; $dp(k)$ has the feature given by

$$\sum_{k=1}^{N_{ms}} dp(k) \times dp(k+c) = \begin{cases} N_{ms} & c = n.N_{ms} : n = 1,2,\ldots \\ -1 & \text{otherwise} \end{cases} \tag{17.2}$$

Figure 17.1(a) also indicates the configuration of the information unit. Next, $N1 \times (N2 + 1) \times N_{ms}$ bits of one information unit are fed into a guard bit insertion circuit. In the guard bit insertion circuit, the guard bits, which consist of N_{gb} bits, are added before and after this parallel data with the period of PN sequence to avoid ISI and to realize easy synchronization of wireless ATM frame and DFT start point at the receiver. The guard bits are extended by repeating the same sequence [6]. Therefore, the period of one extend information unit changes from N_{ms} to $N_{mms} = N_{ms} + 2\,N_{gb}$. The data-inserted guard bit is given as follows: ($n = 1, 2, \ldots$)

$$d'_{i,L}(k) = \begin{cases} d_{i,L}\left(k - (n-1)2N_{gb}\right) & (n-1)N_{mms} \le k < (n-1)N_{mms} + N_{ms} \\ d_{i,L}\left(k - (n-1)2N_{gb} + N_{ms}\right) & (n-1)N_{mms} - N_{gb} \le k < (n-1)N_{mms} \\ d_{i,L}\left(k + (n-1)2N_{gb} - N_{ms}\right) & (n-1)N_{mms} + N_{ms} \le k < nN_{mms} - N_{gb} \end{cases} \tag{17.3}$$

Figure 17.1(b) indicates the configuration of guard bits inserted in the information unit. These are fed into the IDFT circuit and an OFDM signal is generated. The transmitted data are given by

$$s(t) = \sum_{k=-\infty}^{+\infty} \sum_{i=1}^{N_1} \sum_{L=1}^{N_2+1} d'_{i,L}(k) \exp\left(j2\pi f_{i,L}(t - kT_s)\right) f(t - kT_s) \qquad (17.4)$$

where T_S is the symbol duration, and $f_{i,L}$ is transmission frequency of Lth subchannel in ith ATM cell-maker and given as follows

$$f_{i,L} = f_0 + \frac{(i-1)(N_2+1) + L - 1}{T_s} \qquad (17.5)$$

with f_0 as the lowest transmission frequency.

Moreover, $f(t)$ is the pulse waveform of each symbol defined as

$$f(t) = \begin{cases} 1 & 0 \le t \le T_s \\ 0 & \text{otherwise} \end{cases} \qquad (17.6)$$

The transmitted data $s(t)$ is contaminated by multipath fading and AWGN. In the receiver, the received signal is given by

$$r(t) = \int_0^{\infty} h(\tau; t) s(t - \tau) d\tau + n(t) \qquad (17.7)$$

where $h(\tau;\, t)$ is the impulse response of the radio channel at time t, and $n(t)$ is the complex AWGN.

17.2.2 Receiver Configuration

Figure 17.2 shows a block diagram of the proposed OFDM-based wireless ATM cell receiver. At the receiver, the received signal is downconverted to low frequency and quasi-coherently detected at the quadrature demodulator. The detected signal is N_p times oversampled at the A/D converter, and fed to frame and DFT start timing estimator. $\tau_{samp}(k)$: $k = 1, 2, \ldots, (N_p \times M \times N_{mns} \times N_{fft})$ are the stored discrete data, where M is the number of observation symbols to estimate frame and DFT start timing and N_{fft} is the length of the DFT.

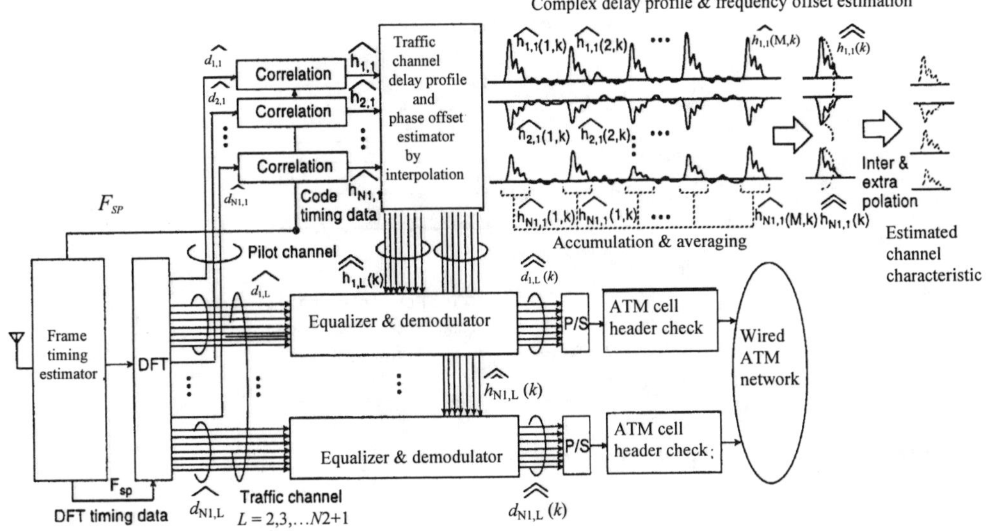

Figure 17.2 Configuration of the receiver .

In this estimator, we decide the frame and DFT start timing by utilizing the guard bits extended by repeating the information unit. Figure 17.3 explains the principle of this estimation method. As shown in Figure 17.3, the estimation method is, first of all, to correlate between received data and delayed received data with the delay time of the period of M sequence (see Figure 17.3).

$$R(k) = r_{\text{samp}}(k)r^{*}_{\text{samp}}(k + N_p N_{\text{ms}} N_{\text{fft}}) \tag{17.8}$$

where * indicates the complex conjugate.

In this case, during guard interval, the correlation is higher than the other interval because the guard bits are extended by repeating the information unit. To increase signal-to-noise power ratio (S/N) of this correlation results, we coherently sum and average the products obtained from M symbols with the interval ($N_p.N_{\text{mns}}$) samples. The value of $R_{\text{av}}(k)$: $k = 1, 2, \ldots, (N_{\text{p}}.\ N_{\text{mns}})$ is given by

$$R_{\text{av}}(k) = \frac{1}{M}\sum_{i=0}^{M-1} R(k + iN_p N_{\text{mms}} N_{\text{fft}}) \tag{17.9}$$

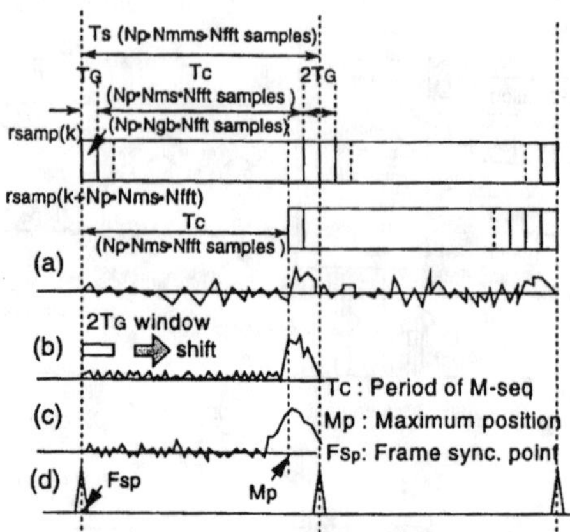

Figure 17.3 Frame and DFT timing estimation method.

The OFDM signal waveform, which includes many kinds of signals, are regarded as one of Gaussian noise waveform by using central limit approximation. If the statistic distribution of signal $g(t)$ is regarded as a Gaussian wave with zero mean and unit variance,

$$\frac{1}{2}E\left[g(t)g^*(t)\right]=1, \qquad \frac{1}{2}E\left[g(t)g^*(t+t_s)\right]=0 \tag{17.10}$$

$E[.]$ means average operation and t_S is the random time but not $t_S = 0$. The guard interval of the OFDM signal is cyclically extended within $T_G = N_{gb}$, that is, if $t_g \in T_G$, $g(t_g) = g(t_g + T_C)$, where T_C is the period of one M sequence. Therefore, the MC-CDM signal $g(t)$ has special feature shown as follows,

$$\frac{1}{2}E\left[g(t)g^*(t+T_c)\right]=\begin{cases}1 & (t \in \text{guard period} \\ 0 & (\text{otherwise})\end{cases} \tag{17.11}$$

That is why $R_{av}(k)$ will be shown in Figure 17.3(b). After that, we set a window with the width of $(N_p.2.N_{gb}.N_{fft})$. We integrate $R_{av}(k)$ during the interval of window and shift this window to one sample. The value of $R_{iav}(1)$: $1 = 1, 2,\ldots,N_p.N_{mns}.N_{fft}$ is

$$R_{\mathrm{iav}}(l) = \sum_{k=l}^{l+N_p 2N_{\mathrm{gb}}N_{\mathrm{fft}}-1} R_{\mathrm{av}}(k) \qquad (if \;\; l + N_p 2N_{\mathrm{gb}} N_{fft} < N_p N_{\mathrm{mms}} N_{\mathrm{fft}})$$

$$= \sum_{k=l}^{N_p N_{\mathrm{mms}} N_{fft}} R_{\mathrm{av}}(k) + \sum_{k=1}^{l+N_p(2N_{\mathrm{gb}}-N_{\mathrm{mms}})N_{\mathrm{fft}}-1} R_{av}(k) \qquad \text{(otherwise)} \tag{17.12}$$

If the window interval is equal to guard interval of $2T_G$ (in this case, we set l = M_p in Figure 17.3(c), R_{iav}(l) is maximum value. So, we decide the position of the guard interval and get the frame synchronization point by defining (see Figure 17.3(d))

$$F_{\mathrm{sp}} = M_p - N_p N_{\mathrm{ms}} N_{fft} \tag{17.13}$$

This synchronization method is based on [6–10]. After deciding the start point of DFT, the received data are resampled with the interval of the number of oversample from the DFT start point F_{sp}. $r_{\mathrm{res}}(k) = r_{\mathrm{samp}}(F_{\mathrm{sp}} + N_p.(k - 1)$: $k = 1, 2, \ldots N_{\mathrm{fft}} \times M \times N_{\mathrm{mms}}$ are the resampled data. After that, the resampled data are fed into the DFT circuit and detected as subchannel data, and we obtain the DFT-detected data $\hat{d}_{i,L}(k)$ described as

$$\hat{d}_{i,L}(k) = \frac{1}{N_{fft}} \sum_{n=(k-1)N_{fft}+1}^{kN_{\mathrm{fft}}} r_{\mathrm{res}}(n) \exp\left(j \frac{2\pi}{N_{\mathrm{fft}}} \big((i-1)(N_2+1) + L\big)n \right) \tag{17.14}$$

After that, we divide the DFT-detected signal by two parts: one is pilot channel, which is expressed as $\hat{d}_{i,L}(k)$: $i = 1, 2, \ldots, N_1$, and another is traffic channel, which is expressed as $\hat{d}_{i,L}(k)$: $i = 1, 2, \ldots, N_1$, $L = 2, 3, \ldots, N_2 + 1$. The DFT-detected pilot data are correlated with M sequence used in transmitter and a complex delay profile is obtained. One example of correlation value is in Figure 17.4. As shown in Figure 17.4, we obtain the delay profile with high quality within guard bits. In addition, F_{sp} is not only the start point of DFT but also the start point of a modified M sequence. Therefore, at the receiver, we only obtain the correlation value within the observation area shown in Figure 17.4 by using F_{sp}. The delay profile within observation area is given by

$$\hat{h}_{i,l}(n,k) = \sum_{l=(n-1)N_{\mathrm{mms}}+k+F_{\mathrm{sp}}+N_{\mathrm{gb}}}^{(n-1)N_{\mathrm{mms}}+N_{\mathrm{ms}}+k+F_{\mathrm{sp}}+N_{\mathrm{gb}}-1} \hat{d}_{i,l}(l) d_p\big(l - (n-1)N_{\mathrm{mms}} - k - F_{\mathrm{sp}} - N_{\mathrm{gb}}\big) \tag{17.15}$$

where $n = 1, 2, \ldots, M$ and $k = -N_{gb} \ldots N_{gb}$. This means that we estimate the characteristics of not only the direct wave but also the delayed wave, and in the case of $k = 0$, we obtain the direct wave. However, this complex delay profile does not accurately estimate the propagation and frequency offset characteristics because of low S/N. Therefore, we also coherently sum and average the complex delay profiles measured in successive symbols.

$$\hat{\hat{h}}_{i,j}(k) = \frac{1}{M}\sum_{n=1}^{M}\hat{h}_{i,l}(n,k) \quad : \quad (i = 1,2,\ldots N_1) \tag{17.16}$$

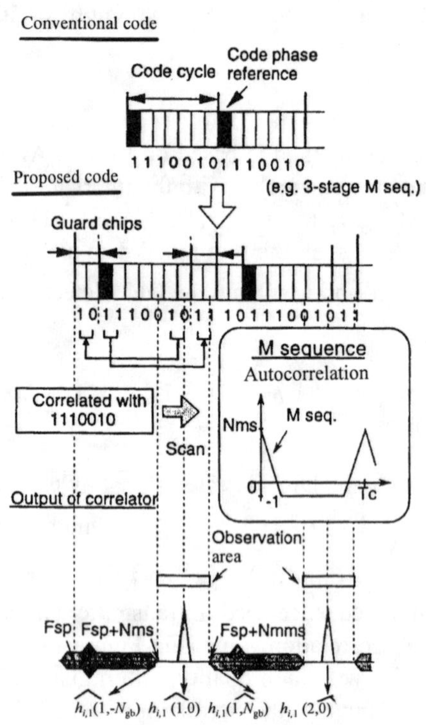

Figure 17.4 Correlation characteristics of guard bit-inserted M seqence.

Until now, we estimated the delay profile of the pilot channel. However, we must estimate the delay profile of the traffic channel, which is expressed as $\hat{\hat{h}}_{i,L}(k)$ $i = 1,2,\ldots N_1$, L=2, 3, ..., $N_2 + 1$. To estimate this propagation and frequency offset characteristics of the OFDM traffic subchannel, we utilize spline interpolating and extrapolating for the delayed profiles obtained by pilot channel. By using the estimated delay profiles of the traffic channel and the DFT-detected data of traffic channel, we equalize, demodulate, and decode these data. If we equalize only direct wave, the equalized data are given by,

$$\hat{\hat{d}}_{i,L}(k) = \frac{\hat{h}^{\wedge *}_{i,L}(0)}{\sum_{l=-N_{gb}}^{N_{gb}} \hat{\hat{h}}_{i,L}(l)\hat{\hat{h}}_{i,L}(l)^*} \hat{d}_{i,L}(k) \tag{17.17}$$

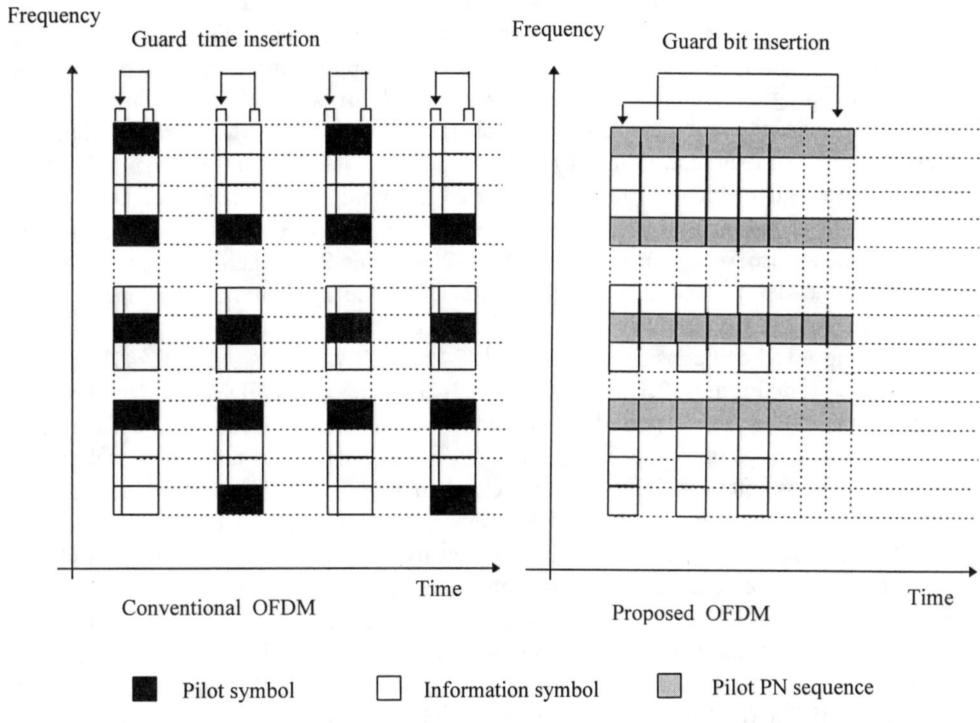

Figure 17.5 Difference between conventional and proposed OFDM schemes.

After that, the receiver reconstructs the ATM cell and checks its header. If there is no header error, this ATM cell transmits to the wired ATM network. This proposed system has several advantages. First, the guard bits are already inserted "unit by unit" on the procedure of information unit generation. At the receiver we need only the DFT circuit and code correlator. Therefore, this system does not need to equip the "bit by bit" guard interval insertion and removal circuits needed in a conventional OFDM scheme [11]. This difference between "unit by unit" and "bit by bit" guard interval insertion method is shown in Figure 17.5. In addition, it is very busy for proposed transmission scheme to estimate delay characteristics, synchronization point of frame, and start point of DFT without complex circuits. Also, there is no feedback circuit: it consists of only a feed-forward circuit. This is a very good effect for DSP design.

17.3 NUMERICAL RESULTS

In this section, the system performance of the proposed OFDM-based wireless ATM transmission scheme by computer simulation is discussed. In the computer simulations, QPSK-OFDM-based wireless ATM transmission is simulated under Rayleigh fading environment with Rayleigh distributed amplifier and the random distributed phase. This Rayleigh fading is evaluated by the parameter f_dT, in which f_d and T are Doppler frequency and the reciprocal of symbol rate per one OFDM parallel channel, respectively. For example, in the case of $f_dT = 6.25e-4$ and $f_d = 80$ Hz, symbol rate at the output of guard-bit insertion circuit is 128 ksymbol/s and if length of DFT N_{fft} is 128 and QPSK transmission, total transmission rate is 32.768 Mbps. Moreover, in our simulation, we use 5-stage M sequence ($N_{ms} = 2^5 - 1$) and 2 guard chips ($N_{gb} = 2$), which estimates the delayed signals within 15 μs when the symbol rate per one OFDM subchannel is 128 ksymbols/s.

One ATM cell consists of 53 bytes (424 bits). If we set $N_1 = 16$, $N_2 = 7$, and $N_{ms} = 2^5 - 1$. From Figure 17.1(a) we obtain 434 bits in each OFDM-ATM cell-maker during one PN sequence period. This means we can transmit 16 ATM cells in one information unit. With these N_1, N_2, and N_{ms} values, BEP and cell error probability (CEP) of the wireless ATM cell is indicated in Figure 17.6. The ATM cell header consists of the cell control data and header error control (HEC) bits. The cell error occurs when we divide the ATM cell header bit by the generation function of HEC and the reminder is not equal to 0 [12]. As shown in Figure 17.6, BEP distributes with the range from 10^{-3} to 10^{-2}. Since these simulations are under the condition of no coding, BEP performance less than 10^{-6} is expected when using channel coding. In addition, as with CEP performance, we compare with the results of [13], which has already simulated BEP and CEP with their own wireless ATM system when several values of delay spread are equal to 0 in [13]. As compared with both results, our system always obtains 2- to 3-dB better than [13],because by using accumulation of delay profile, we obtain a high-quality estimation value of delay profile. Therefore, this transmission scheme is a good candidate to realize a wireless ATM system. But this performance depends on both the interval of pilot PN sequence insertion and the number of accumulation blocks to decide frame and DFT timing points.

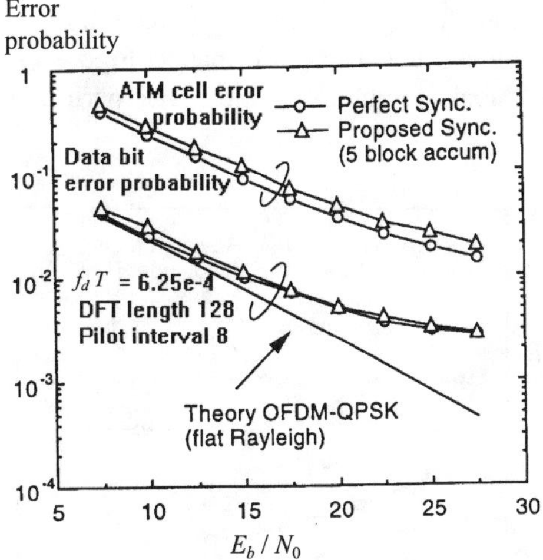

Figure 17.6 BEP and CEP performances of a wireless ATM cell.

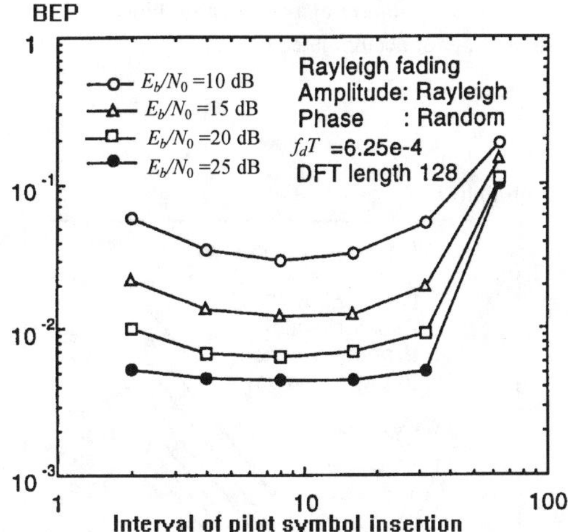

Figure 17.7 Optimum interval of pilot symbol insertion .

The relationship between the insertion interval of the pilot symbol and BEP performance is indicated in Figure 17.7. In Figure 17.7, as the insertion interval of the pilot symbol is smaller, BEP performance is worse. In this case, the ability to estimate for phase rotation is better because the number of pilot symbols increases. However, the energy per data bit decreases as the number of pilot symbols increases when the transmission power is constant. That is why BEP gets worse. On the other hand, as the interval becomes larger, the ability to estimate for phase rotation is worse, therefore,

BEP performance is also worse. As shown in Figure 17.7, the optimum interval of the pilot symbol is 8. Moreover, in Figure 17.8, we indicate the relationship between the number of accumulation blocks and BEP performance. Figure 17.8 shows that we need more than 4 accumulated blocks to obtain its stable BEP performance.

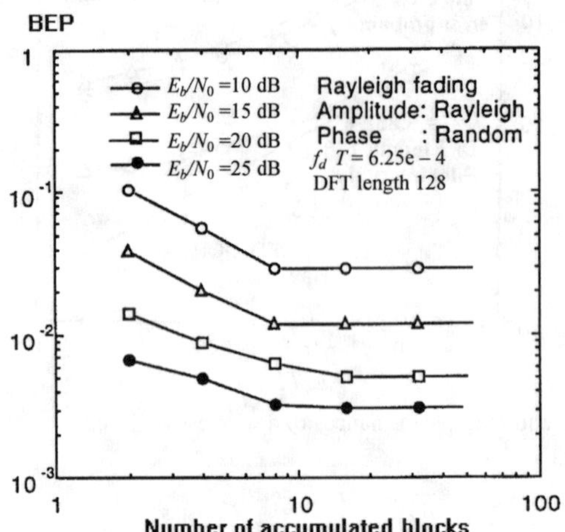

Figure 17.8 Optimum number of accumulated blocks.

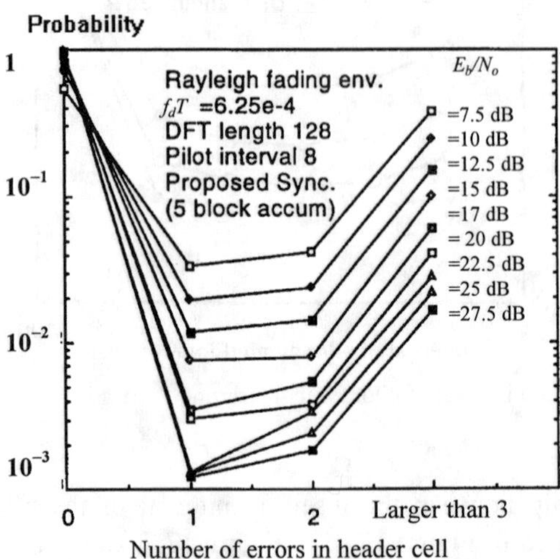

Figure 17.9 The probability of the number of errors occurring in one ATM cell header.

We now discuss the CEP of ATM in detail. Figure 17.9 shows the probability of the number of errors occurred in one ATM cell header. As shown in Figure 17.9, the probability occurred when the errors larger than three bits is larger than 10^{-2}; therefore, we avoid these errors by channel coding as with convolutional code.

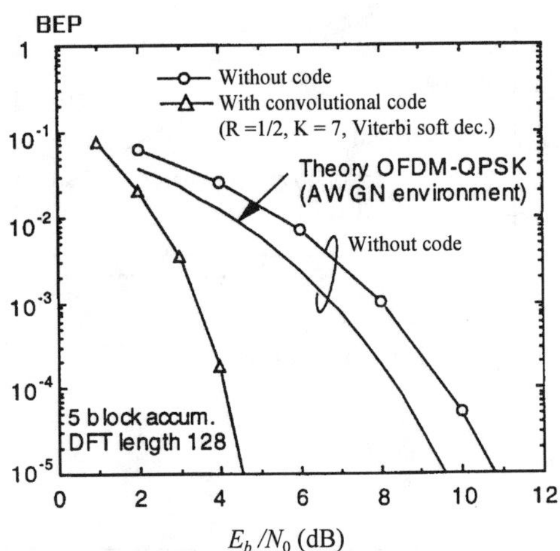

Figure 17.10 BEP performance of the proposed scheme with convolutional code.

In Figure 17.10, we simulate BEP performance with $R = ½$, $K = 7$ convolutional code and Viterbi soft decision decoder under only AWGN environment, because we must discuss interleave method if we simulate under fading environment. Figure 17.10 shows that we can realize less than BEP = 10^{-5} in the area of more than $E_b/N_0 = 5$ dB, and also indicates that we can obtain BEP performance less than 10^{-6} if BEP is between 10^{-2} and 10^{-3} without coding.

Moreover, we discuss BEP performance in several fast-fading environments. Figure 17.11 indicates BEP performance. As shown in Figure 17.11, the proposed method obtains good BEP performance from 10^{-3} to 10^{-2} in the case of f_dT less than 1.25e −3 .

Finally, we discuss the performance comparison between bit by bit and unit by unit guard interval insertion in Figure 17.12. As for bit by bit guard interval insertion, we use the pilot insertion method shown in Figure 17.4 and [11]. In addition, in both systems, we fixed the number of symbols used as pilot, with the parameter "pilot share rate." This parameter is the proportion between the available number of transmission bits of OFDM and the number of bits for pilot symbols or pilot sequence.

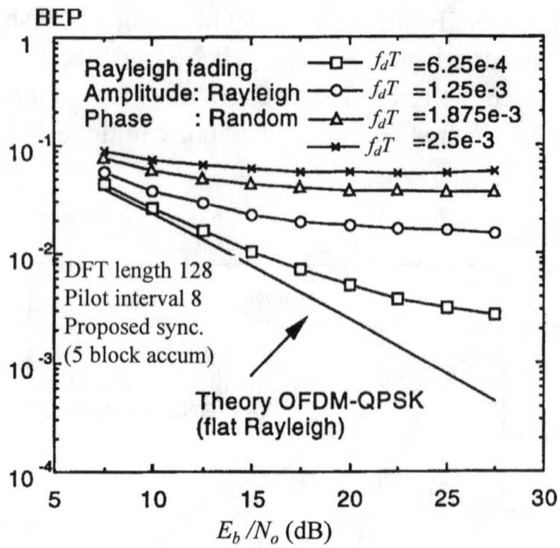

Figure 17.11 BEP performance in fast-fading channels.

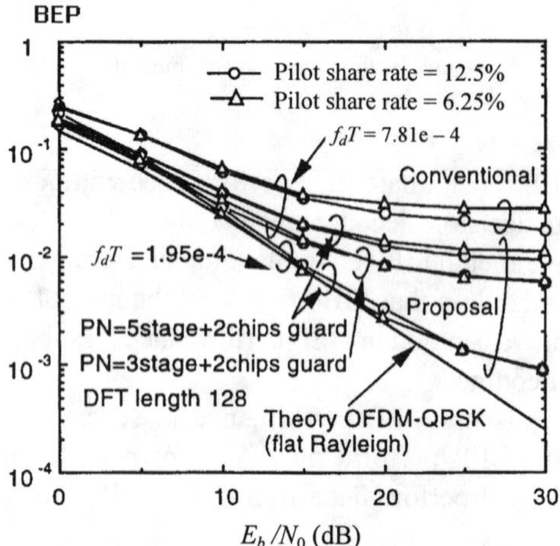

Figure 17.12 Performance comparison between conventional and proposed OFDM transmission schemes.

In Figure 17.12, we use f_dT = 7.8e –4, which means that symbol rate at the output of guard-bit insertion circuit is 128 ksymbol/s in the case of f_d = 100 Hz and f_dT = 1.95e –4. As shown in Figure 17.12, the performance of pilot insertion method (conventional method) is worse than that of PN sequence insertion method. That is because the conventional method cannot chase phase rotation caused by Rayleigh fading in comparison with the proposed system. Moreover, in Figure 17.12, we change the length of the PN sequence that are the cases of 5-stage and 3-stage *M* sequence with 2 guard bits. The shorter the length of the PN sequence is, we can chase fast fading instantaneously. That is why the performance of shorter code is better. Moreover, Figure 17.12 also shows that we obtain good BEP performance close to the theoretical value if we use this scheme under a slow a Doppler frequency environment like $f_d T$ = 1.95e –4.

17.4 CONCLUSIONS

This chapter proposed an OFDM-based wireless ATM transmission system for the future broadband mobile multimedia communications. The proposed system has the following features: (1) The guard bits, which are cyclically extended by repeating information bits, are inserted before and after serial-parallel converted wireless ATM data bits for OFDM transmission, and these bits are applied for the wireless ATM frame synchronization as well as for the decision of the DFT start point at the receiver; (2) the cyclically extended PN sequence is inserted in some of the proposed OFDM transmission subchannels with the fixed interval to estimate phase rotation, frequency offset, and delay characteristics; (3) it does not insert the bit by bit guard interval insertion and removal circuits needed in a conventional OFDM system; and (4) it is easy to estimate phase rotation, frequency offset, and delay characteristics and the synchronization of wireless ATM cell. In this chapter, the performance of the proposed system, in terms of BEP and CEP was evaluated. As a result, the proposed system realized BEP performance from 10^{-3} to 10^{-2}, the optimum number of accumulated symbols is more than 4 information units, and the optimum interval of pilot symbol insertion is 8. This proposed transmission scheme is one of the candidates to realize wireless ATM.

REFERENCES

[1] W. Houcharenko, J.P. Kruys, D.Y. Lee, and N.J. Shah, "Broadband wireless access," *IEEE Comm. Mag.*, Vol. 35, pp. 20–26, January 1997.

[2] L.M. Correia and R. Prasad, "An overview of wireless broadband communications," *IEEE Comm. Mag.*, Vol. 35, pp. 28–33, January 1997.

[3] N. Morinaga, M. Nakagawa, and R. Kohno, "New concepts and technologies for achieving highly reliable and high capacity multimedia wireless communications systems," *IEEE Comm. Mag.*, Vol. 35, pp. 34–40, January 1997.

[4] M. Chelouche, S. Hethuin, and L. Ramel, "Digital wireless broadband corporate and private networks: RNET concepts and applications," *IEEE Comm. Mag.*, Vol. 35, pp. 42–46, January 1997.

[5] R. Prasad, "Wireless broadband communication Systems," *IEEE Comm. Mag.*, Vol. 35, p.18, January 1997.

[6] H. Harada, G. Wu, K. Taira, Y. Hase, and H. Sasaoka, "A new multicode high speed mobile radio transmission scheme using cyclic modified M-sequence," *Proc. VTC'97,* pp. 1709–1713, May 1997.

[7] R. Prasad and H. Harada, "A novel OFDM based wireless ATM system for future broadband multimedia communication," *Proc. ACTS Mobile Communications Summit'97,* Aaalborg, Denmark, pp. 757–762, October 1997.

[8] H. Harada and R. Prasad, "Performance analysis of an OFDM based wireless ATM communication system," *Proc. PIMRC'97,* Helsinki, Finland, pp. 1095–1099, September 1997.

[9] J.-J. Beek, M. Sandell, M. Isaksson, and P.O. Borjesson, "Low complex frame synchronization in OFDM system," *Proc. ICUPC'95,* pp. 982–986, November 1995.

[10] M. Okada, S. Hara, S. Komaki, and N. Morinaga, "Optimum synchronization of orthogonal multicarrier modulated signals," *Proc. PIMRC'96,* pp. 863–867, October 1996.

[11] S. Hara, M. Mouri, M. Okada, and N. Morinaga, "Transmission performance analysis of multicarrier modulation in frequency selective fast Rayleigh fading channel," *Wireless Personal Communications,* (Kluwer journal) pp. 335–356, 1996.

[12] R. Handel and M.N. Huber, *Integrated Broadband Networks,* Addison-Wesley, 1991.

[13] R.D.J. van Nee, "OFDM codes for peak to average power reduction and error correction," *Proc. GLOBECOM'96,* pp. 740–744, November 1996.

Chapter 18
SATCOM

18.1 INTRODUCTION

Information services are becoming more bandwidth-intensive and traffic is rapidly increasing on the Internet. Most, if not all, users of the World Wide Web (WWW), have experienced delays due to traffic congestion at one time or another. Long delays are simply an inconvenience to individual home users, but they may cost time and money to businesses. The demand for more bandwidth and more efficient use of existing capacity is steadily growing.

SATCOM can provide wide area coverage, which makes it well-suited to facilitate the implementation of wide area networks (WANs). SATCOM can be employed to provide connectivity where the terrestrial infrastructure is thin or lacking, as well as enhance user mobility over large geographic regions. In this chapter, the wide area coverage of SATCOM is coupled with the broadcast and multicast capabilities of ATM to facilitate the implementation of a WAN with a high-rate forward link and a low-rate return.

The satellite coverage in this concept is intended to support communication anytime and nearly anywhere. Less than global coverage is envisioned. The primary users are expected to be businesses and the coverage is constrained to the latitudes in which the demand for business communication is greatest. The intent is to keep the constellation to a minimum and the overall system cost down. Although the primary modes of operation are intended to be broadcast and multicast, the use of a multiple beam antenna can facilitate both high-rate and low-rate operation. A constellation of five satellites in geosynchronous orbit is postulated as sufficient to satisfy the coverage requirements and meet a link availability requirement of 99%, with an acceptable margin for rain and other atmospheric effects.

A "bent-pipe" satellite transponder relay is proposed for simplicity of design and ease of interoperability. In the future, the concept is enhanced through the use of an ATM "switch in the sky." The use of CDMA is envisioned to enhance flexibility and facilitate network access and frequency management. Operation on separate transmit and receive frequencies is assumed to reduce interference at the receiver.

Broadband bearer services are considered to begin in the range of hundreds of kilobits per second to a few megabits per second [1,2]. The focus of this chapter is on the ability to support broadcast or multicast transmission of data and video in the range of about 64 Kbps to 8.448 Mbps. However, access rates from tens of kilobits per second to 34 Mbps, consistent with the requirements for services ranging from low-rate data and voice to the broadcast of high-quality video signals, are also considered.

The potential benefits of ATM over SATCOM are examined in this chapter. Key issues with respect to the operation of ATM over EHF SATCOM are identified and the relationship between the traffic profile and design of an ATM-based SATCOM system is discussed. A concept is then presented for broadband wide area networking via IP/ATM over K_a-band SATCOM. The key aspects of the concept addressed in this chapter include the space segment and ground segment, CDMA for ATM-SATCOM links, and error control and correction for ATM-SATCOM links. Performance issues, limitations, and requirements for reliable operation of ATM over SATCOM are examined. Emphasis is placed on error control and correction, and efficiency.

18.1.1 Potential Benefits of ATM Over SATCOM

ATM is rapidly emerging as the backbone of future information networks. The ATM protocol is based on international standards and is designed to support multimedia information services. A combination of features including flexible bandwidth allocation, statistical multiplexing, priority queuing, and multicasting make ATM potentially more efficient in terms of bandwidth utilization than conventional switching technologies, despite the additional overhead of 9.4% due the 5-byte header per 53-byte ATM cell. Flexible bandwidth allocation is a key reason.

The ability to allocate any and all of the available bandwidth on demand can outweigh the reduction in peak throughput. In essence, the peak throughput is sacrificed in favor of flexibility with the goal of greater overall bandwidth efficiency and the ability to accommodate multimedia information services on demand. Multicasting is another key reason. The ability to multicast information and avoid unnecessary transmission of duplicate information, such as copies of e-mail messages or video transmissions, is especially relevant over bandwidth-constrained satellite links.

However, ATM is a new technology. Although it is available now, the real benefits are only beginning to emerge and may not be obvious from a conventional SATCOM perspective. ATM may have a major impact on SATCOM access, efficiency, interoperability, and cost when the technology is mature and the full range of features are available. A summary of the key features of ATM and their potential benefits from a SATCOM perspective is provided in Table 18.1. For these and other reasons to be discussed in this chapter, ATM over SATCOM, where bandwidth is often at a premium, is considered to be an effective combination.

Table 18.1

Key Features of ATM and Potential Benefits From a SATCOM Perspective

Feature	Potential Benefits
Flexible bandwidth allocation	Bandwidth efficiency Only use the required capacity
Bandwidth on demand	Bandwidth efficiency Only use the capacity when required
High speed	Low delay
Nonproprietary standard	Interoperability and lower cost
Support for multimedia	Efficiency, convenience, and potentially lower cost
Multicasting	Bandwidth and power efficiency No unnecessary repetitions
Priority queuing	Accommodation of traffic surge potential for graceful degradation

Table 18.2

Key Issues With Respect to ATM Over SATCOM

Category	Key Issues
ATM protocol	Maturity of the technology and standards maturity of applications
LAN-WAN integration	Traffic control and congestion -QoS negotiation -Flow control -Priority queuing IP over ATM (legacy networks) Physical interface and rate conversion (SATCOM)
ATM over SATCOM	Degradation due to increased delay, delay jitter and error rate mitigation via error control and correction Availability of bandwidth Rain margin and recovery from outage SATCOM access technique Dynamic resource allocation

18.1.2 Key Issues With Respect to the Operation of ATM Over SATCOM

ATM protocol maturity. A summary of the key issues with respect to the operation of ATM over SATCOM is provided in Table 18.2. At present, ATM technology is not fully mature. Future developments are likely to be driven by the proliferation of new and increasingly bandwidth-intensive information services and the requirements imposed by these services. The requirements for wideband services and applications are evolving and ATM standards also continue to evolve. Traffic and congestion control remains an area of development; integration with existing networks is another.

Traffic and congestion control. The role of ATM traffic and congestion control is to help achieve performance objectives in terms of cell-loss ratio, cell delay, and cell delay variation [3]. A connection is refused if the appropriate capacity and QoS are not available. However, it is not always easy to know what capacity is available and at what QoS over a satellite link. Moreover, satisfaction of the QoS requirements over a SATCOM link is only part of the problem. The broader issue concerns satisfaction of the QoS end to end and systemwide.

The problems associated with traffic and congestion control and QoS negotiation are complicated by the need to integrate ATM with legacy networks that may not support priority queuing by service type or QoS-based traffic and congestion control. The lack of information concerning priority and QoS is especially significant over satellite links due to the combination of generally lower transmission rates and longer delay than over (fiberoptic) F-O links. Even a simple "all or nothing" algorithm, where the priority is simply high or low, and flow control is little more than turning the transmission on or off, is useful. Similarly, it is especially important to understand the traffic profile. The solution must be tailored to the information flows, which is point-to-point, point-to-multipoint, or a combination of both. A solution that works well in one case is not necessarily the best for all. The flexibility of ATM not only facilitates tailoring of the solution to the user requirements, it demands so, if the full measure of benefit is to be obtained.

Integration of IP/ATM over SATCOM. In the foreseeable future, ATM is expected to coexist with conventional networks based on IP. Logically, IP over ATM (IP/ATM) is one way to integrate ATM with legacy networks. The integration of ATM into a WAN via SATCOM provides obvious advantages in terms of the ability to extend communications to geographically distant or remote locations. It also poses challenges due to the longer delay, delay jitter, and potentially higher error rates over the SATCOM link, than over the terrestrial F-O links for which the ATM protocol was originally conceived. This raises the issues of error control and correction. Efficient error control and recovery, tailored to the ATM-SATCOM protocol stack, are required. Additionally, the integration of ATM with SATCOM requires interface and rate conversion, as SATCOM links typically do not support rates as high as those of terrestrial F-O links.

ATM over SATCOM. Even if SATCOM links did support data rates commensurate with terrestrial F-O links, broadband networking via SATCOM would place significant demands on already limited spectral resources at 14/12-GHz K_u-band and lower frequencies. Fortunately, more bandwidth is available at extremely high frequency (EHF) in the commercial 30/20-GHz K_a-band. However, losses due to rain and atmospheric attenuation are considerable at EHF. This raises issues concerning recovery from outage and availability, which may place limitations on the type of services that can be supported.

Resource allocation. Flexible resource or bandwidth allocation is one of the key features of ATM, but with flexible resource allocation comes a need for more sophisticated resource allocation techniques that take into consideration user QoS and bandwidth requirements [4]. From an ATM perspective, both static assignment of the required bandwidth and dynamic bandwidth allocation have merit. Issues associated with static and dynamic resource allocation are described in [4]. The question of resource allocation becomes more complex over satellite links, where access is often limited to a few predetermined rates.

18.1.3 Understanding the Traffic Profile and Its Relationship to ATM-SATCOM Links

The performance of the ATM protocol is heavily dependent on the traffic profile. The traffic profile defines the traffic mix and intensity by service type. It also defines the traffic flow, which determines the type of satellite link or links that are required.

Sensitivity to service type and traffic intensity. ATM cells are thought of as packets and performance is evaluated by applying queuing theory [3].

An M/D/1/N queue [3] provides a simple performance model, where M stands for Poisson arrivals, D for packets of fixed size, 1 for a single server and N for the available buffer or queue length. Delay, delay jitter, error rate, and queue length are key factors in determining the cell-loss rate. Assuming that an acceptably low error rate can be achieved and maintained, then delay jitter become the primary source of degradation. In general, the delay increases with traffic intensity, as illustrated in Figure 18.1 [5] for either an M/M/1/∞ queue, with delay normalized to the cell size.

SATCOM implications. Satellite links do not ordinarily at transmission rates as high as those of terrestrial F-O networks. Although they can facilitate LAN-LAN integration into a WAN, they tend to constrain the end to end transmission rate between interconnected LAN, as shown in Figure 18.2, which illustrates the rate conversion problem associated with, for example, interconnection of a 155-Mbps OC3 carrier to a 2.048-Mbps E1 SATCOM carrier. Idle cells can be deleted, but user traffic cannot.

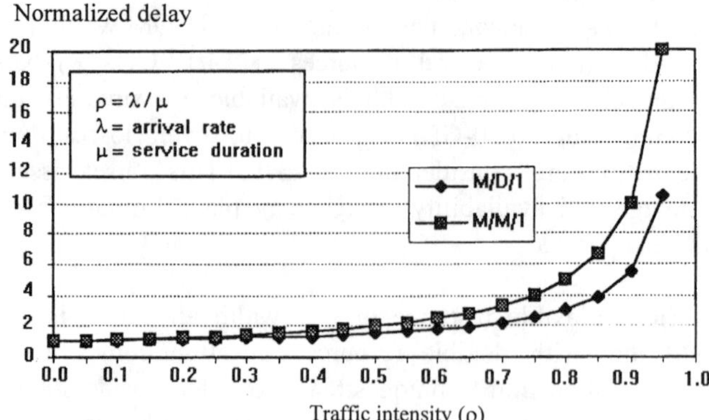

Figure 18.1 The impact of traffic intensity and the need for realistic traffic profiles.

Bandwidth constriction is viewed as analogous to an increase in traffic intensity given a fixed arrival rate. This corresponds to a shift in the operating point on the curves of Figure 18.1. Delay rapidly increases. However, a reduction of the transmission rate does not necessarily imply a significant degradation in performance. For example, if the reduction in data rate is accompanied by a proportional reduction in the arrival rate, then the traffic intensity remains unchanged. Although the focus of this chapter is on ATM over SATCOM, many of the issues raised also apply to ATM over wireless communication in general. Delay resulting from conversion to lower transmission rates is such a case in point.

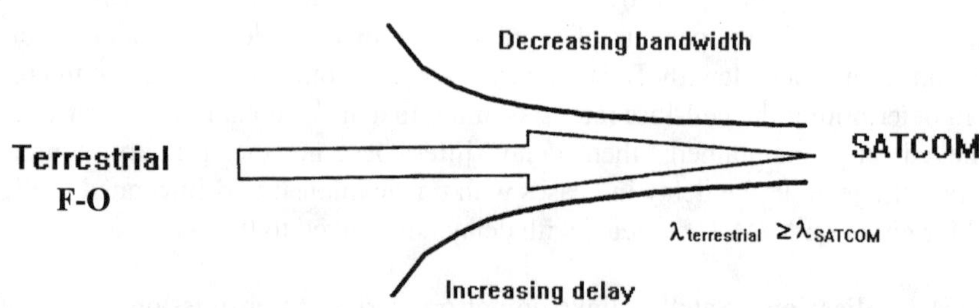

Figure 18.2 Rate conversion and bandwidth constriction.

The performance, as measured in terms of delay, is dependent on ρ_{SATCOM}, which represents the fraction of the total traffic carried over the satellite link. An illustration is provided in Figure 18.3, where native ATM traffic is combined with traffic from legacy systems for transmission over a satellite link. Packet switched ρ_{ps}, circuit switched ρ_{cs}, and native ATM traffic ρ_{ATM} are combined into ρ_{SATCOM} for transmission over the satellite link. This applies to both permanent virtual circuits (PVC) and switched virtual circuits (SVC) through the ATM network. Knowledge of ρ_{SATCOM} is needed to understand the traffic profile.

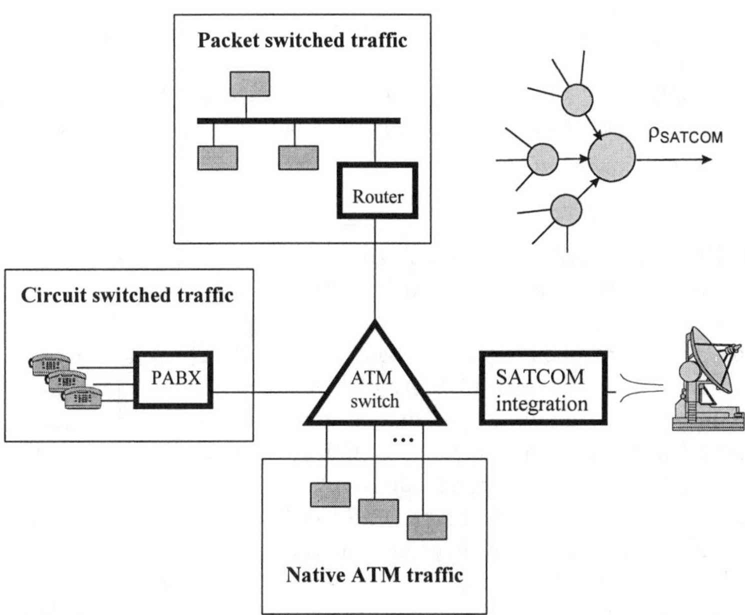

Figure 18.3 General traffic model.

Traffic flow. Another important aspect of the traffic profile is traffic flow. This is more than merely the connectivity. The traffic flow on some links is symmetric in nature. Full duplex 64-Kbps voice provides an example. The transmission rate in each direction is 64 Kbps. Logically, a satellite ground terminal (SGT) of the same size is expected on either end of a symmetric link. TDM or FDMA carriers are commonly employed on such links.

However, a system that is optimized to support primarily symmetric traffic flow is not necessarily the best when it comes to supporting asymmetric traffic flow or mixed traffic flows. Examples of potentially asymmetric traffic include broadcast or multicast of video or data to one or more groups of users. In the case of asymmetric traffic, a large SGT on one end of the link communicates with one or more small SGTs on the other end. This scenario is advantageous in a system where large numbers of small, low-cost, mobile terminals are envisioned. This is the case of interest in this chapter, where the focus is on high-rate wideband communication in the forward direction, with a low-rate return. Typically, the maximum transmission rate in the forward direction will be "downlink-" limited by noise in the small terminal. As will be shown shortly, this is employed to advantage.

A thorough understanding of the user traffic profile is always important if the communication system is to be designed effectively. It is especially important in ATM-SATCOM based systems given the sensitivity to the traffic mix, intensity, and flow. Although ATM provides an efficient mechanism to combine voice, data, and video traffic, knowledge of the traffic mix and intensity is necessary in order to allocate the available capacity equitably and efficiently and to size the SATCOM system. The traffic profile is a key factor in determining whether ATM over SATCOM is an appropriate solution.

18.2 CONCEPTUAL ANALYSIS OF BROADBAND NETWORKING VIA IP/ATM OVER EHF SATCOM USING CDMA

The key advantages of operation at K_a-band (i.e., 30/20 GHz), as opposed to lower frequencies, are the increased availability of bandwidth and the potential for smaller terminals and greater mobility. The major disadvantages are increased rain and atmospheric attenuation and the potential for link outages. The objective is to capitalize on the increased availability of bandwidth, without compromising the requisite QoS requirements of the system. Certain types of SATCOM systems are more suitable than others for implementation of ATM over SATCOM in general and K_a-band in particular.

Multicast via ATM over SATCOM is a natural combination that matches the wide area coverage of SATCOM with transmission of common information to groups of geographically separate users. An illustration is provided in Figure 18.4. In principle, only one copy of an e-mail message is transmitted to a group of users in a multicast via ATM over SATCOM. This eliminates wasteful transmission of duplicate data over the already bandwidth-constrained satellite link.

A broadcast transmission is one that is sent to all users. A multicast transmission is sent to all users or any subset thereof. Thus, broadcasting is a subset of multicasting. Flexibility is a key difference. The benefits of multicasting increase with the bandwidth of the service to be multicast and are very significant when the amount of common user traffic, such as e-mail, is a substantial percentage of the total over the satellite link. Clearly, broadcast or multicast of broadband video transmissions to a user group is much more efficient than sending each user in the group a copy. A broadcast or multicast transmission on the forward link is coupled with a low-rate return link from individual users, which is employed for error recovery of data transmissions. Ideally, multicasting is accomplished via ATM. IP multicasting over ATM networks is more practical at this time. Issues associated with IP multicasting over ATM networks are described in [6]. In the future, multicasting is supported by either ATM or IP version 6.

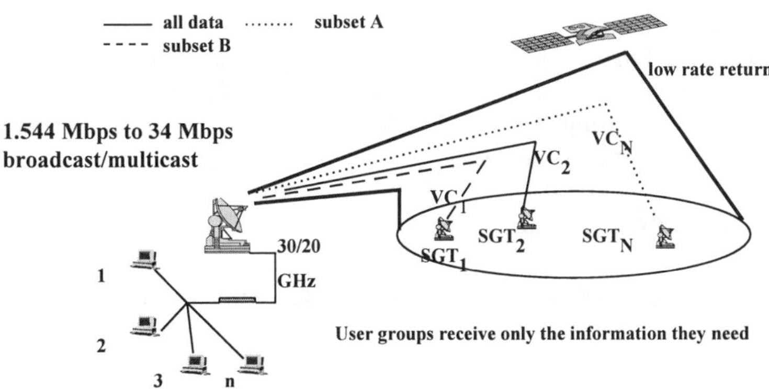

Figure 18.4 High level view of broadband networking via ATM over EHF SATCOM.

A more interesting application of ATM multicast over SATCOM, with a high-rate forward link and a low-rate return link, is to support the full broadband multimedia service on the forward link and only a subset of the components of the broadband multimedia service on the return link. A multimedia service could conceivably be decomposed into voice, data, video, and whiteboard. This requires separate VCIs per component service as described in [3]. Thus, multicasting provides an extra dimension of flexibility. Potentially, it is possible to transmit only a subset of multimedia services to different user groups. If so, this would allow privileged users with larger SGTs to receive the "big picture" (i.e., voice, data, video, and whiteboard), while disadvantaged users with small SGTs could still participate in a multiparty conference via, for example, whiteboard or voice service.

The idea is to facilitate interactive communication between users of all shapes and sizes. For the sake of discussion, a corporate headquarters with a large terminal could transmit broadband multimedia with high-quality video to other users with large terminals and, at the same time, transmit a subset of the component services to users at remote sites with very small or mobile terminals. Similarly, a low-rate return could facilitate remote database access. Thus, the SATCOM link need not be symmetric and, hopefully, the transmission rate over the satellite link can be kept to the minimum necessary to support the essential information transfer. This can help conserve capacity, as well as facilitate the use of smaller, less expensive SGTs.

Priority queuing is another feature of ATM that makes it a potentially good match with SATCOM. Priority queuing helps accommodate surges in the traffic offered to the satellite link or, conversely, it facilitates graceful degradation, by service, of the capacity over the satellite, in the sense that priority users and services are less affected by short-term channel disturbances. The last proposal is especially interesting from the perspective of EHF K_a-band operation.

The concept introduced in this section is idealistic. It assumes that ATM technology is mature and that the full measure of capability exists from end to end. This is not currently the case. An examination of the ATM-SATCOM protocol stack is useful for the purposes of refining the concept in the sections that follow, as well as to help clarify current practical considerations.

18.2.1 The ATM-SATCOM Protocol Stack

An illustration of the basic ATM-SATCOM protocol stack is provided in Figure 18.5 [3,7]. TCP/IP over AAL 5 is employed as an example case because it is relatively common today. The ATM layer is analogous to the link layer in the open systems interconnect (OSI) model [3]. It is divided into the ATM layer itself and the AAL. The AAL is further subdivided into the convergence sublayer and the segmentation and reassembly sublayer. Four AAL protocols have been defined: AAL 1 for CBR service (e.g., voice), AAL 2 for VBR service, AAL 3/4 for data sensitive to loss but not to delay, and AAL 5 for high-speed, connection-oriented data service [3].

For satellite links operating over either a standard E1 2.048-Mbps carrier or a T1 1.544-Mbps carrier, G.704 framing applies [8], and the ATM cells are mapped into the SATCOM carrier via a physical layer convergence procedure (PLCP). In this case, the ATM cells are mapped byte-by-byte into the G.704 frames via the G.804 PLCP [9]. G.704 framing, also applies to higher-order carriers, such as the E2 8.448 Mbps carrier or E3 34 Mbps carrier. Other PLCPs are possible and rates are not limited to T1 or E1 rates; however, these are the only standard options for relatively low-rate ATM available at this time.

Functionally, the PLCP is considered part of the upper portion of the physical layer [3]. However, additional capabilities for error correction is incorporated into the PLCP in the future, which push at least part of the functionality into the lower levels of layer 2 of the OSI model. At present, there is little in the way of error correction and control within the ATM layer. Options for error correction and control will be examined in Section 18.3.

Currently, there are very few applications or information services available that operate in native-mode ATM. The vast majority of information services employed with ATM today actually operate via TCP/IP over ATM or some combination thereof. This is a practical consideration based on the availability of equipment and is expected to continue well into the future. The result is that AAL 5 has become a de facto standard. Other AALs are not widely available. Until they are, the benefits of ATM over

SATCOM is artificially limited. The discussion of ATM in the remainder of this chapter will focus on the use of AAL 5.

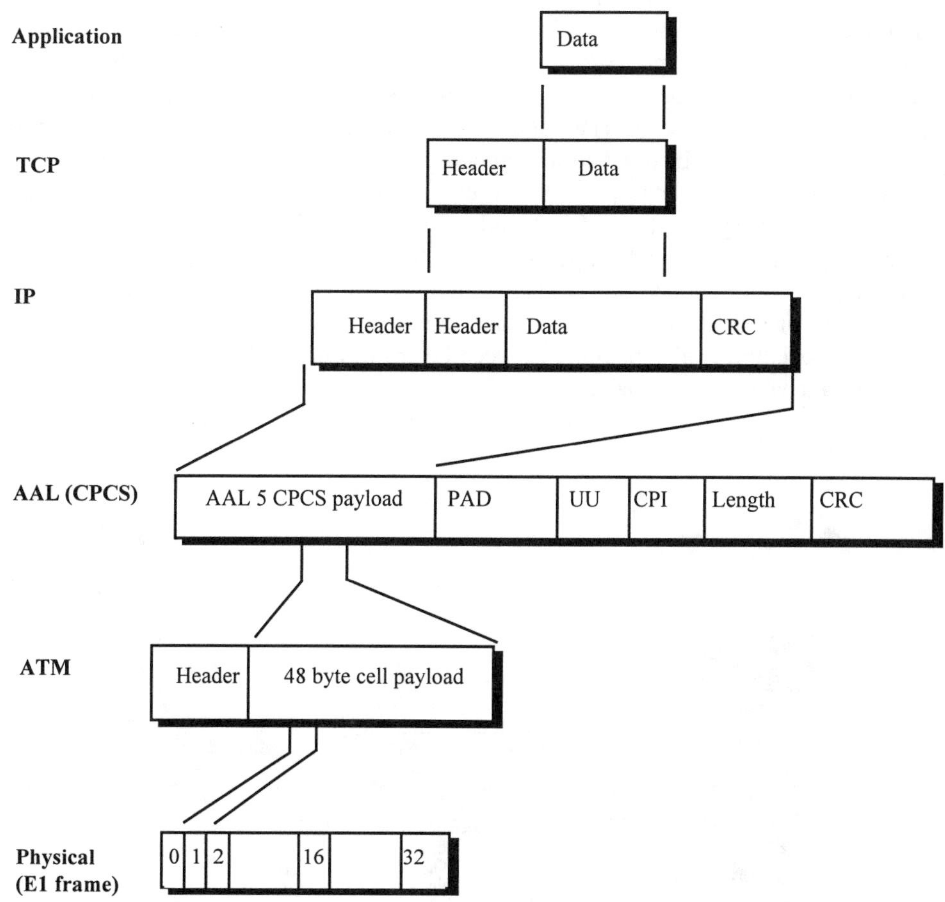

Figure 18.5 Example ATM-SATCOM protocol stack.

18.2.2 CDMA for ATM-SATCOM Links

The basic concept for broadband networking via ATM over EHF SATCOM is enhanced by combining ATM with CDMA. The reasons include the efficiency of

CDMA, the potential to support flexible resource allocation, the potential to support multicasting, and the potential to facilitate ATM traffic and congestion control.

An illustration of the concept is provided in Figure 18.6, from where we see that an association is made between connections and CDMA codes. Ideally, connections to a common user group are made by service and, to the extent possible, multimedia services are separated into their component parts [3]. For example, consider multicasting of a multimedia service from one to many users. The multimedia service is separated into voice, data, and video services and a separate virtual channel, or virtual path, is established over the satellite for each component service. A circuit might carry multiple connections in order to accommodate the different multimedia services and their components from multiple users, where each connection has a unique VCI and/or VPI, which corresponds to a CDMA code. Some of the connections are point-to-point, while others are point-to-multipoint, within the same multiparty conference. Users receiving common information, whether it is a complete multimedia service, or a component of a multimedia service, share one or more common CDMA codes. Note that an association is also possible between beams of a multiple beam antenna (MBA) and virtual channels or virtual paths.

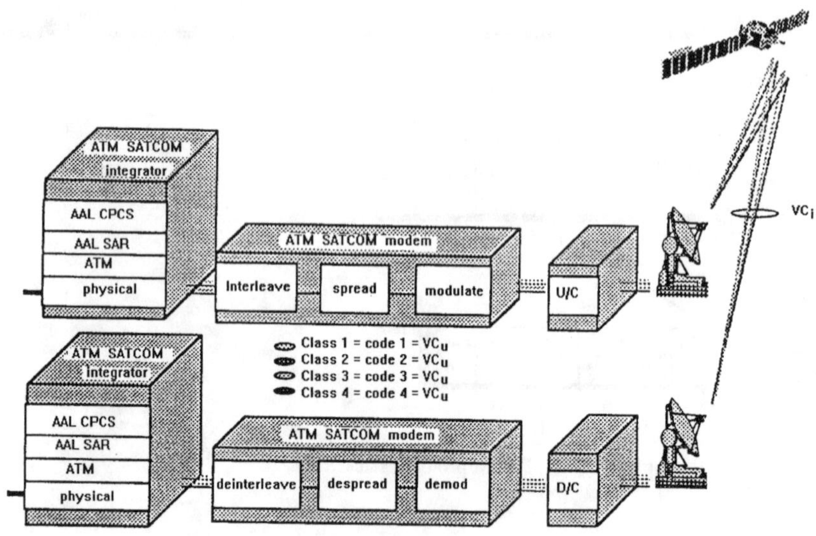

Figure 18.6 ATM over a CDMA-based SATCOM link.

The concept allows for separate and very flexible handling of voice, data, and video traffic over the bandwidth-limited satellite link. Communication requirements are driven more closely by information flow between users and less by the requirements of the most demanding of the services to be supported. To the extent that the information flow is asymmetric, the use of smaller terminals is possible on one end of the satellite link. Aside from the obvious advantages of smaller terminals, the concept would facilitate interactive multiparty communication between a wider variety of users. The concept is also bandwidth-efficient in that the composite transmission rate over the satellite link is reduced, since only the essential information is transmitted to each user. However, the concept favors native mode ATM operation and it may not be possible to fully integrate ATM with CDMA over SATCOM until ATM applications are more mature. A possible interim solution is to assign a CDMA code per service class. The concept is also potentially power efficient in that it allows for averaging of the QoS over the satellite.

Additionally, the fact that all users receive the information in a CDMA broadcast or multicast over a satellite, is exploited to facilitate QoS-based traffic and congestion control. As long as the system is noise limited by the interference from other CDMA users and the traffic can be identified by service class, then it is possible to evaluate the capacity by service class at any given time. Specifically, information about the energy-to-noise-density ratio is employed for the purposes of connection admission control. It should also be possible to estimate the QoS in terms of the error rate. Although decisions concerning congestion control is most relevant from the perspective of the satellite, for example, on a future processed satellite with an on-board ATM switch, information can also be obtained for the purposes of congestion control at ground terminals that are similarly noise limited. An assumption in this case is that the noise is user limited not thermal-noise limited. This is typical of CDMA systems. By comparison, it is much more complicated to discern similar information about an MC-FDMA-based system.

The obvious disadvantage of the concept is that multiple CDMA codes are required to communicate to a single user or group of users (e.g., one for each service class in a multimedia transmission). However, this is considered to be a digital signal processing trade-off and not a technical limitation. Furthermore, the burden is on the larger terminals, or hubs. The smallest terminals need only support operation on one frequency using one code for a single service.

CDMA comes the closest to matching the flexibility of ATM with regard to only using the capacity required for as long as it is required. In fact, any data rate is supported, up to the capacity of the system, at any time, by any user, in support of any service. Where higher rate transmission is required, multiple carriers are considered. Only single carrier operation is considered herein.

Additionally, the ability to support voice, data, and video on separate virtual channels or codes facilitates the graceful degradation in the event of an intermittent outage, as may occur due to rain at EHF. In a multimedia teleconference, the whiteboard could remain active albeit at low-rates even though video and even voice are unavailable, as long as a marginal capacity remains. It also facilitates averaging of the QoS requirements. A conventional satellite link supports all services at the same level (e.g., BEP and delay). ATM over SATCOM with separate virtual circuits for each service, or at least service class, each with their own QoS requirements, allows the QoS to be averaged over all connections and services, according to the traffic mix and intensity. Thus, only the necessary QoS is supported over the satellite link.

In summary, combining the multicast capability of ATM with CDMA allows the voice, data, and video components of a multimedia transmission to be transmitted separately over the satellite link. As a result, not all terminals need to be sized to handle the full multimedia service in order to participate in at least some aspect of the same conference. Without combining ATM multicast with CDMA, multimedia service and component services might have to be carried on separate channels or time-slots. A conventional multiplexing hierarchy based on TDM on FDMA for point-to-point links provides an example.

Point-to-multipoint communication requires that voice, data, and video be placed on either separate static carriers or physical links with FDMA, or burst at a fixed rate into time-slots in a TDMA-based system. Without the ability to separate information flows by service class, and preferably per connection, many of the aforementioned potential benefits of ATM are lost from a SATCOM perspective.

18.2.3 Capacity as a Function of the User Service Distribution

Capacity is a key measure of system performance. It is dependent on many factors. A link budget is useful to help in understanding the problem and one is provided in Table 18.3. A pair of steerable MBAs are employed to provide flexibility and gain: one for reception and one for transmission.

The receive MBA in this conceptual design is 0.75m in diameter and has 169 elements (Ne). The field of view (FOV) is 16 deg, which provides near-Earth coverage. The transmit MBA is 1.0m in diameter. Again, the FOV is 16 deg and the number of elements is 169. Coverage is limited in the far northern and southern latitudes. Additional gain is possible through beamshaping.

A rain margin of 13.2 dB is considered, based on 99% availability in rain region D2 at K_a-band and an elevation angle of 30 deg [11,12]. The vast majority of users are expected to be within a band from about 55 deg north to 55 deg south. Communication outside this band is possible, but coverage is limited and the satellite antenna gain decreases. However, elevation angles less than 30 deg generally correspond to regions that are cooler and drier than rain region D2 (e.g., regions B and C), in this concept. As a result, 13.2 dB of rain margin is considered to be more than adequate for the intended purpose.

A normal constellation of five satellites in geosynchronous orbit is projected. An illustration of the satellite constellation and geometry is provided in Figure 18.7. Cross-links is required to extend the coverage around the world in the absence of terrestrial interconnections, but are not considered in this chapter.

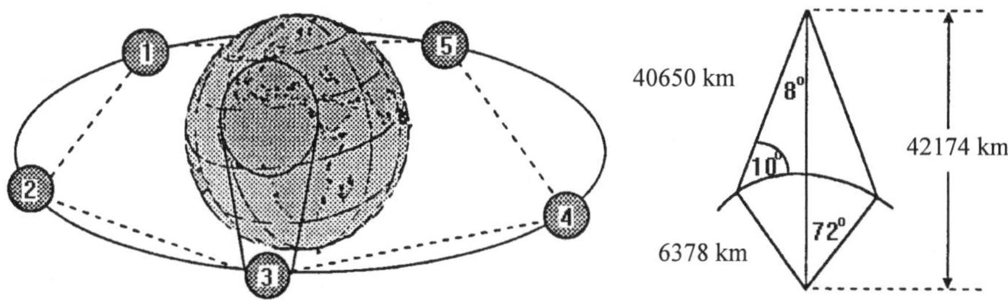

Figure 18.7 Satellite constellation and geometry.

The link budget provides a reference point for the maximum capacity in bits per second that is supported between a 4.8m dish and a 2.4m dish. A capacity of 371.5 Mbps is estimated for a single point-to-point link assuming the use of QPSK modulation with rate ½ FEC and a normal operating point of 10^{-8} based on TCP/IP over an ATM SATCOM link [13]. The bandwidth needed to support 371.5 Mbps with the aforementioned modulation a code is 371.5 MHz, which is well within the coherence bandwidth for K_a-band of 1 GHz or more [12].

When a narrowbeam is used on the uplink (1.1deg) and a 1.6deg regional spot beam is employed on the downlink, 163.8 Mbps are supported. Similarly, a 3.5deg time-zone coverage beam could support 34.4 Mbps and an 8deg continental U.S. coverage beam could support 6.6 Mbps, which is sufficient for broadcast video.

Bits per second is not the only measure of capacity and it is arguably not the best suited to evaluate the performance of an ATM based system, where future users are allocated any portion of the available capacity, on demand, based on priority. A means of evaluating capacity in terms of users and traffic is needed. Erlangs can provide a measure of how efficiently capacity is used. The maximum number of users and capacity in erlangs can be evaluated based on the methodology of [14].

Let the maximum number of users in a CDMA system be M. Each user is transmitting at a bit rate R_j where $j = (1, 2, \ldots, M)$. The total equivalent signal-to-noise density ratio at any receiver is

$$S_T / N_0 = \sum_{j=1}^{M} S_j / N_0 \tag{18.1}$$

where S_j / N_0 is the signal-to-noise density ratio from any given user. All signals are spread over the full spread bandwidth W. The total (composite) data rate over the satellite link is [15]

Table 18.3
Forward Link Budget From a Large Terminal to a Smaller Terminal

Forward Link	Parameter	Units	Comments
General			
1. Satellite type			Geostationary
2. Uplink beam			MBA spot
3. Downlink beam			MBA spot
4. Frequency (up)	30.0	GHz	K_a-band
5. Frequency (down)	20.0	GHz	K_a-band
6. Transponder bandwidth	400.0	MHz	Nominal value
Uplink			
7. Earth station EIRP (peak)	82.6	dBW	4.8m dish, 150W
8. Path loss	–214.2	dB	30deg elevation, 40,650 km
9. Gain of 1 m^2	51.0	dB/m^2	$4\pi/\lambda^2$
10. Operating flux density	–80.6	dBW/m^2	
11. Saturation flux density	–76.0	dBW/m^2	Estimate per Intelsat VII, K_u-band
(beam edge)	–4.6	dB	
12. Input backoff	16.1	dB/K	Ne=169, D=0.675m, NF=3 dB (300deg K)
13. Spacecraft G/T (beam edge)	–115.5	dBW/K	
14. C/T thermal (up)	–228.6	dBW/K-	1.38×10^{-23}
15. Boltzman's constant	113.1	Hz	
16. C/N_0 (up)		dB-Hz	
Downlink			
17. Saturated EIRP (beam edge)	60.9	dBW	Ne=169, D=1.01m, 100W (coupled cavity)
18. Output backoff	–1.0	dB	[10]
19. Path Loss	–210.6	dB	30deg elevation, 40650 km
20. Earth station G/T	34.4	dB/K	2.4m, 50deg K
21. C/T thermal (down)	–116.3	dBW/K-	
22. Boltzman's constant	–228.6	Hz	1.38×10^{-23}
23. C/N_0 (down)	112.3	dBW/K-	
		Hz	
		dB-Hz	
Margin			
24. Modem implementation	1.0	dB	Implementation losses
25. Gaseous attenuation	2.4	dB	1.3 dB (up), 1.1 dB (down), [11]
26. Rain margin	13.2	dB	99%, 6=8.6 dB (up), 4.6 dB (down), D2,
27. Scintilation	1.0	dB	[11]
28. depolarization	0.0	dB	Hot and humid, [11]
29. Sand and dust or clouds	1.0	dB	[11]
30. System margin	2.0	dB	[11]
			Tracking, interference
Performance			
31. E_b/N_0 required	6.0	dB	$P_b < 10^{-8}$, r = ½, QPSK
32. E_c/N_0	6.0	dB	$E_c = rkE_b$, r = ½, k = 2
33. channel symbol rate R_s	85.7	dB-Hz	371.5 Msps
34. Maximum data rate R	85.7	dB-Hz	$R = R_s/rk$ = 371.5 Mbps (maximum)

$$R_T = \sum_{j=1}^{M} R_j \tag{18.2}$$

From [15], using Shannon's formula, with all users operating at equal data rates, such that R_j = R, the maximum number of users in a power-controlled CDMA system are upper-bounded

$$M \leq \frac{W}{R}\log_2\left(1+\frac{S_T}{N_0 W}\right) \tag{18.3}$$

Equation (18.3) is converted to a more useful form [15]

$$M \approx \frac{(W / R)F}{E_b / I_0} \tag{18.4}$$

The term *W/R* is known as the processing gain, F is a reduction factor to account for backoff (e.g., $F = 0.8$), which is already considered in the link budget, and I_0 is the total noise from all sources

$$I_0 = N_0 + \sum_{j=k} E_{cj} \tag{18.5}$$

Based on the assumption that an E_b/N_0 of 6 dB is needed to meet the QoS requirements and assuming a spread bandwidth of 204.8 MHz and a data rate of 2.048 Mbps per user, the maximum number of users is approximated from

$$M \approx \frac{(204{,}800{,}000 / 2{,}048{,}000)}{4} \cong 25 \tag{18.6}$$

Thus, in this example, about 25 users could simultaneously access at 2.048 Mbps rates. However, this result is heavily dependent on the actual size of the satellite ground terminals. It does not consider the possibility of interference cancellation. Another important point is that the approximation for the maximum number of users does not reflect the traffic that is supported in a "busy hour" over the satellite, or the nature of the service supported. Conversion to Erlangs or some other measure of the traffic being offered is necessary. Knowledge of the services and traffic distribution is also required.

The proposed approach is to consider the satellite ground stations as if they were individual users and to model the arrival of service requests by the ground stations as a Poisson process, where the service requests are exponentially distributed in duration and where the capacity requested by the ground stations per service request is either fixed, which is more similar to traditional CDMA, or varies according to a distribution. The latter case is considered more germane to ATM on CDMA. The problem of evaluating the capacity of an ATM-based CDMA system in terms of Erlangs is now revisited. Using the Erlang-B formula, the blocking probability of an M/M/S/S queue is computed [5,14],

$$P_{\text{blocking}} = \frac{(\lambda / \mu)^s / S!}{\sum_{k=0}^{S} (\lambda / \mu)^k / k!} \tag{18.7}$$

Here, S is the number of servers, which corresponds with the number of users. The quantity λ/μ represents the traffic in Erlangs, where λ is the arrival rate and μ is the service duration.

The blocking probability of a CDMA system is expressed as the probability of exceeding a threshold that is a function of the number of users, the processing gain, and the total equivalent noise density due to AWGN and the interference from all other CDMA users.

$$P_{\text{blocking}} = Pr[Z > \text{Threshold}] \tag{18.8}$$

The blocking probability of a CDMA-based cellular radio system is given in [14]. The number of potential users is assumed to be large. In the case of satellite communications, the impact of other cell interference is disregarded, in which case,

$$P_{\text{blocking}} \approx Q\left(\frac{A - E(Z')}{\sqrt{\sigma_z^2}}\right) \tag{18.9}$$

$$E(Z') = (\lambda / \mu) \exp\left[(\beta\sigma)^2 / 2\right] \tag{18.10}$$

$$\sigma_z^2 = (\lambda / \mu) \exp\left[(\beta\sigma)^2\right] \tag{18.11}$$

where $\beta = (\ln 10)/10$ and σ is the standard deviation in decibels of the power control loop. For $\sigma = 0$ dB, power control is ideal. The blocking probability is solved and the capacity in Erlangs is evaluated. Note that the blocking probability is dependent on the requisite QoS, through the E_b/I_0.

$$\frac{\lambda}{\mu} = \frac{(1 - N_0 / I_0)(W / R)\exp[(\beta\sigma)^2 / 2]}{(E_b / I_0)_{\text{median}}}$$

$$\times \left\{1 + (B/2)\exp[3(\beta\sigma)^2 / 2]\left(1 - \sqrt{1 + 4\exp[-3(\beta\sigma)^2 / 2] / B}\right)\right\} \tag{18.12}$$

where,

$$B = \frac{[Q^{-1}(P_{\text{blocking}})]}{A} = \frac{(E_b / I_0)_{\text{median}}[Q^{-1}(P_{\text{blocking}})]^2}{(W / R)(1 - N_0 / I_0)} \tag{18.13}$$

The capacity in Erlangs is evaluated assuming fixed SATCOM access rates and compared with the average capacity in Erlangs based on variable access rates, modeled by a geometric distribution, with a mean access rate equal to the selected fixed rate. Data concerning individual service rates in [3] suggest that the geometric distribution is a reasonable starting point. The actual composite SATCOM access rate is a function of the services supported. Operation at fixed predetermined rates is typical of many, if not most, communication systems. Variable rate access is a better match with the ability of an ATM-based system to flexibly allocate the available capacity. It is also a good match with CDMA from the perspective that the processing gain is allowed to vary while the spread bandwidth remains fixed.

For the purposes of the example that follows, E_b/N_0 is assumed to be the same for all users. In practice, it will vary for reasons including the service to be supported and the accuracy of the power control. In terms of power control, note that SATCOM is relatively immune to the near-far problem associated with terrestrial wireless communication.

The results are provided in Figure 18.8, where we see that the capacity in Erlangs is very sensitive to the access rate. For a fixed access rate of 512 Kbps, the capacity is 268 Erlangs. The "average" capacity when the access rate is geometrically distributed with a mean of 512 Kbps is 762 Erlangs. There is a gain when flexible access rates are considered. This is due to the fact that the relationship between the CDMA capacity in Erlangs and the user access rate is not linear, as seen by examining (18.1). From Figure 18.8, we also see that the relative advantage increases with the mean access rate. The results may vary with the user service distribution, QoS requirements, and access rates, but they have general application.

Intuitively speaking, as long as users in practice do not all require the same access rates, then it is better to provide them with exactly what they need. Any leftover capacity is allocated to other SATCOM users on a priority basis. Conversely, the cost associated with users operating at data rates in excess of the minimum required is greater than merely the overhead associated with the unused portion of the SATCOM carrier. The dependency on QoS will now be examined in terms of P_b and the blocking probability.

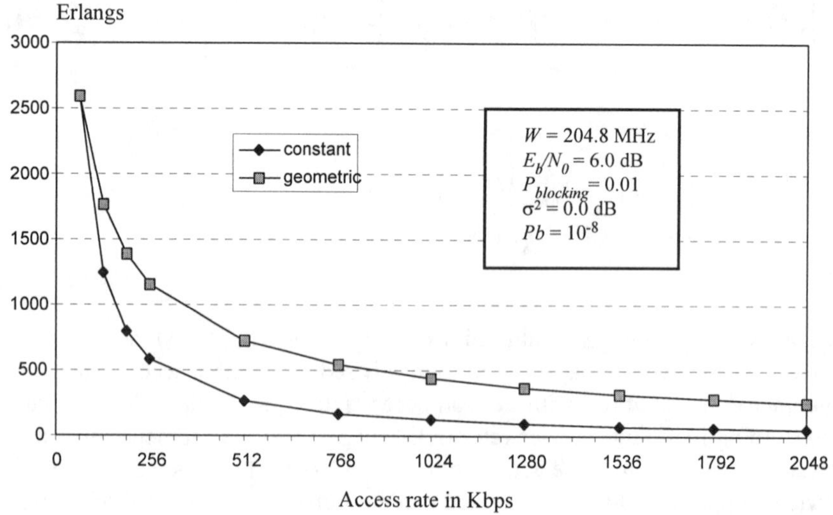

Figure 18.8 Constant versus flexible rate access (Erlangs vs. access rate in Kbps).

In telephony, a busy signal indicates when a call is blocked. The concept of blocking is clear and a blocking probability of 1% to 2% is generally considered acceptable [14]. The concept of blocking is not as clear when we consider data communication. Delay becomes more important. However, blocking can still occur if, for example, service is denied due to lack of capacity anywhere in the system. Blocking due to a lack of capacity over the satellite link is considered herein. The data service is assumed to be a large file transfer via TCP/IP over SATCOM (e.g., e-mail with large enclosures). For ease of discussion, "blocking" is considered to occur upon refusal of a TCP connection by the SATCOM server due to lack of resources on the first try. Blocking can also occur if the end to end delay becomes excessive or, in the case of ATM, if a users' QoS requirements cannot be met.

From Figure 18.9, we see that the capacity in Erlangs is not very sensitive to the probability of blocking for values of P_{blocking} greater than about 1%. However, the capacity degrades quickly when a blocking probability of less than 1% is required.

The probability of blocking is related to the QoS requirements. The probability of bit error is a common measure of the QoS. The available capacity of the system at any given time will vary dependent on the distribution of access rates, as well as the QoS users require. Figure 18.10 illustrates the sensitivity of the system to QoS in terms of BEP in the range of 10^{-3} to 10^{-9}. Over this range, the requisite E_b/N_0 needed to close the link may vary by several decibels dependent on the modulation and coding employed. However, the capacity varies only about 5% for over this range, given a blocking probability of 1%, as seen from Figure 18.10. Thus, the capacity of the system is not very sensitive to changes of a few decibels in terms of E_b/N_0. By comparison, the impact of a few decibels variation in the mean access rate, or processing gain, has a much more dramatic effect on capacity, as seen in comparison with Figure 18.8, which suggests that the capacity is more dependent on bandwidth than on power.

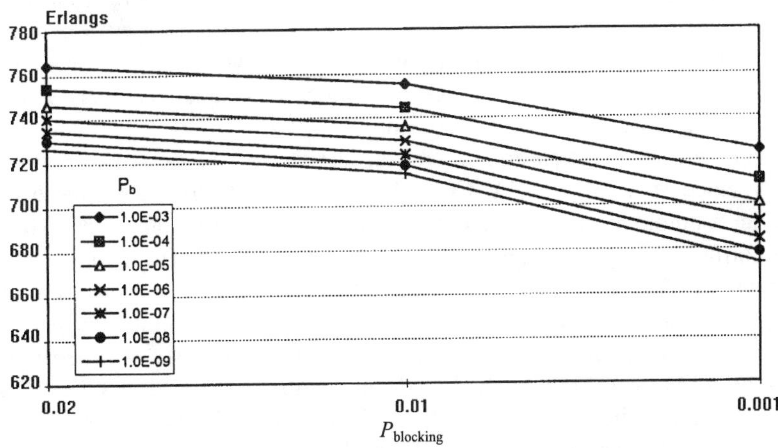

Figure 18.9 Sensitivity to the blocking probability (Erlangs vs. P_{blocking} with BEP P_b as a parameter).

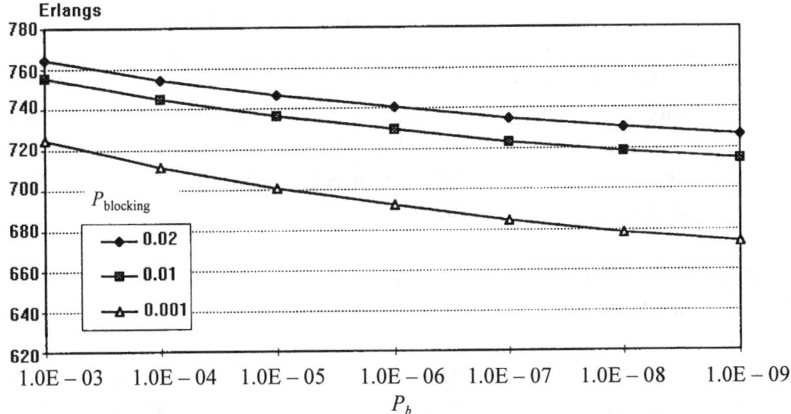

Figure 18.10 Sensitivity to the BEP requirement (Erlangs vs. P_b with P_{blocking} as a parameter).

The sensitivity to the access rate and distribution was shown in this section. The sensitivity to key measures of QoS was also demonstrated. Error correction and control can help achieve and maintain the requisite QoS and improve end to end efficiency. Techniques for error correction and control over ATM-SATCOM links are considered in the next section.

18.3 ERROR CORRECTION AND CONTROL FOR ATM-SATCOM LINKS

Error correction and control are very limited within the ATM protocol, but then ATM was not designed for operation over SATCOM links. The need for error correction and control is typically greater than over a satellite link or over a terrestrial F-O link. It is also dependent on the service, or at least the service class, and as already shown, there is more to efficiency than protocol overhead.

Three error correction and control techniques will be examined for improving the performance of ATM over SATCOM links: FEC, a selective repeat protocol (SRQ) and multicast automatic repeat request (ARQ) with diversity combining. The relative merits, shortfalls, and potential role of each will be discussed. The error correction and control techniques considered in this chapter are intended to be transparent to the ATM switch and augment the existing capabilities provided by commercial off-the-shelf (COTS) satellite modems. From a practical perspective, the use of a separate device between the ATM switch and the SATCOM modem is considered, in order to facilitate integration with existing equipment with minimal change. Special error correction and control techniques for operation over satellite links would reside primarily within the ATM layer of an ATM-SATCOM integration device. However, the error control and correction capabilities could also be implemented within the ATM switch or the SATCOM modem. Additional performance is possible by integrating specific techniques with the ATM switch and SATCOM modem, but this implies modification or replacement of existing equipment and is not considered in this chapter.

The purpose of error correction and control is to maintain the error probability within acceptable limits in terms of proper operation of the system, as well as the end-to-end QoS. Before examining potential error correction and control techniques for ATM over SATCOM, it is useful to consider the requirements for error correction and control imposed by operation of ATM over a SATCOM link. How sensitive is the operation of ATM over SATCOM to errors? To delay?

18.3.1 Performance of TCP/IP Over ATM-SATCOM and Importance of Error Control and Correction

The performance of transmission control protocol (TCP)/IP over ATM-SATCOM links is TCP-limited. Specifically, the throughput per TCP connection is limited by the acknowledgment window size and mechanism [13, 16].

Performance dependency on TCP window size - Figure 18.11 illustrates the dependency of performance on the size of the acknowledgment window. The results were obtained based on the methodology of [17] for evaluating the performance of TCP, which was modified for the case of an E1 2.048 Mbps ATM-SATCOM carrier per [13, 16]. The bit error rate is assumed to be negligible.

The results clearly show the dependency on the transmit window size (W) in bytes. The required window size is determined by the product of the transmission rate and the roundtrip delay (i.e., satellite, plus end to end delay from any other sources). The transmit window must be large enough to store all of the transmitted data until it is acknowledged by the receiver; otherwise, the throughput is degraded. The maximum available is TCP implementation dependent.

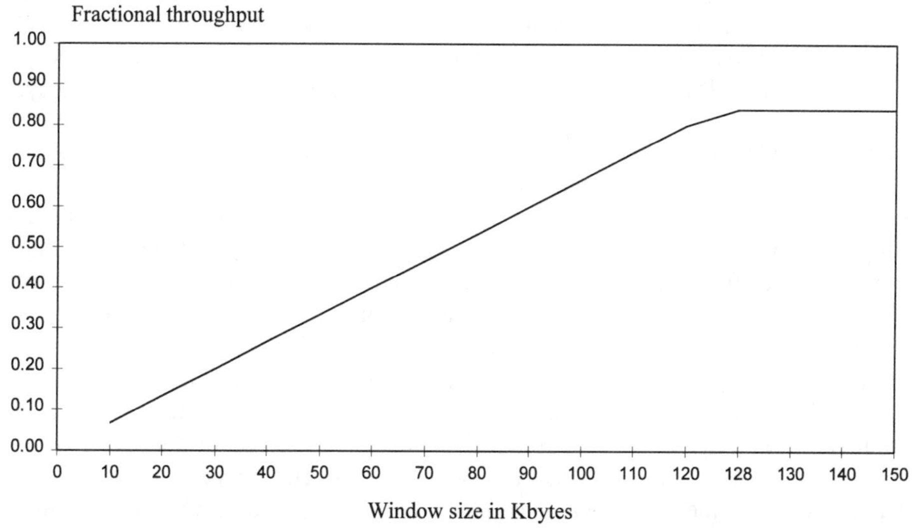

Figure 18.11 Dependency of throughput on TCP window size E1 2.048 Mbps carrier, 9188 byte MTU, 0.25 second one-way delay.

Given a roundtrip delay over the satellite link of about 0.5 seconds and a transmission rate of 2.048 Mbps, the required window size for continuous transmission without stopping to wait for an acknowledgment is 1.024 Mbits. Thus, between the overhead of 20% and the fact that the window size is only 1/2 of the required 1.024 Mbits, the peak throughput is only about 0.8 Mbps, per TCP connection.

Test and analysis indicate that the throughput of a TCP connection over an E1 2.048 Mbps ATM-SATCOM carrier is actually to about 0.8 Mbps due to a combination of protocol overhead (i.e., approximately 20%) and a maximum window size of 64 Kbytes, which is not sufficient to support a TCP connection at rates greater than 512 Kbits over a SATCOM link [13, 16]. This is before consideration of errors.

The combined effects of errors, delay and window size - Figure 18.12 illustrates the combined effects of delay, errors and window size. Controlled measurements were made via a SATCOM delay and error simulator and modeled analytically [13,16]. As can be seen from Figure 18.12, the analytical results are a close match with the measurements obtained using the simulator. In either case, the throughput, given a one-way delay of 0.25 seconds (i.e., roundtrip delay of 0.5 seconds) and a typical TCP implementation, with a window size limitation of 51 Kbytes per connection, is about 750 Kbps. For additional details about simulated and on-air performance of ATM-SATCOM links, refer to [13, 16].

There are two key points to be discerned from Figure 18.12. First, the throughput in Mbps drops off very rapidly when the probability of bit error exceeds a threshold of about 10^{-8}. Second, as long as the window size is large enough and the error rate is low enough, the performance is limited by the capacity of the ATM-SATCOM carrier, less overhead, as opposed to the TCP window. The second point should be clear. It is much less clear as to what is sensitive to the error rate and why.

TCP acknowledgment mechanism - One would not ordinarily expect the 53 byte (424 bit cell) to be catastrophically sensitive to an error rate of 1 bit in 10^6. Nor would one expect the 8192 byte (65536 bit) maximum transmission unit (MTU), which is roughly analogous to an IP packet, to be sensitive. However, the 51 Kbyte TCP transmit window (over 400 Kbits) might be. For example, on average, one out of every 2 to 3 windows is negatively acknowledged, necessitating retransmission, at an error rate of 10^{-6}. Furthermore, it would not matter whether it was a single bit error, or a burst of 51 Kbytes, the entire window's worth of data would need to be retransmitted.

Thus, the performance of TCP/IP over the long delay satellite link is TCP limited and sensitive to delay, errors and the combined effects of delay and errors. If operation of TCP/IP over ATM-SATCOM is to be effective, especially at rates higher than 2.048 Mbps, then additional measures of error control and correction is required, if only to ease the burden on TCP. By comparison, UDP is much less sensitive to errors over the SATCOM channel and insensitive to the delay. It is also unreliable by definition.

The window must be large enough to store transmitted data until it is acknowledged by the receiver. In practice, throughput is limited to about 0.8 Mbps due to a combination of protocol overhead (i.e., approximately 20%) and a maximum window size of 64 KBytes (512 Kb). The maximum widow size is TCP-implementation-dependent. Given a roundtrip delay over the satellite link of about 0.5 seconds and a transmission rate of 2.048 Mbps, the required window size for continuous transmission without stopping to wait for an acknowledgment is 1.024 Mb. Thus, between the overhead of 20% and the fact that the window size was only 1/2 of the required 1.024 Mb, peak throughput was only about 0.8 Mbps, per TCP connection. This is before consideration of errors.

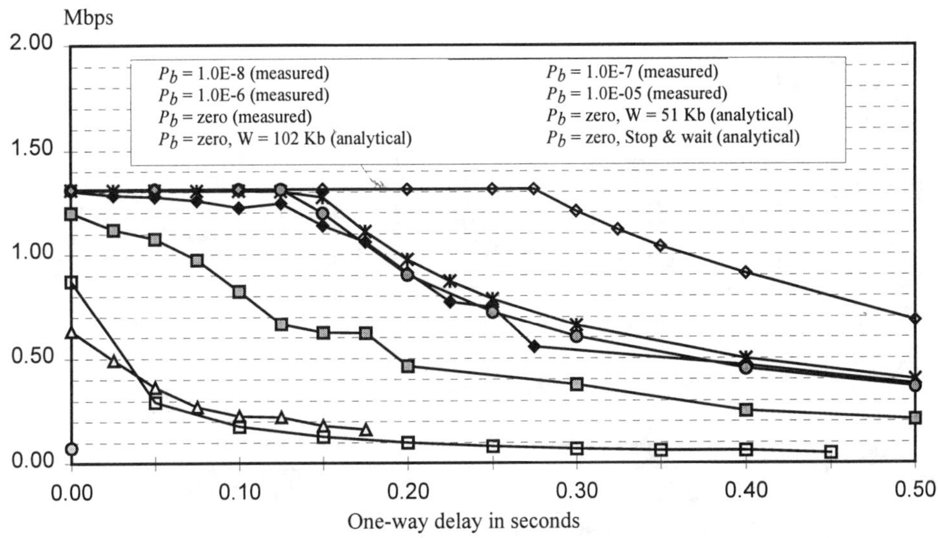

Figure 18.12 The combined effects of errors, delay and window size over an E1 2.048 Mbps ATM-SATCOM carrier.

18.3.2 A Concept for Error Control and Correction Tailored to Service Class

An illustration of a concept for error control and correction over ATM-SATCOM links is provided in Figure 18.13. The illustration depicts the use of an E1 carrier with the G.804 PLCP. Other rates, interfaces and PLCPs are possible. Separate queues per service class are required.

The concept allows for a different error correction scheme per service type. For the purposes of this chapter, the four ATM service classes have been associated with the four AALs (i.e., class 1 = AAL 1, class 2 = AAL 2, class 3 = AAL 3/4 and class 4 = AAL 5). This is only an approximation made to simplify the discussion. Tailoring of the error control and correction technique to the service type, as opposed to applying the same technique to all data, allows the overhead associated with error control to be reduced. This is because the overhead associated with error control and correction is effectively averaged in accordance with the traffic profile over component services on a virtual circuit, the services per SATCOM link and the links over the satellite. The implication is that services that require more power-efficient FEC may do so without necessitating additional overhead for all services.

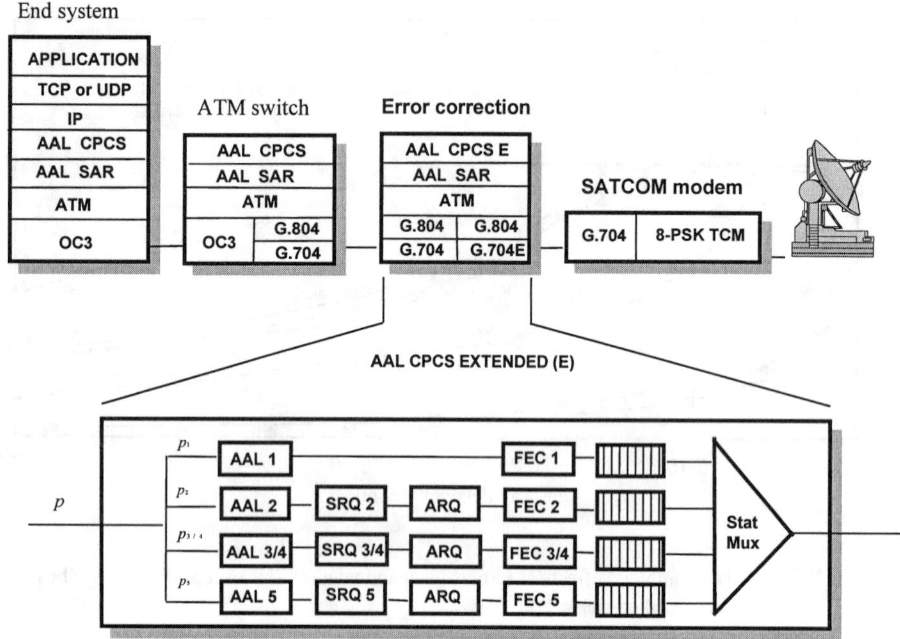

Figure 18.13 Error correction and control for ATM over SATCOM.

18.3.3 A CPCS-Based Truncated SRQ Protocol For AAL 5

The use of an SRQ protocol is intended to support data communication. This includes broadcast data, file transfer, and signaling services running over TCP/IP through AAL 5. TCP/IP over AAL 5 can also be considered to support video. The proposed SRQ protocol is not intended to guarantee error-free communication. End to end reliability remains the responsibility of TCP. Rather, the SRQ is intended to reduce errors seen by the higher-level TCP protocol. In this way, it eases the burden on TCP/IP. TCP is a sliding window protocol. It is more sensitive to the combined effects of delay and errors than an SRQ protocol because it repeats an entire "windows" worth of data in the event of an error. The amount of data to be repeated in the event of an error can quickly become large as the combination of data rate and delay increases. Although there is an SRQ capability in a future version of TCP, none exists at present.

The primary benefit of an SRQ protocol is that only the protocol data unit (PDU) in error is actually retransmitted. This saves bandwidth over the satellite link when errors occur as the channel begins to degrade. Additionally, SRQ protocols are less sensitive to delay than sliding window protocols [17]. The disadvantage of SRQ protocols is that they can be complicated to implement and are memory intensive. An ideal SRQ protocol would continue to retry transmitting a PDU until it was finally successful. We can show that the number of attempts required before success is geometrically distributed, with a mean dependent on the mean PDU error rate [17]. At high error rates, the memory required to implement an SRQ protocol can easily become very large, and the delay is increasingly long and variable as the protocol repeatedly tries to send the PDU. The delay variability makes SRQ protocols unsuitable for CBR services.

The SRQ protocols examined in this chapter are limited to a maximum of three retransmissions. The memory requirement is three times the round-trip delay in the absence of delay from other sources. We will show that three attempts are sufficient to significantly improve the overall performance.

The proposed SRQ protocol is implemented in the convergence sublayer portion of the AAL sublayer. Error correction is consistent with the intended use of the convergence sublayer of AAL 5 [3]. An illustration of the proposed CPCS-SRQ protocol frame is provided in Figure 18.14, from which it can be seen that the protocol is implemented using only 2 bytes out of the CPCS-PDU payload. The CPCS-PDU payload is from 1 to a maximum of 65,353 bytes instead of 1 to 65,355 bytes. Otherwise, the CPCS-PDU remains unchanged. The proposed SRQ protocol would rely on the existing CRC of the CPCS-PDU to determine if retransmission is necessary. The padding is employed to round off the CPCS-PDU to a multiple of 48 bytes, which is the payload of the ATM cell. Ideally, the size of the CPCS-PDU payload is matched in size with the higher-layer TCP/IP packets so as to minimize segmentation.

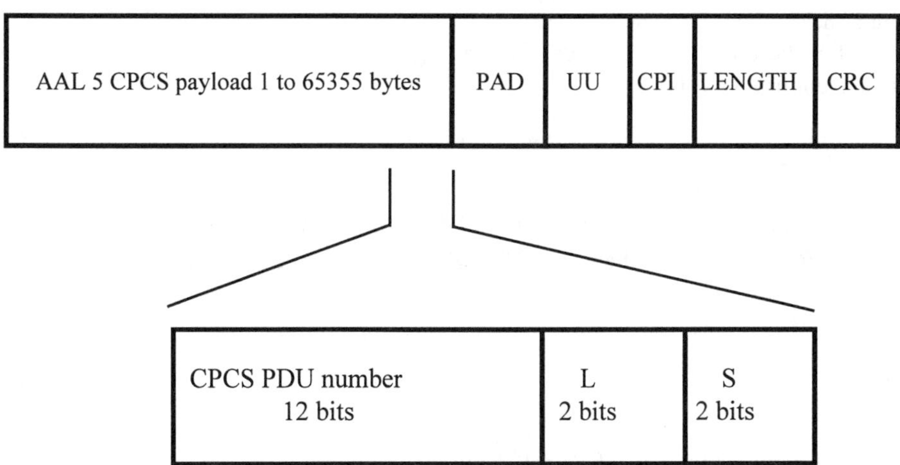

Figure 18.14 ATM CPCS-SRQ protocol frame.

Testing suggests that an IP packet of 1024 bytes (8192 bits) is a reasonable value for operation over a satellite link [16]. Regardless of the exact value, the point is that the CPCS-PDU is expected to be very large compared with the 2 bytes required to implement the SRQ protocol. Consequently, the SRQ protocol has a minimal impact on the total overhead.

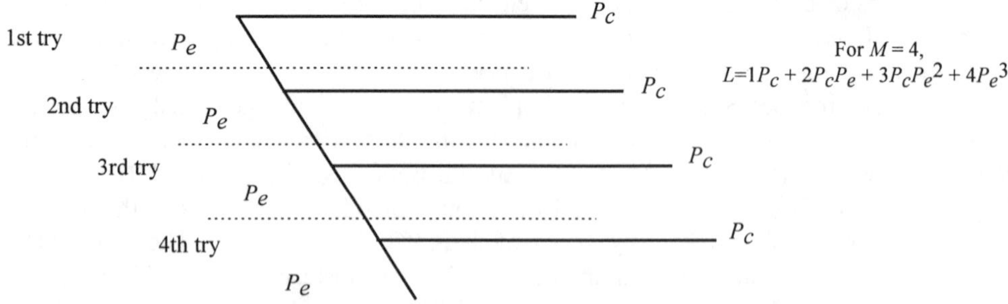

Figure 18.15 ATM CPCS-SRQ protocol performance.

The 2-byte SRQ header is allocated as follows: 2 bits for control, 2 bits for the number of retries (M), 8 bits for CPCS-PDU numbering, and 4 spare bits, which may also be used for numbering. Note that a retry is initiated when the 32-bit CPVS CRC detects an error. The performance of the SRQ protocol is evaluated as illustrated in Figure 18.15.

Let the underlying BEP be given as P_b. The CPCS-PDU error probability per independent trial (P_e) is

$$P_e = 1 - (1 - P_b)^N \tag{18.14}$$

where N is the size in bits of the CPCS-PDU. The probability that the CPCS-PDU is delivered free of error is then $P_c = 1 - P_e$. The efficiency of the SRQ protocol is evaluated from [17]

$$SRQ_{efficiency} = \frac{1}{L} \tag{18.15}$$

where L = average number of CPCS-PDU transmissions

$$L = \sum_{n=0}^{M-1} nP_cP_e^{n-1} + MP_e^{M-1} \tag{18.16}$$

The average probability of successful delivery of a CPCS-PDU is found from

$$P_{\text{success}} = \sum_{n=0}^{M-1} P_c P_e^n \tag{18.17}$$

and the probability that the CPCS-PDU is not correctly received after M tries is

$$P_{\text{fail}} = 1 - P_{\text{success}} \tag{18.18}$$

This is the uncorrected CPCS-PDU error probability as viewed by higher layers (e.g., TCP/IP). The probability of failure to correctly deliver the CPCS-PDU is translated into an average residual BEP; however, the performance of higher-layer protocols is dependent on P_{fail} regardless of whether the failure is due to a single bit error within the CPCS-PDU or a burst of errors. For example, the failure to deliver any CPCS-PDU correctly to TCP necessitates the retransmission over the satellite of a TCP transmit window worth of data.

As we see from the preceding analysis, the performance of the SRQ protocol is not dependent on the round-trip delay as it was for TCP. This is an important feature with respect to operation over SATCOM links. However, delay does factor into the memory required to implement the protocol. The memory must be sufficient to save the CPCS-PDUs for a period of time equal to M-1 times the round-trip delay. For the case of $M_{\text{max}} = 4$, this allows for the possibility that a CPCS-PDU is not received correctly by the receiver on the first attempt or either the first or second retry. Following the third and final retransmission (i.e., M_{max} -1), a CPCS-PDU is dropped by the SRQ, in which case, recovery is via TCP.

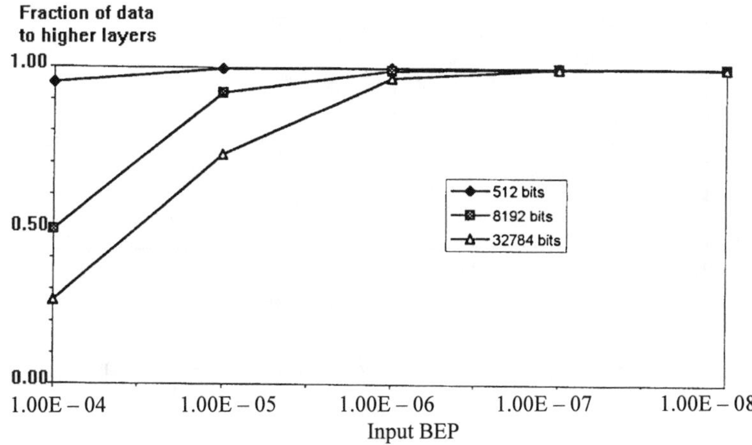

Figure 18.16 CPCS-SRQ protocol efficiency.

From Figure 18.16, we see that the larger the CPCS-SRQ-PDU, the more efficient it is in terms of the fraction of the data passed to higher layers when error probability is low. However, the opposite is true when error probability is high, as larger PDUs are more sensitive to bit errors. In the case of an 8192-bit CPCS-SRQ-PDU, the greatest benefits occur when the input BEP is greater than about 10^{-5} and there is little additional benefit when the input BEP is less than about 10^{-6}.

Figure 18.17 illustrates the performance benefit of the truncated CPCS-SRQ protocol for the case of a CPCS-PDU of 8192 bits. The maximum number of attempts to deliver the data is M_{max}, which means that $M \in \{1, 2, 3, 4\}$. The benefit as expressed in terms of the residual probability of bit error is nearly 4 orders of magnitude at an input BEP of 10^{-5}. As a practical reference, consider the case of TCP/IP over ATM AAL 5, operating via a 2.048 Mbps carrier over a satellite link. As shown in [16], the throughput of TCP/IP over ATM AAL 5 degrades over 750 Kbps (TCP window limited) to less than 10 Kbps as P_b increases from 10^{-8} to 10^{-6} and becomes virtually negligible at 10^{-5}. With the SRQ protocol ($M_{max} = 4$), the residual probability of bit error for an input error probability of 10^{-5} is 4.7×10^{-9}. As we see in Figure 18.16, there is no degradation in the throughput at the ATM AAL 5 CPCS layer, apart from 0.2% overhead to implement the CPCS SRQ header.

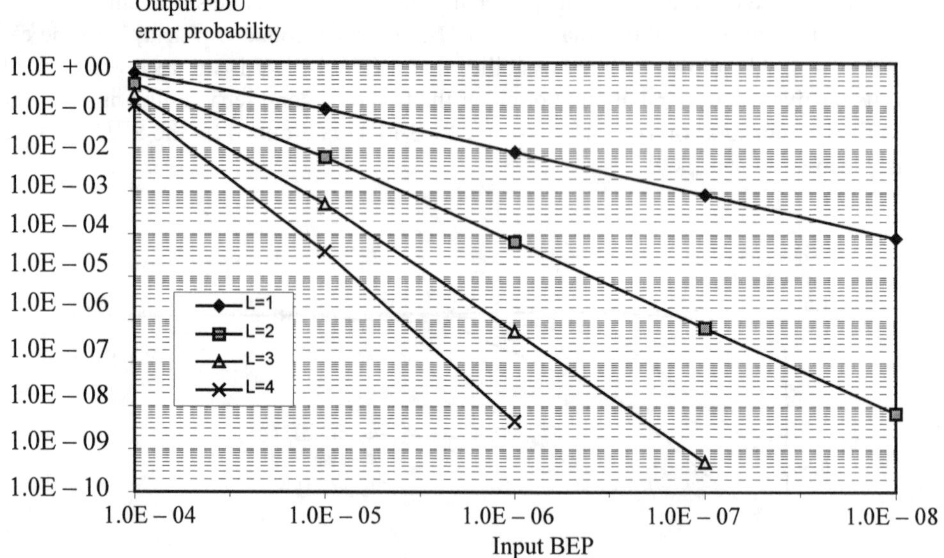

Figure 18.17 CPCS-SRQ protocol I/O characteristic (8,192-bit CPCS-PDU).

18.3.4 FEC Options for ATM Over SATCOM

FEC provides another means of improving the performance of ATM over SATCOM or wireless links. The improvement is substantial. A possible disadvantage is that the overhead required to implement FEC is typically greater than required to implement an SRQ protocol, and the overhead to implement the ATM protocol is already large. An advantage relative to the SRQ protocol is that the delay tends to be less, and at least in the case of block codes, the delay is fixed in duration. Low delay and delay variation are particularly desirable with respect to CBR services, which is why FEC is more suitable for use with voice or other real-time services than an SRQ protocol.

In this section, high-rate block codes are considered as a means of improving communication of ATM over low-rate SATCOM links. A block code is implemented within the ATM switch, the SATCOM modem, or in a separate external box, which is the preferred approach. The block code is effectively concatenated with existing FEC within the SATCOM modem. Block codes operate on hard decisions about the channel bits. Soft-decision decoding is more efficient than hard-decision decoding; however, hard-decision decoding is the most practical here since decisions have already been made within the SATCOM modem on the inner channel bits and the outer code is implemented externally.

The block code is implemented within the physical layer or the ATM layer. Physical layer FEC provides additional protection for the ATM header not afforded by the codes implemented at the ATM layer or by CPCS-SRQ protocol described earlier in this chapter. On the other hand, ATM-layer FEC is tailored to the service class, or even by component services, which is potentially more efficient. BCH codes and shortened block codes, matched to the ATM protocol, will be examined.

As bit errors over a SATCOM link may come in bursts either due to channel effects or the operation of the SATCOM modem (e.g., code or differential encoding), burst error correction through the use of interleaved BCH codes and Reed-Solomon (RS) codes will be considered. Burst error correction is expected to be most applicable for codes implemented at the physical layer, or at the ATM layer in support of information services other than those running over TCP, where the effect of error bursts tend to be masked by the effects of the sliding window retransmission mechanism.

The special case of a high-rate BCH code mapped into spare bits in the popular E1 2.048 Mbps G.704 frame will also be examined. Although not as effective as lower rate codes, any improvement in performance is "free" from the perspective of the cost in terms of overhead.

A cross-section of BCH and RS block codes are considered. A summary is provided in Table 18.4. For the (n,k) BCH codes, n is the number of bits per block and k is the number of information bits per block. The code rate is defined as rate $= k/n$. In the case of RS codes, n represents the number of symbols per block and k represents the number of information symbols per block. Each symbol contains $Q = 2^m$ bits, where m is the related to the block size by the equation $m = \log_2(n)$ and m represents the burst-error-correcting capability of the RS code. Ordinary BCH codes do not correct burst errors ($\lambda = 1$); however, interleaved BCH codes [18] can correct bursts of errors ($\lambda > 1$). The BCH codes will be examined first.

Table 18.4

Summary of Codes

Code	**Code Type**	***(n, k)***	λ	**OH(%)**	*t*	**Burst Bits**	**Delay (bits)**
1	BCH	(1023,1013)	1	0.98	1	1	1024
2	BCH	(63,53)	1	18.9	1	1	64
3	Short BCH	(443,423)	1	4.7	2	1	581
4	BCH	(511,457)	1	11.8	6	1	512
5	BCH	(255,223)	1	14.3	4	1	256
6	BCH	(2046,2026)	2	0.98	1	2	2048
7	BCH	(4092,4052)	4	0.98	1	4	4096
8	Short BCH	(886,846)	2	4.7	2	2	2048
9	RS	(255,223)	N/A	14.3	16	8	2048
10	Short RS	(61,53)	N/A	15.1	1	8	3600

Nonburst error-correcting block codes

The performance of a *t* error-correcting BCH code is evaluated from [18]

$$P_{b_{\text{BCH}}} = \sum_{i=t+1}^{n} (1 - P_b)^{n-1} P_b^i \tag{18.19}$$

A summary of the performance of the nonburst error-correcting block codes is provided in Figure 18.18. The performance is expressed in terms of an I/O characteristic as it was for the SRQ protocol. The baseline is the uncoded performance.

Testing [13,16] indicates that the performance of TCP/IP over ATM-based systems begins to deteriorate very quickly at error rates above 10^{-8}. The (1023,1013) BCH code begins to be effective at an input error probability of about an input BEP of 10^{-6}. Although not nearly as effective as the lower rate BCH codes, even this high-rate BCH code reduces the BEP seen by higher layers by more than an order of magnitude at an input error probability of 10^{-7}, at a cost of only 1% overhead.

The best of the BCH codes of Figure 18.18 is code 4, the 6-error-correcting (511,451) code. An improvement of more than 4 orders of magnitude is achieved at an input BEP of 7.10^{-4}, to the desired output error probability of 10^{-8}, at a cost of 11.7% overhead.

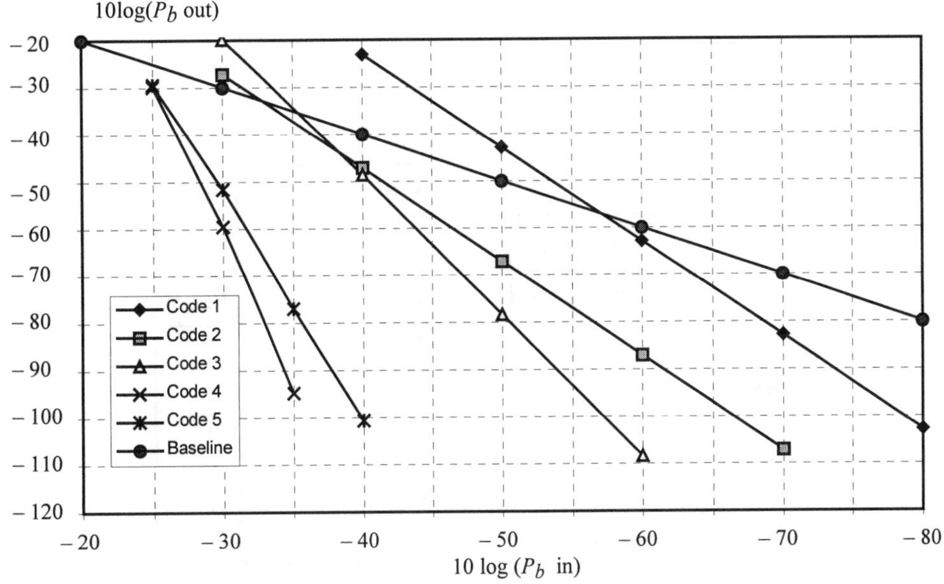

Figure 18.18 Nonburst error-correcting block code performance.

Burst error-correcting block codes

Burst error correction is possible with BCH codes by constructing an interleaved code [18]. An interleaved BCH code is constructed from a BCH code as illustrated by Figure 18.19. The burst error-correction capability is equal to the interleaving degree λ. Alternatively, RS codes are powerful, burst error-correcting codes, but they do not perform as well at low values of E_b/N_0.

Figure 18.20 summarizes the performance of the burst error-correcting codes defined in Table 18.4. Code 6 is an interleaved BCH code constructed from code 1, the (1023,1013) BCH code, with $\lambda = 2$. Code 7 is also constructed from code 1, with $\lambda = 4$. Code 8 is a shortened BCH code constructed from code 3. Code 9 is a 16-error-correcting (255,223) RS code with a burst-error correction capability of 8 bits. Code 8 is a shortened RS code constructed from the (255,223) RS code, which is also capable of correcting 16 errors of burst length 8 bits.

The best code, from the perspective of the lowest output bit error rate for a given input BEP, was code 9. But the overhead with this code is 12.6%, which comes off of the peak throughput achievable at higher layers. Dependent on the actual burst error-correction requirements, code 8 is of interest. This interleaved double-error-correcting BCH code is less effective than code 8 or code 10, but it is matched to the cell size and the overhead is only 4.5%. Code 10 is a short RS code, which is also matched to the cell size, but the overhead is 13.1%.

***n* bits**

λ rows				
c_{00}	c_{01}	c_{02}	• • •	c_{0n-1}
c_{10}	c_{11}	c_{12}	• • •	c_{1n-1}
c_{20}	c_{21}	c_{22}	• • •	c_{2n-1}
⋮	⋮	⋮		⋮
$c_{\lambda-10}$	$c_{\lambda-11}$	$c_{\lambda-12}$	• • •	$c_{\lambda-1n-1}$
$c_{\lambda 0}$	$c_{\lambda 0}$	$c_{\lambda 0}$	• • •	$c_{\lambda 0}$

Figure 18.19 $(\lambda n, \lambda k)$ interleaved BCH code.

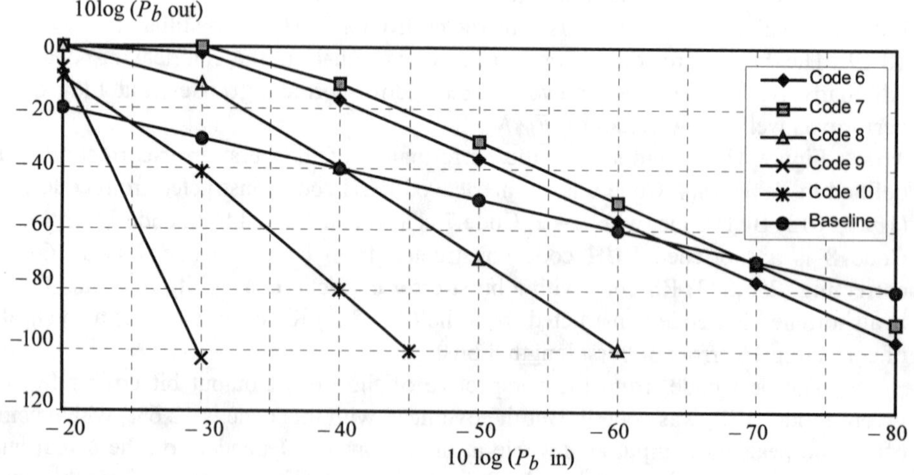

Figure 18.20 Burst error-correcting block code performance.

FEC matched to the spare bits in the E1 carrier

The (1023,1013) BCH code and the interleaved codes constructed from it were the least effective of the codes examined from the perspective of the I/O performance. However, this class of codes has the special quality that they can be implemented using only the spare bits of the E1 G.704 frame, or other framing structures based on the E1 (e.g., the E3 8.448 Mbps framing or Intelsat IBS framing). This means that any coding gain, however limited, is essentially free with respect to the cost in overhead. An illustration of how the (4092,4052) interleaved BCH code is mapped into the E1 CRC-4 extended super frame (ESF) is provided in Figure 18.21. The mapping occurs over a 16 frame multiframe [8]. The 40 parity check bits of the code are mapped I-to-I into the 40 spare bits of the ESF, protecting 4052 information bits, which is sufficient to accommodate the 3840-bit (16 × 30 byte) payload load in the ESF. Like other physical layer coding options, it applies to all traffic on the link.

	Submultiframe	Frame	Bits 1 to 8 of the Frame							
			1	2	3	4	5	6	7	8
Multiframe	I	0	C_1	0	0	1	1	1	1	1
		1	0	1	A	P_0	P_1	P_2	P_3	P_4
		2	C_2	0	0	1	1	1	1	1
		3	0	1	A	P_5	P_6	P_7	P_8	P_9
		4	C_3	0	0	1	1	1	1	1
		5	0	1	A	P_{10}	P_{11}	P_{12}	P_{13}	P_{14}
		6	C_4	0	0	1	1	1	1	1
		7	0	1	A	P_{15}	P_{16}	P_{17}	P_{18}	P_{19}
	II	8	C_1	0	0	1	1	1	1	1
		9	1	1	A	P_{20}	P_{21}	P_{22}	P_{23}	P_{24}
		10	C_2	0	0	1	1	1	1	1
		11	1	1	A	P_{25}	P_{26}	P_{27}	P_{28}	P_{29}
		12	C_3	0	0	1	1	1	1	1
		13	E	1	A	P_{30}	P_{31}	P_{32}	P_{33}	P_{34}
		14	C_4	0	0	1	1	1	1	1
		15	E	1	A	P_{35}	P_{36}	P_{37}	P_{38}	P_{39}

E = CRC-4 error indication bits, P_n = parity bits in the (λ 1023, λ 1013) BCH code (i.e., $\lambda = 4$), C_i = CRC-4 bits and A = remote alarm indication bits

Figure 18.21 Mapping of the FEC into the E1 CRC-4 ESF.

Multicast with ARQ diversity for local error recovery

From a SATCOM perspective, diversity is most commonly employed to improve link availability and reduce power requirements, for example, in terms of rain margin. However, under the appropriate channel conditions, diversity is also a powerful error correction and control technique.

Broadcast or multicast of data via ATM over SATCOM is employed to provide a diversity gain. There is gain due to spatial separation where outages due to rain are concerned over EHF, but there is also the potential for diversity gain where the information is transmitted to multiple ground terminals. A key assumption is that the receiver is noise limited on the downlink. This is typical of small terminals and, to the extent that the AWGN process is independent from one terminal to another, diversity gain is possible. An examination of the link budget in Table 18.2 reveals that the forward link from the large terminal to the small terminal is limited by the receiver noise in the small terminal. The limitation is expected to be more pronounced given smaller terminals on the receive end and the use of CDMA.

To realize a diversity gain in this context, multiple ground terminals must be interconnected. An illustration is provided in Figure 18.22. Interconnection via a combination of the Internet and phone lines is shown. Other possibilities exist. Since the primary purpose of the interconnection is to support local error recovery as opposed to recovery over the satellite, operation at very low data rates over the terrestrial interconnections is conceivable, as long as the need for retransmission remains infrequent.

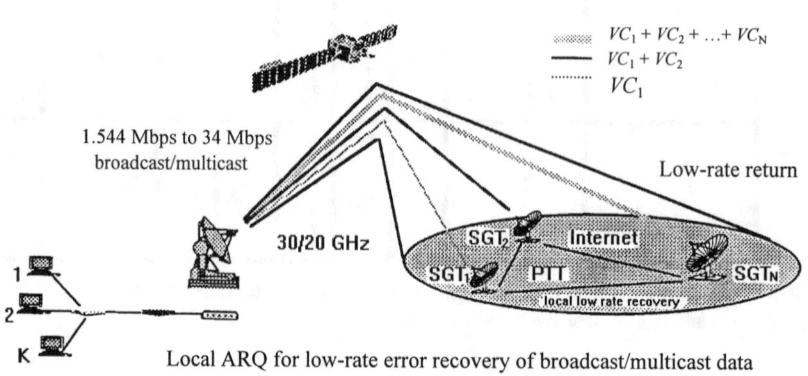

Figure 18.22 ATM-SATCOM multicast with ARQ diversity.

The aforementioned diversity technique applies most directly to multicast data transmissions over ATM (i.e., TCP/IP over ATM AAL 5), but may have broader application dependent on the round-trip delay required for local error recovery between ground terminals receiving the multicast transmission. The ability to rely on a local ground-based neighbor for limited error recovery relaxes the need for error recovery over the satellite. As already shown, this is important because of TCP limitations over long-delay, high-data-rate links. It is also important since error recovery via local terrestrial links is more desirable for some information services than recovery over the satellite link in terms of the delay.

Local error recovery is implemented using an ARQ protocol similar to the SRQ protocol previously described. Upon detection of an error in the CPCS-PDU CRC, a one-time ARQ for error recovery is multicast to "terrestrial neighbors" in the SATCOM multicast. In the special case of fixed rate access using standard E1 2.048 Mbps carriers, the ARQ protocol is implemented in the spare bits of the E1 carrier, like the (4092,4052) block code. Although the access rate is inflexible, no additional overhead is required and protection of the cell header as well as the CPCS-PDU is possible.

The concept is probably best suited to regular scheduled use of SATCOM, which is why virtual paths are shown in Figure 18.22. However, it is neither constrained to pre-scheduled transmissions, nor to virtual paths, and is dynamic as long as terrestrial interconnectivity can be counted on between the SGTs.

The diversity gain is evaluated as follows. Given N links, where each is independently limited by noise in the receiver, diversity gain is estimated using the simple relationship

$$P_e = \left[1-\left(1-P_b\right)^{\mathrm{MTU}}\right]^N \tag{18.20}$$

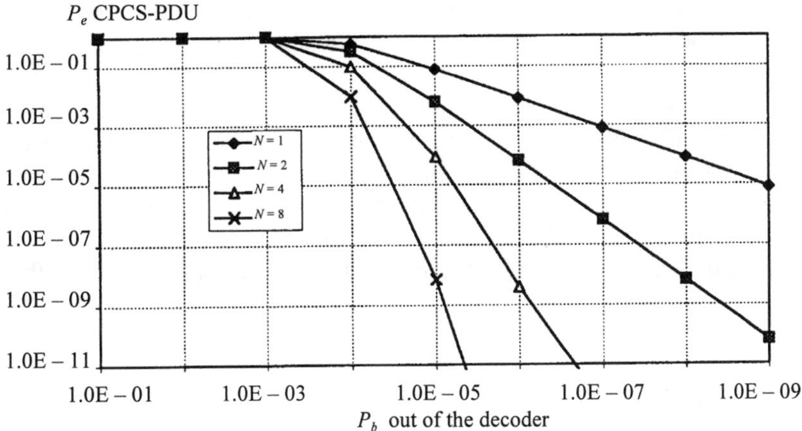

Figure 18.23 The performance benefit with diversity.

A summary of the potential performance benefit is provided in Figure 18.23. The benefit was evaluated assuming a nominal MTU of 8192 bits (1024 bytes) where the MTU is viewed as essentially an IP packet and it is assumed that the IP packet and the CPCS-PDU are matched. A reduction of the CPCS-PDU error probability of several orders of magnitude is possible with relatively few interconnected SGTs. A reduction in CPCS-PDU error probability has end to end significance since a single erroneous CPCS-PDU passed to TCP/IP will necessitate retransmission over the satellite of a potentially much larger block of data. Note that the multicast gain is actually twofold: (1) N-level diversity gain dependent on the number of sites participating in the multicast, and (2) a reduction of the data rate over the satellite by up to a factor of N.

Summary of error control and correction

Error control and correction are very important where TCP/IP over ATM links are concerned; especially over SATCOM and wireless communication links where error probability and/or delay is elevated relative to the terrestrial F-O links for which ATM was conceived.

A simple SRQ protocol, with limited memory requirements, was proposed to improve the throughput and take the burden off of TCP, where TCP/IP over ATM data links are concerned. The cost to implement the protocol in terms of overhead was found to be 2 bytes, which is less than 1% for CPCS-PDU sizes of 1024 bytes. An SRQ protocol is most applicable to ATM service class 4 (AAL 5), as well as class 3 (AAL 3/4).

FEC is applicable to all four ATM service classes. Several FEC alternatives were examined, including BCH codes, shortened BCH codes, interleaved BCH codes, Reed-Solomon codes, and a special BCH code tailored to fit in the spare bits of an E1 carrier.

An ARQ multicast/diversity concept was proposed to improve performance when multicasting is considered for multiple, small, interconnected terminals. Diversity is used in combination with the SRQ protocol and the FEC alternatives examined. As with the SRQ protocol, this technique is most applicable to data and possibly video services. Since error recovery is "local" and delay is much less than over the satellite link, this technique may have broader application to information services than the SRQ protocol.

18.4 SATCOM EFFICIENCY

The primary measures of efficiency considered in this chapter thus far are capacity in Erlangs and throughput in bits per second. The overall system capacity was evaluated in terms of the Erlangs of traffic that is supported in Section 18.2. The end to end throughput per connection was discussed relative to TCP/IP over ATM in Section 18.3.

The potential bandwidth and power efficiency of several of the more promising error control and correction techniques examined in Section 18.3 are now considered.

The performance is evaluated in terms of the throughput in error-free bits delivered to protocols above the ATM layer normalized by the signal bandwidth occupied over the satellite. The results are expressed in terms of bits per hertz (bits/Hz) of carrier bandwidth over the satellite link. A "snapshot" of the results is provided in Figure 18.24, to illustrate the relative potential of the techniques. For the purposes of discussion, a normal 8192-byte CPCS-PDU is assumed. The performance may vary somewhat with CPCS-PDU size. High-speed file transfer via TCP/IP over ATM AAL 5 is considered and QoS requirement $P_b \leq 10^{-8}$ is assumed [13,16].

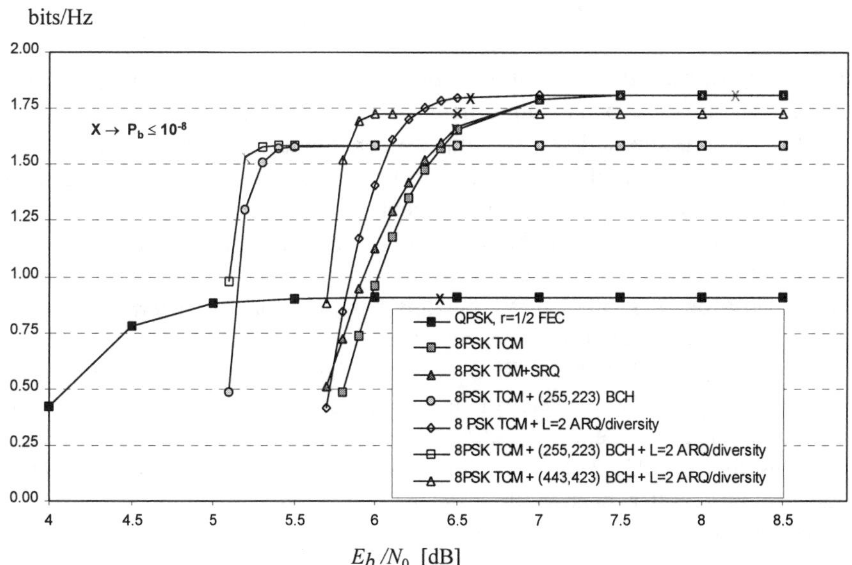

Figure 18.24 ATM-SATCOM protocol stack efficiency 8192-bit AAL 5 CPCS-PDU.

The baseline is QPSK with rate ½ FEC, on a FDMA carrier with a carrier spacing equal to the channel symbol rate. A carrier spacing of 1.0 times the channel symbol rate is optimistic for FDMA, where a carrier spacing of 1.4 times the channel symbol rate is typical[19]; however, it is reasonable for single-carrier CDMA. An 8-PSK TCM on a CDMA carrier is considered for all other cases. The baseline provides the best performance when E_b/N_0 is less than about 5.5 dB. The peak efficiency is 0.91 bits/Hz as seen at the boundary between the ATM layer and TCP/IP. Although full performance is achieved from the perspective of the ATM layer at about an E_b/N_0 of 5.5 dB, the QoS requirement of $P_b \leq 10^{-8}$ requires an E_b/N_0 of 6.4 dB. The reason for the discrepancy is that the link in this example case is TCP-limited at the higher layer, not by the ATM layer.

An 8-PSK TCM without additional error correction and control outperforms the baseline in terms of the bandwidth efficiency starting at about 6.0 dB E_b/N_0, achieving a maximum of 1.81 bits/Hz at 7.2 dB E_b/N_0. However, 8.2 dB E_b/N_0 is needed to meet the QoS requirement.

The maximum efficiency of the SRQ protocol with 8-PSK TCM is just under 1.81 bits/Hz due the overhead associated with the SRQ protocol. At first glance the performance improvement appears to be very small. However, this technique is more bandwidth-efficient than the baseline beginning at 5.8-dB E_b/N_0. The QoS requirement is satisfied at 6.5-dB E_b/N_0, which means that it is only 0.1 dB less power-efficient than the baseline.

The FEC (255,223) BCH code is 0.5 dB more power-efficient than the baseline meeting the QoS requirement at E_b/N_0 of 5.9 dB, and is much more effective overall. A maximum bandwidth efficiency of 1.58 bits/Hz is achieved at an E_b/N_0 of 5.4 dB; however, it is 12.6% less bandwidth-efficient than SRQ technique for 6.5-dB E_b/N_0.

The ARQ multicast/diversity with two interconnected ground terminals (2-level diversity) is as bandwidth-efficient as the SRQ protocol, but 0.1 dB less power-efficient bases on 6.6 dB to meet the QoS requirement. Note that the diversity here is in the availability of a good copy of the CPCS-PDU from a neighbor also receiving the multicast transmission. Additional improvement is possible if diversity combining is performed on a bit or symbol basis, but this has implications in terms of additional system complexity, as well as the rates over the terrestrial links.

An 8-PSK TCM with the SRQ protocol, the (255,223) BCH code, and ARQ multicast/diversity (2-level diversity), is the most power-efficient in this example, meeting the QoS requirement at 5.2-dB E_b/N_0. The bandwidth efficiency is essentially the same as for the (255,223) code alone. Alternatively, the (443,423) BCH codes provides slightly better bandwidth efficiency at a cost of about 1 dB in power efficiency.

18.5 CONCLUSIONS

Techniques for broadband wide area networking based on IP over ATM-SATCOM links at K_a-band were examined in this chapter. The key technical issues with respect to operation over SATCOM are: delay, delay variation, and error probability over the satellite link, as well as outages due to rain and atmospheric affects. The latter takes on particular significance where operation of high-speed ATM networks is concerned. Current TCP limitations also affect performance at high speeds; especially over long-delay satellite links.

From a SATCOM perspective, the efficiency of ATM lies in its ability to flexibly allocate capacity and, potentially, on the manner in which multicasting and priority queuing are exploited. ATM is very sensitive to traffic intensity, type, and mix.

Previous generations of SATCOM systems were designed in terms of voice circuit capacity available to any other user on a priority basis.

In a broader context, the key question is cost, which is viewed in terms of bandwidth, power, and ultimately, economics. Efficiency is critical and is only achieved with a system-level solution tailored to meet the user's specific information exchange requirements, which include service type, QoS requirements, traffic intensity and mix, as well as users' geographic considerations. Providing communication to anyone, anywhere, anytime is an admirable goal. However, there is a niche for a lower-cost system that only attempts to satisfy the 80% of users that impose only 20% of the system design cost.

A concept was proposed for combining the benefits of ATM with those of SATCOM at K_a-band SATCOM. A variety of different techniques for integrating ATM with SATCOM were considered. These include matching the flexibility of ATM with that of CDMA, multicasting, and diversity combining for small terminals that are downlink-receiver-noise limited, priority queuing for graceful degradation of service at EHF in fades due to rain and error correction concepts tailored to ATM-SATCOM.

There is a common perception that broad acceptance of ATM in the market is dependent on the need for information services that require the flexibility, performance, and speed of ATM. It is difficult to know what these services will be, but perhaps broadband wide area networking via ATM over SATCOM is such a service.

REFERENCES

[1] L.M. Correia and R. Prasad, "An overview of wireless broadband communication," *IEEE Comm. Mag.*, Vol.35, pp. 28–33, January 1997.

[2] W. Honcharenko, J. Kruys, D. Lee, and N. Shah, "Broadband wireless access," *IEEE Comm. Mag.*, Vol.35, pp.20–26, January 1997.

[3] M. de Prycker, *Asynchronous Transfer Mode Solution for Broadband ISDN*, Third Edition, London, Prentice-Hall, 1995.

[4] H. Saito, "Dynamic resource allocation in ATM networks," *IEEE Comm. Mag.*, May 1997.

[5] J. Hayes, *Modeling and Analysis of Computer Communication Networks*, New York, Plenum Press, 1984.

[6] G. Armitage, "IP multicasting over ATM networks," *IEEE J. Selected Areas in Comm.*, April 1997.

[7] ITU-T Recomendation I.363, *B-ISDN ATM Adaptation Layer (AAL) Specification*, ITU, March 1993.

[8] International Telecommunication Union (ITU), *CCITT Blue Book*, Volume III, Fascicle III.4, "General aspects of digital transmission systems; terminal equipments, recommendation G.704, synchronous frame structures used at

primary and secondary hierarchical levels" (Malaga-Tormelinos, 1984; amended Melbourne, 1988).

[9] ITU-T G.804, *ATM Cells Mapping into Plesiochronous Digital Hierarchy*, International Telecommunication Union, Telecommunication Standardization Sector (ITU-T), November 1993.

[10] C. Feher, *Digital Communications Satellite/Earth Station Engineering*, Englewood Cliffs, NJ: Prentice-Hall, 1981.

[11] F. Davarian (Ed.), *Proceedings of the Eighteenth NASA Propagation Experimenters Meeting (NAPEX XVIII) and the Advanced Communications Technology Satellite (ACTS) Propagation Studies Mini workshop*, Vancouver, Canada, June 16–17, 1994. Propagation characteristics of 20/30 GHz links with a 40 degree masking angle, F. Davarian, A. Kantak, C. Le, Jet Propulsion Laboratory.

[12] L. Ippolito, *Radiowave Propagation in Satellite Communications*, New York: Van Nostrand Reinhold Co., 1986.

[13] J. Farserotu and A. Tu, "TCP/IP over low-rate ATM-SATCOM links," *IEEE Milcom'96*, McLean VA, 1996.

[14] A.M. Viterbi and A.J. Viterbi, "Erlang capacity of a power-controlled CDMA system," *IEEE J. Selected Areas in Comm.*, August 1993.

[15] A. Viterbi, "Error correcting coding for CDMA systems," *Proc. IEEE 3rd International Symposium on Spread Spectrum Techniques and Applications (ISSSTA'94)*, Oulu, Finland, July 4–6, 1994.

[16] J. Farserotu and A. Tu, "Test and analysis of low-rate asynchronous transfer mode (ATM) cell relay over NATO satellite communications: limitations of transmission control protocol (TCP)/internet protocol (IP) over ATM," SHAPE Technical Center, STC-TN-640, The Hague, The Netherlands, 1996.

[17] A. Tannenbaum, *Computer Networks*, Englewood Cliffs, NJ, Prentice-Hall, 1988.

[18] S. Lin and D. Costello, *Error Control Coding: Fundamentals and Applications*, Englewood Cliffs, NJ: Prentice-Hall, 1983.

[19] Intelsat Earth Station Standard (IESS) 309, Document IESS-309 (Rev. 2), "QPSK/FDMA performance characteristics for Intelsat business service (IBS)," March 9, 1990.

About the Author

Ramjee Prasad was born in Babhnaur (Gaya), Bihar, India, on July 1, 1946. He is now a Dutch Citizen. He received a B.Sc. (Eng) degree from Bihar Institute of Technology, Sindri, India and an M.Sc. (Eng) and Ph. D. degrees from Birla Institute of Technology (BIT), Ranchi, India, in 1968, 1970 and 1979, respectively.

He joined BIT as a Senior Research Fellow in 1970 and became associate professor in 1980. While he was with BIT, he supervised a number of research projects in the areas of microwave communications and plasma engineering. During 1983 to 1988, he was with the University of Dar es Salaam (UDSM), Tanzania, where he rose to the level of professor of telecommunications at the Department of Electrical Engineering in 1986. At USDM, he was responsible for the collaborative project "Satellite Communications for Rural Zones" with Eindhoven University of Technology, The Netherlands. Since February 1988, he has been with Telecommunications and Traffic-Control Systems Group, Delft University of Technology (DUT), The Netherlands, where he is actively involved in the area of wireless personal and multimedia communications (WPMC). He is head of the Transmission Research Section of IRCTR (International Research Center for Telecommunications - Transmission and Radar) and also Program Director of a newly established Center for Wireless Personal Communications (CEWPC). He is currently involved in the European ACTS project FRAMES (Future Radio Wideband Multiple Access System) as a Project Leader of DUT. He is Project Leader of several international industrial funded projects. He has published over 200 technical papers and authored a book "*CDMA for Wireless Personal Communications*," published by Artech House, Boston. His current research interest lies in wireless networks, packet communications, multiple access protocols, adaptive equalizers, spread-spectrum CDMA systems, and multimedia communications.

He has served as a member of advisory and program committees of several IEEE international conferences. He has also presented keynote speeches, invited papers, and tutorials on WPMC at various universities, technical institutions, and IEEE conferences. He was the Organizer and Interim Chairman of the IEEE Vehicular Technology / Communications Society Joint Chapter, Benelux Section. He is now the Elected Chairman of the joint chapter. He is also founder of the IEEE Symposium on Communications and Vehicular Technology (SCVT) in the Benelux and he was the Symposium Chairman of SCVT'93.

He is the Co-ordinating Editor and Editor-in-chief of the Kluwer international journal on *Wireless Personal Communications* and also a member of the editorial board of other international journals, including the IEEE Communications Magazine and the IEE Electronics Communication Engineering Journal. He was the Technical Program Chairman of the PIMRC'94 International Symposium held in The Hague, The Netherlands, during September 19–23, 1994, and also of the Third Communication Theory Mini-Conference in conjunction with the GLOBECOM'94 held in San Francisco, CA, November 27–30, 1994. He is the Conference Chairman of IEEE Vehicular Technology Conference, VTC'99 (Fall), Amsterdam, The Netherlands to be held on September 19–22, 1999.

He is listed in the US Who's Who in the World. He is a fellow of the IEE, a fellow of the Institution of Electronics & Telecommunication engineers, a senior member of IEEE and a member of NERG (The Netherlands Electronics and Radio Society).

Index

The Artech House Mobile Communications Series

John Walker, Series Editor

Advanced Technology for Road Transport: IVHS and ATT, Ian Catling, editor

An Introduction to GSM, Siegmund M. Redl, Matthias K. Weber, Malcolm W. Oliphant

CDMA for Wireless Personal Communications, Ramjee Prasad

Cellular Communications: Worldwide Market Development, Garry A. Garrard

Cellular Digital Packet Data, Muthuthamby Sreetharan, Rajiv Kumar

Cellular Mobile Systems Engineering, Saleh Faruque

Cellular Radio: Analog and Digital Systems, Asha Mehrotra

Cellular Radio: Performance Engineering, Asha Mehrotra

Cellular Radio Systems, D. M. Balston, R. C. V. Macario, editors

Digital Beamforming in Wireless Communications, John Litva, Titus Kwok-Yeung Lo

GSM System Engineering, Asha Mehrotra

Handbook of Land-Mobile Radio System Coverage, Garry C. Hess

Introduction to Wireless Local Loop, William Webb

Introduction to Radio Propagation for Fixed and Mobile Communications, John Doble

Land-Mobile Radio System Engineering, Garry C. Hess

Low Earth Orbital Satellites for Personal Communication Networks, Abbas Jamalipour

Mobile Antenna Systems Handbook, K. Fujimoto, J. R. James

Mobile Communications in the U.S. and Europe: Regulation, Technology, and Markets, Michael Paetsch

Mobile Data Communications Systems, Peter Wong, David Britland

Mobile Information Systems, John Walker, editor

Personal Communications Networks, Alan David Hadden

RF and Microwave Circuit Design for Wireless Communications, Lawrence E. Larson, editor

Smart Highways, Smart Cars, Richard Whelan

Spread Spectrum CDMA Systems for Wireless Communications, Savo G. Glisic, Branka Vucetic

Transport in Europe, Christian Gerondeau

Understanding GPS: Principles and Applications, Elliott D. Kaplan, editor

Universal Wireless Personal Communications, Ramjee Prasad

Vehicle Location and Navigation Systems, Yilin Zhao

Wireless Communications for Intelligent Transportation Systems, Scott D. Elliott, Daniel J. Dailey

Wireless Communications in Developing Countries: Cellular and Satellite Systems, Rachael E. Schwartz

Wireless Data Networking, Nathan J. Muller

Wireless: The Revolution in Personal Telecommunications, Ira Brodsky